Handbook for III-V High Electron Mobility Transistor Technologies

Handbook for III-V High Electron Mobility Transistor Technologies

Edited by
D. Nirmal
J. Ajayan

CRC Press
Taylor & Francis Group
Boca Raton London New York

CRC Press is an imprint of the
Taylor & Francis Group, an **informa** business

CRC Press
Taylor & Francis Group
6000 Broken Sound Parkway NW, Suite 300
Boca Raton, FL 33487-2742

First issued in paperback 2020

© 2019 by Taylor & Francis Group, LLC
CRC Press is an imprint of Taylor & Francis Group, an Informa business

No claim to original U.S. Government works

ISBN-13: 978-1-138-62527-3 (hbk)
ISBN-13: 978-0-367-72924-0 (pbk)

Visit the Taylor & Francis Web site at
http://www.taylorandfrancis.com

and the CRC Press Web site at
http://www.crcpress.com

Contents

Preface

The last four decades have witnessed an explosion of interest in solid state transistor technologies for ultra-high-speed and high-power applications. A wide variety of integrated circuits incorporating millions and billions of solid state transistors on a single crystalline semiconductor have been demonstrated. The last decade has also witnessed a tremendous improvement in speed of solid state transistors and integrated circuits with amplification demonstrated at 1 THz in high electron mobility transistor (HEMT)-based IC amplifiers. HEMTs have been considered as the most suitable solid state transistor technology for providing amplification and frequency conversion up to the millimeter wave, sub-millimeter wave and THz frequency regime for high-speed wide-band communication networks, sensing and imaging systems and deep space applications. HEMTs are also an attractive choice for cryogenic applications. The dynamic behavior of HEMTs depends on their material systems, geometric structures and device fabrication techniques. This handbook deals with the DC/RF performance of HEMTs on different material systems, analytical modelling and their applications. This book is intended for under graduate and post graduate engineering courses in the fields of electrical and electronics. Beginning with the motivation behind the HEMTs, the book also covers the various types of HEMTs on different material systems and their analytical modelling. This book has been organized in 16 chapters. Chapter 1 presents the motivation behind the HHEMTs, current and future trends in transistor technologies, advanced MMICs and TMICs and their applications at millimeter wave, sub-millimeter wave and THz frequency applications. Chapter 2 gives a brief description of history and background of HEMTs, basic structure and principle of operation of HEMTs. Chapter 3 covers the material systems and epitaxial deposition techniques for HEMTs. Chapter 4 explains source/drain engineering, gate engineering and channel engineering techniques in HEMTs. Chapter 5 introduces AlGaN/GaNHEMTs for high power applications. Chapter 6 deals with the fabrication challenges in AlGaN/GaNHEMTs and Chapter 7 is devoted to the analytical modeling of HEMTs. Chapter 8 explains the polarization effects in AlGaN/GaNHEMTs. Chapter 9 concentrates on the current collapse effect in AlGaN/GaNHEMTs and Chapter 10 deals with the modelling and simulation of AlGaN/GaNHEMTs. Chapter 11 describes the breakdown voltage improvement techniques in AlGaN/GaNHEMTs for high power applications. Chapter 12 includes HEMTs on InP/InAlAs/InGaAs material system for high-speed low-noise and low-power applications. Chapter 13 studies the elemental and surface characterization of AlGaN/GaNHEMTs by magnetron sputtering systems. Chapter 14 introduces metamorphic HEMTs for sub-millimeter wave applications. Chapter 15 describes the structure and operation of metal oxide semiconductor high electron mobility transistors (MOSHEMTs) for high-speed, low-power applications. Finally, Chapter 16 covers the geometric structure and working principle of double gate high electron mobility transistors (DGHEMTs).

We are grateful to the contributors of each chapter from renowned institutes and industries, editorial and production teams for their unconditional support to publish this handbook.

Editors

D. Nirmal (M'08 – SM'15) is currently an associate professor in the School of Electrical sciences, Karunya University, India. He received the PhD degree in Information and Communication Engineering from Anna University, India. His research interest includes nano electronics, optoelectronics, microelectronics, VLSI design, device fabrication and modelling. He is the author of many refereed international journals and conferences. He is a chair of IEEE ED Coimbatore Chapter. He has been awarded as Shri. P. K. Das Memorial Best Faculty Award in the year 2013. He has received best high impact factor journal publication award and best researcher award from Karunya University in 2012 and 2014, respectively. He has delivered many lecture and invited to chair several conference/workshop in national and international level. He is currently an editor in microelectronics journal. He is a senior member of IEEE, member of IETE, SSI, ISTE, and IEI societies.

J. Ajayan received the BTech degree in electronics and communication engineering from Kerala University, Trivandrum, India, in 2009 and the MTech degree in VLSI Design from Karunya University, Coimbatore, India, in 2012 and the PhD degree in electronics and communication engineering from Karunya University, Coimbatore, India, in 2017. He is a senior assistant professor in the department of electronics and communication engineering at SNS College of Technology, Coimbatore, India. He has presented papers in many international conferences and also he is an author of many refereed international journals (Elsevier, Taylor & Francis Group, Springer, IOP Science).

Contributors

J. Ajayan
Department of Electronics and
   Communication Engineering
SNS College of Technology
Coimbatore, India

G. Amarnath
Department of Electronics and
   Communication Engineering
National Institute of Technology Silchar
Silchar, India

N. B. Balamurugan
Department of Electronics and
   Communication Engineering
Thiagarajar College of Engineering
Madurai, India

Jorge Castillo
The University of Texas Rio Grande Valley
Edinburg, Texas

Mayank Chakraverty
Maxim India Integrated Circuit Design
   Pvt Ltd.
Bangalore, India

Palash Das
National Institute of Science and
   Technology
Berhampur, India

Amitava DasGupta
Department of Electrical Engineering
Indian Institute of Technology Madras
Chennai, India

Nandita DasGupta
Department of Electrical Engineering
Indian Institute of Technology Madras
Chennai, India

Gourab Dutta
Department of Electronics and Electrical
   Communication Engineering
Indian Institute of Technology Kharagpur
Kharagpur, India

Roman Garcia-Perez
The University of Texas Rio Grande Valley
Edinburg, Texas

Mridula Gupta
Department of Electronic Science
University of Delhi South Campus
New Delhi, India

Hasina F. Huq
Electrical Engineering Department
The University of Texas Rio Grande Valley
   (UTRGV)
Brownsville, Texas

Nilesh Kumar Jaiswal
Department of Electrical and Electronics
   Engineering
Vellore Institute of Technology
   (VIT University)
Vellore, India

Sneha Kabra
Department of Instrumentation
Shaheed Rajguru College of Applied
   Sciences for Women
University of Delhi
New Delhi, India

Srikanth Kanaga
Department of Electrical Engineering
Indian Institute of Technology Madras
Chennai, India

Rama Komaragiri
Department of Electronics and
    Communication Engineering
Bennet University
Greater Noida, India

Atanu Kundu
Department of Electronics and
    Communication Engineering
Heritage Institute of Technology

T. R. Lenka
Department of Electronics and
    Communication Engineering
National Institute of Technology, Silchar
Silchar, India

and

Department of Electronics and
    Communication Engineering
National Institute of Science and
    Technology
Berhampur, India

Karen Lozano
The University of Texas Rio Grande Valley
Edinburg, Texas

Satya Sopan Mahato
National Institute of Science and
    Technology
Berhampur, India

Anuja Menokey
Indian Institute of Technology Palakkad
Pudussery East, India

D. Nirmal
Department of Electronics and
    Communication Engineering
Karunya Institute of Technology and
    Sciences (Deemed-to-be-University)
Coimbatore, India

Vimala Palanichamy
Department of Electronics and
    Communication Engineering
Dayananda Sagar College of Engineering
Bangalore, India

A. K. Panda
National Institute of Science and
    Technology
Berhampur (Odisha), India

D. K. Panda
Department of Electronics and
    Communication Engineering
National Institute of Science and
    Technology
Berhampur, India

P. Prajoon
Department of Electronics and
    Communication Engineering
Jyothi Engineering College
Thrissur, India

V. N. Ramakrishnan
Department of Electrical and Electronics
    Engineering
Vellore Institute of Technology
    (VIT University)
Vellore, India

Ajith Ravindran
Department of Electronics and
    Communication Engineering
Saintgits College of Engineering
Kottayam, India

Binit Syamal
Department of Electronics and
    Communication Engineering
Heritage Institute of Technology
Kolkata, India

1

Motivation Behind High Electron Mobility Transistors

Mayank Chakraverty

CONTENTS

1.1 Introduction

Compounds like gallium nitride (GaN), aluminum nitride (AIN, indium nitride (InN) and their alloys form a unique material system termed as nitride semiconductors. When compared to most of the other material systems, a much wider spectrum of bandgaps is covered by the nitrides. This is illustrated in Figure 1.1 with the help of a bandgap-lattice constant plot. Nitrides have been researched by several research groups for more than three decades; it started when the first GaN-based light-emitting diode (LED) was reported by Pankove et al. The intrinsic material properties did not drive the performance of these early devices; instead, high-defect density and poor surface morphology of the heteroepitaxial films dictated the performance of these nitride semiconductor devices. The work done by Isamu Akasaki at Nagoya and Meijo Universities and Shuji Nakamura at Nichia Chemical Company in Japan during the mid-1980s did go a long way in mitigating such performance issues. Using AIN or GaN nucleation layers, metal-organic chemical vapor deposition (MOCVD) has been used to grow GaN films of high quality on sapphire substrates. This has led to an explosion in the field of GaN research in optoelectronics and gradually moved into electronics. Using the nitride semiconductor family of materials, blue, white, green and violet LEDs (along with blue-light semiconductor) lasers were fabricated by Shuji Nakamura, who currently a professor at the University of California at Santa Barbara. Owing to all the work that has happened to date, a very large number of applications starting from traffic light to large displays and high-definition DVD players make wide use of such nitride-based optoelectronics.

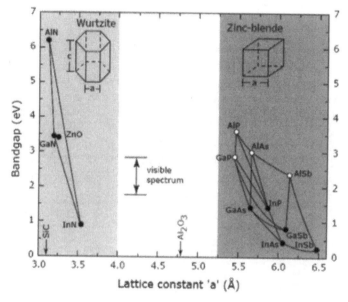

FIGURE 1.1

The bandgap-lattice constant plot. All bandgaps plotted are the direct gaps, indirect gap semiconductors are shown by open circles. Wurtzite crystals are characterized (see insets) by two lattice constants (a, c) of which the a-lattice constant is used for the figure. The a-lattice constants of the common substrates for growth of III-V Nitrides-SiC and Al_2O_3 (sapphire) are indicated by arrows.

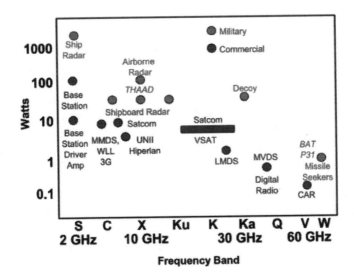

FIGURE 1.2
Some of the most important potential applications for GaN-based power transistors.

It was in the same decade that nitride semiconductors found their use in electronic devices as well, apart from optoelectronics. GaN-based power transistors have been used in multiple applications, as summarized in Figure 1.2. Cell phone base stations have already been using GaN for commercial applications. Digital radio, satellite communications and high-speed digital communications are the other important applications that will use nitrides in the near future. These materials and devices made from them are suited to military applications as well, where GaN amplifiers find apt use in radar and 94 GHz direct energy weapons.

It was in the early 1990s that high-quality AlN films were managed to be grown on sapphire substrates by several groups, and this marked the use of GaN for electronic applications. For the gate insulating layer, AlN proved to be a good choice as it is highly insulating in nature, similar to oxides in silicon MOSFET. The high electron mobility values in GaN heterojunctions grown by Khan et al. were attributed to the conduction by two-dimensional electron gas. It was in 1993 that the group of Khan et al. demonstrated the first GaN transistors. These transistors did not show good electrical characteristics at that time as the transconductance was limited to only 28 mS/mm at 300 K and no significant frequency response was exhibited by the devices. However, there has been a tremendous evolution in the area of nitride semiconductor-based transistors in the last two decades in terms of frequency as well as output power.

As many other discoveries, the idea for a HEMT structure was a product of a research with different purposes and there were several factors superimposed. The late 1970s saw the evolution of the molecular beam epitaxy growth technique and modulation doping together with a vivid interest in the behavior of quantum well structures [1] (the latter peaking in the work of Klitzing, Laughlin, Stomer, and Tsui).

At this time, Mimura and his colleagues at Fujitsu were working on GaAs MESFETs. Facing problems with a high density of the surface states near the interface, they decided to use a modulation-doped heterojunction superlattice and were able to produce depletion type MOSFETs [2]. While those structures were still plagued by several issues, the idea to control the electrons in the superlattice occurred to him. He achieved this by introducing a Schottky gate contact over a single heterojunction. Thus, the AlGaAs/GaAs HEMT was born [3]. Subsequently the first HEMT-based integrated circuit was reported [4]. Alongside Fujitsu a number of other research facilities joined on the further development of the new structures: Bell Labs, Thomson CSF, Honeywell, IBM [5]. In order to counter different problems, several designs were proposed: AlGaAs/GaAs HEMTs, AlGaAs/InGaAs pseudomorphic HEMTs (pHEMTs), AlInAs/ InGaAs/InP HEMTs (ordered by increasing) [6]. However, until the end of the decade HEMTs mainly found military and space applications [7]. Only in the 1990s did the technology enter the consumer market in satellite receivers and emerging mobile phone systems.

In the beginning of the last decade new methods for deposition of GaN on sapphire by MOCVD were developed. Thus, the production of AlGaN/GaN-based HEMTs was possible [8]. GaN has a wide band gap, which brings the advantages of higher breakdown voltages and higher operational temperature. Due to the large lattice mismatch between AlN and GaN a strain in the AlGaN layer is induced, which generates a piezoelectric field. Together with the large conduction band offset and the spontaneous polarization, this leads to very high values for the electron sheet charge density [9]. This large potential of AlGaN/GaN structures (and the indirect advantage of excellent thermal conductivity of the sapphire substrates) was realized very soon, and the research focus partially shifted from AlGaAs/GaAs to AlGaN/GaN devices.

In the course of further development and optimization various techniques were adopted. An approach previously used in high-voltage p–n junctions [10], the field-plate electrode, significantly improved device performance by reducing the peak values of the electric field in the device. Thus, the breakdown voltage could be further increased. This technique was further refined to T-shaped [11] and subsequently Y-shaped gate electrodes [12]. Another step-in optimization of the structure is the addition of a thin AlN barrier between the GaN channel and the AlGaN layer. It increases the conduction band offset and the two-dimensional electron gas (2DEG) density and decreases the alloy disorder scattering, thereby increasing the mobility [13]. An additional option to enhance the electron gas transport properties is the double-heterojunction structure [14]. The InGaN layer under the channel introduces a negative polarization charge at the interface, and thereby improves the carrier confinement in the channel.

While the depletion mode (D-mode) technology has been significantly improved, no comparable progress on the enhancement counterparts can be noted. However, such devices have advantages in certain applications and are therefore getting in the focus of research activities in the recent years. Several groups have proposed interesting approaches. Devices featuring very thin AlGaN layers [15] and fluoride-based plasma treatment [16] have been proposed, however certain stability concerns remain. A very promising method is the recess gate structure reported by Kumar et al. [17]. Also, recently, excellent results have been achieved with InGaN-cap devices [18].

Owing to their high electron saturation velocity, high breakdown field and high operating temperatures have instigated increased research and development work on GaN-based electron devices, and they have been widely investigated for RF and microwave power amplifiers over the last decade. Especially for cell phones, satellites, TV broadcasting and similar RF power electronics applications, AlGaN/GaN high-electron mobility transistors (HEMTs) have recently caught considerable attention as the next generation device. Wider bandwidth and higher efficiency are absolute requirements for the next generation cell phones in the mobile communication applications. Similarly, there is a need for amplifiers that can operate at higher frequencies and higher power for the development of satellite communication and TV broadcasting. It is for these needs that the remarkable property of AlGaN/GaN HEMTs prove to be a promising candidate for microwave power application in wireless communication [19,20].

The HEMT or High Electron Mobility Transistor is a type of field effect transistor (FET), that is used to offer a combination of low noise figure and very high levels of performance at microwave frequencies. This is an important device for high-speed, high-frequency, digital circuits and microwave circuits with low-noise applications. These applications include computing, telecommunications, and instrumentation. And the device is also used in RF design, where high performance is required at very high RF frequencies.

The key element that is used to construct an HEMT is the specialised PN junction. It is known as a hetero-junction and consists of a junction that uses different materials either side of the junction. Instead of the *p–n* junction, a metal-semiconductor junction (reverse-biased Schottky barrier) is used, where the simplicity of Schottky barriers allows fabrication to close geometrical tolerances.

The most common materials used Aluminium Gallium Arsenide (AlGaAs) and Gallium Arsenide (GaAs). Gallium Arsenide is generally used because it provides a high level of basic electron mobility which has higher mobilities and carrier drift velocities than Si [21].

The future HEMT devices based on two-dimensional carrier confinement seem very bright in electronics, communications, physics, and other disciplines. GaAs, InP, and GaN-based HEMTs will continue their journey toward higher integration, higher frequency, higher power, higher efficiency, lower noise, and lower cost. GaN, in particular, offers high-power, high-frequency territory of vacuum tubes and leads to lighter, more efficient, and more reliable communication systems. HEMTs will continue to mold themselves into other kinds of FETs that will exploit the unique properties of 2DEG in various materials systems. In power electronics, GaN-based HEMTs can create a great impact on consumer, industrial, transportation, communication, and military systems. On the other hand, MOS-HEMT or MISFET structures are likely to be operated in enhancement mode with very low leakage current. Si CMOS technology is rapidly advancing toward 10 nm gate regime. To achieve this, power dissipation management in future generation ultra-dense chips will be a significant challenge. Operating voltage reduction may be a solution to meet this challenge. However, currently, it is difficult to accomplish this with Si CMOS while maintaining quality performance. Quantum well-based devices such as InGaAs or InAs HEMTs offer very high potential. Therefore, HEMTs may extend the Moore's law for several more years which will be gigantic for the society [22].

From the past, it can be anticipated that research on new device models and structures of HEMTs will definitely result in new insights into the often-bizarre physics of quantized electrons. ZnO, SiGe, and GaN have shown fractional quantum Hall effect (FQHE), the greatest exponent for impeccable purity and atomic order, which ensures the bright future of HEMT devices [23]. The concept of different kinds of physical and biosensors are still very new to these kinds of devices. The ultra-high mobility that is possible in InAlSb/InAsSb-based system enables high-sensitivity micro-Hall sensors for many applications including scanning Hall probe microscopy and biorecognition [24]. Three-axis Hall magnetic sensors have been reported in micromachined AlGaAs/GaAs-based HEMTs [25]. These devices may be used in future electronic compasses and navigation. THz detection, mixing and frequency multiplication can also be used by 2DEG-based devices [26]. GaN and related materials have strong piezoelectric polarization, and they are also chemically stable semiconductors. Combining functionalized GaN-based 2DEG structures with free-standing resonators, there is a possibility of designing sophisticated sensors [27]. These can offer methods of measurements of several properties such as viscosity, pH, and temperature. Without references, expansion of this technology in the machine-to-machine

(M2M) field is expected to be used in cloud networking-based various sensing functions. Diverse applications such as environmental research, biotechnology, and structural analysis can greatly benefit with the help of newly emerged sensing technology which has high-speed, high-mobility, and high-sensitivity characteristics. HEMT technology is expected to make a great change in the intelligent social infrastructure from the device level. A smart city system, transport system, food industry, logistics, agriculture, health welfare, environmental science, and education systems are examples where this technology can make exceptions [28]. The rise of III-N-based solid-state lighting will lead to a continuous development of materials, substrates, and technologies pushed by a strong consumer market. In an analogy, III-N optoelectronics will challenge the light bulbs, while III-N electronics will challenge the electronic equivalent, the tubes [29].

1.2 Current and Future Device Technologies

Our modern world is based on semiconductors. In addition to your computer, cell phones and digital cameras, semiconductors are a critical component of a growing number of devices. Think of the high-efficiency LED lights you are putting in your house, along with everything with a lit display or control circuit: cars, refrigerators, ovens, coffee makers and more. You would be hard-pressed to find a modern device that uses electricity that does not have semiconductor circuits in it.

While most people have heard of silicon and Silicon Valley, they do not realize that this is just one example of a whole class of materials.

But the workhorse silicon—used in all manner of computers and electronic gadgets—has its technical limits, particularly as engineers look to use electronic devices for producing or processing light. The search for new semiconductors is on. Where will these materials innovations come from?

As the name suggests, semiconductors are materials that conduct electricity at some temperatures but not others—unlike most metals, which are conductive at any temperature, and insulators like glass, plastic and stone, which usually don't conduct electricity.

However, this is not their most important trait. When constructed properly, these materials can modify the electricity moving through them, including limiting the directions it flows and amplifying a signal.

The combination of these properties is the basis of diodes and transistors, which make up all our modern gadgets. These circuit elements perform a multitude of tasks, including converting the electricity from your wall socket to something usable by the devices, and processing information in the form of zeros and ones.

Light can also be absorbed into semiconductors and turned into electrical current and voltage. The process works in reverse as well, allowing for the emission of light. Using this property, we make lasers, LED lights, digital cameras and many other devices.

While this all seems very modern, the original discoveries of semiconductors date back to the 1830s. By the 1880s, Alexander Graham Bell experimented with using selenium to transmit sound over a beam of light. Selenium was also used to make some of the first solar cells in the 1880s.

A key limitation was the inability to purify the elements being used. Tiny impurities—as small as one in a trillion, or 0.0000000001%—could fundamentally change the way a semiconductor behaved. As technology evolved to make purer materials, better semiconductors followed.

The first semiconducting transistor was made of germanium in 1948, but silicon quickly rose to become the dominant semiconductor material. Silicon is mechanically strong, relatively easy to purify, and has reasonable electrical properties.

It is also incredibly abundant: 28.2% of the Earth's crust is silicon. That makes it literally dirt cheap. This almost-perfect semiconductor worked well for making diodes and transistors and is still the basis of almost every computer chip out there. There was one problem: silicon is very inefficient at converting light into an electrical signal or turning electricity back into light.

When the primary use of semiconductors was in computer processors connected by metal wires, this wasn't much of a problem. But, as we moved toward using semiconductors in solar panels, camera sensors and other light-related applications, this weakness of silicon became a real obstacle to progress.

The search for new semiconductors begins on the periodic table of the elements, a portion of which is in the figure at right.

In the column labeled IV, each element forms bonds by sharing four of its electrons with four neighbors. The strongest of these "group IV" elements bonds is for carbon (C), forming diamonds. Diamonds are good insulators (and transparent) because carbon holds on to these electrons so tightly. Generally, a diamond would burn before you could force an electrical current through it.

The elements at the bottom of the column, tin (Sn) and lead (Pb), are much more metallic. Like most metals, they hold their bonding electrons so loosely that when a small amount of energy is applied the electrons are free to break their bonds and flow through the material.

Silicon (Si) and germanium (Ge) are in between and accordingly are semiconductors. Due to a quirk in the way both of them are structured, however, they are inefficient at exchanging electricity with light.

To find materials that work well with light, we have to step to either side of the group IV column. Combining elements from the "group III" and "group V" columns results in materials with semiconducting properties. These "III-V" materials, such as gallium arsenide (GaAs), are used to make lasers, LED lights, photodetectors (as found in cameras) and many other devices. They do what silicon does not do well.

But why is silicon used for solar panels if it is so bad at converting the light into electricity? Cost. Silicon could be refined from a shovel full of dirt scooped up from anywhere on the Earth's surface; the III-V compounds' constituent elements are far rarer.

A standard silicon solar panel converts the sunlight with an efficiency of 10%–15%. A III-V panel can be three times as efficient, but often costs more than three times as much. The III-V materials are also more brittle than silicon, making them hard to work with in wide panels.

However, the III-V materials' increased electron speeds enable construction of much faster transistors, with speeds hundreds of times faster than the ones you find in your computers. They may pave the way for wires inside computers to be replaced with beams of light, significantly improving the speed of data flow.

In addition to III-V materials, there are also II-VI materials in use. These materials include some of the sulfides and oxides researched in the 1800s. Combinations of zinc, cadmium, and mercury with tellurium have been used to create infrared cameras as well as solar cells from companies such as First Solar. These materials are notoriously brittle and very challenging to fabricate.

High power III-V (gallium-nitride) semiconductor electronics will be the backbone of our electrical grid system, converting power for high-voltage transmission and back again. New III-V materials (antimonides and bismuthides) are leading the way for infrared sensing for medical, military, other civilian uses, as well new telecommunication possibilities. And what of the old standby, silicon? Its inability to harness light efficiently does not mean that it is destined for the dust bin of history. Researchers are giving new life to silicon, creating "silicon photonics" to better handle light, rather than just shuttling electrons [30].

1.2.1 Current Semiconductor Device Technologies

1.2.1.1 Bipolar Junction Transistors (BJTs)

This type of transistor is one of the most important of the semiconductor devices. It is a bipolar device in that both electrons and holes are involved in the conduction process. The bipolar transistor delivers a change in output current in response to a change in input voltage at the base. The ratio of these two changes has resistance dimensions and is a "transfer" property (input-to-output), hence the name transistor.

A perspective view of a silicon *p–n–p* bipolar transistor is shown in Figure 1.3. Basically, the bipolar transistor is fabricated by first forming an *n*-type region in the *p*-type substrate; subsequently a *p*+ region (very heavily doped *p*-type) is formed in the n region. Ohmic contacts are made to the top *p*+ and *n* regions through the windows opened in the oxide layer (an insulator) and to the p region at the bottom.

An idealized, one-dimensional structure of the bipolar transistor, shown in Figure 1.3a, can be considered as a section of the device along the dashed lines in Figure 1.3a. The heavily doped *p*+ region is called the emitter, the narrow central *n* region is the base, and the *p* region is the collector. The circuit arrangement in Figure 1.3b is known as a common-base configuration. The arrows indicate the directions of current flow under normal operating

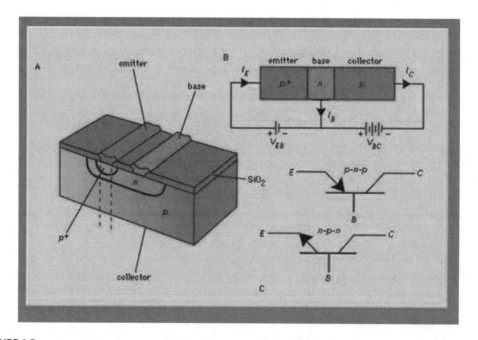

FIGURE 1.3
(a) Perspective of a *p-n-p* bipolar transistor; (b) idealized one-dimensional transistor; (c) symbols for *p-n-p* and *n-p-n* bipolar transistors (E is an emitter, B is a base, and C is a collector).

conditions—namely, the emitter-base junction is forward-biased and the base-collector junction is reverse-biased. The complementary structure of the *p–n–p* bipolar transistor is the *n–p–n* bipolar transistor, which is obtained by interchanging *p* for *n* and *n* for *p* in Figure 1.3a. The current flow and voltage polarity are all reversed. The circuit symbols for *p–n–p* and *n–p–n* transistors are given in Figure 1.3c.

The bipolar transistor is composed of two closely coupled *p–n* junctions. The emitter-base p^+–n junction is forward-biased and has low resistance. The majority carriers (holes) in the p^+-emitter are injected (or emitted) into the base region. The base-collector *n–p* junction is reverse-biased. It has high resistance, and only a small leakage current will flow across the junction. If the base width is sufficiently narrow, however, most of the holes injected from the emitter can flow through the base and reach the collector. This transport mechanism gives rise to the prevailing nomenclature: emitter, which emits or injects carriers, and collector, which collects these carriers injected from a nearby junction.

The current gain for the common-base configuration is defined as the change in collector current divided by the change in emitter current when the base-to-collector voltage is constant. Typical common-base current gain in a well-designed bipolar transistor is very close to unity. The most useful amplifier circuit is the common-emitter configuration, as shown in Figure 1.4a, in which a small change in the input current to the base requires little power but can result in much greater current in the output circuit. A typical output current-voltage characteristic for the common-emitter configuration is shown in Figure 1.4b, where the collector current I_C is plotted against the emitter-collector voltage V_{EC} for various base currents. A numerical example is provided using Figure 1.4b. If V_{EC} is fixed at five volts and the base current I_B is varied from 10 to 15 microamperes (μA; 1 μA = 10^{-6} A), the collector current I_C will change from about four to six milliamperes (mA; 1 mA = 10^{-3} A), as can be read from the left axis. Therefore, an increment of 5 μA in the input-base current gives rise to an increment of 2 mA in the output circuit—an increase of 400 times, with the input signal thus being substantially amplified. In addition to their use as amplifiers, bipolar transistors are key components for oscillators and pulse and analog circuits, as well as for high-speed integrated circuits. There are more than 45,000 types of bipolar transistors

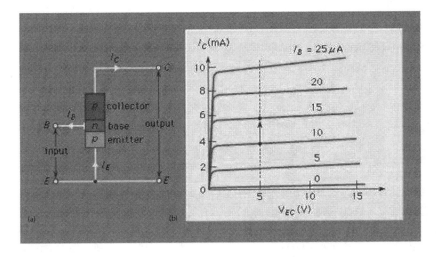

FIGURE 1.4
(a) Common-emitter configuration of a *p-n-p* transistor; (b) output characteristics for a *p-n-p* transistor in the common-emitter configuration.

for low-frequency operation, with power outputs up to 3,000 watts and a current rating of more than 1,000 amperes. At microwave frequencies, bipolar transistors have power outputs of more than 200 watts at 1 gigahertz and about 10 watts at 10 gigahertz.

1.2.1.2 Metal Semiconductor Field-Effect Transistors (MESFETs)

The metal-semiconductor field-effect transistor (MESFET) is a unipolar device, because its conduction process involves predominantly only one kind of carrier. The MESFET offers many attractive features for applications in both analog and digital circuits. It is particularly useful for microwave amplifications and high-speed integrated circuits, since it can be made from semiconductors with high electron mobilities (e.g., gallium arsenide, whose mobility is five times that of silicon). Because the MESFET is a unipolar device, it does not suffer from minority-carrier effects and so has higher switching speeds and higher operating frequencies than do bipolar transistors.

A perspective view of a MESFET is given in Figure 1.5. It consists of a conductive channel with two ohmic contacts, one acting as the source and the other as the drain. The conductive channel is formed in a thin n-type layer supported by a high-resistivity semi-insulating (nonconducting) substrate. When a positive voltage is applied to the drain with respect to the source, electrons flow from the source to the drain. Hence, the source serves as the origin of the carriers, and the drain serves as the sink. The third electrode, the gate, forms a rectifying metal-semiconductor contact with the channel. The shaded area underneath the gate electrode is the depletion region of the metal-semiconductor contact. An increase or decrease of the gate voltage with respect to the source causes the depletion region to expand or shrink; this in turn changes the cross-sectional area available for current flow from source to drain. The MESFET thus can be considered a voltage-controlled resistor.

A typical current-voltage characteristic of a MESFET has the drain current I_D plotted against the drain voltage V_D for various gate voltages. For a given gate voltage (e.g., $V_G = 0$), the drain current initially increases linearly with drain voltage, indicating that the conductive channel acts as a constant resistor. As the drain voltage increases, however, the cross-sectional area of the conductive channel is reduced, causing an increase in the channel resistance. As a result, the current increases at a slower rate and eventually saturates. At a given drain voltage, the current can be varied by varying the gate voltage. For example, for $V_D = 5$ V, one can increase the current from 0.6 to 0.9 mA by forward-biasing the gate to 0.5 V, or one can reduce the current from 0.6 to 0.2 mA by reverse-biasing the gate to −1.0 V.

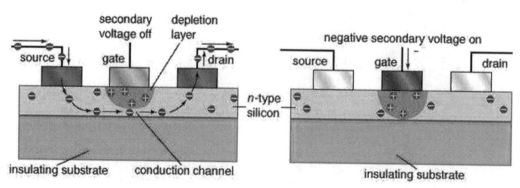

FIGURE 1.5
Perspective of a MESFET.

A device related to the MESFET is the junction field-effect transistor (JFET). The JFET, however, has a *p–n* junction instead of a metal-semiconductor contact for the gate electrode. The operation of a JFET is identical to that of a MESFET.

There are basically four different types of MESFET (or JFET), depending on the type of conductive channel. If, at zero gate bias, a conductive *n*-channel exists and a negative voltage has to be applied to the gate to reduce the channel conductance, then the device is an *n*-channel "normally on" MESFET. If the channel conductance is very low at zero gate bias and a positive voltage must be applied to the gate to form an *n* channel, then the device is an *n*-channel "normally off" MESFET. Similarly, *p*-channel normally on and *p*-channel normally off MESFETs are available.

To improve the performance of the MESFET, various heterojunction field-effect transistors (FETs) have been developed. A heterojunction is a junction formed between two dissimilar semiconductors, such as the binary compound GaAs and the ternary compound $Al_xGa_{1-x}As$. Such junctions have many unique features that are not readily available in the conventional *p–n* junctions discussed previously.

Figure 1.6 shows a cross section of a heterojunction FET. The heterojunction is formed between a high-bandgap semiconductor (e.g., $Al_{0.4}Ga_{0.6}As$, with a bandgap of 1.9 eV) and one of a lower bandgap (e.g., GaAs, with a bandgap of 1.42 eV). By proper control of the bandgaps and the impurity concentrations of these two materials, a conductive channel can be formed at the interface of the two semiconductors. Because of the high conductivity in the conductive channel, a large current can flow through it from source to drain. When a gate voltage is applied, the conductivity of the channel will be changed by the gate bias, which results in a change of drain current. The current-voltage characteristics are similar to those of the MESFET shown in Figure 1.5b. If the lower-bandgap semiconductor is a high-purity material, the mobility in the conductive channel will be high. This in turn can give rise to higher operating speed.

1.2.1.3 Metal Oxide Semiconductor Field Effect Transistor (MOSFETs)

The most important device for very-large-scale integrated circuits (those that contain more than 100,000 semiconductor devices such as diodes and transistors) is the metal-oxide-semiconductor field-effect transistor (MOSFET). The MOSFET is a member of the family of field-effect transistors, which includes the MESFET and JFET.

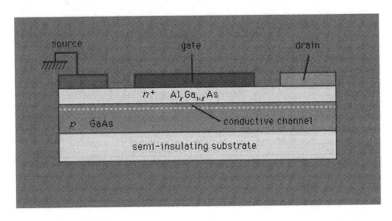

FIGURE 1.6
Cross section of a heterojunction FET having a conductive channel at the heterojunction interface.

The basic principle of this kind of transistor (JFET) was first patented by Julius Edgar Lilienfeld in 1925. In 1959, Dawon Kahng and Martin M. (John) Atalla at Bell Labs invented the metal oxide semiconductor field-effect transistor (MOSFET) as an offshoot to the patented FET design. The earliest microprocessors starting in 1970 were all MOS microprocessors, i.e., fabricated entirely from PMOS logic or fabricated entirely from NMOS logic. The most widely used device in integrated circuit technology, the MOSFET has profound impact on semiconductor electronics superseded JFET making JFET obsolete.

A perspective view for an *n*-channel MOSFET is shown in Figure 1.7. Although it looks similar to a MESFET, there are four major differences: (1) the source and drain of a MOSFET are rectifying *p–n* junctions instead of ohmic contacts; (2) the gate is a metal-oxide-semiconductor structure, meaning that there is an insulator—silicon dioxide (SiO_2)—sandwiched between the metal electrode and the semiconductor substrate, while for the MESFET the gate electrode forms a metal-semiconductor contact; (3) the left edge of the gate electrode must be aligned or overlapped with the source contact to facilitate device operation, while in a MESFET there is no overlapping of gate and source contact; and (4) the MOSFET is a four-terminal device, so that there is a fourth substrate contact in addition to the source, drain, and gate electrode, as in the case of a MESFET.

One of the key device parameters is the channel length, *L*, which is the distance between the two n^+–*p* junctions, as indicated in Figure 1.7. When the MOSFET was first developed, in 1960, the channel length was longer than 20 micrometres (µm). Today, channel lengths less than 1 µm have been fabricated in volume production, and lengths less than 0.1 µm have been created in research laboratories.

The source is generally used as the voltage reference and is grounded. When no voltage is applied to the gate, the source-to-drain electrodes correspond to two *p–n* junctions connected back to back. The only current that can flow from source to drain is a small leakage current. When a high positive bias is applied to the gate, a large number of electrons will

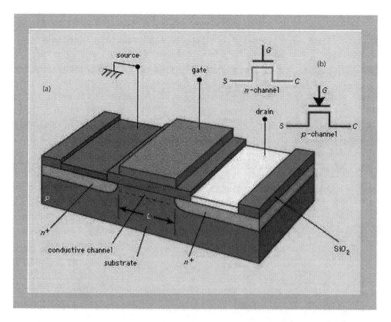

FIGURE 1.7
(a) Perspective of a MOSFET with (b) symbols for *n*- and *p*-channel devices. *Encyclopædia Britannica, Inc.*

be attracted to the semiconductor surface and form a conductive layer just underneath the oxide. The n^+ source and n^+ drain are now connected by a conducting surface n layer (or channel) through which a large current can flow. The conductance of this channel can be modulated by varying the gate voltages; the conductance also can be changed by the substrate bias.

Metal Oxide Field Effect transistors are of mainly of two types based on constructional and operational features of the device.

1. Enhancement type MOSFET
   - *N*-channel Enhancement type
   - *P*-channel Enhancement type

2. Depletion type MOSFET
   - *N*-channel Depletion type
   - *P*-channel Depletion type

In Enhancement mode MOSFET channel is induced by applying gate voltage contrary to the depletion mode MOSFET in which already existing channel is modulated by applying gate voltage like JFET.

Enhancement-type mode MOSFET will be off for gate to source 0 V as there exists no channel to conduct. Depletion-type MOSFET conducts at 0 V has positive cut off gate voltage so less preferred. Depletion MOS also conducts at 0 V, therefore has less useful application. Since the logic operations of depletion MOSFET is the opposite to the enhancement MOSFET, the depletion MOSFET produces positive logic circuits, such as buffer, AND, and OR. The depletion MOSFET is free from sub-threshold leakage current and gate oxide leakage current [31]. As an enhancement MOSFET shrinks in size, there is no way to stop the sub threshold leakage current diffused across from source to drain because the drain and source terminals are closer physically. This is not the problem with depletion type MOSFET because a pinched channel stops the diffusion current completely.

The current-voltage characteristic of a MOSFET is similar to that shown in Figure 1.5b. There are also four different kinds of MOSFETs, depending on the type of conducting layer. The four are *n*-channel normally off, *n*-channel normally on, *p*-channel normally off, and *p*-channel normally on MOSFETs. They are similar to MESFET varieties.

The main reasons why the MOSFET has surpassed the bipolar transistor and become the dominant device for very-large-scale integrated circuits are: (1) the MOSFET can be easily scaled down to smaller dimensions, (2) it consumes much less power, and (3) it has relatively simple processing steps, and this results in a high manufacturing yield (i.e., the ratio of good devices to the total) [31].

1.2.1.4 Dual-Gate Metal Oxide Semiconductor Field Effect Transistors (DG-MOSFETs)

One form of MOSFET that is particularly popular in many RF applications is the dual-gate MOSFET.

The dual-gate MOSFET is used in many RF and other applications where two control gates are required in series. The dual-gate MOSFET is essentially a form of MOSFET where two gates are fabricated along the length of the channel—one after the other. In this way, both gates affect the level of current flowing between the source and drain. In effect, the dual-gate MOSFET operation can be considered the same as two MOSFET devices in series.

FIGURE 1.8
Dual-gate MOSFET circuit symbol.

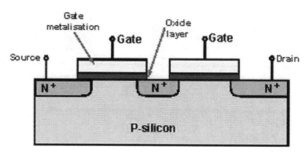

FIGURE 1.9
Dual-gate MOSFET structure.

Both gates affect the overall MOSFET operation and hence the output. Figure 1.8 depicts the circuit symbol of DG-MOSFET while Figure 1.9 shows the cross-sectional structure of the same device. The dual-gate MOSFET can be used in a number of applications including RF mixers/multipliers, RF amplifiers, amplifiers with gain control and the like.

The dual-gate MOSFET has what may be referred to as a tetrode construction where the two grids control the current through the channel. The different gates control different sections of the channel which are in series with each other.

Dual-gate MOSFETs are able to operate with improved performance as amplifiers over single gated FETs. The dual-gate MOSFET enables a cascade two-stage amplifier to be constructed using a single device.

The cascade amplifier helps overcome the Miller effect where capacitance is present between the input and output stages. Although the Miller effect can relate to any impedance between the input and output, normally the most critical is capacitance [32]. This capacitance can lead to an increase in the level of input capacitance experienced and in high frequency (e.g., VHF & UHF) amplifiers it can also lead to instability.

The effect is overcome by using a cascade amplifier using a single dual-gate FET. In this configuration, biasing the drain-side gate at constant potential reduces the gain loss caused by Miller effect. The effects of capacitive coupling between the input and output are virtually eliminated.

The dual-gate MOSFET is used in many RF applications, both as a mixer and as an RF amplifier. In these and other applications the use of two gates to a single device makes it particularly useful. Although not as widely used as its single gate relative, the dual-gate MOSFET is able to provide improved performance in several niche areas [32].

1.2.1.5 Schottky Barrier Metal Oxide Semiconductor Field Effect Transistors (SB-MOSFETs)

As transistors are continuously scaled, parasitic effects, so-called short-channel effects (SCE), begin to diminish performance improvements and can lead to device failures. The most prevalent limitations observed to date are tunneling of carriers though the thin gate oxide, tunneling of carriers from drain to source and substrate, loss of control over doping profiles in channel, source and drain, leading to consequent reduction in ON-OFF current ratio, and finite subthreshold slope.

Moreover, once these problems are solved, the next, and, as most scientists working in the area agree, final limit of fundamental atomic sizes looms. Scaling devices to sub-nm dimensions means that atomic scales are being approached and classical physics gives way to quantum mechanics. Given that the operation of a MOSFET is based on classical physics, this essentially spells the end of the conventional transistor. Current technology operates at scales of 45–90 nm, still well above the quantum physics regime and estimates predict another 5–10 years' worth of scaling capability left in MOSFETs.

One proposed solution to the problematics relating to source and drain junctions is the introduction of metallic materials in place of conventional doped semiconductor regions. Rectifying metal-semiconductor junctions, so-called Schottky barriers, have very similar electrical characteristics to doped pn junctions and thus represent a simple replacement for the source and drain regions of a MOSFET. These so-called Schottky barrier MOSFETs provide a number of scaling benefits. Throughout this text, the acronym SB-MOSFET will be used when referring to Schottky barrier source/drain MOSFETs. As junction depths are scaled to below 50 nm, source and drain series resistances become increasingly significant due to the reduction in cross-sectional area. This reduces drive current and counteracts benefits offered by scaling. Thompson et al. (1998) showed that scaling junctions to below 30 nm results in little or no performance benefits, since increases in source/drain resistance offset any improvements gained by scaling device. By replacing the doped source and drain regions with a metal, their resistance is significantly reduced, even for very shallow junctions. Doing so makes the metal-semiconductor junction an integral part of the transistor. The reduction of source/drain resistances was the original factor for investigating Schottky barrier MOSFETs for highly scaled devices, and subsequently other benefits were also discovered.

As gate lengths are scaled down, the requirements for source/drain to channel junction abruptness become more stringent and require ever-increasing doping concentrations. Silicide-silicon junctions are inherently atomically abrupt, thus directly solving this problem and allowing very short physical channel lengths to be defined. As well as that, the Schottky barrier present at the interface between metallic source/drain and semiconducting channel provides a potential barrier to carriers in the OFF state, thus providing greater control over the OFF-state leakage current in short-channel devices. Together with the low resistivity, these two factors represent the main benefits of Schottky barrier metal sources and drains to highly scaled MOSFET devices, though there are additional benefits, as described in the following. Conventional doped source/drain (DSD) MOSFETs require a gate-to-source/drain overlap to prevent current from spreading to lower doping locations in the source/drain extensions, thereby increasing accumulation and spreading resistance. It has been reported in literature that reducing this overlap results in a degradation of saturation current. SB-MOSFETs do not require these overlaps—in fact, the presence of an underlap, a gap between the edge of the gate and source/drain electrodes, is beneficial to SBMOSFET performance. The absence of gate-to-source/drain overlaps eliminates corresponding parasitic capacitances, which is of great importance for high-frequency applications. Because of this, SB-MOSFETs are thought to promise performance benefits for radio-frequency operation.

One problem in DSD MOSFETs is parasitic bipolar latchup between adjacent devices. For example, *p*-type source or drain and the *n*-type substrate of a *p*-channel device and the *p*-well of an adjacent *n*-channel transistor form a bipolar transistor and lead to parasitic conduction. Both npn and pnp parasitic bipolar latchup is entirely eliminated in SB-MOSFETs due to the presence of metallic materials in source and drain. Doped source/drain devices require high temperature RTAs of about 900°C for dopant activation in source and drain implants. This is incompatible with proposed high-κ gate dielectrics required for further gate oxide scaling, which are damaged by such high temperature treatment. Silicides, such as PtSi or ErSi, form at much lower temperatures of 500°C and lower, thereby maintaining compatibility with fabrication requirements for high-κ dielectrics. In addition to lowering thermal budgets, SB-MOSFETs require fewer processing steps by not requiring source/drain extension and halo implants, dopant activation anneals and associated masking and cleaning steps. This is achieved with processes fully compatible with current CMOS fabrication technologies.

Despite aforementioned benefits, there are a few issues preventing widespread adoption of SB-MOSFETs. The first concerns n-channel devices. While most studies of SB-MOSFETs have centred around p-channel devices, relatively few investigations into n-channel devices have been made. Currently, the major obstacle for n-channel SB-MOSFETs is the lack of a suitable source/drain silicide for n-channel devices due to Fermi level pinning at the metal-semiconductor interface, a result of interface states.

The concept of the neutral level $\Phi 0$, around which the Fermi level in a metal-semiconductor junction is pinned. Experimentally, it has been determined that $\Phi 0$ is generally located about one-third of the band gap above the valence band. While this results in relatively low hole Schottky barrier heights, it conversely also pins the electron Schottky barrier height at considerably higher values, generally about twice the hole barrier height. Consequently, electron transport across a contact between a metal and a n-type semiconductor is more restricted than hole transport for a metal on a p-type semiconductor. Despite the use of a wide variety of metals as suitable silicides, such as Iridium, Erbium or Ybitterium, the performance of n-channel SB-MOSFETs has remained inferior to p-channel devices. The recently proposed use of interfacial layers to de-pin the Fermi level in the semiconductor holds some promise to solving this problem, though at the expense of additional fabrication steps. In addition, the use of dopant segregation techniques to create a layer of highly doped semiconductor close to the metal-semiconductor interface has been shown to reduce the effective Schottky barrier height and may be able to sufficiently reduce the electron Schottky barrier height. Secondly, the silicidation process is a very energetic one. Semiconductor bonds are broken and replaced with metal-semiconductor bonds that, although being covalent, are not entirely non-polar [33]. Differences in electronegativity between metal and semiconductor create slightly polarised bonds that act as a dipole layer at the interface. Furthermore, the crystal structures differ, for example, silicon and germanium are arranged in a diamond lattice, whereas silicides generally have a different crystal lattice, such as an orthorhombic arrangement for PtSi. Depending on their alignment during the silicidation process, the properties of the Schottky contact, specifically the barrier height, may vary from junction to junction [33].

1.2.1.6 Tunnel Field Effect Transistors (TFETs)

The term *TFET* stands for tunneling field effect transistor. It was developed in 1992 by Baba as one of the capable changes to the conventional MOSFETs based on numerous performance factors, including the possibility for above the 60 mV/decade, sub-threshold swing, ultra-low power and ultra-low voltage, the effects of short channel, leakage current reduction, speed requirement exceeding due to the effects of tunneling, capability to work on sub-threshold and super-threshold voltage, similarity in the assembly process as equated with a MOSFET. Taking these factors into account, the MOSFET could be changed by a potential substitute in terms of tunneling field effect transistor for the purpose of high-speed, energy-efficient, and ultra-low-power applications in the area of integrated circuits.

Tunneling field effect transistor (TFET) is a one type of emerging device. Generally, a MOSFET is used for low-energy electronic devices. The structure of the tunneling field effect transistor is closer to the MOSFET, but with a different important switching mechanism. The switching mechanism of TFET is done by modulating quantum tunneling through a barrier in its place of modulating thermionic emission over a barrier as in traditional MOSFETs.

This transistor is a three-terminal or four-terminal device built in Si (silicon). The working principle of this transistor is gate-controlled band-to-band tunneling, and its basic structure is a gated PIN diode. Compared to the MOSFET, it has numerous advantages like apt for low power applications due to lower outflow current, better immunity to short-channel effects, sub threshold swing is not restricted to 60 mV/decade, greater operating speed due to tunneling, the threshold voltage is much smaller, and the current ratio is low off and higher on/off. Thus, TFET can be thought of as a capable alternative to the MOSFET for low-power and high-speed applications.

The basic construction of TFET is similar to a MOSFET excluding that both the source and drain terminals of a TFET are doped of reverse type. A common tunneling field effect transistor device structure consists of a PIN junction (*p*-type, intrinsic, *n*-type), in which the electrostatic potential of the intrinsic area is controlled by a gate terminal.

The TFET device is functioned by applying gate bias so that electron buildup occurs in the intrinsic section. At ample gate bias, BTBT (band-to-band tunneling) happens when the conduction band of the intrinsic region brings into line with the valence band of the *P*-region. This can be illustrated using Figure 1.10.

In valance band, the flow of electrons in the *p*-type region channel into the conduction band of the intrinsic region and the flow of current across the device. As the gate terminal bias is reduced, the bands develop some sort of misalignment, and the flow of current is no longer there.

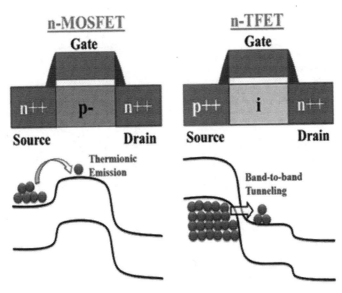

FIGURE 1.10
Illustration of conduction in MOSFET and TFET.

After having a wider research and study on the TFET (tunneling field effect transistor) it can be decided that the source channel tunneling process in doping less TFET can be measured by a gate voltage, and the similar idea is also applied in case of other transistors.

Off-to-on state transition speed of a MOSFET is represented by a parameter, subthreshold swing (SS), which is defined as the minimum gate-voltage required for a decade (a factor of 10) increase of the drain-current. Thermionic emission being the main carrier injection mechanism, thermal broadening of Fermi distribution sets the classical limit of 60 mV/dec (ln10*kT/q at room temperature) SS in conventional MOSFET. A typical MOSFET must be so designed that the current at on-state is at least 10,000 times as much high as the current at off-state; which means a minimum gate-voltage of 240 mV (four decades with each of them requires 60 mV) is required to turn on the MOSFET. Therefore, if the supply voltage is scaled down to maintain the trend of reduced energy consumption; the minimum required gate-voltage, i.e., the SS should be reduced below its classical limit for the future nanoscale MOSFETs. Unfortunately, as long as the carriers are injected by means of thermionic emission, the SS is impossible to scale down below its classical limit.

All credits must go to the quantum mechanical tunneling for which the device engineers can now really think about a device that may break the SS barrier. Tunnel FETs (TFETs), which make use of tunneling for carrier injection, have been demonstrated to achieve SS much lower than its classical limit. As shown in Figures 1.10 and 1.11, a TFET is effectively a reverse biased *p–i–n* structure with a gate-stack similar to MOSFET. Figure 1.12 illustrates the operating principle of a TFET. Unlike the MOSFET where the gate voltage controls source-to-drain barrier, the voltage applied at the gate of a TFET controls the tunneling barrier width to turn on or turn off the device.

Drain current versus gate voltage characteristics for theoretical TFET and MOSFET devices is shown in Figure 1.13. The TFET is able to reach higher drain current for small voltages. Last but not the least the TFET is absolutely protected to random dopant variations and it has been a significant feature for this transistor. The add on point of this transistor is that it does not need very high thermal resources and it can achieve the thermal budget at a very slight one. From the current features of the TFET, it can be detected and estimated that in upcoming a lot work and progress can be expected from this.

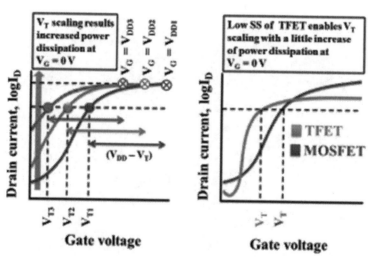

FIGURE 1.11
Problems with conventional MOSFET.

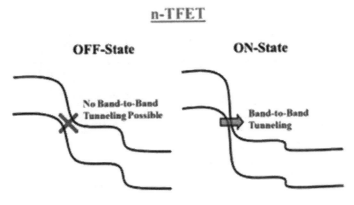

FIGURE 1.12
Operating principle of TFET.

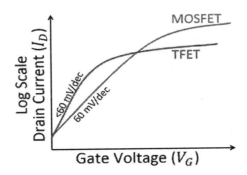

FIGURE 1.13
Transfer characteristics of TFET.

TFET or tunnel FETs are similar to MOSFETs, and the applications of these two are similar to a digital switch, etc. The working principle of TFETs is quite different than MOSFETs. In MOSFETs, the flow of current is due to diffusion phenomenon, while in Tunnel FETs, the conduction mechanism is allied to Zener Tunneling.

The TFET belongs to the family of so-called steep slope devices that are presently being examined for ultra-low-power electronic applications. Because of their low-off currents, they are perfectly suitable for low-standby-power logic and low-power applications which are functioning at moderate frequencies. Other applications of tunnel FETS include ultra-low-power specific analog ICs (integrated circuits) with better temperature strength and low power SRAM.

The main advantages of TFETs include the following:

- Less SS < 60 mV/decade
- Low power requirement

1.2.1.7 *Fin Field Effect Transistors (FinFETs)*

FinFET, also known as Fin Field Effect Transistor, is a type of non-planar or 3D transistor used in the design of modern processors. As in earlier planar designs, it is built on an SOI (silicon on insulator) substrate. However, FinFET designs also use a conducting channel

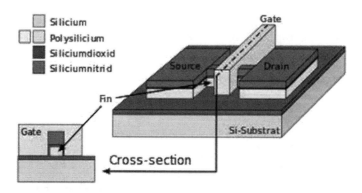

FIGURE 1.14
FinFET device schematic.

that rises above the level of the insulator, creating a thin silicon structure, shaped like a fin, which is called a gate electrode. This fin-shaped electrode allows multiple gates to operate on a single transistor [34]. A schematic of a FinFET device is shown in Figure 1.14.

This type of multi-gate process extends Moore's law, allowing semiconductor manufacturers to create CPUs and memory modules that are smaller, perform faster, and consume less energy. Intel began releasing FinFET CPU technology in 2012 with its 22-nm Ivy Bridge processors [34].

FinFET technology has recently seen a major increase in adoption for use within integrated circuits. The FinFET technology promises to provide the deliver superior levels of scalability needed to ensure that the current progress with increased levels of integration within integrated circuits can be maintained.

The FinFET offers many advantages in terms of IC processing that mean that it has been adopted as a major way forwards for incorporation within IC technology.

This type of FET gained its name from Profs. Chenming Hu, Tsu-Jae King-Liu and Jeffrey Bokor at the University of California, Berkeley who were the first to coin the term as a result of the shape of the structure.

FinFET technology has been born as a result of the relentless increase in the levels of integration. The basic tenet of Moore's law has held true for many years from the earliest years of integrated circuit technology. Essentially it states that the number of transistors on a given area of silicon doubles every two years. Some of the landmark chips of the relatively early integrated circuit era had a low transistor count even though they were advanced for the time. The 6800 microprocessor for example had just 5000 transistors. Todays have many orders of magnitude more. To achieve the large increases in levels of integration, many parameters have changed. Fundamentally the feature sizes have reduced to enable more devices to be fabricated within a given area. However other figures such as power dissipation, and line voltage have reduced along with increased frequency performance. There are limits to the scalability of the individual devices and as process technologies continued to shrink towards 20 nm, it became impossible to achieve the proper scaling of various device parameters. Those like the power supply voltage, which is the dominant factor in determining dynamic power were particularly affected. It was found that optimizing for one variable such as performance resulted in unwanted compromises in other areas like power. It was therefore necessary to look at other more revolutionary options like a change in transistor structure from the traditional planar transistor.

FinFET technology takes its name from the fact that the FET structure used looks like a set of fins when viewed. The main characteristic of the FinFET is that it has a conducting channel wrapped by a thin silicon "fin" from which it gains its name. The thickness of the fin determines the effective channel length of the device. In terms of its structure, it typically has a vertical fin on a substrate which runs between a larger drain and source area. This protrudes vertically above the substrate as a fin. The gate orientation is at right angles to the vertical fin. And to traverse from one side of the fin to the other it wraps over the fin, enabling it to interface with three side of the fin or channel. This form of gate structure provides improved electrical control over the channel conduction and it helps reduce leakage current levels and overcomes some other short-channel effects. The term FinFET is used somewhat generically. Sometimes it is used to describe any fin-based, multigate transistor architecture regardless of number of gates.

There are a number of subtly different forms of trigate transistor structure that are being described as FinFETs. The architecture typically takes advantage of self-aligned process steps to produce extremely narrow features that are much smaller than the wavelength of light generally used to pattern devices on a silicon wafer. It is possible to create very thin fins—of 20 nm in width or less—on the surface of a silicon wafer using selective-etching processes, although they typically cannot currently be made less than 20–30 nm because of the limits of lithographic resolution. The fin is used to form the raised channel. The gate is then deposited so that it wraps around the fin to form the trigate structure. As the channel is extremely thin the gate has much greater control over the carriers within it but, when the device is switched on, the shape limits the current through it to a low level. So, multiple fins are used in parallel to provide higher drive strengths.

Originally, the FinFET was developed for use on silicon-on-insulator (SOI) wafers. Recent developments have made it possible to produce working FinFETs on bulk silicon wafers and improve the performance of certain parameters. The steep doping profile used to control leakage into the bulk substrate has a beneficial impact on DIBL, although increased doping has a negative impact on variability.

Fully depleted SOI transistors have been shown to offer comparable or better performance than FinFETs. However, the relative compatibility of the bulk-silicon FinFET with existing wafer fabrication processes and today's wafer-supply chain favors the FinFET for high-volume IC production at 22 nm and below.

FinFETs have key advantages over planar bulk devices. They exhibit more drive current per unit area than planar devices, largely because the height of the fin can be used to create a channel with a larger effective volume but still take advantage of a wraparound gate.

The added performance capability of FinFETs can be used to achieve higher frequency numbers compared to bulk for a given power budget or lower power. The power reduction can come from two sources: reduced need for wide, high-drive standard cells; and the ability to operate with a lower supply voltage for a given amount of leakage.

A key difference between FinFET-based design and that using conventional planar devices is that the freedom to choose the device's drive strength is reduced, especially for devices that are close to the minimum size. Drive strength can only be improved during layout by adding more fins. The effective width of the device becomes quantized, and the quantization effect is worse for smaller transistors for which the next step up from the minimum-size device is one that is twice as wide. In addition, the minimum number of fins may be two in practical manufacturing processes. This is due to the self-aligned spacer processes that are used to create fins at tight pitches—each sacrificial spacer element that is deposited creates a pair of fins.

The FinFET is a technology that is used within ICs. FinFETs are not available as discrete devices. However, FinFET technology is becoming more widespread as feature sizes within integrated circuits fall and there is a growing need to provide very much higher levels of integration with less power consumption within integrated circuits [35].

Other than the difficulties of dealing with a new, 3D transistor design in terms of parasitic extraction and physical behavior, the major issue is cost: building a FinFET uses a number of additional steps in a manufacturing flow that is already struggling to contain the cost of advanced lithography: double patterning in the next few years, and possibly a move to EUV lithography in the second half of the decade. Figures presented by Qualcomm at IEDM 2013 indicated that the jump in cost to FinFET was lower than that caused by the shift to double patterning from 28 to 20 nm. The back-end of line (BEOL) processes are more or less the same for 20, 16, and 14 nm technologies provided by the foundries.

Real-world FinFETs do not have the same profile as research devices, at least not yet. Analysis by reverse-engineering specialist Chipworks found that device-level simulation by GSS, a spinout from research at the University of Glasgow, indicated that the shape of the fin yields a device with less favorable operating behavior than the ideal FinFET profile, in which the fin walls are parallel to each other. The tapered fin tends to force most of the current into the very top of the fin at higher voltages, which could lead to reliability issues, and into the bulk of the fin in the off-state, which can lead to short-channel effects.

A further issue with bulk FinFETs is higher variability than most research devices. Originally, the FinFET architecture was conceived as being built on SOI wafers: the oxide layer provides a "stop" for the etch processes used to define the raised channel fins. On a bulk-silicon process, control over fin depth is more difficult. The manufacturing issues appear to be manageable but leads to greater variability in transistor behavior versus SOI-based implementations [36].

1.2.2 Future Semiconductor Device Technologies

1.2.2.1 Junctionless Transistor (JLT)

Since all the conventional transistors have junctions that limit their scaling as it requires very abrupt junctions, achieving high-concentration gradient becomes challenging with every technology node. The problems come when there is a junction in such a device like MOSFET, which has source junction and drain junction formed with oppositely doped substrate. A junction-less transistor is a uniformly doped nanowire without junctions and no doping concentration gradient exists. It shows better short-channel effect and less degradation in mobility with temperature, small DIBL, subthreshold swing, and higher voltage gain, which shows good scalability below 10 nm and reduces the fabrication complexity [37]. Junctionless transistors are very promising candidates for future nanoscale MOSFET devices.

The scaling of MOSFETs have the worth-mentioning feature that as they become smaller, they become faster, and at the same time they consume less power and also become cheaper. So, due to scaling, the number of functionalities of a given integrated circuit (ICs) increases on a given silicon area, but as the source and drain come so closer to each other, the gate loses control over the channel and results in short-channel effects (SCEs), which also result in large leakage current due to source-drain coupling, sub-threshold conduction, drain-induced barrier lowering (DIBL), and threshold voltage roll off. To overcome these issues, multiple gate FET (MUGFET), such as double gate and triple gate FinFETs and gate all around (GAA) FETs where the gate is wrapped around the thin fin channel. In these devices, more than one gate controls the channel, which provides better electrostatic control. In a very short-channel device, it puts great difficulties on lithography for the formation of an ultra-sharp source and drain junction and is quite challenging, which imposes thermal budget. A junction-less transistor is an attractive electronic device due to its simple process architecture without concerning thermal budget for the formation of source and drain junctions. Since a junction-less transistor (JLT) is a uniformly doped device throughout with N+ silicon nanowire for *n*-channel MOSFET or P+ silicon nanowire film for *p*-channel MOSFET, it has no source and drain junctions. Hence, it is an accumulation mode device with heavy dopant concentration of either donor type for *n*-channel JLT or acceptor type for *p*-channel JLT.

In junction-less transistors (JLTs) the semiconductor layer is thin and narrow enough to allow for full depletion of carriers when the device is turned off. The device is turned off because of work function differences between the gate electrode and thin silicon film. Since metals and semiconductors have different work functions, this causes distortions in the band diagram, because Fermi levels have to align and hence bands bend up. Electrons get depleted as the thickness of silicon is between 5 and 10 nm. Hence, the entire nanowire gets fully depleted in the OFF condition, which is not because of the reverse bias applied to the gate as in case of conventional MOSFETs. Therefore, in JLTs, polysilicon gate is oppositely doped (*p*+ doping for *n*-channel JLT). As a result, under flat band conditions the device is turned on.

When a positive bias is applied to the gate, the channel is formed within the bulk of the device and small current starts flowing and under a flat band condition the channel becomes neutral. The energy band diagram is shown in Figure 1.15a and b. The electric field is maximum under fully depletion condition when the device is off, and it becomes zero when the device is turned on because of the absence of the junction (electric field exist only when the junction is present). The interesting point here in JLT is that under the on state the effective channel length become equal to that of gate length, so short-channel effect is less in case of JLT compared to conventional MOSFET where the effective channel length reduces as the device is turned on. The spacer formation is not necessary in JLT, however using a high-k spacer on either side enhances the electrostatic integrity, and under off condition it depletes the device up to the spacers.

In a very short channel, device leakage current also increases which increases the standby power of device. Silicon compatible with bulk type compound JLT using epitaxial growth of GaAs on Si substrate with a thick buffer layer of P+ germanium (Ge) is used for *n*-JLT. Since GaAs is grown epitaxially on buffer layer of Ge with Silicon substrate and have large energy bandgap between GaAs and Ge reduces leakage path to the substrate and is compatible with Si technology further enhance the performance by reducing the leakage current. The current driven in JLT is controlled by doping concentration and not by the gate capacitance. The use of a high-k spacer in either side causes increase in effective length in off state but in on state effective length is unaffected and hence the off-state leakage current is reduced using a high-k spacer.

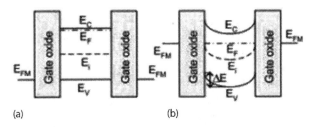

(a)   (b)

FIGURE 1.15
Energy band diagram (a) Under on state (b) Under off state.

Compared to inversion mode (IM) transistor, SCE of a JLT is better because in these IM devices, there exists a concentration gradient between source-channel and drain-channel, whereas the concentration gradient is absent in JLTs. In case of IM device, the depletion charge is balanced by source as well as drain. Hence, when the channel length becomes, shorter, SCE increases whereas in JLTs, charge is controlled by the gate alone [8]. An important point here is that the effective channel in a JLT under off-state is greater than the physical gate length, and under on-state it is equal to the physical gate length. This results in an improvement in SCEs. For shorter gate length, DIBL of bulk junction-less transistor shows similar performance compared to its SOI counterpart whereas SS performance degrades. The threshold voltage of JLT is almost insensitive to the channel length.

The threshold voltage of an IM transistor with a lightly doped channel, V_T, decreases slightly with W_{si} but as the doping concentration increases, V_T increases. Whereas in case of JLTs, since it is a heavily doped device to ensure high drive current, it shows large V_T variations with W_{si} for high channel doping. Thus, the threshold voltage of JLTs is sensitive to the variations in channel doping concentration. As a result, increasing silicon channel width (W_{si}) causes a reduction in the threshold voltage. The threshold voltage is also affected by temperature although there is no direct relation of V_T with temperature; it is related to surface potential which is proportional to temperature. Fully depleted silicon on insulator (FDSOI) MOSFET shows smaller V_T variation compared to bulk MOSFET and using multiple gate can further reduce V_T variation where as a JLT shows large V_T variation compared to IM device. There is a V_T variation with Lg for IM and JL bulk FinFET, small V_T variation with Lg for JL device compared to IM device is because the effective channel length of a JLT under on-state condition is equal to that of gate length where as in case of IM transistor effective channel length is smaller than the gate length but it shows large variation in threshold voltage compared to inversion mode device if the channel doping concentration is increased so threshold voltage of a JLT is more sensitive to channel doping, silicon film width and less sensitive to channel length variation.

1.2.2.2 III-V Tri-gate Quantum Well MOSFET

With Si CMOS technology gradually reaching its ultimate limit, integration of III-V materials in multigate device structures has become a topic of intensive research and study. With excellent carrier transport properties and lower effective mass, III-V semiconductors outperform Si in many aspects of nanoscale device applications [38]. With sub10nm node technology approaching fast, counteracting short-channel effects (SCEs), sub-threshold conduction, improving electrostatic behavior, reduced source/drain tunneling and gate leakage current have become aspects of research and investigation. Multigate device architectures like FinFETs, GAA FETs provide better control over channel carrier accumulation and improved short-channel performance over planar structures. Integration of III-V semiconductors with multigate device architectures can certainly meet the demands of future technology nodes [39]. First experimental demonstration of III-V semiconductors in non-planar tri-gate architecture took place in 2009 [40]. Since then, various III-V semiconductors and their alloys have been employed successfully in multigate architectures like FinFET, Tri-Gate, GAAFETs with increased scaling and improved short-channel performance [40–46]. A sub-100 nm III-V tri-gate device that incorporates bi-layer high-k dielectric material (Al_2O_3/HfO_2 with EOT by lowering carrier velocity in the channel and consequently degrading device "on" current. Presence of trapped charges at the high-k oxide and semiconductor interface could introduce Fermi-level pinning and degrade device performance significantly. Recent studies have shown, the effect of surface roughness scattering becomes lower at wider channel devices [47]. However, as device dimensions are scaled, surface roughness becomes an important issue and degrades device mobility in InGaAs material system. Apart from surface roughness scattering, alloy disorder scattering, impurity scattering, phonon scattering also degrades device performance in III-V materials [48]. In NEGF formalism, the effects of both elastic and inelastic scattering events can be taken into account in ballistic simulation using Büttiker probe method mentioned in [49,50]. Inclusion of non-ideal scattering would decrease effective mobility in the channel, lower carrier velocity, increase channel resistance and degrade device performance.

1.2.2.3 Silicon Nanowire Transistors (SNWTs)

Silicon Nano Wire Transistors (SNWT) represent a promising alternative architecture to the conventional planar technology for devices at the end of the ITRS roadmap, because of the improved electrostatic control of the channel via the gate voltage and the consequent suppression of short-channel effects [51–54]. Since electrostatic plays an important role in these devices, simulations can represent a valid tool in order to understand device behavior and to give design guidelines. Three-dimensional drift-diffusion simulations have been performed in Refs. [51,52] to study the optimum configuration that reduces short-channel effects, and in [55–57], where quantum corrections to the electron density have been considered. In [58] instead, a simulation study of ballistic SNWTs with different cross sections, has shown advantages with respect to Double Gate MOSFETs, as far as downscaling is concerned.

Silicon MOSFETs with effective channel length smaller than 50 nm will not be fully ballistic, so it is reasonable to assume that even for SNWTs scattering events can occur and a fully ballistic assumption can be considered as a limiting case. Transport in SNWT is then likely to be in an intermediate regime between ballistic and drift-diffusion, and its simulation would require the detailed knowledge of the scattering rates of electrons in the 1D sub-bands. Relevant information can be obtained by considering the two limiting cases (ballistic and drift-diffusion), assuming that a partially ballistic transistor would have an intermediate behavior. Another important aspect of interest is source-to-drain tunneling, that is expected to have a significant impact on the shortest devices.

Nanowire transistors made from silicon and germanium have been found to outperform conventional silicon ones. As transistors are the crucial switches used to control electronic circuitry this marks a key step towards super-fast nanoscale computing.

Each nanowire transistor is about half the size of the smallest silicon transistor and could theoretically be used to pack more processing power onto the same area of microchip. Lieber's team found that the nanowire components also performed better than conventional MOSFETs in several important benchmarks.

These included measures of the amount of current the wires could conduct, the speed with which they could be switched on and off and their sensitivity to the applied electric field. The NWFETs were two to three times better than MOSFETs in each case.

Silicon nanowires have a multitude of potential applications, including transistors [51,52], semiconductor memories [53], photovoltaics [54], thermoelectric generators [55], biosensors [56], colour selective photodetectors [57], and qubits [58]. The use of nanowires in commercial products, however, has to date been limited. A major challenge for transistor nanoelectronic applications is that, as transistor dimensions are reduced, it is difficult to maintain a low off-current (I_{off}) whilst simultaneously maintaining a high on-current (Ion). High Ion is fundamental for high gain and/or high speed in any transistor technology and is therefore one of the key parameters requiring optimization.

Reducing I_{off} is significantly harder when transistor critical dimensions reach levels where quantum mechanical tunnelling, short-channel effects [59] and statistical variability [60,61] can be significant. A single gate in a MOSFET transistor becomes unable to provide sufficient electrostatic control to fully deplete carriers in the transistor channel, resulting in increased I_{off} values [62]. A variety of new architectures, including ultra-thin silicon-on-insulator (SOI) [63–65], double gate [63,66], FinFETs [67–71], π- [72]/Ω-gate [73], tri-gate [69], junctionless [52], and gate all-around (GAA) nanowire transistors [74,75] have therefore been developed to improve the electrostatic control of the conducting channel. This is essential since a low I_{off} implies low static power dissipation, and will therefore improve power management in the multibillion transistor circuits employed globally in microprocessors, sensors and memory [76].

Conventional MOSFETs running in inversion have a drain current, I_D, that improves with reduced gate-length Lg, since $I_D \propto \mu Lg(Vg-V_T)^2$ where μ is the mobility, Vg is the gate voltage and V_T is the threshold voltage. As the dimensions of these conventional transistors are reduced, however, higher doping in the channel is required to suppress short-channel effects, which in turn reduces the mobility, thus reducing Ion [77]. The large vertical electric field required to form an inversion layer also significantly reduces the mobility, through interface roughness scattering [52]. A substantial volume of research is therefore focused on investigating new high-mobility channel materials to improve the drive current at lower voltages [78,79]. Alternatively, the problem can be circumvented by developing a range of flat-band devices, such as the junctionless transistor [52]. This has a 3D wire-like channel rather than planar channel of MOSFETs, and acts as a gated resistor that pinches-off the carrier density of the wire by the application of a gate voltage. It is a normally-on device but by selecting a gate metal with an appropriate work-function, it can become a depleted, normally-off device. When switched on, and assuming flat-band conditions, ID is due to the resistive behavior of the channel and is given by

$$I_D = q\mu N_D A V_D / Lg$$

where q is the electronic charge, N_D is the channel doping density, A is the channel conducting area and V_D is the drain voltage. Thus, $I_D(=I_{on})$ again improves with reduced Lg. The channel doping can also be increased to improve Ion as the drive current is directly proportional to the electronic conductivity of the channel, given by $\sigma = q\mu N_D$. However, this cannot be increased arbitrarily because the higher the doping the closer the semiconductor will be to a nearly metallic system, making the channel depletion for particular cross section very difficult. For P-doped Si, this implies a doping limit of 3.5×10^{18} cm^{-3} [80], although in small devices such as nanowires, surface state traps and donor deactivation [81] can actually reduce the activated carrier density, pushing the critical doping limit above the Mott criteria.

1.2.2.4 Carbon Nanotube Field Effect Transistors (CNTFETs)

The increasing demand for ultra-high-speed processors, smaller dimensions and lower power consumption of integrated circuits has made the technology scaling of the electronic components a challenging issue for device designers. In the past few decades, miniaturization of transistors has always obeyed Moore's law: the number of transistors that can be placed inexpensively on an integrated circuit has doubled approximately every 2 years. Nanoscale field effect transistors in the sub-10 nm regime suffer from short-channel effects such as direct tunneling from source to drain, increase in gate-leakage current and punch-through effect. These effects have posed severe problems for miniaturized transistors and directed the recent research toward better alternative semiconductors than silicon. Semiconducting carbon nanotubes (CNTs) because of their properties like large mean free path, excellent carrier mobility and improved electrostatics at nanoscales as the result of their non-planar structure, have been known as the best ideal replacement for silicon. In particular, they exhibit ballistic transport over length scales of several hundred nanometers. Absence of the dangling bond states at the surface of CNTs and purely one-dimensional transport properties improve gate control while meeting gate leakage constrains and allows for a wide choice of gate insulators. That's why CNTs suppress the short-channel effects in transistor devices. Symmetry of the conduction and valence bands makes CNTs advantageous for complementary applications. CNTs are very attractive for nanoelectronic applications and can be used to achieve high speed ballistic carbon nanotube field effect transistors (CNTFETs). Theoretically, CNTFETs could reach a higher frequency domain (terahertz regime) than conventional semiconductor technologies. CNTFETs, only a few years after the initial discovery of CNTs in 1991 by Sumio Iijima were first demonstrated in 1998 and soon after by groups at IBM and Stanford University. Intensive research has led to significant progress in understanding the fundamental properties of CNTs. By using a single-wall CNT as the channel between two electrodes, which work as the source and drain contacts of a FET, a coaxial CNTFET can be fabricated. Coaxial devices are of special interest because their geometry allows for better electrostatics because the gate contact wraps all around the channel (CNT) and has a very good control on carrier transport. Type of metal-CNT contacts plays crucial role in the output characteristics of the transistor. Heavily doped semiconductors, because of the ability to form Ohmic contacts, can be used as ideal electrodes but they suffer from high parasitic resistance. Existence of potential barrier at the metal-CNT interface, changes the device to a CNTFET resembling to Schottky barrier MOSFETs. However, heavily doped CNT contacts can be used to get to a behavior similar to conventional MOSFETs. Understanding CNTFETs from electronic point of view requires a deep insight for mesoscopic physics. For modelling a CNTFET, a powerful methodology with the abilities of solving Schrödinger equation under non-equilibrium conditions in the presence of self-consistent electrostatics and treating coupling of the channel to contacts is needed. The nonequilibrium Green's function (NEGF) formalism provides a sound basis for quantum device simulations.

There are two main types of CNTFETs that are currently being studied, differing by their current injection methods. CNTFETs can be fabricated with Ohmic or Schottky contacts. The type of the contact determines the dominant mechanism of current transport and device output characteristics. CNTFETs are mainly divided into Schottky barrier CNTFETs (SBCNTFETs) with metallic electrodes which form Schottky contacts and MOSFET-like CNTFETs with doped CNT electrodes, which form Ohmic contacts. In SB-CNTFETs, tunneling of electrons and holes from the potential barriers at the source and drain junctions constitutes the current. The barrier width is modulated by the application of gate voltage, and thus, the transconductance of the device is dependent on the gate voltage. The other type of the CNTFETs takes advantage of the n-doped CNT as the contact [82]. Potassium doped source and drain regions have been demonstrated and the behavior like MOSFETs have been experimentally verified. In this type of transistor, a potential barrier is formed at the middle of the channel and modulation of the barrier height by the gate voltage controls the current [82].

1.2.2.5 *Spintronics Based Field Effect Transistors*

Spintronics, which relies on the magnetic moment or the spin of an electron, could revolutionize the entire electronics field. Much of its benefit goes to computing since it can create ultra-speed PCs. Due to its high speed, density and low power consumption its more efficient than the traditional technologies. Until now research was more concentrated on downsizing the transistors. Transistors are based on charge-based technology while spintronics is aimed at replacing this with the spin-based technology and it seems that the Moore's law is nearing its end.

Moore's law describes a long-term trend in the history of computing hardware whereby the number of transistors that can be placed inexpensively on an integrated circuit doubles approximately every 2 years.

There might come a time when transistors would shrunk to the size of an atom and cannot be shrunk any further and this was expected by 2015. So, the researchers have been concentrating on an alternative to the conventional transistor and they have already succeeded in developing spin-based electronics. GMR (Giant Magneto-Resistance effect) was the stepping stone to this. GMR effect brought about a breakthrough in gigabyte hard disk drives and has also been a key in the development of portable electronic devices such as the iPod.

The unique property of spintronics is that spins can be transferred without the actual flow of charge. This is called spin current, and it can transfer information without much loss of energy in the form of heat. The only hurdle that remains now is the generation of a large volume of spin current, which could support the electronic devices. In order to create enhanced spin currents, the researchers used the collective motion of spins called spin waves. According to the research one of the spin wave interaction generates spin current ten times more efficiently than using pre-interacting spin waves [83].

The magnetically sensitive transistor (also known as the spin transistor or spintronic transistor—named for spintronics, the technology which this development spawned) originally proposed in 1990 by Supriyo Datta and Biswajit Das, [84] currently still being developed, [85] is an improved design on the common transistor invented in the 1940s. The spin transistor comes about as a result of research on the ability of electrons (and other fermions) to naturally exhibit one of two (and only two) states of spin: known as "spin up" and "spin down." Unlike its namesake predecessor, which operates on an electric current, spin transistors operate on electrons on a more fundamental level; it is essentially the application of electrons set in particular states of spin to store information.

One advantage over regular transistors is that these spin states can be detected and altered without necessarily requiring the application of an electric current. This allows for detection hardware (such as hard drive heads) that are much smaller but even more sensitive than today's devices, which rely on noisy amplifiers to detect the minute charges used on today's data storage devices. The potential end result is devices that can store more data in less space and consume less power, using less costly materials. The increased sensitivity of spin transistors is also being researched in creating more sensitive automotive sensors, a move being encouraged by a push for environmentally friendlier vehicles.

A second advantage of a spin transistor is that the spin of an electron is semi-permanent and can be used as means of creating cost-effective non-volatile solid-state storage that does not require the constant application of current to sustain. It is one of the technologies being explored for Magnetic Random Access Memory (MRAM).

Because of its high potential for practical use in the computer world, spin transistors are currently being researched in various firms throughout the world, such as in

England and in Sweden. Recent breakthroughs have allowed the production of spin transistors, using readily available substances, that can operate at room temperature: a precursor to commercial viability.

1.2.2.6 High Electron Mobility Transistors (HEMTs)

The HEMT, or High Electron Mobility Transistor, is a form of field effect transistor, FET, that is used to provide very high levels of performance at microwave frequencies.

The HEMT offers a combination of low noise figure combined with the ability to operate at the very high microwave frequencies. Accordingly, the device is used in areas of RF design where high performance is required at very high RF frequencies.

The development of the HEMT took many years. It was not until many years after the basic FET was established as a standard electronics component that the HEMT appeared on the market. The specific mode of carrier transport used in HEMTs was first investigated in 1969, but it was not until 1980 that the first experimental devices were available for the latest RF design projects. During the 1980s they started to be used, but in view of their initial very high cost their use was considerably limited. Now with their cost somewhat less, they are more widely used, even finding uses in the mobile telecommunications as well as a variety of microwave radio communications links, and many other RF design applications.

The key element within a HEMT is the specialised PN junction that it uses. It is known as a hetero-junction and consists of a junction that uses different materials either side of the junction. The most common materials used aluminium gallium arsenide (AlGaAs) and gallium arsenide (GaAs) [113]. Gallium arsenide is generally used because it provides a high level of basic electron mobility, and this is crucial to the operation of the device. Silicon has a much lower level of electron mobility and as a result it is never used in a HEMT.

There is a variety of different structures that can be used within a HEMT, but all use basically the same manufacturing processes.

In the manufacture of a HEMT, an intrinsic layer of gallium arsenide is first set down on the semi-insulating gallium arsenide layer. This is only about one micron thick. Next, a very thin layer (between 30 and 60 Å) of intrinsic aluminum gallium arsenide is set down on top of this. Its purpose is to ensure the separation of the hetero-junction interface from the doped aluminum gallium arsenide region. This is critical if the high electron mobility is to be achieved. The doped layer of aluminum gallium arsenide (about 500 Å thick) is set down above this. Precise control of the thickness of this layer is required and special techniques are used to achieve the required precision in control.

There are two main structures that are used. These are the self-aligned ion implanted structure and the recess gate structure. In the case of the self-aligned ion implanted structure the gate, drain and source are set down and are generally metallic contacts, although source and drain contacts may sometimes be made from germanium. The gate is generally made from titanium, and it forms a minute reverse biased junction similar to that of the GaAs FET.

For the recess gate structure another layer of *n*-type gallium arsenide is set down to enable the drain and source contacts to be made. Areas are etched as shown in the diagram. The thickness under the gate is also very critical since the threshold voltage of the FET is determined by this. The size of the gate, and hence the channel is very small. Typically the gate is only 0.25 microns or less, enabling the device to have a very good high frequency performance.

The operation of the HEMT is somewhat different to other types of FET and as a result it is able to give a very much improved performance over the standard junction or MOSFETs, and in particular in microwave radio applications.

Electrons from the *n*-type region move through the crystal lattice and many remain close to the hetero-junction. These electrons form a layer that is only one layer thick, forming what is known as a two-dimensional electron gas. Within this region the electrons are able to move freely because there are no other donor electrons or other items with which electrons will collide, and the mobility of the electrons in the gas is very high.

A bias applied to the gate formed as a Schottky barrier diode is used to modulate the number of electrons in the channel formed from the 2-D electron gas, and in turn this controls the conductivity of the device. This can be compared to the more traditional types of FET where the width of the channel is changed by the gate bias.

The HEMT was originally developed for high speed applications. It was only when the first devices were fabricated that it was discovered they exhibited a very low noise figure. This is related to the nature of the two dimensional electron gas and the fact that there are less electron collisions.

As a result of their noise performance they are widely used in low noise small-signal amplifiers, power amplifiers, oscillators and mixers operating at frequencies up to 60 GHz and more and it is anticipated that ultimately devices will be widely available for frequencies up to about 100 GHz. In fact, HEMT devices are used in a wide range of RF design applications including cellular telecommunications, Direct broadcast receivers—DBS, radar, radio astronomy, and any RF design application that requires a combination of low noise and very-high-frequency performance.

HEMTs are manufactured by many semiconductor device manufacturers around the globe. They may be in the form of discrete transistors, but nowadays they are more usually incorporated into integrated circuits. These Monolithic Microwave Integrated Circuit chips, or MMICs are widely used for RF design applications, and HEMT-based MMICs are widely used to provide the required level of performance in many areas.

A further development of the HEMT is known as the PHEMT. PHEMTs, or Pseudomorphic High Electron Mobility Transistors, are extensively used in wireless communications and LNA applications. PHEMT transistors find wide market acceptance because of their high power added efficiencies and excellent low noise figures and performance. As a result, PHEMTs are widely used in satellite communication systems of all forms including direct broadcast satellite television, DBS-TV, where they are used in the low noise boxes, LNBs used with the satellite antennas. They are also widely used in general satellite communication systems as well as radar and microwave radio communications systems. PHEMT technology is also used in high-speed analogue and digital IC technology where exceedingly high speed is required [86].

1.3 Advanced MMIC and TMIC Technologies

Wide bandgap electronics show great potential for becoming the next generation of solid-state technology for high-power microwave applications. With power densities far beyond that of traditional technologies, i.e., GaAs pseudomorphic HEMT (pHEMT) and Si LDMOS, and high efficiencies [87–89], AlGaN/GaN-based circuits are well suited for radar and base-station systems. AlGaN/GaN HEMT-based high-power microwave electronics have been extensively researched over the past years. Power amplifiers in code division multiple access (CDMA) operation at 2 GHz [90,91], as well as monolithic microwave integrated circuits (MMICs) operating at band [88,89,92], have shown excellent performance. The high-power-handling capabilities also makes GaN suitable for receiver electronics where the high component breakdown voltage makes the circuits inherently robust [93–96]. The high linearity of circuits based on AlGaN/GaN HEMTs also ensures a high dynamic range of the receiver electronics [97]. The demonstration of low-loss switches in AlGaN/GaN technology [98] indicates the possibility to fabricate entire transceivers on one chip. The combination of high-power density, high efficiency, moderate noise figure, and robustness opens up new possibilities in the design of radar transceivers. For example, even though the noise figure of GaN technology is inferior to that of an optimized GaAs pHEMT, the overall transceiver noise performance may be much improved over the GaAs implementation. This is possible since the robustness of a GaN-based low-noise amplifier (LNA) may facilitate the omission of limiter/protection circuitry in the receiver path. To realize multifunction chips, process repeatability and uniformity are essential. Demonstrator circuits have been fabricated using an in-house process, which represent transceiver functions such as transmit/receive (T/R) switches, wideband gain blocks, and mixers. The output power and low noise performance of the in-house technology are demonstrated by load–pull and noise-figure measurements. The results indicate that a fully integrated transceiver is feasible.

The transistor technology is suitable for transceiver applications with an epi-structure and a layout appropriate for both in low-noise and switch designs, as well as high power amplifiers. The layouts are adjusted to each application (e.g., minimizing gate–source distance for low-noise application, etc.). The standard HEMT is a 0.25 µm gate-length device with a source–drain spacing of 3 µm giving a typical on resistance of 2.2 ohms.

1.4 Submillimeter Wave Applications

The performance of GaN-based transistors for high frequency RF power amplifiers has been improved tremendously in the last few years through extensive research. The maximum f_T has been increased from a few tens of GHz to about 450 GHz, reducing the gap with low bandgap semiconductor technologies. Moreover, in addition to new technologies to increase the peak frequency performance, novel approaches to maintain the high speed under a wide bias range have been developed. It is expected that the high frequency performance of these devices, in combination with the large breakdown voltage of GaN HEMTs, will open a new frontier for RF amplifiers and mixed signal electronics in the near future.

1.5 Terahertz Applications

The HEMT was formerly developed for high-speed applications. Because of their low noise performance, they are widely used in small-signal amplifiers, power amplifiers, oscillators and mixers operating at frequencies up to 60 GHz.

HEMT devices are used in a wide range of RF design applications including cellular telecommunications, direct broadcast receivers—DBS, radio astronomy, RADAR (Radio Detection and Ranging System), and majorly used in any RF design application that requires both low noise performance and very high-frequency operations.

Nowadays HEMTs are more usually incorporated into integrated circuits. These Monolithic Microwave Integrated Circuit chips (MMIC) are widely used for RF design applications.

A further development of the HEMT is PHEMT (Pseudomorphic High Electron Mobility Transistor). The PHEMTs are extensively used in wireless communications and LNA (Low Noise Amplifier) applications. They offer high power added efficiencies and excellent low noise figures and performance [99].

1.6 Transistor Technology for 4G/5G Communications

Although the standards for 5G have not been finalized yet, general expectations for this fifth generation of mobile networks are high. 5G is expected to enable extreme mobile broadband with data rates up to 10 Gbps, in order to meet, for example, the future demand for video streaming. It also promises to enable machine-to-machine communication in support of the Internet-of-Things Platform. And it is expected to allow for critical machine communication—such as driverless cars communicating with each other and with neighboring base stations. These applications typically require extremely high reliability, and low latencies, below 1 ms.

To allow for this almost unlimited experience, innovations are required in the overall network infrastructure (including base stations and small cells) as well as in the technologies for mobile devices.

In the first phase of the 5G deployment, wireless communication radios will most probably operate in the sub-6GHz radio frequency (RF) bands. But to cope with the upcoming spectrum scarcity within these bands, bandwidth is being sought at millimeter-wave (mm-wave) bands—more specifically the RF bands within the 24–100 GHz range. The introduction of these mm-wave frequencies will have a significant impact on the overall 5G network infrastructure. For mobile handsets such as smartphones, this translates into an increasing complexity (Figure 1.16) of the RF front-end modules—that contain, e.g., the transmitter/receiver, bandpass filters, power amplifiers and local oscillators. Both sub-6 GHz bands and mm-wave bands will now need to be enabled in one common architecture, and devices will probably need to access several bands simultaneously. Therefore, higher speed front-end devices than currently used in 4G-LTE are needed. Also, the mm-wave functionality will have to be implemented in battery powered mobile devices, which will put severe restrictions on the power consumption of the mm-wave circuits. To meet all these challenges, we will need high-speed devices that have both a high output power and a high-power efficiency.

Today, several device technologies are being used for RF applications, including for example RF SOI and SiGe technologies. Of particular interest is the use of III-V circuits. III-V high electron mobility transistors (HEMTs, GaAs or InP based) are already in standard use for high-frequency applications. The RF performance of these devices significantly outperforms that of standard Si CMOS devices, especially when considering FinFETs which suffer from intrinsically higher parasitics. Next to that, III-V heterojunction

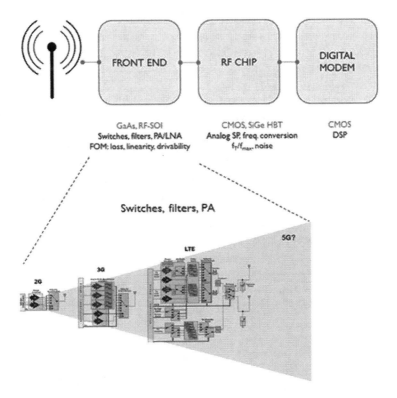

FIGURE 1.16
Schematic representation of the increased RF-front-end-module complexity with every generation of mobile communication.

bipolar transistors (HBTs) have also shown great potential when high-speed requirements need to be fulfilled. And, although originally designed for high-power applications, III-N devices (such as GaN-based HEMT devices) have demonstrated high-frequency performance exceeding 400 GHz.

So far, Si and III-V (or III-N) circuits have been fabricated and packaged separately, and then later assembled on the same carrier substrate. This approach however does not really allow for the optimization of the performance, the reduction of power, cost and form factor, and the increase in the complexity of the circuits. Reducing the form factor will however be essential, as many different dies will be needed to fabricate the RF front-end module, and space within the mobile handset is limited. The fabrication of III-V-based devices presents other challenges. In general, (lab-like) fabrication processes and materials are being used that are not compatible with cost-effective high-volume Si manufacturing. In addition, these III-V HEMT and HBT devices are mostly fabricated on smaller size (2–3 inch) non-Si substrates [100].

The 5th generation (5G) mobile communication service is expected to provide the significant benefits of the huge traffic capacity, the mobility and flexibility, the multi-dimensional connection over human to machines, and so on. Especially in enhancing the traffic capacity responding to the explosive growth of the traffic demand, the strategic technologies development has been proposed consists of the network density enhancement, the spectrum expansion and the spectrum efficiency enhancement [101].

The power amplifier (PA) is one of the key components dominating the system power consumption, the output power and quality of downlink signal on the base station. And using the complex QAM scheme requires the larger output back at the amplifiers, and it causes the increase of consumption. The crest factor reduction (CFR) technique which is used with the adaptive digital pre-distortion (DPD) technique is today in the main stream to achieve the high efficiency by less back-off and the compliance to the 3GPP specification simultaneously.

Numerous studies have been done and are also ongoing toward the PA's consumption reduction [102]. A big progress in the last decade is employing the Doherty PA (DPA) into the base station transmitter [103]. This revived solution successfully achieved the average drain efficiency of over 40%, which was nearly twice efficient compared with the conventional class-AB amplifier in 2003 around. After a few year experiences of the DPA, highly efficient GaN HEMT devices are introduced in earnest and combined with the advanced architectures such as asymmetrical/multi-way Doherty and Envelope Tracking (ET) technique [104]. These techniques have been greatly pushing the PA final stage efficiency up to 60% or more over the extended power back off range. But the amount of enhanced efficiency itself seems to be slightly stagnated in the last few years.

The other recent progress of Outphasing/Linear-amplification-using-Nonlinear-Components (LINC) PA and the switch-mode PA based on Class-D/Class-S amplifiers have actually broken the 70% efficiency barrier [105,106]. Since the amplifier efficiency and the compactness are still major competitiveness for 4G/5G systems, those PA techniques could be potentially next generation main stream. Because the network density enhancement by Small-Cell is assumed to use the large number of radio equipment (RE) to overlay the existing macro cell, and the spectrum efficiency enhancement using the Massive-MIMO technique requires huge number of RF front-end for a system.

The GaN HEMT is one of the most suitable devices for both the Outphasing/LINC and switch-mode PA. Because its potentially high peak drain-efficiency is pushing the PA efficiency toward the theoretical limitation. And the higher f-max than Si LDMOS enables the higher frequency operation which is planned to use 5G service as well as broadband capability and based on the wide band gap properties and high supply voltage operation. We discuss the design methodologies and realizing technique of the highly efficient PA architectures having a broad/multi-band capability and compactness suitable for the shortly coming 5G service and try to lead the future requirements and expectation for GaN HEMT or the compound semiconductor devices.

1.7 Need for Enhancing the Data Capacity of Advanced Wireless Communication

As demand for mobile broadband services continues to explode, mobile wireless networks must expand greatly their capacities. This paper describes and quantifies the economic and technical challenges associated with deepening wireless networks to meet this growing demand. Methods of capacity expansion divide into three general categories: the deployment of more radio spectrum; more intensive geographic reuse of spectrum; and increasing the throughput capacity of each MHz of spectrum within a given geographic area. Two significant throughput-improving advances associated with LTE-Advanced are not achieved just through software tweaks. Higher-order MIMO requires new antennas at base stations, and completely new user devices that incorporate both the increased number of antennas and the chipsets necessary for the more complex processing of received and sent signals. New chipsets will also be needed for advanced versions of CoMP [107].

Cellular communication has got the most important nonmilitary applications of HEMT devices by replacing Si transistors. For such broadband/multiband communication applications, we get a lot of advantages. The increase in relative bandwidth for a given power level is one of those. Some new circuit and system concepts provide bandwidth with increased efficiency. Linearity has been improved for the same output power. Reduction of memory effects is also found by using GaN HEMT devices.

1.8 Space Applications

High gain and low noise amplifiers are the main characteristics for making radar components. GaN HEMTs are one of the first choices for such components. Active electronic sensor arrays are built from GaN-based HEMTs, which are used for airborne radars, ground-based air defense radars, and naval radars [108]. Ka-band missile applications at 35 GHz are also being discussed in literature [109]. Discrete HEMTs are almost always used as the preamplifier in a typical DBS receiver, followed by one or more GaAs MESFET monolithic microwave integrated circuits (MMICs) due to their excellent low-noise characteristics. The use of the low-noise HEMT preamplifier has resulted in substantial improvements in system performance at little additional cost. A low-noise down-converter consisting of a 0.25 pm HEMT and three GaAs MMIC chips has shown a system noise figure less than 1.3 dB with a gain of about 62 dB from 11.7 to 12.2 GHz, which is phenomenal for a commercial, system [110]. Microwave equipment used for space applications are very expensive as they need extra protection from harsh environment in space to survive. Moreover, spacecraft shall be launched, and this implies that the equipment should also sustain without damage at high levels of vibrations and shocks. HEMTs can be fabricated to survive these conditions and have been extensively used in various fields. Generally, a microwave component for space applications is ten to hundred times more expensive than for commercial applications. Workers at the National Radio Astronomy Observatory (NRAO) have used the excellent cryogenic performance of HEMTs to receive signals during the Neptune flyby of the voyager spacecraft.

Phased array systems are also used for remote sensing and earth observation for civilian and scientific purposes. Several such programs have been started [111]. Due to the immense costs of delivering cargo in near-earth orbits, reduced volume and weight, as well as low power consumption are the main goals. The high radiation levels and extreme temperatures must also be considered.

The GaN-based high electron mobility transistor (GaN HEMT) is generally considered to be the best choice in order to meet the requirements for many current HPA designs. HPAs have recently become desirable for high linearity and wider bandwidth operation at lower power consumption. While GaAs HEMT or LDMOS have traditionally been widely used as the HPA devices, GaN HEMT offers the following advantages:

- Higher PAE, which not only saves electrical power usage (OPEX), but also can reduce the size and cost of HPAs, due to the lower amount of heat dissipated (CAPEX).

- High operating voltage—GaN HEMT operates with a power supply voltage of up to 50 V, similar to the range of the power feeder voltage of 48 V, which is commonly used for communication equipment. Furthermore, for any given output power and supply voltage, the operating current can be reduced, when compared to other technologies.

In general, the amplifier design becomes more challenging as the transistor impedances become lower. GaN HEMT devices have a higher impedance than other technologies.

Hence, the HPA design engineer can use the benefits of GaN to enhance HPA performance, such as wider frequency band coverage and higher PAE, depending on the required performance of the HPA. In space applications, the vacuum-tube-based traveling wave tube amplifier (TWTA) is still used, because of a high PAE. However, because a TWTA needs an extremely high voltage, of the order of several thousands of volts, and reliability is considered not ideal due to the hot electrons in the vacuum tube, the solid-state power amplifier (SSPA) is often considered to be the favored solution. GaN-based SSPAs are now in development, in order to replace TWTAs in many space applications and plans are in place to soon launch GaN SSPAs into space.

In space applications, such as satellite communication systems, the RF power amplifier is one of the key components. In the amplifiers, a high PAE is important in order to reduce the launch cost of a satellite. C-band amplifiers, in the frequency range of 3.7–4.2 GHz, are often used for satellite downlinks. In spite of the tremendous merits of SSPAs, GaAs based amplifiers cannot typically offer an acceptable PAE for many satellite applications and the GaN HEMT SSPA, with its higher PAE, is expected to be the best candidate for replacing TWTAs. Many organizations have recently developed GaN HPAs at C-band. However, there are very few reports for HPAs with more than 60% PAE.

GaN HEMT is the best transistor technology for producing high performance HPAs in many applications, so it can satisfy the market demands for high power, high PAE devices.

1.9 Defense Applications

The gallium nitride (GaN) high-electron-mobility transistor (HEMT) has emerged as the dominant force in high-frequency solid-state power amplifiers (PAs)—not that it does not have competition. Silicon (Si) bipolar junction transistors (BJTs) and Si laterally diffused metal–oxide–semiconductor (LDMOS) field-effect transistors (FETs) are still commercially available. They are viable alternatives to GaN HEMTs in aerospace/defense applications such as L-band transponders/interrogators for the identification friend or foe; Link 16 data links; electronic warfare; and surveillance radar; and, in the case of Si LDMOSs, commercial cellular base stations. These older technologies can be favored due to their mature heritage, good performance, and low cost. The pseudomorphic HEMT (PHEMT) is ubiquitous in microwave and millimeter-wave power applications. Vacuum electron devices (VEDs) still reign in the regime of brute power. The GaN HEMT has been displacing these technologies as it has matured, and costs have come down.

Based on the platform constraints three main application areas for GaN-based HEMTs are distinguished (depending on the application the available power and/or volume are limited) [96–99]: ground-based applications, airborne applications, and maritime applications. Phased array systems for these applications are characterized by their power-aperture product and high reliability. The frequency range includes the S-band, C-band, and X-band, with ultra-broadband applications reaching into the Ku-band.

Discrete gallium arsenide (GaAs) devices have been used in an increasing number of microwave systems for about 15 years. The emergence of monolithic GaAs integrated circuits from the laboratory to use in fielded systems has taken place only within the past five years. Almost all the current systems applications make use of ion-implanted metal-semiconductor field effect transistors (MESFETs). The improvements in basic materials, computer aided design (CAD) accuracy, patterning resolution, processing control, test speed, and packaging efficiency that have made the application of MESFET technology possible are, with very little further delay, being applied to the design and manufacture of high electron mobility transistor (HEMT) integrated circuits (ICs). Many applications being targeted for HEMT technology are in the millimeter wave range from 35 to 94 GHz where MESFET performance begins to fall off. However, even in the frequency range from about 5 to 35 GHz, particularly for those applications that require the lowest noise or highest efficiency circuits, HEMT technology has advantages. The particular tradeoff between cost and performance required by a system will determine which technology is chosen. As a result of the progress that has been made in the manufacture of HEMT components, they are being designed for use in a large number of specific systems in the United States for both military and commercial applications—sometimes as a cost-effective replacement for MESFETs that have already been designed into the system, and sometimes as original, enabling system components. In order to merit consideration for insertion into the hardware design of any system, whether military or commercial, each technology under consideration must meet a number of criteria:

- Performance: It must meet the form, fit, and functional performance requirements of the system with adequate margin to ensure acceptable operation in the field.
- Availability: Circuits must be available in assured quantities from reputable vendors, preferably with second-source availability.
- Reliability: A level of reliability must be demonstrated which is compatible with the operational mission of the system.
- Affordability: Components must be affordable within the context of the system use.

The most prominent parameters of interest and measures of performance are the noise figure, the power output, and the efficiency as a function of frequency. In all three factors, HEMT ICs are competitive with other GaAs technologies below 35 GHz and generally superior above that frequency. At all frequencies, HEMT technology currently provides the lowest noise figure of any GaAs device. The availability of HEMT IC components from merchant foundries within the past two years is largely a result of the stimulus provided by the U.S. Microwave and Millimeter Wave Monolithic Integrated Circuits (MIMIC) Program. It also is the result of the prospects for a rapidly developing market in commercial systems both at microwave and millimeter wave frequencies that require performance characteristics that can only be achieved readily through the use of HEMT monolithic microwave/millimeter wave ICs (MMICs). A number of foundries are now offering HEMT fabrication and/or design capabilities.

1.10 Medical and Military Applications

In the recent decade, chemical sensors have gained importance for applications that include homeland security, medical and environmental monitoring, and food safety. The desirable goal is the ability to simultaneously analyze a wide variety of environmental and biological gases and liquids in the field and be able to selectively detect a target analyte with high specificity and sensitivity. The conducting 2DEG channel of HEMTs is very close to the surface and very sensitive to adsorption of analytes. Hence, HEMT sensors can be a good alternative for detecting gases, ions, and chemicals [111].

Au-gated AlGaN/GaN HEMTs functionalized in the gate region with label free 3'-thiol modified oligonucleotides serves as a binding layer to the AlGaN surface, which can detect the hybridization of matched target DNAs. XPS shows immobilization of thiol modified DNA covalently bonded with gold on the gated region. Drain-source current shows a clear decrease of 115 μA as this matched target DNA is introduced to the probe DNA on the surface, showing the promise of the DNA sequence detection for biological sensing [112].

Using amino-propyl silane in the gate region, ungated AlGaN/GaN HEMT structures can be activated, which can serve as a binding layer to the AlGaN surface for attachment of biotin. Biotin has a very high affinity to streptavidin proteins. When the chemicals are attached to AlGaN/GaN HEMTs, the charges on the attached chemicals affect the current of the device. The device shows a clear decrease of 4 μA as soon as this protein is collected at the surface, showing indication of protein sensing [112].

The use of Sc_2O_3 gate dielectric produces superior results to either native oxide or UV ozone- induced oxide in the gate region. The ungated HEMTs with Sc_2O_3 in the gate region exhibit a linear change in current between pH 3–10 of 37 μA/pH. The HEMT pH sensors show stable operation with a resolution of <0.1 pH over the entire pH range. The results indicate that HEMTs may have application in monitoring pH solution changes between 7 and 8, the range of interest for testing human blood [111].

References

1. R. Ross, S. Svensson, and P. Lugli, *Pseudomorphic HEMT Technology and Applications*, Dordrecht, the Netherlands: Kluwer Academic Publisher, 1996.
2. T. Mimura, "The Early History of the High Electron Mobility Transistor (HEMT)," *IEEE Trans. Microw. Theory Tech.*, vol. 50, no. 3, pp. 780–782, 2002.
3. T. Mimura, S. Hiyamizu, T. Fujii, and K. Nanbu, "A New Field-Effect Transistor with Selectively Doped GaAs/n-$Al_xGa_{1-x}As$ Heterojunctions," *Jpn. J. Appl. Phys.*, vol. 19, no. 5, pp. L225–L227, 1980.
4. T. Mimura, K. Joshin, S. Hiyamizu, K. Hikosaka, and M. Abe, "High Electron Mobility Transitor Logic," *Jpn. J. Appl. Phys.*, vol. 20, no. 8, pp. L598–L600, 1981.
5. J. Orton, *The Story of Semiconductors*. Oxford, UK: Oxford University Press, 2004.
6. C. Weisbuch and B. Vinter, *Quantum Semiconductor Structures*. San Diego, CA: Academic Press, 1991.
7. K. Duh, C. Chao, P. Ho, A. Tessmer, J. Liu, Y. Kao, M. Smith, and M. Ballingall, "W-Band InGaAs HEMT Low Noise Amplifiers," *IEEE Intl. Microw. Symp. Dig.*, vol. 1, pp. 595–598, 1990.
8. M. Khan, A. Bhattarai, J. Kuznia, and D. Olson, "High Electron Mobility Transistor Based on a GaN-$Al_xGa_{1-x}N$ Heterojunction," *Appl. Phys. Lett.*, vol. 63, no. 9, pp. 1214–1215, 1993.

9. S. Mohammad and H. Morkoc, "Base Transit Time of GaN/InGaN Heterojunction Bipolar Transistors," *J. Appl. Phys.*, vol. 78, no. 6, pp. 4200–4205, 1995.
10. F. Conti and M. Conti, "Surface Breakdown in Silicon Planar Diodes Equipped with Field Plate," *Solid-State Electron.*, vol. 15, no. 1, pp. 93–105, 1972.
11. R. Thompson, T. Prunty, V. Kaper, and J. Shealy, "Performance of the AlGaN HEMT Structure With a Gate Extension," *IEEE Trans. Electron Devices*, vol. 51, no. 2, pp. 292–295, 2004.
12. K. Makiyama, T. Ohki, M. Kanamura, K. Imanishi, N. Hara, and T. Kikkawa, "High-f_{max} GaN HEMT with High Breakdown Voltage over 100 V for Millimeter-Wave Applications," *Phys. Stat. Sol. (a)*, vol. 204, no. 6, pp. 2054–2058, 2007.
13. L. Shen, S. Heikman, B. Moran, R. Coffie, N. Zhang, D. Buttari, I. Smorchkova, S. Keller, S. DenBaars, and U. Mishra, "AlGaN/AlN/GaN High-Power Microwave HEMT," *IEEE Electron Device Lett.*, vol. 22, no. 10, pp. 457–459, 2001.
14. N. Maeda, T. Saito, K. Tsubaki, T. Nishida, and N. Kobayashi, "Two-Dimensional Electron Gas Transport Properties in AlGaN/GaN Single- and Double-Heterostructure Field Effect Transistors," *Mat. Sci. Eng. B*, vol. 82, no. 1–3, pp. 232–237, 2001.
15. M. Khan, Q. Chen, C. Sun, J. Yang, M. Blasingame, M. Shur, and H. Park, "Enhancement and Depletion Mode GaN/AlGaN Heterostructure Field Effect Transistors," *Appl. Phys. Lett.*, vol. 68, no. 4, pp. 514–516, 1996.
16. Y. Cai, Y. Zhou, K. Chen, and K. Lau, "High-Performance Enhancement-Mode AlGaN/GaN HEMTs Using Fluoride-Based Plasma Treatment," *IEEE Electron Device Lett.*, vol. 26, no. 7, pp. 435–438, 2005.
17. V. Kumar, A. Kuliev, T. Tanaka, Y. Otoki, and I. Adesida, "High Transconductance Enhancement-Mode AlGaN/GaN HEMTs on SiC Substrate," *Electron. Lett.*, vol. 39, no. 24, pp. 1758–1760, 2003.
18. T. Mizutani, M. Ito, S. Kishimoto, and F. Nakamura, "AlGaN/GaN HEMTs With Thin InGaN Cap Layer for Normally Off Operation," *IEEE Electron Device Lett.*, vol. 28, no. 7, pp. 549–552, 2007.
19. J.-C. Gerbedoen, A. Soltani, S. Joblot, J.-C. De jaeger, C. Gaquiere, Y. Cordier, and F. Semond, "AlGaN/GaN HEMT on (001) Silicon Substrate With Powe Density Performance of 2.9 W/mm at 10GHz," *IEEE Trans. Electron Device*, vol. 57, p. 7, 2010.
20. T. Palacio, A. Chakrabotry, S. Rajan, C. Poblenz, S. Keller, S.P. DEnBaars, J.S. Speck, and U.K. Mishra, "High-Power AlGaN/GaN HEMTs for Ka-Band Application," *IEEE Electron Device Lett.*, vol. 26, p. 11, 2005.
21. T. Agarwal, Tutorial on High Electron Mobility Transistor (HEMT), https://www.elprocus.com/high-electron-mobility-transistor-hemt-construction-applications/.
22. J.A. Del Alamo, The High-Electron Mobility Transistor at 30: Impressive Accomplishments and Exciting Prospects. In: *2011 International Conference on Compound Semiconductor Manufacturing Technology*, May 2011, (pp. 16–19), Indian Wells, CA: IEEE.
23. Y.Z. Chen, F. Trier, T. Wijnands, R.J. Green, N. Gauquelin, R. Egoavil, D.V. Christensen, G. Koster, M. Huijben, N. Bovet, and S. Macke, Extreme mobility enhancement of two-dimensional electron gases at oxide interfaces by charge-transfer-induced modulation doping. *Nat. Mater.*, vol. 14, no. 8, pp. 801–806, 2015.
24. M. Bando, T. Ohashi, M. Dede, R. Akram, A. Oral, S.Y. Park, I. Shibasaki, H. Handa, and A. Sandhu, "High Sensitivity and Multifunctional Micro-Hall Sensors Fabricated using InAlSb/InAsSb/InAlSb Heterostructures." *J. Appl. Phys.*, vol. 105, no. 7, pp. 07E909, 2009.
25. A. Vorob'ev, A. Chesnitskiy, A. Toropov, and V. Prinz, "Three-axis Hall Transducer based on Semiconductor Microtubes." *Appl. Phys. Lett.*, vol. 103, no. 17, pp. 173513.
26. M. Dyakonov and M. Shur, "Detection, Mixing, and Frequency Multiplication of Terahertz Radiation by Two-dimensional Electronic Fluid." *IEEE Trans. Electron Devices.*, vol. 43, no. 3, pp. 380–387, 1996.
27. K. Brueckner, F. Niebelschuetz, K. Tonisch, S. Michael, A. Dadgar, A. Krost, V. Cimalla, O. Ambacher, R. Stephan, and M.A. Hein, "Two-dimensional Electron Gas Based Actuation of Piezoelectric AlGaN/GaN Microelectromechanical Resonators." *Appl. Phys. Lett.*, vol. 93, no. 17, p. 173504.

28. K. Joshin, T. Kikkawa, S. Masuda, and K. Watanabe, "Outlook for GaN HEMT Technology." *Fujitsu Sci. Tech. J.*, vol. 50, no. 1, pp. 138–143.
29. R. Quay, *Gallium Nitride Electronics*. Springer Science & Business Media: Berlin, Germany, 2008.
30. T. Vandervelde, Beyond silicon: The search for new semiconductors, https://theconversation.com/beyond-silicon-the-search-for-new-semiconductors-55795.
31. S.M. Sze and W.C. Holton, Semiconductor device, Encyclopædia Britannica, Encyclopædia Britannica, inc., April 10, 2016, https://www.britannica.com/technology/semiconductor-device.
32. I. Poole, Dual Gate MOSFET, Radio-Electronics.com, http://www.radio-electronics.com/info/data/semicond/fet-field-effect-transistor/dual-gate-mosfet.php.
33. https://warwick.ac.uk/fac/sci/physics/research/condensedmatt/silicon/papers/theses/pearman_thesis.pdf.
34. https://www.computerhope.com/jargon/f/finfet.htm.
35. Ian Poole, http://www.radio-electronics.com/info/data/semicond/fet-field-effect-transistor/finfet-technology-basics.php.
36. http://www.techdesignforums.com/practice/guides/finfets/.
37. J.A. del Alamo, Nanometre-scale Electronics With III–V Compound Semiconductors, *Nature* vol. 479, pp. 317–323. doi:10.1038/nature10677.
38. ITRS Home n.d. http://www.itrs.net/.
39. Y.Q. Wu, R.S. Wang, T. Shen, J.J. Gu, and P.D. Ye, "First Experimental Demonstration of 100 nm Inversion-mode In Ga As F in FET Through Damage-Free Sidewall Etching." *Tech. Dig. Int. Electron Devices Meet*, pp. 331–334, 2009. doi:10.1109/IEDM.2009.5424356.
40. M. Radosavljevic, G. Dewey, J.M. Fastenau, J. Kavalieros, R. Kotlyar, B. Chu-Kung et al., "Non-planar, Multigate In Ga As Quantum well Field Effect Transistors with High-K gate Dielectric and Ultra-scaled Gate-to drain/Gate-to-Source separation for Low Power Logic Applications." *Tech Dig. Int. Electron Devices Meet*, vol. 6, pp. 126–129, 2010. doi:10.1109/IEDM.2010.5703306.
41. J.J. Gu, Y.Q. Liu, Y.Q. Wu, R. Colby, R.G. Gordon, and P.D. Ye, "First Experimental Demonstration of Gate-all-around III #x2013;V MOSFETs by Top-down Approach." *Electron Devices Meet (IEDM)*, pp. 33.2.1–33.2.4. doi:10.1109/IEDM.2011.6131662.
42. M. Radosavljevic, G. Dewey, D. Basu, J. Boardman, B. Chu-Kung, J.M. Fastenau et al., "Electrostatics improvement in 3-D tri-gate over ultra-thin body planar In Ga As quantum well field effect transistors with high-K gate dielectric and scaled gate-to-drain/gate-to-source separation." *Tech. Dig. Int Electron Devices Meet*, pp. 765–768. doi:10.1109/IEDM.2011.6131661.
43. J.J. Gu, X.W. Wang, J. Shao, T. Neal, M.J. Manfra, R.G. Gordon et al., "III–V gate-all-around nanowire MOSFET process technology: From 3D to 4D." *Tech. Dig. Int. Electron Devices Meet*, pp. 529–532, 2012. doi:10.1109/IEDM.2012.6479091.
44. J.J. Gu, X.W. Wang, H. Wu, J. Shao, T. Neal, M.J. Manfra et al., "20–80 nm Channel length In Ga As gate-all around nanowire MOSFETs with EOT = 1.2 nm and lowest SS = 63 mV/dec." *Tech. Dig. Int. Electron Devices Meet*, pp. 633–636, 2012. doi:10.1109/IEDM.2012.6479117.
45. S.H. Kim, M. Yokoyama, R. Nakane, O. Ichikawa, T. Osada, M. Hata et al., "High performance sub-20-nmchannel-lengthextremely-thin body InAs-on-insulator tri-gate MOSFETs with high short channel effect immunity and Vth tunability." *Tech. Dig. Int. Electron Devices Meet*, pp. 429–432, 2013. doi:10.1109/IEDM.2013.6724642.
46. T.-W. Kim, D.-H. Kim, D.H. Koh, H.M. Kwon, R.H. Baek, D. Veksler et al., "Sub-100 nm In Ga As Quantum-Well (QW)Tri-Gate MOSFETs with Al2O3/HfO2 (EOT < 1 nm) for Low-Power Logic Applications." *Echnical. Dig. Int. Electron Devices Meet*, pp. 16.3.1–16.3.4, 2013. doi:10.1109/IEDM.2013.6724641.
47. E.G. Marin, F.G. Ruiz, A. Godoy, I.M. Tienda-luna, F. Gámiz, S. Member et al., "Mobility and Capacitance Comparison in Scaled In Ga As Versus Si Trigate MOSFETs." *Electron Device Lett IEEE*, vol. 36, pp. 114–116. doi:10.1109/LED.2014.2380434.
48. S. Oktyabrsky, P. Nagaiah, V. Tokranov, S. Koveshnikov, M. Yakimov, R. Kambhampati, R. Moore, and W. Tsai, "Electron Scattering in Buried In Ga As MOSFET Channel with HfO2 Gate Oxide." *MRS Proc.*, vol. 1155, Cambridge University Press; 2009, pp. 2–7. doi:10.1557/PROC-1155-C02-03.

49. J. Wang, Device physics and simulation of silicon nanowire transistors. PhD Thesis 2005.
50. Z. Ren, Nanoscale MOSFETs: Physics, Simulation and Design. PhD Thesis 2001.
51. J. Appenzeller, J. Knoch, M. Bjork, H. Riel, H. Schmid, and W. Riess, "Toward nanowire electronics," *IEEE Trans. Electron Devices*, vol. 55, no. 11, pp. 2827–2845, 2008.
52. J.-P. Colinge, C.W. Lee, A. Afzalian, N.D. Akhavan, R. Yan, I. Ferain, P. Razavi, B. O'neill, A. Blake, M. White, and A.M. Kelleher, "Nanowire transistors without junctions," *Nature Nanotechnol.*, vol. 5, no. 3, pp. 225–229, 2010. doi:10.1038/nnano.2010.15.
53. C. Busche, L. Vilà-Nadal, J. Yan, H.N. Miras, D.L. Long, V.P. Georgiev, A. Asenov, R.H. Pedersen, N. Gadegaard, M.M. Mirza, and D.J. Paul, "Design and fabrication of memory devices based on nanoscale polyoxometalate clusters," *Nature*, vol. 515, no. 7528, pp. 545–549, 2014. doi:10.1038/nature13951.
54. B. Tian, X. Zheng, T.J. Kempa, Y. Fang, N. Yu, G. Yu, J. Huang, and C.M. Lieber, "Coaxial silicon nanowires as solar cells and nanoelectronics power sources," *Nature*, vol. 449, no. 7164, pp. 885–889, 2007. doi:10.1038/nature06181.
55. A.I. Boukai, Y. Bunimovich, J. Tahir-Kheli, J.K. Yu, W.A. Goddard, and J.R. Heath, "Silicon nanowires as efficient thermoelectric materials," *Nature*, vol. 451, no. 7175, pp. 168–171, 2008.
56. G. Zheng, F. Patolsky, Y. Cui, W.U. Wang, and C.M. Lieber, "Multiplexed electrical detection of cancer markers with nanowire sensor arrays," *Nature Biotechnol.*, vol. 23, no. 10, pp. 1294–1301, 2005. doi:10.1038/nbt1138.
57. H. Park, Y. Dan, K. Seo, Y.J. Yu, P.K. Duane, M. Wober, and K.B. Crozier, "Filter-free image sensor pixels comprising silicon nanowires with selective colour absorption," *Nano Lett.*, vol. 14, no. 4, pp. 1804–1809, 2014.
58. S. Nadj-Perge, S.M. Frolov, E.P.A.M. Bakkers, and L.P. Kouwenhoven, "Spin-orbit qubit in a semiconductor nanowire," *Nature*, vol. 468, no. 7327, pp. 1084–1087, 2010. doi:10.1038/nature09682.
59. Y. Taur, D.A. Buchanan, W. Chen, D.J. Frank, K.E. Ismail, S.H. Lo, G.A. Sai-Halasz, R.G. Viswanathan, H.J. Wann, S.J. Wind, and H.S. Wong, "CMOS scaling into the nanometer regime," *Proc. SPIE*, vol. 85, no. 4, pp. 486–504, 1997.
60. A. Asenov, A.R. Brown, J.H. Davies, S. Kaya, and G. Slavcheva, "Simulation of intrinsic parameter fluctuations in decananometer and nanometer scale MOSFETs," *IEEE Trans. Electron Devices*, vol. 50, no. 9, pp. 1837–1852, 2003.
61. A. Asenov, B. Cheng, X. Wang, A.R. Brown, C. Millar, C. Alexander, S.M. Amoroso, J.B. Kuang, and S.R. Nassif, "Variability aware simulation based design- technology cooptimization (DTCO) flow in 14 nm Finfet/sramcooptimization," *IEEE Trans. Electron. Devices*, vol. 62, no. 6, pp. 1682–1690, 2015.
62. K. J. Kuhn, "Considerations for ultimate CMOS scaling," *IEEE Trans. Electron. Devices*, vol. 59, no. 7, pp. 1813–1828, 2012.
63. H. Wong, D. Frank, and P. Solomon, "Device design considerations for double-gate, ground-plane, and single-gated ultra-thin SOI MOSFET's at the 25 nm channel length generation," *Proc. Int. Electron Devices Meet*, vol. 98, pp. 407–410, 1998.
64. D. Esseni, M. Mastrapasqua, G.K. Celler, C. Fiegna, L. Selmi, and E. Sangiorgi, "An experimental study of mobility enhancement in ultrathin SOI transistors operated in double-gate mode," *IEEE Trans. Electron Devices*, vol. 50, no. 3, pp. 802–808, 2003.
65. J.-P. Colinge, *The SOI MOSFET: From Single Gate to Multigate*. Boston, MA: Springer, 2008, pp. 1–48.
66. F. Gamiz, A. Godoy, C. Sampedro, N. Rodriguez, and F. Ruiz, "Monte carlo simulation of low-field mobility in strained double gate SOI transistors," *J. Comput. Electron.*, vol. 7, no. 3, pp. 205–208, 2008. doi:10.1007/s10825-007-0163-5.
67. D. Hisamoto, W.C. Lee, J. Kedzierski, H. Takeuchi, K. Asano, C. Kuo, E. Anderson, T.J. King, J. Bokor, and C. Hu, "FinFET-a self-aligned double-gate MOSFET scalable to 20 nm," *IEEE Trans. Electron Devices*, vol. 47, no. 12, pp. 2320–2325, 2000.
68. X. Wang, A.R. Brown, B. Cheng, and A. Asenov, "Statistical variability and reliability in nanoscale FinFETs," *Proc. 2011 IEEE Int. Electron Devices Meet*, pp. 5.4.1–5.4.4, 2011.

69. C.-H. Jan, "A 22 nm SoC platform technology featuring 3-D tri-gate and high-k/metal gate, optimized for ultra low power, high performance and high density SoC applications," *Proc. Int. Electron Devices Meet*, vol. 12, pp. 44–47, 2012.

70. X. Wang, B. Cheng, A.R. Brown, C. Millar, J.B. Kuang, S. Nassif, and A. Asenov, "Interplay between process-induced and statistical variability in 14-nm cmos technology double-gate SOI finfets," *IEEE Trans. Electron Devices*, vol. 60, no. 8, pp. 2485–2492, 2013.

71. P.L. Yang, T.B. Hook, P.J. Oldiges, and B.B. Doris, "Vertical slit fet at 7-nm node and beyond," *IEEE Trans. Electron Devices*, vol. 63, no. 8, pp. 3327–3334, 2016.

72. J.-T. Park, J. Colinge, and C. Diaz, "Pi-gate SOI MOSFET," *IEEE Electron Device Lett.*, vol. 22, no. 8, pp. 405–406, 2001.

73. F. Yang, H.Y. Chen, F.C. Chen, C.C. Huang, C.Y. Chang, H.K. Chiu, C.C. Lee, C.C. Chen, H.T. Huang, C.J. Chen, and H.J. Tao, "25 nm CMOS omega FETs," *Proc. Int. Electron Devices Meet.* vol. 2, pp. 255–258, 2002.

74. N. Singh, A. Agarwal, L.K. Bera, T.Y. Liow, R. Yang, S.C. Rustagi, C.H. Tung, R. Kumar, G.Q. Lo, N. Balasubramanian, and D.L. Kwong, "High-performance fully depleted silicon nanowire (diameter ≤ 5 nm) gate-all-around CMOS devices," *IEEE Electron Device Lett.*, vol. 27, no. 5, pp. 383–386, 2006.

75. A. Asenov, Y. Wang, B. Cheng, X. Wang, P. Asenov, T. Al-Ameri, and V.P. Georgiev, "Nanowire transistor solutions for 5nm and beyond," *Proc. 2016 17th Int. Symp. Qual. Electron. Design.*, pp. 269–274, 2016.

76. W. Lu and C.M. Lieber, "Semiconductor nanowires," *J. Phys. D Appl. Phys.*, vol. 39, no. 21, pp. R387–R406, 2006.

77. E. Abrahams, P.W. Anderson, D.C. Licciardello, and T.V. Ramakrishnan, "Scaling theory of localization: Absence of quantum diffusion in two dimensions," *Phys. Rev. Lett.*, vol. 42, pp. 673–676, 1979. doi:10.1103/PhysRevLett.42.673.

78. J.J. Gu, Y.Q. Liu, Y.Q. Wu, R. Colby, R.G. Gordon, and P.D. Ye, "First experimental demonstration of gate-all-around III–V MOSFETs by top-down approach," *Proc. Int. Electron Devices Meet.*, vol. 11, pp. 769–773, 2011.

79. S.W. Chang, X. Li, R. Oxland, S.W. Wang, C.H. Wang, R. Contreras-Guerrero, K.K. Bhuwalka, G. Doornbos, T. Vasen, M.C. Holland, and G. Vellianitis, "InAs N-MOSFETs with record performance of Io n = 600 µA/µm at Ioff = 100 nA/µm (Vd = 0.5 V)," *Proc. Int. Electron Devices Meet*, vol. 13, pp. 417–420, 2013.

80. P.P. Edwards and M.J. Sienko, "Universality aspects of the metal-nonmetal transition in condensed media," *Phys. Rev. B*, vol. 17, pp. 2575–2581, 1978. doi:10.1103/PhysRevB.17.2575.

81. V.P. Georgiev, M.M. Mirza, A.I. Dochioiu, F. Adamu-Lema, S.M. Amoroso, E. Towie, C. Riddet, D.A. MacLaren, A. Asenov, and D.J. Paul, "Experimental and simulation study of silicon nanowire transistors using heavily doped channels," *IEEE Trans. Nanotechnol.*, vol. 16, no. 5, pp. 727–735, 2017. doi:10.1109/TNANO.2017.2665691.

82. https://www.intechopen.com/books/carbon-nanotubes/fundamental-physical-aspects-of-carbon-nanotube-transistors.

83. https://electrosome.com/spinronics-devices-transistor/.

84. S. Datta and B. Das, "Electronic analog of the electrooptic modulator". *Applied Physics Letters.* vol. 56, pp. 665–667, 1990.doi:10.1063/1.102730.

85. P. Chuang, S.-H. Ho, L.W. Smith, F. Sfigakis, M. Pepper, C.-H. Chen, J.C. Fan, J.P. Griffiths, I. Farrer, "All-electric all-semiconductor spin field-effect transistors". *Nature Nanotechnol.*, vol. 10, no. 1, pp. 35–39.

86. I. Poole, http://www.radio-electronics.com/info/data/semicond/fet-field-effect-transistor/hemt-phemt-transistor.php.

87. Y.-F. Wu, M. Moore, A. Saxler, T. Wisleder, and P. Parikh, "40 W/mm double field-plated GaN HEMTs," *Device Res. Conf.*, University Park, PA, 2006, pp. 151–152.

88. U. Mishra, S. Likun, T. Kazior, and Y.-F. Wu, "GaN-based RF power devices and amplifiers," *Proc. IEEE*, vol. 96, no. 2, pp. 287–305, 2008. doi:10.1109/JPROC.2007. 911060.

89. M.-Y. Kao, C. Lee, R. Hajji, P. Saunier, and H.-Q. Tserng, "AlGaN/GaN HEMTs with PAE of 53% at 35 GHz for HPA and multi-function MMIC applications," *IEEE MTT-S International Microwave Symposium Digest*, pp. 627–630, 2007.

90. A. Wakejima, K. Matsunaga, Y. Okamoto, K. Ota, Y. Ando, T. Nakayama, and H. Miyamoto, "370-W output power GaN-FET amplifier with low distortion for W-CDMA base stations," *IEEE MTT-S International Microwave Symposium Digest*, San Francisco, CA, pp. 1360–1363, 2006.

91. D. Kimball, P. Draxler, J. Jeong, C. Hsia, S. Lanfranco, W. Nagy, K. Linthicum, L. Larson, and P. Asbeck, "50% PAE WCDMA basestation amplifier implemented with GaN HFETs," *IEEE Compound Semiconduct. Integrated Circuit Symp.*, Palm Springs, CA, pp. 89–92, 2005.

92. A. Darwish, K. Boutros, B. Luo, B. Huebschman, E. Viveiros, and H. Hung, "AlGaN/GaN Ka-band 5 W MMIC amplifier," *IEEE Trans. Microw. Theory Tech.*, vol. 54, no. 12, pp. 4456–4463, 2006. doi:10.1109/TMTT.2006.883599.

93. M. Rudolph, R. Behtash, K. Hirche, J. Wurfl, W. Heinrich, and G. Trankle, "A highly survivable 3–7 GHz GaN low-noise amplifier," *IEEE MTT-S International Microwave Symposium Digest*, San Francisco, CA, pp. 1899–1902, 2006.

94. M. Micovic, A. Kurdoghlian, H. Moyer, P. Hashimoto, A. Schmitz, I. Milosavljevic, P. Willadsen et al., "GaN MMIC technology for microwave and millimeter-wave applications," *IEEE Compound Semiconduct. Integr. Circuit Symp.*, pp. 173–176, 2005. doi:10.1109/CSICS.2005.1531801.

95. M. Micovic, A. Kurdoghlian, T. Lee, R. O. Hiramoto, P. Hashimoto, A. Schmitz, I. Milosavljevic et al., "Robust broadband (4 GHz–16 GHz) GaN MMIC LNA," *IEEE Compound Semiconduct. Integr. Circuit Symp.*, pp. 1–4, 2007. doi:10.1109/CSICS07.2007.54.

96. M. Rudolph, R. Behtash, R. Doerner, K. Hirche, J. Wurfl, W. Heinrich, and G. Trankle, "Analysis of the survivability of GaN low-noise amplifiers," *IEEE Trans. Microw. Theory Tech.*, vol. 55, no. 1, pp. 37–43, 2007. doi:10.1109/TMTT.2006. 886907.

97. P. Schuh, R. Leberer, H. Sledzik, D. Schmidt, M. Oppermann, B. Adelseck, H. Brugger et al., "Linear broadband GaN MMICs for Ku-band applications," *IEEE MTT-S International Microwave Symposium Digest*, pp. 1324–1326, 2006. doi:10.1109/MWSYM.2006.249475.

98. V.S. Kaper, R.M. Thompson, T.R. Prunty, and J.R. Shealy, "Signal generation, control, and frequency conversion AlGaN/GaN HEMT MMICs," *IEEE Trans. Microw. Theory Tech.*, vol. 53, no. 1, pp. 55–65, 2005. doi:10.1109/TMTT.2004. 839336.

99. https://www.elprocus.com/high-electron-mobility-transistor-hemt-construction-applications/.

100. https://www.imec-int.com/en/imec-magazine/imec-magazine-february-2018/tackling-the-challenges-of-5g-with-hybrid-iii-v-si-technology.

101. T. Nakamura "Future LTE Enhancements toward Future Radio Access "Workshop on IoT/M2M and Future Radio Access 2012.

102. F.H. Raab, "Power Amplifiers and Transmitters for RF and Microwave," *IEEE Trans. Microw. Theory Tech.*, vol. 50, no. 3, pp. 814–826, 2002.

103. A. Grebennikov, "High-Efficiency Doherty Power Amplifiers: Histrical Aspects and Modern Trends," *Proc. IEEE*, vol. 100, no. 12, pp. 3190–3219, 2012.

104. R.S. Pengelly, "A Review of GaN on SiC High Electron-Mobility Power Transistors and MMICs," *IEEE Trans. Microw. Theory Tech.*, vol. 60, no. 6, pp. 1764–1783, 2012.

105. S. Kimura, "Efficiency Improvement of Transmitter for Mobile Phone Base Station using Outphasing Amplifier," *IEICE Technical Report*, pp. 53–58, 2013.

106. K. Motoi, A. Wentzel, M. Tanio, S. Hori, M. Hayakawa, W. Heinrich, and K. Kunihiro, "Digital Doherty Transmitter with Envelope A Σ Modulated Class-D GaN Power Amplifier for 800 MHz band," *IEEE MTT-S Microwave International Symposium Digest*, 2014.

107. C.P. Wen, J. Wang, and Y. Hao, "Current collapse, memory effect free GaN HEMT." In: *2010 IEEE MTT-S International Microwave Symposium Digest (MTT)*, May 23, 2010, pp. 149–152, Anaheim, CA: IEEE.

108. Y. Mancuso, P. Gremillet, and P. Lacomme, T/R-modules technological and technical trends for phased array antennas. In: *2005 European Microwave Conference*, October 4–6, 2005, pp. 614–617, Paris, France: IEEE.

109. H. Hommel and H. Feldle, Current status of airborne active phased array (AESA) radar systems and future trends. In: *34th European Microwave Conference*, October 12–14, 2004, pp. 121–124, Amsterdam, the Netherlands: IEEE. High Electron Mobility Transistors: Performance Analysis, Research Trend. doi:10.5772/6779663.
110. T. Kiehl and R. Sollner, *High Speed Heterostructure Devices*. 1st ed. Boston, MA: Academic Press, 1994.
111. B.H. Chu, B.S. Kang, H.T. Wang, C.Y. Chang, Y. Tseng, A. Goh, A. Sciullo et al., AlGaN/ GaN HEMT and ZnO nanorod-based sensors for chemical and bio-applications. In: *SPIE OPTO: Integrated Optoelectronic Devices*, February 12, 2009, (pp. 72162A–72162A), San Jose, CA: International Society for Optics and Photonics.
112. B.S. Kang, Fabrication and Characterization of Compound Semiconductor Sensors for Pressure, Gas, Chemical, And Biomaterial Sensing (Doctoral dissertation) University of Florida.
113. Muhammad Navid Anjum Aadit, Sharadindu Gopal Kirtania, Farhana Afrin, Md. KawsarAlam and QuaziDeenMohdKhosru (June 7th 2017). High Electron Mobility Transistors: Performance Analysis, Research Trend and Applications, Different Types of Field-Effect Transistors MomčiloPejović, IntechOpen, doi:10.5772/67796. Available from: https://www. intechopen.com/books/different-types-of-field-effect-transistors-theory-and-applications/ high-electron-mobility-transistors-performance-analysis-research-trend-and-applications.

2

Introduction to High Electron Mobility Transistors

Rama Komaragiri

CONTENTS

2.1 History and Background of HEMTs

Around 1979 Dr. Takashi Mimura, with Fujitsu Laboratories Ltd., Kawasaki, Japan, deeply immersed researching on III-V compound semiconductor high-frequency, high-speed devices invented the heterostructure high electron mobility transistor (HEMT). The proposed energy band diagram of the HEMT, as in [1] is shown in Figure 2.1. The first HEMT was developed using the AlGaAs/GaAs material system. The working of HEMT uses the concept of modulation doping, which was first demonstrated in 1978 by Ray Dingle and his collaborators at Bell Labs [2,3]. During the same time period, Daniel Delagebeaudeuf and Trong L. Nuyenfrom Thomson-CSF (France) filed a patent for the same device, with the title "Field effect transistor with high cut-off frequency and process for forming same" [4]. From the early device research, the high-speed switching characteristics of HEMT devices became evident in 1981when Fujitsu demonstrated a ring oscillator switching at a delay as low as 17.1 ps [5]. This first HEMT integrated circuit used both enhancement and depletion-mode logic.

With more stabilized technology and the improvements in manufacturing technologies, HEMTs became essential components of the devices with lowest noise characteristics [6].

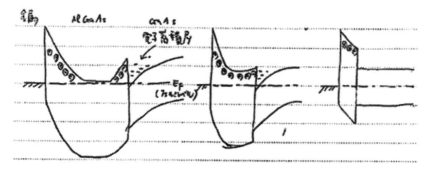

FIGURE 2.1
Energy band diagram as proposed by Dr. Takashi Mimura in the landmark patent. (From Mimura, T., Japan Patent 1409643, 1987.)

As a result of the matured manufacturing process, GaAs MESFET were replaced by HEMTs, which resulted in a drastic size reduction of antennas.

With applications ranging from hi-tech applications like radio astronomy to space and military applications and to entertainment applications in satellite broadcasting receivers, HEMT quickly gained the popularity. Advancements in digital applications like multibit data registers 0.5 µm GaAs/AlGaAs heterojunction device HEMT in LSI technology have paved the path for using HEMTs in development of super computers [7].

In early 1990s, HEMTs were first demonstrated using heterojunctions based on nitride semiconductors AlGaN/GaN HEMT. HEMTs have been demonstrated in several material systems in (and the most notable are) the AlGaAs/GaAs and AlGaN/GaN systems [8,9]. An interesting monograph on HEMT development in the own words of Dr. Mimura is available in the literature [10]. Nowadays, HEMT is omnipresent in applications ranging from cryogenic lownoiseamplifiers, radio telescope to detect microwavesignals from a dark nebula, broadcasting satellite receivers, cell-phone handsets and automotive radars.

2.2 Basic Structure of HEMTs

The structural cross section of a HEMT is shown in Figure 2.2. The heart of HEMT is the formation of two-dimensional electron gas (2DEG) in a quantum well. The 2DEG is formed by appropriately choosing the material system. In a HEMT, the barrier layer is a wide bandgap material and the buffer layer constitute a narrow bandgap material. The barrier layer and buffer layer may have the same doping type, n-type, forming a heterojunction. Generally, the barrier layer doping is greater than buffer layer doping. When the barrier layer is brought in contact with the buffer layer, to acquire minimum energy configuration, electrons diffuse from the wide bandgap barrier layer to narrow bandgap buffer layer. This electron transfer continues till the diffusion is counterbalanced by a built-in electric field (similar to the process that takes place in a PN junction). Thus, the Fermi level on both the sides gets aligned and inside the structure, and the spatial change of Fermi level across the junction is zero. Under the equilibrium condition, conduction band, valence bands bend suitably. In the channel

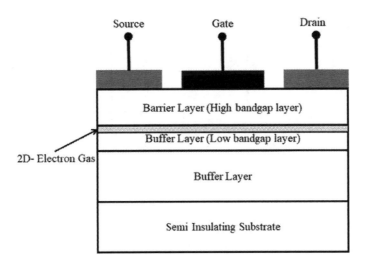

FIGURE 2.2
Basic cross-section of a HEMT device.

region, the band bending facilitates formation of a two-dimensional quantum well with a finite energy barrier height and width. Due to quantum mechanical phenomenon, the electrons reside in quantized energy levels. The electrons present in the channel are confined to the quantum well and reside in the energy levels thus forming a 2DEG. As the n-doped barrier layer supplies electrons to the channel, the charge carriers (electrons) in the channel are thus spatially separated from their ionized donor atoms, thus reducing the impurity scattering of mobile charges. Further, as the current carrying region is away from the surface, the surface scattering is also reduced. Both these phenomena improve the carrier mobility and hence the name high electron mobility transistor. An undoped channel further increases the carrier mobility. Thus, it is a standard practice to minimize the doping of the buffer layer and hence the buffer layer is undoped. The gate forms a Schottky contact with the barrier layer while drain and source form an ohmic contact with the barrier layer.

To understand the principle of operation of a HEMT, various types of heterostructures, formation of 2DEG and different effects involved in the HEMT are explained later in the section.

2.2.1 Heterojunctions

A heterojunction is formed when two different semiconductors are brought into physical contact. Depending on the line-ups of conduction and valance bands of two semiconductors and forbidden gap, heterojunctions can be classified into three distinct categories. Figure 2.3a shows the most common alignment, which is called Type-I alignment or straddled alignment. $Al_xGa_{(1-x)}As/GaAs$ is an example of straddled alignment. The Type-II alignment consist of staggered alignment. In the staggered alignment, the difference in alignment of conduction band and valance band is in the same direction. As shown in Figure 2.3b, $GaAs_ySb_{(1-y)}/Ga_xIn_{(1-x)}As$ material system is an example of staggered alignment. The broken gap alignment shown in Figure 2.3c occurs in GaSb/InAs material systems. The broken gap alignment is also called as Type-III alignment. Type-I band alignment is discussed further as it is suitable for a HEMT device.

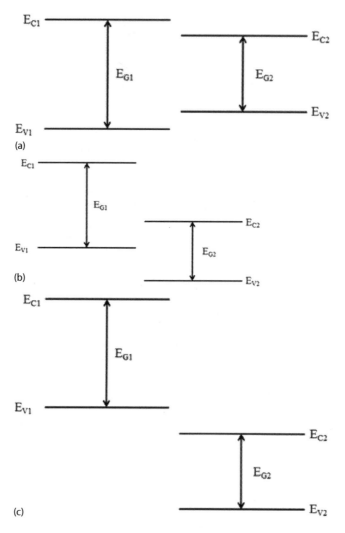

FIGURE 2.3
(a) Type-I or straddled heterojunction; (b) Type-II or staggered heterojunction; and (c) Type-III or broken gap heterojunction.

2.2.2 Equilibrium Band Diagram of a Type-I Heterojunction

Consider that the two semiconductor materials shown in Figure 2.3a are brought into contact. Let χ_1 and χ_2 be the electron affinity, E_{G1} and E_{G2} be the bandgap of the two semiconductor materials respectively, as shown in Figure 2.4. When the effect of image forces is neglected, the electron affinity model assumes that the energy balance of an electron moved from the vacuum level to semiconductor-1 to vacuum level of semiconductor-2. After bringing the two materials into contact, in the field free region, the conduction band edge and valance band edge differences are illustrated in Figure 2.4. Thus, the conduction band edge difference can be calculated using (2.1).

$$\Delta E_C = \chi_1 - \chi_2 \tag{2.1}$$

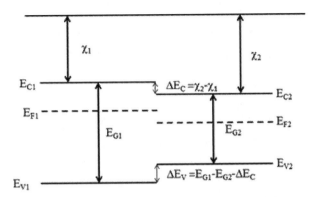

FIGURE 2.4
Description of energy levels in a Type-I heterojunction material system.

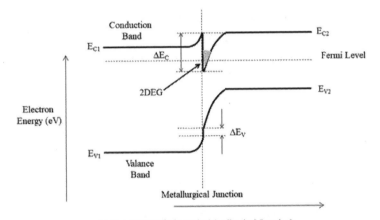

FIGURE 2.5
Band diagram and 2DEG formation in a Type-I heterojunction.

Hence the valance band edge difference is given by (2.2).

$$\Delta E_V = E_{g2} - E_{g1} - \Delta E_C \tag{2.2}$$

Once the wideband gap and narrow band gap materials are brought to from a heterojunction, the bands align so that the system reaches equilibrium. The equilibrium band diagram of a Type-I heterojunction is shown in Figure 2.5 indicating the 2DEG. Please note that the 2DEG region in Figure 2.5 is not to scale, but rather only for illustration purposes.

2.2.3 Electrostatics of a Heterojunction

Consider the material description of heterojunction shown in Figure 2.6. The aim is to study electrostatics of a $n-n$ heterojunction. The following assumptions are made: (i) the semiconductors regions are uniformly doped and form an abrupt junction, and (ii) depletion approximation is valid.

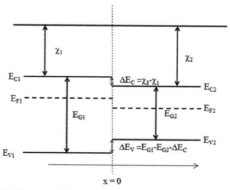

FIGURE 2.6
Description of system to find built-in potential of a n-n heterojunction.

The built-in potential is given by (2.3).

$$qV_{bi} = E_{F1} - E_{F2} \qquad (2.3)$$

Let the potential drop across the semiconductor region-1 and semiconductor region-2 be ϕ_1 and ϕ_2 respectively. Thus, the built-in potential can be written as (2.4).

$$V_{bi} = \phi_1 + \phi_2 \qquad (2.4)$$

The potential drop in semiconductor region-1 is related to the thickness of the space charge region and is given by (2.5). Here q is unit charge and x_n is the space charge region width in the semiconductor region-1.

$$\phi_1 = \frac{qN_{d1}x_n^2}{2\epsilon_1} \qquad (2.5)$$

In the region $0 \le x \le \infty$, assume that the potential is zero at ∞. The carrier concentration in the above-said region is given by (2.6).

$$n(x) = N_{C2}e^{\left(\frac{E_f - E_{C1}(x) + q\phi(x)}{kT}\right)} \qquad (2.6)$$

In (2.6), k is Boltzmann's constant and T is absolute temperature. By using the relation between N_C and N_D, as given by (2.7), (2.6) can be simplified into (2.8).

$$N_D = N_C e^{\left(\frac{E_F - E_C(\infty)}{kT}\right)} \qquad (2.7)$$

$$n(x) = N_{D2}e^{\frac{q\phi(x)}{kT}} \qquad (2.8)$$

Writing Poisson's equation

$$-\epsilon_2 \frac{\partial^2 \phi(x)}{\partial x^2} = q\left[N_{D2} - n(x)\right] \tag{2.9}$$

By substituting (2.8) into (2.9) and simplifying results in

$$\frac{\partial}{\partial x}\left[\left(\frac{\partial \phi}{\partial x}\right)^2\right] = 2q\frac{N_{D2}}{\epsilon_2}\left[e^{\left(\frac{q\phi(x)}{kT}\right)} - 1\right]\frac{\partial \phi}{\partial x} \tag{2.10}$$

using the boundary conditions specified in (2.11) and integrating from 0 to ∞, results in

$$\phi(\infty) = 0$$
$$\phi(0) = \phi_2 \tag{2.11}$$

$$\left(\frac{\partial \phi}{\partial x}\right)^2\bigg|_{x=0} = \frac{2N_{D2}kT}{\epsilon_2}\left[\left(e^{\left(\frac{q\phi_2}{kT}\right)} - 1\right) - \frac{q\phi_2}{kT}\right] \tag{2.12}$$

thus, the electric field at the interface is given by

$$E(x)\big|_{x=0} = -\sqrt{\frac{2N_{D2}kT}{\epsilon_2}}\sqrt{\left[\left(e^{\left(\frac{q\phi_2}{kT}\right)} - 1\right) - \frac{q\phi_2}{kT}\right]} \tag{2.13}$$

At the interface, $x = 0$, Gauss's law can be written as

$$\epsilon_2 E(x)\big|_{x=0^+} = \epsilon_1 E(x)\big|_{x=0^-} = -qN_{D1}x_n \tag{2.14}$$

Using (2.13) and (2.14),

$$\sqrt{\frac{2N_{D2}kT}{\epsilon_2}}\sqrt{\left[\left(e^{\left(\frac{q\phi_2}{kT}\right)} - 1\right) - \frac{q\phi_2}{kT}\right]} = qN_{D1}x_n \tag{2.15}$$

Finally, V_{bi} can be written as

$$V_{bi} = \phi_1 + \frac{qN_{D2}x_n^2}{2\epsilon_2} \tag{2.16}$$

To find x_n, ϕ_1 and ϕ_2, (2.15) and (2.16) can be solved simultaneously.

2.2.4 Potential Well with Infinite Depth

A potential well with infinite energy depth with a width of a is shown in Figure 2.7. The potential is infinite in Region-I and Region-III, while potential is zero in Region-II. The potential is described as (2.17).

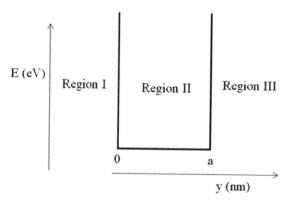

FIGURE 2.7
Square well potential of infinite depth.

$$V(y) = \begin{cases} 0 & \text{for } 0 \le y \le a \\ \infty & \text{for } y < 0, \ y > a \end{cases} \tag{2.17}$$

The time-independent Schrödinger equation that needs to be solved is given by (2.18). Here, m denotes effective mass of particles.

$$-\frac{\hbar^2}{2m}\frac{d^2\psi}{dy^2}\psi(y) = E\psi(y) \tag{2.18}$$

The boundary conditions that are applicable to this problem are

$$\psi(0) = 0$$
$$\psi(a) = 0 \tag{2.19}$$

The solutions to (2.18) are given by

$$\psi(y) = A\sin(ky) + B\cos(ky) \tag{2.20}$$

where

$$k = \sqrt{\frac{2mE}{\hbar^2}} \tag{2.21}$$

The boundary condition $\psi(0) = 0$ results in $B = 0$ and the boundary condition $\psi(a) = 0$ results in $ka = n\pi, n = 1,2,3,\ldots$.

Thus, the bound state energy of particle is given by

$$E_n = \frac{n^2 h^2}{8ma^2}, \ n = 1,2,3,\ldots \tag{2.22}$$

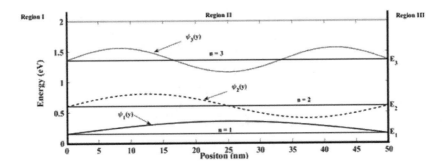

FIGURE 2.8
Energy levels and wavefunction description in a square well of infinite depth.

and wave function of particle in each bound state is given by

$$\psi(y) = A\sin(ky), n = 1, 2, 3, \dots \tag{2.23}$$

There is a possibility of infinitely many states in a potential well of infinite depth. The first few energy levels and wavefunctions are shown in Figure 2.8. In a real system the difference between conduction band minimum and to next energy level (at the best vacuum level) is finite. Hence the infinite potential is a very idealistic approximation of a real system.

2.2.5 Square Well Potential of Finite Depth

A square well potential of finite depth is shown in Figure 2.9. The square well potential of the finite depth can be described as (2.24).

$$V(y) = \begin{cases} -V_0 & \text{for } -\dfrac{a}{2} \le y \le \dfrac{a}{2} \\[2mm] 0 & \text{for } \dfrac{a}{2} \le |y| \end{cases} \tag{2.24}$$

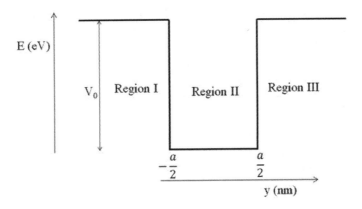

FIGURE 2.9
Description of square well potential of finite depth.

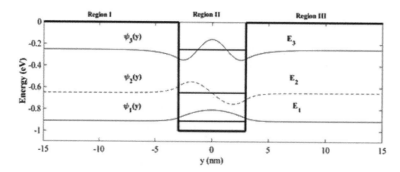

FIGURE 2.10
Energy levels and wavefunction description in a square well of infinite depth.

The time-independent Schrödinger equation that needs to be solved is given by (2.25).

$$\left(-\frac{\hbar^2}{2m}\frac{d^2}{dy^2}+V(y)\right)\psi(y)=E\psi(y) \tag{2.25}$$

In Region-II, the solutions are of the form

$$\psi(y)=A\sin(ky)+B\cos(ky) \tag{2.26}$$

Outside the well, the solutions are of the form

$$\psi(y)=Ce^{\pm\kappa y} \tag{2.27}$$

The solutions to the system are solved by using (2.26) and (2.27). The boundary conditions applicable are: $\psi(y)$ is well-behaved over the entire region, across the region boundaries, $y=-\frac{a}{2}$ and $y=\frac{a}{2}$ the wave function and its derivative are continuous. The bound state energy levels and corresponding wavefunctions are shown in Figure 2.10.

2.2.6 Triangular Potential Well

Consider the triangular potential well shown in Figure 2.11. The potential distribution is given by (2.28). The potential variation is linear with an infinite barrier at $y = 0$. The infinite barrier is assumed to eliminate the probability of tunneling in to the regions $y \leq 0$ to zero.

$$V(y)=\begin{cases}qFy & y>0\\ \infty & y\leq 0\end{cases} \tag{2.28}$$

The time-independent Schrödinger equation for the potential distribution in (2.28) is given by (2.29). In (2.29), F denotes electric field.

$$\left[-\frac{\hbar^2}{2m}\frac{d^2}{dy^2}+qFy\right]\psi(y)=E\psi(y) \tag{2.29}$$

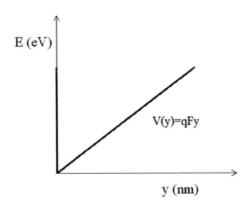

FIGURE 2.11
Schematic description of a triangular potential well.

The energy levels and wavefunctions are obtained by solving the Schrödinger equation subject to the boundary condition $\psi(y=0)=0$. Let

$$y_0 = \left(\frac{\hbar^2}{2mqF}\right)^{\frac{1}{3}}$$

$$E_0 = \left[\frac{(qF\hbar)^2}{2m}\right]^{\frac{1}{3}} = qFy_0$$

(2.30)

Using (2.30), the Schrödinger equation in (2.29) can be rewritten as

$$\frac{d^2\psi}{d\bar{y}^2} = (\bar{y} - \bar{E})\psi(\bar{y})$$

(2.31)

By changing the variable $\bar{y} - \bar{E} = s$, the Schrödinger equation can be written as Stoke's or Airy equation as shown in (2.32).

$$\frac{d^2\psi}{ds^2} = s\psi$$

(2.32)

The Schrödinger equation has two independent solutions $Ai(s)$ and $Bi(s)$, which are plotted in Figure 2.12. $Ai(s)$ and $Bi(s)$ are Airy functions. As the wavefunction needs to be bounded for $y \rightarrow \infty$, which is same as $s \rightarrow \infty$, the solutions constituted by $Bi(s)$ can be rejected. As shown in Figure 2.8, there are infinite values of s for which $Ai(s) = 0$. The allowed energy levels are given by (2.33). Here m indicates effective mass of particle and can be different in each bound state.

$$E_n = c_n \left[\frac{(qF\hbar)^2}{2m}\right]^{\frac{1}{3}} \quad n = 1,2,3,\ldots$$

(2.33)

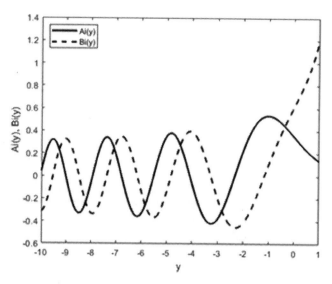

FIGURE 2.12
Airy functions $Ai(y)$ and $Bi(y)$.

where c_n is given by

$$c_n = \left[\frac{3}{2}\pi \left(n - \frac{1}{4} \right) \right]^{\frac{2}{3}} \tag{2.34}$$

The corresponding wave functions are given by

$$\psi_n(y) = Ai\left(\frac{qFy - E}{E_0} \right) \tag{2.35}$$

The wavefunctions and energy levels in a triangular potential well are depicted in Figure 2.13. Under suitable bias conditions, each energy level can accommodate electrons, which are free from scattering mechanisms. The electrons can move freely in energy level and are confined to move in the energy level to which they belongs to.

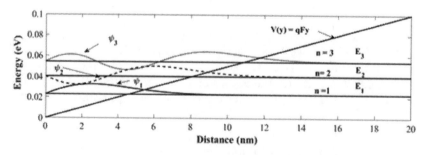

FIGURE 2.13
Description of energy levels and wavefunctions in a finite triangular potential well.

2.3 Principle of Operation of HEMTs

The structure used for HEMT's is shown in Figure 2.14. In AlGaAs/GaAs system, the HEMT structure consists of AlGaAs/GaAs layers grown on a semi-insulating GaAs substrate. Typical widths of these layers are shown in Figure 2.14. The transistor is finally realized by depositing a metal layer which forms a Schottky barrier and serves as gate. Aluminum can be used as a gate metal layer. Two ohmic contacts serve as contacts to the source and drain regions. Quantum well is formed in the GaAs layer near the interface with the AlGaAs. Carriers are supplied by the doped AlGaAs layer.

The transistors can be operated in two modes, normally-on, called as depletion mode and normally-off mode, called as enchantment mode. The mode of operation is determined by the thickness of the AlGaAs layers. When AlGaAs layers are thick enough, charge is supplied by the layer to fill up the surface states at the interface between AlGaAs layers and the gate metal and also to the GaAs layer for the alignment of the Fermi levels. In this case, the transistor is normally-on. Application of a suitable voltage on gate increases the depletion layer width and redistribution of the charges and in effect alters the carrier density and hence the current through the quantum well. In the GaAs layer near the heterojunction interface and the bound state electrons are removed. As a result, the current flow seizes, and the device is switched off.

On the other hand, if the barrier layer is made thin, then the charge available in the layer is not enough to cause alignment of the Fermi levels, and the GaAs layer is required to supply additional charge. The layer being depleted, the transistor is normally off. Application of a suitable bias on the gate results in redistribution of the charges, and in effect alters the carrier density and hence the current through the channel formed by the quantum well. In the GaAs layer near the heterojunction interface, electrons occupy the energy levels and HEMT conducts current. Both modes of operation are required for different kinds of transistor operations and different functional circuits. Realizing enhancement mode and depletion mode HEMT on the same chip is considered to be a great advantage.

Band diagram of an enhancement mode HEMT on application of a positive and negative gate voltage are shown in Figures 2.15 and 2.16, respectively.

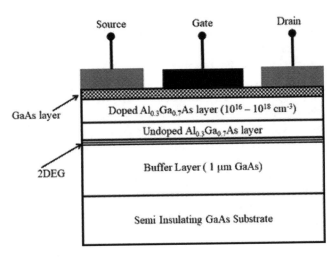

FIGURE 2.14
HEMT structure in AlGaAs/GaAs system.

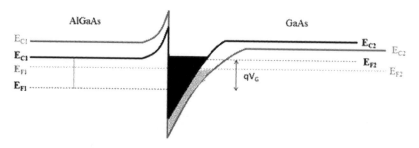

FIGURE 2.15
Band diagram depicting band bending and 2DEG formation in an AlGaAs/GaAs HEMT when $V_G < 0$. Dark lines indicate $V_G < 0$, and shaded, colored lines indicate thermal equilibrium band diagram when $V_G = 0$.

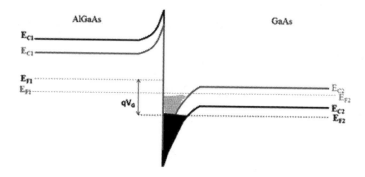

FIGURE 2.16
Band diagram depicting band bending and 2DEG formation in an AlGaAs/GaAs HEMT when $V_G > 0$. Dark lines indicate $V_G < 0$, and shaded, colored lines indicate thermal equilibrium band diagram when $V_G = 0$.

An expression for the current is obtained by first working out the potential distribution in the structure and determining the carrier density, and then using the mobility-field relations to evaluate the current as explained below.

2.3.1 Threshold Voltage

The potential distribution on application of a negative gate voltage (V_G) between the gate metal and small bandgap material is shown in Figure 2.17. For the ease of discussion, assume that the material system used in these calculations is AlGaAs/GaAs and the band alignment is shown in Figure 2.17. The Fermi level is aligned in AlGaAs and GaAs layers while in position of Fermi level in lowered by qV_G in metal. Due to applied bias, two depletion regions are formed—one at the gate-semiconductor interface (region marked as A) and the other at heterojunction (region marked with B). In region-A, electrons get accumulated on the metal surface, which produce an electric field E_s. The positively charged ionized donors in the AlGaAs layer produce a positive field gradient, which counters E_s at some distance t_1 and produces equilibrium. Then there exists a field free region of a thickness ($t_2 - t_1$). It is further assumed that the AlGaAs layer thickness is t_d and the undoped layer thickness is t_u. A detailed discussion of electrostatics can be found in [11].

The threshold voltage for charge control V_{th1} is given by (2.36).

$$V_{th1} = V_s - \frac{\Delta E_{F2}}{q} - a(d - d_0) \tag{2.36}$$

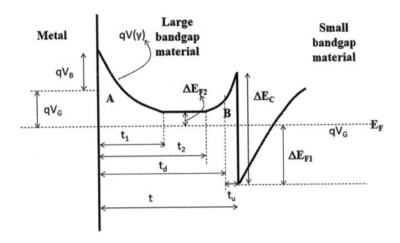

FIGURE 2.17
Potenital distribution in a HEMT on application of gate bias.

where a and d_0 are given by (2.37) and (2.38) respectively. N_D and ϵ_1 are doping concentration and permittivity of AlGaAs layer.

$$a = \frac{qN_D}{\epsilon_1} \tag{2.37}$$

$$ad_0^2 = ad^2 + \frac{\Delta E_C - \Delta E_{F2} - \Delta E_{F1}}{q} \tag{2.38}$$

The threshold voltage to control gate, V_{th2}, is given by (2.39).

$$V_{th2} = V_B - \frac{\Delta E_C}{q} - \frac{E_F}{q} - at_d^2 \tag{2.39}$$

To operate a HEMT, along with the charge accumulation at the interface, the charge needs to be controlled using a gate voltage. Hence, the applied gate bias needs to be between V_{th1} and V_{th2}. For enhancement mode HEMTs, the thickness t of the AlGaAs layer is such that charge accumulates till $V_G > V_{th2}$ and remains controllable up to $V_G < V_{th1}$, thus, $V_{th1} > V_{th2} > 0$. In case of depletion mode HEMTs, the t is chosen such that $V_{th2} < 0$ and $|V_{th2}| > |V_{th1}|$. Thus, the operating conditions of a depletion mode HEMT are $|V_{th2}| > |V_G| > |V_{th1}|$.

2.3.2 Current Voltage Characteristics

Assume that to a HEMT with a channel length L, a gate voltage V_{GS} and drain voltage V_{DS} are applied with respect to the source. Let the current conduction be in x-direction and gate on the source is the origin. The applied voltage between drain and source varies as a function of x and indicated by $V(x)$. The effective voltage V_{eff} at a position x is given by (2.40) [11].

$$V_{eff} = V_G - V(x) \tag{2.40}$$

The surface charge density at x is given by (2.41).

$$qn(x) = \frac{\epsilon_1}{d}\left(V_{eff} - V_{th2}\right) \tag{2.41}$$

If W is the width of gate electrode, then the current through the channel region is given by (2.42).

$$I = \mu(E)qn(x)\frac{dV}{dx}W \tag{2.42}$$

By substituting for $n(x)$, assuming constant mobility and integrating from the source end to drain end yields total current through the channel.

2.4 Modulation Doping and 2DEG

The modulation doping (MD) technique has its origin from semiconductor heterostructures and superlattices [12]. The charge carriers are confined by heterojunctions, and the carriers occupy quantized energy states. The quantization occurs in the dimension perpendicular to the heterostructure, whereas the motion parallel to the interface is free. With only two degrees of free motion available, the charge carriers in each quantum state form a two-dimensional system. The modulation structures are designed to separate donors from mobile electrons into increase the time between scattering events, thus increasing the speed of operation.

Schematic representation of a modulation doped structure is shown in Figure 2.18. Low energy of electron states in the wells relative to electrons residing in the barriers results in a finite well. As a result, a finite depth square well is formed. The charge transfer and carrier confinement which form the basis for most modulation doping

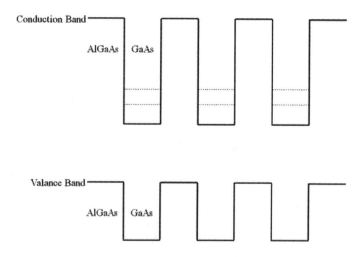

FIGURE 2.18
Schematic representation of modulation doping. Dashed lines indicate energy levels.

effects in heterostructures are produced by the step in the band edge energies occurring at the heterojunction. These band-edge discontinuities form an upper limit for the difference in energy between electrons bound to a donor in the barrier layer near the interface and electrons in the channel layer. The electrons are thus confined to the bound states and form 2DEG. The mobility enhancement is mainly due to the reduction in scattering of the carriers from impurities by means of spatial separation of the impurities and the mobile electrons. The actual energy difference between energy bands are generally far less than the theoretical band edge energy differences and are reduced by several effects.

2.5 HEMT Material Systems

Recent works on HEMTs by various research groups indicates that InP HEMTs shows superior transistor gain, transition frequency and low-noise behavior compared with GaN and GaAs HEMTs due to the high electron mobility, electron saturation velocity and two-dimensional sheet charge densities (2DEG) of the InGaAs or InAs channel materials. Further, in an InP, parasitics can be reduced as aggressive scaling is possible by proper selection of device architecture. Due to low cost and reduced process complexity, GaAs substrate-based HEMTs, MMICs and terahertz monolithic integrated circuits (TMICs) are most cost effective when compared to InP substrate-based devices. On the other hand, GaAs-based HEMTs are more suitable for high-frequency, low-power applications, while GaN-based HEMTs are useful for high-power, high-frequency and low-noise systems. AlGaAs/GaAs HEMT system is shown in Figure 2.19. InAlGaAs/InGaAs HEMT system is shown in Figure 2.20. AlGaN/GaN HEMT system is shown in Figure 2.21.

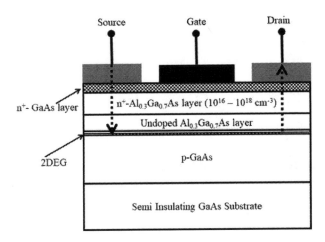

FIGURE 2.19
AlGaAs/GaAs HEMT system. Arrow indicates the direction of electron flow.

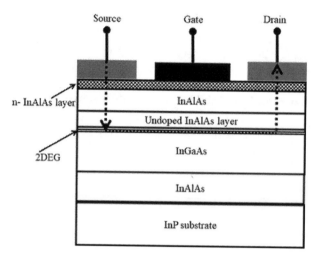

FIGURE 2.20
InAlGaAs/InGaAs HEMT system. Arrow indicates the direction of electron flow.

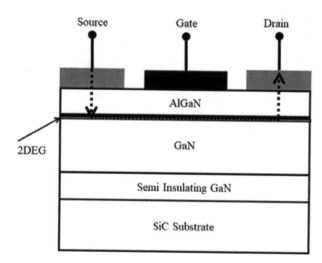

FIGURE 2.21
AlGaN/GaN HEMT system. Arrow indicates the direction of electron flow.

2.6 Classification of HEMTs

HEMTs can be classified based on structure and device operation. To form a quantum well and thus a 2DEG, two different materials are required. An ideal requirement is to have same lattice constant across the heterojunction. However, in practice, the two different materials used to from a heterojunction would have different lattice constants, typically slightly different. The lattice mismatch thus results in crystal defects. In semiconductors, these lattice mismatches result in lattice discontinuities which in turn form deep-level traps and greatly reduce device performance.

Using a thin crystal lattice, the lattice mismatch can be avoided. A very thin layer stretches itself to fit to the other lattice. A HEMT where the lattice mismatch is avoided using thin crystal layer is called as pseudomorphic HEMT (pHEMT). Due to reduced dislocations, pHEMT allows fabrication of transistors with larger bandgap differences than otherwise possible, giving them better performance.

If the lattice mismatch is avoided is using a buffer layer at the heterojunction, the HEMT is called as metamorphic HEMT or mHEMT. The buffer layer is achieved by using a concentration gradient. For example, by varying the indium concentration, the lattice constant of both the GaAs substrate and the GaInAs channel can be matched. Graded doping concentration provides an advantage that a desired concentration in the channel can be realized, so the devices can be optimized for different applications.

An induced HEMT is a device in which charge carriers are "induced" to the 2DEG plane rather than created by dopants. The absence of a doped layer enhances the electron mobility significantly when compared to their modulation-doped counterparts. As the doping is reduced, thermal noises due to dopant ionization are reduced, thus has applications in ultra-stable and ultra-sensitive electronic devices. For example, Unlike AlGaAs/GaAs HEMT, which require intentional doping to form 2DEG, 2DEG in AlGaN/GaN HEMT is due to induced polarization. electrons in 2DEG are due to surface states.

Due to the electronegativity differences, Al-N and Ga-N bonds are highly ionic and each carry a strong dipole. As the electronegativity of N is much higher than that of Ga, the electron wave function around the Ga-N pair is offset to the nitrogen side. The effect is even more exaggerated in the Al-N pair, which is a special feature of the III-nitrides and the degree of spontaneous polarization is more than five times greater than in most III-V semiconductors. In an AlGaN/GaN system, AlGaN layer has two different polarization vectors, due to the piezoelectric and spontaneous polarizations, respectively. The polarization in the AlGaN results in formation of dipole charges with a positive sheet charge at the AlGaN/GaN junction. The polarization in the GaN layer causes a negative sheet charge at the AlGaN/GaN junction. Since the total polarization in the AlGaN is larger, the overall result is a net positive sheet charge at the AlGaN/GaN interface. The presence of charges and polarization induces an electric field in AlGaN resulting in a 2DEG formation without barrier doping.

References

1. T. Mimura, Semiconductor device, Japan Patent 1409643 (1987).
2. R. Dingle et al., Electron mobilities in modulation-doped semiconductor heterojunction superlattices, *Applied Physics Letters*, 33 (1978), 665.
3. E. C. Niehenke, The evolution of transistors for power amplifiers: 1947 to today, *Microwave Symposium (IMS), 2015 IEEE MTT-S International*, IEEE (2015).
4. US 4471366, Field effect transistor with high cut-off frequency and process for forming same, more details at https://patents.justia.com/patent/4471366.
5. T. Mimura, K. Joshin, S. Hiyamizu, K. Hikosaka and M. Abe, High electron mobility transistor logic, *Japanese Journal of Applied Physics*, 20(8) (1981).
6. T. Mimura, Development of high electron mobility transistor, *Japanese Journal of Applied Physics*, 44(12) (2005), 8263–8268.
7. Y. Watanabe, S. Saito, N. Kobayashi, M. Suzuki, T. Yokoyama, E. Mitani, K. Odani, T. Mimura and M. Abe, A HEMT LSI for a multibit data register, *IEEE International Solid-State Circuits Conference, Digest of Technical Papers, ISSCC* (1988).

8. M. A. Khan, A. Bhattarai, J. N. Kuznia and D. T. Olson, High electron mobility transistor based on a GaN-AlxGa1–xN heterojunction, *Applied Physics Letters*, 63 (1993), 1214.

9. H. Xing, S. Keller, Y.-F. Wu, L. McCarthy, I. P. Smorchkova, Gallium nitride based transistors, *Journal of Physics: Condensed Matter* 13 (2001).

10. T. Mimura, The early history of the high electron mobility transistor (HEMT), *IEEE Transactions on Microwave Theory and Techniques*, 50(3) (2002).

11. B. R. Nag, *Physics of Quantum Well Devices*. Dordrecht, the Netherlands: Springer (2002).

12. L. L. Chang and K. Ploog, *Molecular Beam Epitaxy and Heterostructures, NATO Science Series E: Applied Sciences*. Dordrecht, the Netherlands: Springer (1985).

3

HEMT Material Technology and Epitaxial Deposition Techniques

Rama Komaragiri

CONTENTS

3.1 III-V Compound Semiconductors

The elemental conductors belong to Group 14 (earlier, Group IVA) and have four valence electrons, and covalent bonding prevails. Compound semiconductors are formed by chemically combining elements from various groups in the periodic table with an average valency of four. For example, consider the family of III-V semiconductors, developed from cations of Group 13 (earlier, Group IIIA) and anions of Group 15 (earlier, Group VA). Consider a specific example of a compound semiconductor, gallium arsenide, GaAs. Gallium belongs to Group 13 and has three valence electrons, while arsenic belongs to Group 15 and has five valence electrons. The one-unit volume of GaAs semiconductor contains 50% of Ga atoms and 50% of As atoms. Thus, each Ga atom contributes three valence electrons while As atom contributes five valence electrons, thus the compound semiconductor GaAs has an average of four valence electrons.

Electronegativity, the tendency of an atom to attract a shared pair of electrons, is affected by both its atomic number and the distance between the valence electrons and the nucleus. Higher the electronegativity, the more an element attracts electrons. Excluding inert gases, generally, electronegativity increases while passing from left to right along a period, and decreases on descending a group.

The electron affinity of an atom is the amount of energy released when an electron is added to a neutral atom to form a negative ion. Generally, nonmetals have more positive electron affinity values than metals. Electron affinity generally increases across a period the and decreases while going down groups.

The atomic radii decrease while passing from left to right along a period, and increases on descending a group. There is a sharp increase in the atomic radius between the noble gas at the end of each period and the alkali metal at the beginning of the next period.

Generally, low values of ionization energy, electronegativity and electron affinity result in the more metallic character of an element.

If one compares the compound semiconductors, gallium arsenide and gallium phosphide, even though the cation is same, the anion different. Therefore, from the above discussion As is more metallic than P; hence the properties of GaAs and GaP differ significantly.

There are three important categories of III-V semiconductors, namely, binary, ternary and quaternary. An undoped binary semiconductor is made up of two elements, an element each belonging to Group-13 and Group-15. Examples of III-V binary compound semiconductors are GaAs, InAs, GaN, to name a few. A ternary compound semiconductor consists of three elements. Examples of III-V ternary compound semiconductors are $Ga_xIn_{(1-x)}Sb$ and $GaAs_ySb_{(1-y)}$. In a III-V ternary compound semiconductor, 50% of the atoms are constituted from Group-13 elements and remaining 50% is constituted from the Group-15 element. For example, in $Ga_xIn_{(1-x)}Sb$, Ga and In atoms constitute 50% of the total atoms while Sb constitutes the remaining 50% of the total atoms. The percentage fraction of Ga and In add up to one. Thus, Ga and In contribute to a valency of three, and Sb contributes to a valency of five. Thus, the average valency remains at four. Similarly, in the case of $GaAs_ySb_{(1-y)}$, Ga constitutes 50% of atoms while As and Sb constitute remaining 50% of atoms. A quaternary III-V compound semiconductor consists of four elements, in which 50% of the atoms belong to Group-13, and 50% of the elements belong to Group-15. An example of a quaternary semiconductor is $Ga_xIn_{(1-x)}As_ySb_{(1-y)}$.

In ternary and quaternary semiconductors, different values of mole fractions "*x*" or "*y*" result in different constituents. As the variation of mole fraction in a compound semiconductor results in different values of electron affinity, electronegativity, metallic characteristic, distinct and continuous electronic, mechanical and optical properties of a semiconductor can be achieved. Hence, the properties of a compound semiconductor can be tuned as per requirements of the application. Thus, ternary and quaternary semiconductors provide a distinct advantage over binary semiconductors. The major disadvantage of ternary and quaternary semiconductors is their material and fabrication related costs.

3.2 Physical Properties of III-V Compound Semiconductors

At 300 K, Table 3.1 lists various physical properties of binary III-V compound semiconductors, and Table 3.2 lists various physical properties of ternary and quaternary III-V compound semiconductors. The data are compiled from references [2–5] and references therein. The properties of quaternary compounds depend on the lattice matching. In this chapter, the properties of $Ga_xIn_{(1-x)}As_ySb_{(1-y)}$ lattice matched to GaSb are provided.

TABLE 3.1

Physical Properties of Some Binary III-V Semiconductors

Property	Unit	GaAs	GaP	InAs	InP	GaN[a]
Crystal structure		Zinc Blend	Zinc Blend	Zinc Blend	Zinc Blend	Wurzite
Lattice constant	°A	5.65325	5.4505	6.0583	5.8687	a: 3.160 c: 5.125
Density	g/cm^{-3}	5.32	4.14	5.86	4.81	6.15
Group symmetry		$T_d^2 - F\bar{4}3m$	$T_d^2 - F\bar{4}3m$	$T_d^2 - F\bar{4}3m$	$T_d^2 - F\bar{4}3m$	$6_{6v}^4 - P6_33m$
Number of atoms in 1 cm^3	$\times 10^{22}$	4.42	4.94	3.59	3.96	8.9
Electron affinity	eV	4.07	3.8	4.9	4.38	4.1

Source: Levinshtein, M. et al. (Eds.), *Handbook Series on Semiconductor Parameters Volume 1: Si, Ge, C (Diamond), GaAs, GaP, GaSb, InAs, InP, InSb*, World Scientific, Singapore; http://www.ioffe.ru/SVA/NSM/Semicond/; Levinshtein, M. et al., *Handbook Series on Semiconductor Parameters Volume 2: Ternary and Quaternary III–V Compounds*, World Scientific, Singapore.

[a] http://www.ioffe.ru/SVA/NSM/Semicond/GaN/basic.html.

TABLE 3.2

Physical Properties of Some Ternary and Quaternary III-V Semiconductors

Property	Unit	GaAsSb	GaInSb	$Ga_xIn_{(1-x)}As_ySb_{(1-y)}$[b]
Crystal structure		Zinc Blend	Zinc Blend	
Lattice constant	°A	5.6532 – 6.0959[a]	6.479 – 0.383x	6.0959
Density	g/cm^{-3}	5.32 + 0.29x	5.77 – 0.16x	5.69 – 0.08x
Group symmetry		$T_d^2 - F\bar{4}3m$	$T_d^2 - F\bar{4}3m$	
Number of atoms in 1 cm^3	$\times 10^{22}$	4.42 – 0.89x	2.94 + 0.59x	3.53
Electron affinity	eV	4.07	4.59 – 0.53x	4.87 – 0.81x

Source: Levinshtein, M. et al., *Handbook Series on Semiconductor Parameters Volume 2: Ternary and Quaternary III–V Compounds*, World Scientific, Singapore; Adachi, S. (Ed.), *Physical Properties of III–V Semiconductor Compounds: InP, InAs, GaAs, GaP, InGaAs, and InGaAsP*, John Wiley & Sons, Hoboken, NJ.

[a] Non-linear function between $x = 0$ to $x = 1$.
[b] These are approximations, a realistic value depends upon the actual values of x and y.

3.3 Electrical Properties of III-V Compound Semiconductors

At 300 K, Table 3.3 lists various electrical properties of binary III-V compound semiconductors, and Table 3.4 lists various electrical properties of ternary and quaternary III-V compound semiconductors. The data are compiled from references [2–8] and references therein. The properties of quaternary compounds depend on the lattice match. In this chapter, the properties of $Ga_xIn_{(1-x)}As_ySb_{(1-y)}$ lattice matched to GaSb are provided.

TABLE 3.3

Electrical Properties of Some Binary III-V Semiconductors

Property	Unit	GaAs	GaP	InAs	InP	GaN
Bandgap	eV	1.424	2.26	0.354	1.344	3.39
Intrinsic Carrier concentration (n_i)	cm^{-3}	2.1×10^6	2	10^{15}	1.3×10^7	10^{10}
Effective mass of electrons (m_e) (in m_0)		0.063	$m_l = 1.12$ $m_t = 0.22^a$	0.023	0.08	0.20
Heavy effective mass of holes (m_{hh}) (in m_0)		0.51	0.79	0.41	0.6	1.4
Light effective mass of holes (m_{lh}) (in m_0)		0.082	0.14	0.026	0.089	0.3
Conduction band density of states (N_C)	cm^{-3}	4.7×10^{17}	1.8×10^9	8.7×10^{16}	5.7×10^{17}	2.3×10^{18}
Valence band density of states (N_V)	cm^{-3}	9×10^{18}	1.9×10^9	6.6×10^{18}	1.1×10^{19}	4.6×10^{19}
Electron mobility (μ_n)	$cm^2V^{-1}s^{-1}$	≤ 8500	≤ 250	$\leq 4 \times 10^4$	≤ 5400	≤ 1000
Hole mobility (μ_p)	$cm^2V^{-1}s^{-1}$	≤ 400	≤ 150	≤ 500	≤ 200	≤ 350
Diffusion coefficient of electrons (D_e)	cm^2s^{-1}	≤ 200	≤ 6.5	$\leq 10^3$	≤ 130	25
Diffusion coefficient of holes (D_p)	cm^2s^{-1}	≤ 10	≤ 4	≤ 13	≤ 5	9
Thermal velocity of electrons (v_e)	ms^{-1}	4.4×10^5	2×10^5	7.7×10^5	3.9×10^5	2.6×10^5
Thermal velocity of electrons (v_e)	ms^{-1}	1.8×10^5	1.3×10^5	2×10^5	1.7×10^5	9.4×10^4
Breakdown field	Vcm^{-1}	4×10^5	10^6	$\simeq 4 \times 10^4$	5×10^5	5×10^6
Static relative dielectric constant		12.9	11.1	15.15	12.5	10.4
High-frequency relative dielectric constant		10.89	9.11	12.3	9.61	

Source: Adachi, S. (Ed.), *Physical Properties of III-V Semiconductor Compounds: InP, InAs, GaAs, GaP, InGaAs, and InGaAsP*, John Wiley & Sons, Hoboken, NJ; Look, D. C. et al., *Solid State Commun.*, 102, 297–300, 1997; https://www.azom.com/article.aspx?ArticleID=8370; Levinshtein, M. et al. (Eds.), *Handbook Series on Semiconductor Parameters Volume 1: Si, Ge, C (Diamond), GaAs, GaP, GaSb, InAs, InP, InSb*, World Scientific, Singapore; http://www.ioffe.ru/SVA/NSM/Semicond/.

^{a} m_l and m_t indicate longitudinal and traverse effective masses respectively.

TABLE 3.4

Electrical Properties of Some Ternary and Quaternary III-V Semiconductors

Property	Unit	GaAsSb	GaInSb	$Ga_xIn_{(1-x)}As_ySb_{(1-y)}$[a]
Bandgap	eV	$1.42 - 1.9x + 1.2x^2$ (0 < x < 0.3)	$0.014 + 0.0178x + 0.0092x^2$	$0.022 + 0.03x - 0.012x^2$
Intrinsic carrier concentration (n_i)	cm^{-3}			
Effective mass of electrons (m_e) (in m_0)		$0.063 - 0.0495x + 0.0258x^2$	$0.43 - 0.03x$	0.4
Heavy effective mass of holes (m_{hh}) (in m_0)		$0.51 - 0.11x$		
Light effective mass of holes (m_{lh}) (in m_0)		$0.082 - 0.032x$	$0.015 + 0.01x + 0.025x^2$	$0.025(1 + x)$
Conduction band density of states (N_C)	cm^{-3}	$2.5 \times 10^{19}[0.063 - 0.0495x + 0.0258\,x^2]^{3/2}$	$2.5 \times 10^{19}[0.014 + 0.0178x + 0.0092x^2]^{3/2}$	$2.5 \times 10^{19}[0.022 + 0.03x - 0.012x^2]^{3/2}$
Valence band density of states (N_V)	cm^{-3}	Refer [4]	Refer [4]	$2.5 \times 10^{19}[0.41 + 0.16x + 0.23x^2]^{3/2}$
Electron mobility (μ_n)	cm^{2}V^{-1}s^{-1}	Refer [4]	Refer [4]	Refer [4]
Hole mobility (μ_p)	cm^{2}V^{-1}s^{-1}	Refer [4]	≤1000	Refer [4]
Diffusion coefficient of electrons (D_e)	cm^{2}s^{-1}	Refer [4]	Refer [4]	Refer [4]
Diffusion coefficient of holes (D_p)	cm^{2}s^{-1}	Refer [4]	Refer [4]	Refer [4]
Thermal velocity of electrons (v_t)	ms^{-1}	$4.4 \times 10^5 (1 + 0.4x - 0.09x^2)$	$(9.8 - 4x) \times 10^5$	Refer [4]
Thermal velocity of electrons (v_t)	ms^{-1}	Refer [4]	$(1.3 + 0.3x) \times 10^5$	
Breakdown field	Vcm^{-1}	Refer [4]	≃10^3	≃5 × 10^4
Static relative dielectric constant		$12.9 + 2.8x$	$16.8 - 1.1x$	$15.3 + 0.4x$
High-frequency relative dielectric constant		$10.89 + 3.51x$	$15.7 - 1.3x$	$12.6 + 1.8x$

Source: http://www.ioffe.ru/SVA/NSM/Semicond/; Levinshtein, M. et al., *Handbook Series on Semiconductor Parameters Volume 2: Ternary and Quaternary III–V Compounds,* World Scientific, Singapore; Adachi, S. (Ed.), *Physical Properties of III–V Semiconductor Compounds: InP, InAs, GaAs, GaP, InGaAs, and InGaAsP,* John Wiley & Sons, Hoboken, NJ; https://www.azom.com/article.aspx?ArticleID=8370.

a Lattice-matched to GaSb.

3.4 Mechanical Properties of III-V Compound Semiconductors

At 300 K, Table 3.5 lists various mechanical and thermal properties of binary III-V compound semiconductors, and Table 3.6 lists various mechanical and thermal properties of ternary and quaternary III-V compound semiconductors. The data are compiled from references [2–8] and references therein.

TABLE 3.5

Mechanical and Thermal Properties of Some Binary III-V Semiconductors

Property	Unit	GaAs	GaP	InAs	InP	GaN
Bulk modulus	Nm^{-2}	0.753	0.88	0.58	0.71	0.204
Melting point	°C	1240	1457	942	1060	>2250
Specific heat	$Jg^{-1}°C^{-1}$	0.33	0.43	0.25	0.31	0.49
Thermal conductivity	$Wcm^{-1}°C^{-1}$	0.55	1.1	0.27	0.68	1.3
Thermal diffusivity	cm^2s^{-1}	0.31	0.62	0.19	0.372	0.43
Coefficient of linear thermal expansion	$°C^{-1}$	5.73×10^{-6}	4.65×10^{-6}	4.52×10^{-6}	4.60×10^{-6}	3.17×10^{-6}

Source: http://www.ioffe.ru/SVA/NSM/Semicond/; Levinshtein, M. et al., *Handbook Series on Semiconductor Parameters Volume 2: Ternary and Quaternary III–V Compounds*, World Scientific, Singapore; Adachi, S. (Ed.), *Physical Properties of III–V Semiconductor Compounds: InP, InAs, GaAs, GaP, InGaAs, and InGaAsP*, John Wiley & Sons, Hoboken, NJ; Look, D. C. et al., *Solid State Commun.*, 102, 297–300, 1997; https://www.azom.com/article.aspx?ArticleID=8370.

TABLE 3.6

Mechanical and Thermal Properties of Some Ternary and Quaternary III-V Semiconductors

Property	Unit	GaAsSb	GaInSb	$Ga_xIn_{(1-x)}As_ySb_{(1-y)}$[a]
Bulk modulus	Nm^{-2}	$0.753 - 0.019x$	$0.466 + 0.096x$	0.570
Melting point	°C	Refer [4]	Refer [4]	Refer [4]
Specific heat	$Jg^{-1}°C^{-1}$	Refer [4]	Refer [4]	0.25
Thermal conductivity	$Wcm^{-1}°C^{-1}$	Refer [4]	Refer [4]	Refer [4]
Thermal diffusivity	cm^2s^{-1}	Refer [4]	Refer [4]	Refer [4]
Coefficient of linear thermal expansion	$°C^{-1}$	Refer [4]	Refer [4]	Refer [4]

Source: http://www.ioffe.ru/SVA/NSM/Semicond/; Levinshtein, M. et al., *Handbook Series on Semiconductor Parameters Volume 2: Ternary and Quaternary III–V Compounds*, World Scientific, Singapore; Adachi, S. (Ed.), *Physical Properties of III–V Semiconductor Compounds: InP, InAs, GaAs, GaP, InGaAs, and InGaAsP*, John Wiley & Sons, Hoboken, NJ; Look, D. C. et al., *Solid State Commun.*, 102, 297–300, 1997; https://www.azom.com/article.aspx?ArticleID=8370, last accessed on 24th June 2018.
[a] Lattice-matched to GaSb.

TABLE 3.7

Optical Properties of Some Binary III-V Semiconductors

Property	Unit	GaAs	GaP	InAs	InP	GaN
Infrared refractive index		3.3	3.02	3.51	3.1	2.29
Radioactive recombination coefficient	cm^3s^{-1}	7×10^{-10}	10^{-13}	1.2×10^{-10}	1.2×10^{-10}	1.1×10^{-8}

TABLE 3.8

Optical Properties of Some Ternary and Quaternary III-V Semiconductors

Property	Unit	GaAsSb	GaInSb	$Ga_xIn_{(1-x)}As_ySb_{(1-y)}$[a]
Infrared refractive index		$3.3 + 0.5x$	$4 - 0.2x$	Refer [4]
Radioactive recombination coefficient	cm^3s^{-1}	$\sim 10^{-10}$	$\sim 10^{-10}$	$\sim 10^{-10}$

Source: http://www.ioffe.ru/SVA/NSM/Semicond/; Levinshtein, M. et al., *Handbook Series on Semiconductor Parameters Volume 2: Ternary and Quaternary III–V Compounds*, World Scientific, Singapore; Adachi, S. (Ed.), *Physical Properties of III–V Semiconductor Compounds: In P, InAs, GaAs, GaP, InGaAs, and InGaAsP*, John Wiley & Sons, Hoboken, NJ.

[a] Lattice-matched to GaSb.

3.5 Optical Properties of III-V Compound Semiconductors

At 300 K, Table 3.7 lists various optical properties of binary III-V compound semiconductors, and Table 3.8 lists various optical properties of ternary and quaternary III-V compound semiconductors. The data are compiled from references [2–8] and references therein.

3.6 Challenges in III-V MOS Technologies

The ultimate challenge for the III-V MOSFETs is to achieve a comparable channel performance of HEMTs in ON state. A comparison with mature HEMTs [9–15] easily provides the essential requirements for III-V MOSFETs technology suitable for logic circuit applications, which are tabulated in Table 3.9. While addressing individual issues, improvement in one issue results in degradation in another issue, and hence there exists a trade-off between design and technology. Superior electron transport properties of III-V Compound semiconductors are a major reason for advancing CMOS technology beyond silicon and beyond Moore technologies. Similar issues like Coulomb scattering due to interface charges, surface roughness and soft phonon scattering issues faced by

TABLE 3.9

Challenges in III-V MOSFET Technology

Property	Issues	Approaches
High drain current	Low effective mass, high carrier density with low scattering	Buried channel devices, high mobility [16,17]
Low gate leakage current density and high gate capacitance	Electrostatic integrity, scalability	High-k gate [18,19]
High-temperature process stability of gate stack	Passivation, annealing and dopant activation	Preventing inter diffusion [20–23]
Low interface trap density	Large subthreshold swing and reduced speed of the device and high-power operation	Interface passivation with a trap density below 10^{11} cm^{-2} eV^{-1} [24]
Low source/drain resistance	large parasitic resistances and scaling issues	Implanted source /drain structures, modulation doped channel, and crystal regrowth [16,22,25–27]
p-Channel	Low hole mobility and large hole effective mass, difficult to implement CMOS ratioed logic	Strained channel to split heavy/light holes, different channel materials with large mobility [28]
Integration with Si	Lattice and thermal mismatches resulting in high defect densities	Growing of III-V's on Si with thick buffer [29,30], epitaxially grown oxide interlayers [31], wafer bonding [32]
High device integration density	Low defect density	No known solutions yet
3D gate structures	Non-planar structures	Schottky gates [33,34]; surface passivation [35]

Source: Oktyabrsky, S. and Peide, D. Y., *Fundamentals of III–V Semiconductor MOSFETs*, Springer, New York, 2010.

III-V semiconductors [36–42]. Reduction of scattering mechanisms may be achieved by inserting a thick wide-band gap semiconductor barrier spacer between the channel and oxide.

3.7 Molecular Beam Epitaxy

Molecular Beam Epitaxy (MBE) at its root has a basic process of ultra-high vacuum evaporation. The fundamental process is due to the foundation and pioneering work and can be found in the literature [43–46]. The strength of MBE is in understanding the atomic level growth of the epitaxial film. A treatise on MBE is found in the literature [47]. During the process of MBE, the growth of materials takes place under UHV conditions on a heated crystalline substrate by the interaction of adsorbed species supplied by atomic or molecular beams [46]. The deposited layers have the same crystalline structure as the substrate or a structure with a similar symmetry and with a maximum lattice mismatch of 10%. The molecular beams generally have thermal energy and are produced by evaporation or sublimation of suitable materials contained in ultra-pure crucibles.

An MBE system (refer to Figure 3.1 in reference [47]) typically consists of a growth chamber and auxiliary chamber and a load-lock connected to associated pumping

Group Period	Group-13	Group-14	Group-15
2	Boron (B, 5)	Carbon (C, 6)	Nitrogen (N, 7)
3	Aluminum (Al, 13)	Silicon (Si, 14)	Phosphorus (P, 15)
4	Gallium (Ga, 31)	Germanium (Ge, 32)	Arsenic (As, 33)
5	Indium (In, 49)	Tin (Sn, 50)	Antimony (Sb, 51)
6		Led (Pb, 82)	Bismuth (Bi, 83)

FIGURE 3.1
Group 13, Group 14, and Group 15 elements of periodic table. The symbol and number in the bracket indicate IUPAC name and atomic number respectively. An interactive periodic table is available at https://www.ptable.com/#Writeup/Wikipedia [1].

systems. Using the load lock, one can introduce or remove samples or wafer without significantly affecting the vacuum in the growth chamber. Auxiliary chambers contain various tools such as additional deposition equipment, processing equipment, and analytical tools.

A schematic of a growth chamber consists various sources, control systems and gauges (refer to Figure 3.2 in reference [47]). Sources or molecular beams, a hearing system, a setup to translate or rotate sample stage, a cryoshroud surrounding the growth region, shutters to control the molecular beams, different gauges to measure pressure, temperature and molecular beam fluxes, reflection electron diffraction gun, systems to monitor film surface and its structure, a mass analyzer to monitor the background gas species and molecular beam flux compositions are integral parts of a MBE growth chamber.

An auxiliary chamber hosts a wide variety of process and analytical equipment. Further, hearing arrangements and ion bombardment equipment might be required to clean the sample surface. The process equipment further includes sources for deposition.

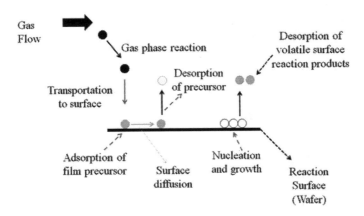

FIGURE 3.2
Representation of a CVD process. (From Jones, A. C. and O'Brien, P., *CVD of Compound Semiconductors*, VCH, Weinheim, CA, 1997.)

TABLE 3.10

Comparison of Various Deposition Techniques

| Parameter | Liquid Phase Epitaxy | Vapor Phase Epitaxy | | Molecular Beam Epitaxy |
		Chemical Vapor Deposition (CVD)	Metal Organic CVD	
Growth rate (μm/min)	~1	~0.1	~0.1	~0.01
Growth Temperature (°C)	800	700	700	550
Thickness Control (nm)	~50	~25	~2.5	~0.5
Interface thickness (nm)	≥ 5	≤ 5	≤ 1	≤ 0.5
Doping range (cm^{-3})	$10^{13}–10^{19}$	$10^{13}–10^{19}$	$10^{14}–10^{19}$	$10^{14}–10^{19}$

The MBE growth process involves controlling the molecular and/or atomic beams directed at a suitable heated single crystal sample to achieve epitaxial growth. The opening or closing of a shutter during the process facilitates control of different molecular beams. A gas background minimizes unintentional contamination. A deposition rate of approximately 1 μm per hour is achievable. The pressure in the growth chamber is maintained approximately in the range of $10^{-12}–10^{-11}$ Torr. At such a small pressure, the mean free path of the molecules/atoms is so high that the beams impinge unreacted on the sample surrounded with a cryoshroud. As a result, the reaction takes place at the sample surface where the source beams initiate the film growth. Thus, the purity of the sample highly depends on the cleanliness of the sample surface. A controlled actuation of the source shutters facilitates a controlled growth at monolayer level. Various deposition techniques are compared in Table 3.10.

3.8 Chemical Vapor Deposition

Chemical vapor deposition (CVD) involves the formation of a thin solid film on a substrate material by a chemical reaction of vapor-phase precursors. In physical vapor deposition (PVD) processes, such as evaporation and reactive sputtering, involves the adsorption of atomic or molecular species on the substrate. In CVD, the chemical reactions of precursor species occur both in the gas phase and on the substrate. In CVD, the reactions can be enhanced by heat (thermal CVD), higher frequency radiation or a plasma (plasma-enhanced CVD) [50–51].

CVD processes involve a series of extremely complex gas-phase and surface reactions and are often summarized by using an overall reaction scheme. Generally, an overall reaction scheme provides little or no information about the physicochemical processes and the gas-phase and surface reactions involved in the process. Steps involved in a basic physicochemical reaction in an overall CVD process is illustrated in Figure 3.2 [47]. An extensive description of CVD process can be found in [50–56]. The key steps involved in a typical CVD process are [50,55]:

1. Evaporation and transport of reagents (i.e., precursors) into the reactor.
2. Gas phase reactions of precursors in the reaction zone to produce reactive intermediates and gaseous by-products.

3. Mass transport of reactants to the substrate surface.
4. Adsorption of the reactants on the substrate surface.
5. Surface diffusion to growth sites, nucleation and surface chemical reactions leading to film formation.
6. Desorption and mass transport of remaining fragments of the decomposition away from the reaction zone.

3.9 Metalorganic Chemical Vapor Deposition

Metal-organic chemical vapor deposition (MOCVD), also known as Metal Organic Vapor Phase Epitaxy (MOVPE) or Organo-Metallic Vapor Phase Epitaxy (OMVPE), is a chemical growth mechanism in which a sample grows on a heated substrate in a pressure regime typically 15–750 Torr. MOVPE uses more complex compound sources, namely, metal-organic sources (e.g., tri-methylaluminum, tri-methyl antimony, triethylgallium, triisopropylgallium, trimethylgallium, etc.), hydrides (e.g., AsH_3, etc.), and other gas sources (e.g., disilane), whereas MBE uses elemental sources. In MOCVD, the reactants are flown across the substrate where they react resulting in epitaxial growth. In contrast to MBE, MOVPE requires the use of carrier gas to transport reagent materials across the substrate surface. Layered structures are achieved by valve actuation for differing injection ports of a gas manifold. A comparison between MOCVD and MBE is shown in Table 3.11 [57]. A detailed survey on MOCVD can be found in [58–60]. A schematic

TABLE 3.11

Comparison Between MOCVD and MBE

Characteristic	MOCVD	MBE
Process development	Complex	Simple
Growth rate	High growth rate for bulk layers	Low
Interface control	Good	Superior
Growth of materials	Facilitates growth of material systems near thermodynamic equilibrium with excellent quality and crystallinity	Facilitates growth of thermodynamically forbidden materials
Process	Ability to explicitly control background doping	Explicit control of doping is possible
	Hydrogen passivation and burn-in are required	Uniformity is set by reactor geometry
	Process development is very complex as there are a number of process parameters	Process development is easy to tune
Commercial aspects	Shorter maintenance periods	Longer maintenance periods
	More flexibility for source and reactor configuration changes	Less setup variability
	High safety risk and environmental issues	Lower material cost/wafer
	Less overhead costs and scale with run rate	High overhead costs do not scale with run rate
	Economic to scale in a manufacturing environment	Economically difficult to scale in a manufacturing environment
		Can become a bottleneck process in the manufacturing line

Source: Pelzel, R., A comparison of MOVPE and MBE growth technologies for III–V epitaxial structures, *CS MANTECH Conference*, New Orleans, LA, May 13–16, 2013.

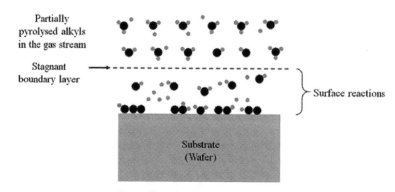

FIGURE 3.3
Schematic representation of an MOCVD process.

representation of MOCVD process is shown in Figure 3.3 [61]. The MOCVD process is extremely complex and involves a series of gas phase and surface reactions. The MOCVD process can be mainly divided into the following steps [61]: (a) evaporation and transport of reagents or precursors, (b) pyrolysis of precursors leading to deposition of the semiconductor materials, and (c) removal of the remaining fragments of the decomposition reactions from the reactor zone.

References

1. https://www.ptable.com/#Writeup/Wikipedia, last accessed on June 24, 2018.
2. M. Levinshtein, S. Rumyantsev, and M. Shur (Eds.), *Handbook Series on Semiconductor Parameters Volume 1: Si, Ge, C (Diamond), GaAs, GaP, GaSb, InAs, InP, InSb*. World Scientific, Singapore.
3. http://www.ioffe.ru/SVA/NSM/Semicond/, last accessed on June 24, 2018.
4. M. Levinshtein, S. Rumyantsev, and M. Shur, *Handbook Series on Semiconductor Parameters Volume 2: Ternary and Quaternary III–V Compounds*. World Scientific, Singapore.
5. S. Adachi (Ed.), *Physical Properties of III–V Semiconductor Compounds: InP, InAs, GaAs, GaP, InGaAs, and InGaAsP*. John Wiley & Sons, Hoboken, NJ.
6. D. C. Look, J. R. Sizelove, S. Keller, Y. F. Wu, U. K. Mishra, and S. P. Den Baars, Accurate mobility and carrier concentration analysis for GaN, *Solid State Communications*, 102(4), 297–300, 1997.
7. https://www.azom.com/article.aspx?ArticleID=8370, last accessed on June 24, 2018.
8. V. B. Gera, R. Gupta, and K. P. Jain, Electronic structure of III–V ternary semiconductors, *Journal of Physics: Condensed Matter*, 1, 4913–4930, 1989.
9. N. Harada, S. Kuroda, T. Katakami, K. Hikosaka, T. Mimura, and M. Abe, Pt-based gate enhancement-mode InAlAs/InGaAs HEMTs for large-scale integration, *Indium Phosphide and Related Materials, Proceeding of 3rd International Conference*, New York, IEEE, pp. 377–380, 1991.
10. A. Mahajan, M. Arafa, P. Fay, C. Caneau, and I. Adesida, 160 GHz enhancement-mode InAlAs/InGaAs/InP high electron mobility transistor, *Device Research Conference, Technical Digest*, pp. 132–133, 1996.
11. D. Moran, E. Boyd, H. McLelland, K. Elgaid, Y. Chen, D. S. Macintyre, S. Thoms, C. R. Stanley, and I. G. Thayne, Novel technologies for the realisation of GaAs pHEMTs with 120 nm self-aligned and nanoimprinted T-gates, *Microelectronic Engineering*, 67–68, 769–774, 2003.

12. A. Cetronio, F. Giannini, and G. Leuzzi, Self-aligned gate technology for analogue and digital GaAs integrated circuits, *Gallium Arsenide applications Symposium*, Paris, France, October 2–6, 2000.

13. T.-W. Kim, D.-H. Kim, S.-H. Shin, S.-J. Jo, J. H. Jang, and J.-I. Song, Characteristics of 0.2 μm depletion and quasi-enhancement mode self-aligned gate capless p-HEMTs, *Electronics Letters*, 42(20), 1178–1180, 2006.

14. I. Watanabe, A. Endoh, T. Mimura, and T. Matsui, 35-nm-gate In0.7Ga0.3As/In0.52Al0.48As HEMT with 520-GHz fT, *International Conference on Indium Phosphide and Related Materials*, pp. 28–31, 2007.

15. S. Oktyabrsky and D. Y. Peide, *Fundamentals of III-V Semiconductor MOSFETs*. New York: Springer, 2010.

16. R. J. W. Hill, D. A. J. Moran, X. Li et al., Enhancement mode GaAs MOSFETs with an In0.3Ga0.7As channel, a mobility of over 5000 cm²/Vs, and transconductance of over 475 μS/μm, *IEEE Electron Device Letters*, 28(12), 1080–1082, 2007.

17. Y. Xuan, Y. Q. Wu, and P. D. Ye, High-performance inversion-type enhancement-mode InGaAs MOSFET with maximum drain current exceeding 1 A/mm, *IEEE Electron Device Letters*, 29(4), 294–296, 2008.

18. S. Koveshnikov, R. Kambhampati, V. Tokranov, M. Yakimov, D. Schlom, M. Warusawithana, C. Adamo, W. Tsai, and S. Oktyabrsky, Thermal stability of electrical and structural properties of GaAs based MOS capacitors with amorphous LaAlO3 gate oxide, *Applied Physics Letter*, 93(1), 012903–1–3, 2008.

19. T. D. Lin, P. Chang, H. C. Chiu, Y. C. Chang, C. A. Lin, W. H. Chang, Y. J. Lee, Y. H. Chang, M. L. Huang, J. Kwo, and M. Hong, Nanoelectronics of high-k dielectrics on InGaAs for key technologies beyond Si CMOS, *Indium Phosphide and Related Materials, Proceedings of International Conference*, Piscataway, NJ, IEEE, pp. 94–99, 2009.

20. S. Oktyabrsky, V. Tokranov, M. Yakimov, R. Moore, S. Koveshnikov, W. Tsai, F. Zhu, and J. C. Lee, High-k gate stack on GaAs and InGaAs using in situ passivation with amorphous silicon, *Materials Science & Engineering B*, 135(3), 272–276, 2006.

21. I. Ok, H. Kim, M. Zhang et al., Metal gate HfO2 metal-oxide-semiconductor structures on InGaAs substrate with varying Si interface passivation layer and post deposition anneal condition, *Journal of Vacuum Science & Technology B*, 25(4), 1491–1494, 2007.

22. Y. Xuan, Y. Q. Wu, H. C. Lin, T. Shen, and P. D. Ye, "Submicrometer inversion-type enhancement-mode InGaAs MOSFET with atomic-layer-deposited as gate dielectric," *IEEE Electron Device Letters*, 28(11), 935–938, 2007.

23. C. P. Chen, T. D. Lin, Y. J. Lee, Y. C. Chang, M. Hong, and J. Kwo, Self-aligned inversion nchannel In0.2Ga0.8As/GaAs metal-oxide-semiconductor field-effect-transistors with TiN gate and Ga2O3(Gd2O3) dielectric, *Solid State Electronics*, 52(10), 1615–1618, 2008.

24. M. Passlack, M. Hong, and J. P. Mannaerts, Quasistatic and high frequency capacitance–voltage characterization of Ga/sub 2/O/sub 3/-GaAs structures fabricated by in situ molecular beam epitaxy, *Applied Physics Letter*, 68(8), 1099–1101, 1996.

25. K. Hiruma, M. Yazawa, H. Matsumoto, O. Kagaya, M. Miyazaki, and Y. Umemoto, Selective growth of ultra-low resistance GaAs/InGaAs for high performance InGaAs FETs, *Journal of Crystal Growth*, 124(1), 255–259, 1992.

26. C. Liao, D. Cheng, C. Cheng, K. Y. Cheng, M. Feng, T. H. Chiang, J. Kwo, and M. Hong, Inversion-channel enhancement-mode GaAs MOSFETs with regrown source and drain contacts, *Journal of Crystal Growth*, 311(7), 1958–1961, 2009.

27. U. Singisetti, M. A. Wistey, G. J. Burek et al., Enhancement mode In0.53Ga0.47As MOSFET with self-aligned epitaxial source/drain regrowth, 2009 *IEEE International Conference on Indium Phosphide & Related Materials*, Piscataway, NJ, IEEE, pp. 120–123, 2009.

28. J. B. Boos, B. R. Bennett, N. A. Papanicolaou, M. G. Ancona, J. G. Champlain, R. Bass, and B. V. Shanabrook, High mobility p-channel HFETs using strained Sb-based materials, *Electronics Letters*, 43(15), 834, 835, 2007.

29. M. K. Hudait and R. Chau, Integrating III–V on silicon for future nanoelectronics, *2008 IEEE Compound Semiconductor Integrated Circuits Symposium*, Piscataway, NJ, IEEE, pp. 1–2, 2008.

30. C. L. Dohrman, K. Chilukuri, D. M. Isaacson, M. L. Lee, and E. A. Fitzgerald, Fabrication of silicon on lattice-engineered substrate (SOLES) as a platform for monolithic integration of CMOS and optoelectronic devices, 2006 *International SiGe Technology and Device Meeting*, Piscataway, NJ, IEEE, pp. 44–45, 2006.
31. Z. Yu, R. Droopad, D. Jordan et al., GaAs-based heterostructures on silicon, *2002 GaAs MANTECH Conference. Technical Digest*, pp. 276–279, 2002.
32. F. Letertre, Formation of III–V semiconductor engineered substrates using smart Cut layer transfer technology, *AIP Conference Proceedings*, 1068, pp. 185–196, 2008.
33. J. W. Lee, Y. W. Ahn, J. H. Song, B. G. Cho, and I. H. Ahn, AlGaAs/InGaAs PHEMT with multiple quantum wire gates, *Microelectronics Journal*, 36(3), 389–391, 2005.
34. S. A. Fortuna and X. Li, GaAs MESFET with a high-mobility self-assembled planar nanowire channel, *IEEE Electron Device Letters*, 30(6), 593–595, 2009.
35. H. Hasegawa and M. Akazawa, Surface passivation technology for III–V semiconductor nanoelectronics, *Applied Surface Science*, 255(3), 628–632, 2008.
36. S. Oktyabrsky, P. Nagaiah, V. Tokranov, S. Koveshnikov, M. Yakimov, R. Kambhampati, R. Moore, and W. Tsai, Electron scattering in buried InGaAs MOSFET channel with HfO2 gate oxide, *MRS Proceedings*, 1155, C02–C03, 2009.
37. T. Matsuoka, E. Kobayashi, K. Taniguchi, C. Hamaguchi, and S. Sasa, Temperature dependence of electron mobility in InGaAs/InAlAs heterostructures, *Japan Journal of Applied Physics*, Pt. 1, 29(10), 2017–2025, 1990.
38. M. A. Negara, K. Cherkaoui, P. Majhi, C. D. Young, W. Tsai, D. Bauza, G. Ghibaudo, and P. K. Hurley, The influence of HfO2 film thickness on the interface state density and low field mobility of n channel HfO2/TiN gate MOSFETs, *Microelectronic Engineering*, 84(9–10), 1874–1877, 2007.
39. K. Maitra, M. M. Frank, V. Narayanan, V. Misra, and E. A. Cartier, Impact of metal gates on remote phonon scattering in titanium nitride/hafnium dioxide n-channel metal-oxide-semiconductor field effect transistors-low temperature electron mobility study, *Journal of Applied Physics*, 102(11), 114507–1141, 2007.
40. S. Barraud, L. Thevenod, M. Casse, O. Bonno, and M. Mouis, Modeling of remote Coulomb scattering limited mobility in MOSFET with HfO2/SiO2 gate stacks, *Microelectronic Engineering*, 84(9–10), 2404–2407, 2007.
41. P. D. Ye, G. D. Wilk, B. Yang et al., GaAs metal-oxide-semiconductor field-effect transistor with nanometer-thin dielectric grown by atomic layer deposition, *Applied Physics Letter*, 83(1), 180, 2003.
42. M. L. Huang, Y. C. Chang, C. H. Chang, Y. J. Lee, P. Chang, J. Kwo, T. B. Wu, and M. Hong, Surface passivation of III–V compound semiconductors using atomic-layer-deposition- grown Al2O3, *Applied Physics Letter*, 87, 252104, 2005.
43. J. R. Arthur Jr., Interaction of Ga and As₂ molecular beams with GaAs surfaces, *Journal of Applied Physics*, 39, 4032, 1968.
44. A. Y. Cho, Citation. Film deposition by molecular-beam techniques, *Journal of Vacuum Science & Technology*, 8, S31, 1971.
45. L. L. Chang, L. Esaki, W. E. Howard, R. Ludeke, and G. Schul, Structures grown by molecular beam epitaxy, *Journal of Vacuum Science & Technology*, 10, 655, 1973.
46. A. Y. Cho, J. R. Arthur, Molecular beam epitaxy, *Progress in Solid State Chemistry*, 10, 157–191, 1975.
47. M. Henini (Ed.), *Molecular Beam Epitaxy from Research to Mass Production*. San Diego, CA, Elsevier, 2010.
48. A. Y. Cho, Recent developments in molecular beam epitaxy (MBE), *Journal of Vacuum Science & Technology*, 16, 275, 1979.
49. I. K. Schuller, New class of layered materials, *Physical Review Letters*, 44, 24, 1980.
50. A. C. Jones and M. L. Hitchman, Overview of chemical vapour deposition, in *Chemical Vapour Deposition: Precursors Processes and Applications*, The Royal Society of Chemistry, Cambridge, UK, pp. 1–36, 2009.
51. J.-O. Carlsson and P. M. Martin, Chemical vapor deposition, in *Handbook of Deposition Technologies for Films and Coatings*, 3rd ed., Boston, MA, William Andrew Publishing, pp. 314–363, 2010.

52. G. B. Stringfellow, Overview of the OMVPE process, in *Organometallic Vapor-Phase Epitaxy*, 2nd ed., San Diego, CA, Academic Press, pp. 1–16, 1999.

53. M. L. Hitchman and K. F. Jensen, Chemical vapour deposition—An overview, in *Chemical Vapor Deposition*, M. L. Hitchman and K. F. Jensen (Eds.), New York, Academic Press, 1989.

54. T. T. Kodas and M. J. Hampden-Smith, An overview of CVD processes, in *The Chemistry of Metal CVD*, Weinheim, Germany, VCH, 1994.

55. A. C. Jones and P. O'Brien, *CVD of Compound Semiconductors*. Weinheim, Germany, VCH, 1997.

56. R. L. Moon and Y.-M. Houng, *In Chemical Vapor Deposition*, M. L. Hitchman and K. F. Jensen (Eds.), New York, Academic Press, 1989.

57. R. Pelzel, A comparison of MOVPE and MBE growth technologies for III–V epitaxial structures, *CS MANTECH Conference*, May 13–16, 2013, New Orleans, LA.

58. A. G. Thompson, MOCVD technology for semiconductors, *Materials Letters*, 30(4), 255–263, 1997.

59. B. Gerald, *String Fellow, Organometallic Vapor-Phase Epitaxy*, 2nd ed. San Diego, CA, Academic Press, 1999.

60. J. L. Zilko, Metal organic chemical vapor deposition: Technology and equipment, in *Handbook of Thin Film Deposition*, 3rd ed., Seshan, K. (Ed.), Cambridge, MA, Elsiver.

61. M. Bochmann, K. J. Webb, J.-E. Hails, and D. Wolverson, Volatile cadmium chalcogenolato complexes as single source precursors for the MOCVD growth of II–VI films, *European Journal of Solid State Inorganic Chemistry*, 29, 155–166, 1992.

4

Source/Drain, Gate and Channel
Engineering in HEMTs

Palash Das, T. R. Lenka, Satya Sopan Mahato, and A. K. Panda

CONTENTS

4.1 Introduction

The rapid down-scaling of semiconductor devices towards gigascale integration level is causing a fundamental change from experiment-dominated development to simulation-dominated development of new electronics technology. Therefore, the high cost and complexity of fabrication is the essence of the gigascale challenge to bring new technology to the market faster and cheaper. In order to accomplish the goal, the device simulation is more and more important to lead new generation development.

Nowadays, device modeling development also meets the gigascale challenge because of rapid progress in materials and device structures incorporating complex-geometry and quantum effects [1,2]. To meet the gigascale challenges, a device simulator needs to be designed, which can be able to provide more extensible device simulations and characteristics in the framework of quantum mechanics (QM). In the nano-scale device simulation of AlGaN/GaN or AlGaAs/GaAs based HEMTs, modeling plays a major role for the first-hand information and hence small-signal model developed is presented here.

4.2 HEMT Modeling

Usually III-V based heterostructure device is subjected to 1 D Poisson's solver and results into various electrical parameters such as charge distribution, electric field, potential function, sheet charge density and various subbands with eigenvalues. The optimized device design parameters are also fixed during this solution. This is used as input to determine small-signal, RF and microwave characteristics of the device.

4.3 Small-Signal Modeling

The parameters like device capacitance, routing capacitance and the ON resistance of device limit the high-speed performance in terms of its high frequency performance and switching speed. The small-signal equivalent model of HEMT with its intrinsic circuit model is shown in Figure 4.1 [2].

The equivalent circuit for the intrinsic HEMT is shown within the dashed rectangular boundary, which includes the gate capacitance (C_g), the charging resistance (r_i), the output conductance (g_0), and the drain-to-gate transconductance (g_m). The element C_i (not shown here), which arises due to the passive coupling between the drain and the channel, is not included. These intrinsic parameters, together with the extrinsic ones, can be used to analyze and predict the AC operation behavior of the HEMT.

Combined with the parameter extraction techniques, the equivalent circuit can be used to characterize the performance of the devices. In the real applications, it is also desirable to predict the ultimate potential of the device performance or make a decision which device should be chosen. Apart from the capacitance effect, the RF performance of device is limited by several other factors such as gate length, source and

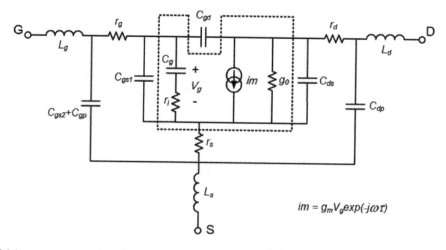

FIGURE 4.1
Small-Signal equivalent circuit model of HEMT, and the intrinsic circuit model is enclosed by the dashed line. (From Snider, G. L. et al., *J. Appl. Phys.*, 68, 1990.)

drain parasitic resistance, gate–drain parasitic capacitance, etc. These parameters play important role for high frequency limitation of HEMT and hence the high frequency parameters like parasitic resistances etc are important and hence discussed below.

4.4 Parasitic Resistances in HEMTs

As shown in figure, r_g, r_s and r_d are parasitics resistances in HEMTs on gate, source and drain side and are shown in Figure 4.1.

4.5 Parasitic Capacitances in HEMTs

Parasitic capacitances are also shown in Figure 4.1.

4.5.1 RF and Microwave Simulation Model

The RF and microwave characteristics are discussed with the help of two-port network analysis as shown in Figure 4.2.

Here the small-signal AC characteristics are analyzed for various frequencies. A mixed-mode environment is created for the AC simulation, i.e., HEMT is embedded in an external circuit which forms a two-port network (as shown in Figure 4.2). Here the voltage sources are attached to the gate (port 1) and drain (port 2) terminals and all other terminals are grounded.

The small-signal output file contains the admittance (A) and capacitance (C) matrices, which are equivalent to the Y-matrix as $Y = A + j\omega C$. The rows and columns of the matrices are given by the nodes in the small-signal analysis. Y-matrix obtained now can be converted to any other matrix such as S, Z or h-matrix and the RF and microwave gain parameters are obtained in the following way.

The Y-matrix describes how the currents in a circuit would react if the applied voltages at different contact nodes of the circuit change on small scales, that is:

$$\delta i = Y \cdot \delta v = \left(A + j\omega C \right) \delta v \qquad (4.1)$$

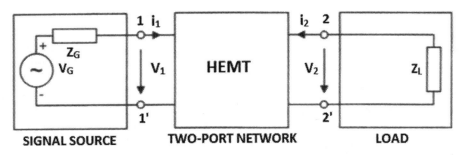

FIGURE 4.2
Two Port Network for Microwave Characteristics.

where the Y-matrix measures the small current changes (δi), in response to a small change in the gate voltages (δv).

The complex Y-matrix can be split into two parts: The real part A is called *conductance matrix*, which measures the in-phase response of the current with the voltage; the imaginary part C is called the *capacitance matrix*, which measures the out-phase response. The symbol j denotes the imaginary unit and ω denotes the frequency of the small-signal change.

For a typical HEMT device with four terminals, gate (g), drain (d), source (s) and bulk (b), the A and C matrices have the following elements:

$$\begin{bmatrix} i(g) \\ i(d) \\ i(s) \\ i(b) \end{bmatrix} = \left(\begin{bmatrix} a(g,g) & a(d,g) & a(s,g) & a(b,g) \\ a(g,d) & a(d,d) & a(s,d) & a(b,d) \\ a(g,s) & a(d,s) & a(s,s) & a(b,s) \\ a(g,b) & a(d,b) & a(s,b) & a(b,b) \end{bmatrix} + j\omega \begin{bmatrix} C(g,g) & C(d,g) & C(s,g) & C(b,g) \\ C(g,d) & C(d,d) & C(s,d) & C(b,d) \\ C(g,s) & C(d,s) & C(s,s) & C(b,s) \\ C(g,b) & C(d,b) & C(s,b) & C(b,b) \end{bmatrix} \right) \begin{bmatrix} v(g) \\ v(d) \\ v(s) \\ v(b) \end{bmatrix}$$

where the $a(d,g)$ element represents the in-phase drain-current response to a small gate voltage change, which is by definition the transconductance of the HEMT device. Similarly, $C(b,g)$ represents the out-phase body current response to a small change in gate voltage. This quantity corresponds to the gate-to-body capacitance. Then Y matrix is converted to the S matrix using the following relations and S-parameters are determined as

$$s_{11} = \frac{\left(1 - \bar{y}_{11}\right)\left(1 + \bar{y}_{22}\right) + \bar{y}_{12}\,\bar{y}_{21}}{N_y}, \quad s_{12} = \frac{-2\bar{y}_{12}}{N_y} \tag{4.2}$$

$$s_{21} = \frac{-2\bar{y}_{21}}{N_y}, \quad s_{22} = \frac{\left(1 - \bar{y}_{22}\right)\left(1 + \bar{y}_{11}\right) + \bar{y}_{12}\,\bar{y}_{21}}{N_y} \tag{4.3}$$

where S_{11} and S_{22} are the reflection coefficients at port 1 and port 2, respectively, and plotted on Smith chart, whereas S_{21} and S_{12} are the voltage transfer ratio from port 1 to port 2 and port 2 to port 1, respectively, and plotted on Polar plot in our computation.

$$N_y = \left(1 + y_{11}\right)\left(1 + y_{22}\right) - y_{12}y_{21} \tag{4.4}$$

Furthermore, $\bar{y}_{ij} = Z_0 y_{ij}$ and Z_0 is the characteristic impedance.

A RF or microwave transistor is capable of power amplification or sustained oscillation. Whether the transistor in a circuit will oscillate or not depends on the transistor itself and on the source and load impedance [3,4]. During amplifier design it is important to check the stability of the device chosen, otherwise the amplifier may well turn into an oscillator. RF transistors, like all other active devices, are unconditionally stable at any operating frequency above a critical frequency (f_k). At operating frequencies below f_k, the transistor is conditionally stable and certain termination conditions can cause oscillations.

As discussed in Carson [5], the stability measure $B > 0$ and Rollett stability factor $K > 1$ are the necessary and sufficient conditions for stability behavior of a device, which is calculated using a set of S-parameters for the device at the frequency of operation. We can calculate the stability parameters such as K, $|\Delta|$ and B to give us an indication to whether a device is likely to oscillate or not or whether it is conditionally/unconditionally stable. The parameters are defined in terms of S-parameters as

$$K = \frac{1 - |S_{11}|^2 - |S_{22}|^2 + |\Delta|^2}{2|S_{12} \cdot S_{21}|} > 1 \tag{4.5}$$

$$B = 1 + |s_{11}|^2 - |s_{22}|^2 - |\Delta|^2 \tag{4.6}$$

where $|\Delta| = |S_{11} \cdot S_{22} - S_{12} \cdot S_{21}| < 1$.

The stability parameters must satisfy, i.e., $K > 1$ and $|\Delta| < 1$ or $B > 0$ for a transistor to be unconditionally stable. The Rollett stability can also be computed from obtained Y-parameters and is given by [6]

$$K = \frac{2\Re(y_{11})\Re(y_{22}) - \Re(y_{12}y_{21})}{|y_{12}y_{21}|} \tag{4.7}$$

where $\Re(y)$ denotes the real part of the complex number y.

When $K > 1$, it indicates conjugate matching between output and input loads. The standard criterion of stability, i.e., when $K > 1$, the circuit is stable unconditionally and when $K < 1$, the circuit is conditionally stable. Then different power gains are obtained. The power gain is the ratio of output power (P_{out}) delivered from the transistor output to the load to the input power (P_{in}) delivered from the signal source to the transistor input. The matching conditions between the signal source and transistor and between the transistor and load influence the power transfer. For a transistor to achieve maximum power gain, the power matching is required. We have taken this into account, and the gain parameters are determined using the following equations.

Once K factor is calculated and found that the device is unconditionally stable then we can calculate the maximum available gain (MAG) and the Maximum Stable Gain (MSG) which can be computed with relating to the Rollett stability factor and is given by [3,4]

$$MAG = \left|\frac{S_{21}}{S_{12}}\right| \left(K - \sqrt{K^2 - 1}\right) \tag{4.8}$$

$$MAG = MSG\left(K - \sqrt{K^2 - 1}\right) \tag{4.9}$$

When K is in the limit of unity and if the input and output of the whole network is conjugately impedance-matched, the maximum stable gain (MSG) is also defined from the S parameters by using the relation,

$$MAG = MSG = \left|\frac{S_{21}}{S_{12}}\right| \tag{4.10}$$

In this case the MAG or G_{MAG} is known as the maximum stable gain (MSG). The MAG can also be expressed as

$$G_{MAG} = \frac{|S_{21}|^2}{\left(1-|S_{11}|^2\right)\left(1-|S_{22}|^2\right)} = G_1|S_{21}|^2 G_2 \tag{4.11}$$

where the maximum input and output gain factors are

$$G_1 = \frac{1}{1-|S_{11}|^2} \text{ and } G_2 = \frac{1}{1-|S_{22}|^2} \tag{4.12}$$

The Mason's unilateral gain is used as a figure of merit across all operating frequencies, and its value at f_{max} is especially very useful. It is the maximum oscillation frequency of a device, where *MUG* (f_{max}) = 1. At this frequency the maximum stable gain (*MSG*) and the maximum available gain (*MAG*) of the device become unity. Therefore, f_{max} is the characteristic of the device, and it has the significance that it is the maximum frequency of oscillation in a circuit where only one active device is embedded in a passive network.

The Mason's Unilateral Power Gain (*MUG*) is defined as a power gain of a two-port having no output-to-input feedback, with input and output conjugately impedance matched to signal source and load respectively. It can be expressed with *S*, *Y* and *Z* parameters, such as:

$$MUG = \frac{S_{12}S_{21}S_{11}^*S_{22}^*}{\left(1-|S_{11}|^2\right)\left(1-|S_{22}|^2\right)} \tag{4.13}$$

$$MUG = \frac{|y_{21}-y_{12}|^2}{4\left[\Re(y_{11})\left[\Re(y_{22})\right]-\Re(y_{12}y_{21})\right]} \tag{4.14}$$

$$MUG = \frac{|z_{21}-z_{12}|^2}{4\left[\Re e(z_{11})\cdot\Re e(z_{22})-\Re e(z_{12})\cdot\Re e(z_{21})\right]} \tag{4.15}$$

where $\Re e(y)$ and $\Re e(z)$ denotes the real part of the complex number y or z, respectively.

The cut-off frequency (f_t) is defined as the frequency at which the magnitude of the short-circuit current gain $|h_{21}|$ rolls off to 1 (0 dB) i.e. ($|h_{21}| = 1$). The current gain can be calculated from the measured *S*-parameters [3,4].

$$|h_{21}| = \frac{2\cdot S_{12}}{\left(1-S_{11}\right)\cdot\left(1+S_{22}\right)+S_{12}\cdot S_{21}} \tag{4.16}$$

If $|h_{21}|$ (dB) is positive in the whole measured frequency range, then $|h_{21}|$ (dB) is linearly extrapolated to zero versus $\log(f)$ with a slope of 20dB/dec.

Again the current gain cut-off frequency can be expressed as a function of transconductance parameter (g_m) and gate-to-source (C_{gs}) and gate-to-drain (C_{gd}) parasitic capacitances.

$$f_t = \frac{g_m}{2\pi\left(C_{gs}+C_{gd}\right)} \tag{4.17}$$

Similarly, the cut-off frequency is also related to the saturation velocity and channel length of the device and can be expressed as

$$f_t = \frac{v_{sat}}{2\pi L_g} \tag{4.18}$$

The maximum frequency of oscillation (f_{max}) can be related to the short circuit current gain (f_t) and the parasitic resistances of the device as

$$f_{max} = \frac{f_t}{2}\sqrt{\frac{R_{ds}}{R_i + R_g}} \tag{4.19}$$

where R_{ds} is the parasitic drain to source resistance, R_i is the intrinsic resistance and R_g is the gate resistance.

If the extrapolated f_{max} is known, then the unilateral power gain at any frequency can be calculated by

$$MUG(f) = -20\log(f) + 20\log(f_{max}) \tag{4.20}$$

Using f_t and f_{max}, the RF potential of different transistors can be realized.

However, the transconductance basically governs the current driving capability and is an extremely important parameter for estimating the RF/microwave performance of the device and is given by

$$g_m = \left.\frac{\partial I_{dsLin}}{\partial V_{gs}}\right|_{V_{ds}} = \frac{\mu_n C_0 W \, V_D}{L_g} \tag{4.21}$$

$$g_m = \left.\frac{\partial I_{dsSat}}{\partial V_{gs}}\right|_{V_{ds}} = \frac{\mu_n C_0 W \, (V_G - V_T)}{L} \tag{4.22}$$

Using these techniques, the RF parameters can be extracted in the following way.

4.5.1.1 RF Parameters Extraction

- g_m: The current carrying capability i.e. transconductance is extracted from $I_d V_{gs}$ plot.
- f_{max}: The maximum frequency of oscillation (f_{max}) is plotted as a function of bias for three different extraction methods: unit-gain-point, extract-at-dBPoint, and extract-at-frequency.
- *Unit-gain-point* $(|h_{21}| = 1)$: With this method where the frequency at which $|h_{21}| = 1$ is returned.
- *Extract-at-dBpoint*: With this method, where the frequency at which $|h_{21}|$ drops by 10 dB. Then, it is assumed that the gain curve exhibits 20 dB/decade decay at this point and, using this assumption, the unit gain-point is computed by extrapolation.

- *Extract-at-frequency*: With this method, the extrapolation of the gain to the unit-gain-point is performed by assuming that the 20 dB/decade decay is established at a given frequency.

- f_t: The cut-off frequency (f_t) is plotted as a function of bias for three different extraction methods: unit-gain-point, extract-at-dBPoint, and extract-at-frequency.

- *MUG*: A family of Mason's unilateral gain curves is plotted as a function of frequency for all bias points. A family of Mason's unilateral gain curves is plotted as a function of bias for all frequencies.

- *MSG*: The maximum stable gain (*MSG*), the maximum available gain (*MAG*), and the Rollett stability factor (*K*) are plotted as a function of frequency for a gate bias of 0 V and as a function of bias for a frequency of 1 GHz.

- *Current Gain* (h_{21}): The RF parameter h_{21} is plotted as a function of frequency for a gate bias of 0 V and as a function of bias for a frequency of 1 GHz. The plot shows the real and imaginary parts of h_{21} as well as the magnitude and phase. Here we have only considered the real parts of h_{21}.

- *Smith Chart*: The RF parameters S_{11} and S_{22} are plotted on a Smith chart for all frequencies and a gate bias of 0 V and for all bias points and a frequency of 1 GHz.

- *Polar Plot*: The RF parameters S_{21} and S_{12} are plotted on a Polar plot for all frequencies and a gate bias of 0 V.

4.6 Gate Engineering

Gate Engineering plays a major role on the device performance and hence different structures of different kind of gates such as conventional Schottky gate contact, T-shaped gate, Gate Recessed are discussed in this section.

4.6.1 Schottky Gate Contact HEMT

Schottky gate contact HEMT structure as shown in Figure 4.3 was proposed by Takuma Nanjo et al. [6]. They have improved the performance of nitride HEMTs by inserting an aluminum nitride (AlN) spacer layer at the interface of AlGaN/GaN and by using silicon (*Si*)

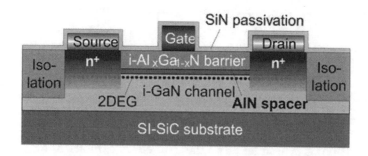

FIGURE 4.3
Schematic cross-sectional structure of AlGaN/AlN/GaN HEMT fabricated by employing Si ion implantation doping. (From Nanjo, T. et al., *Jpn. J. Appl. Phys.*, 50, 064101, 1–7, 2011.)

implanted source–drain regions [6]. AlN spacer layer placed between the GaN buffer and AlGaN barrier layers significantly increases the conductivity of the two-dimensional electron gas (*2DEG*) that is used as the channel in nitride HEMTs which results in current density of 1.3 A/mm. However, AlN is usually highly insulating and thus makes the channel difficult to access through the source–drain regions. The epitaxial structures with layers of GaN buffer, AlN spacer and AlGaN barrier as shown in this figure were grown on silicon carbide (*SiC*) substrates using metal-organic chemical vapor deposition (*MOCVD*).

4.6.2 T-Shaped HEMT

HEMTs are field effect transistors where the current flow between two ohmic contacts, i.e., source and drain is controlled by a third contact, called gate. Most often the gate is a Schottky contact. In contrast to ion implanted MESFETs, HEMTs are based on epitaxially grown layers with different band gaps E_g. A schematic cross section of a HEMT with a T-shaped gate is shown in Figure 4.4.

Depending on the use of GaAs or AlGaAs for the buffer layer the HEMT is called single heterojunction HEMT (SHHEMT) or double heterojunction HEMT (DHHEMT), respectively.

In order to achieve higher frequency, the HEMT gate has to be mushroom shaped in cross section and have the smallest possible width of footprint as shown in Figure 4.5. In terms of fabrication process technology, f_t and f_{max} can be significantly improved with a T-shaped or mushroom shaped gate structure. This is possible by constructing shorter gate length, L_g, while maximizing the cross-sectional area of the gate.

4.6.3 Gate Recessed HEMT

In the AlGaN/GaN heterostructure, high-density 2DEG induced by spontaneous and piezoelectric polarization effects present the conventional AlGaN/GaN HEMT as depletion-mode (D-mode) transistors with a threshold voltage (V_{th}) typically around −4V. Usually,

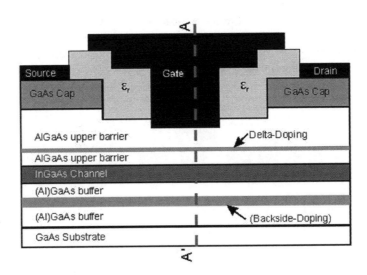

FIGURE 4.4
Schematic cross section of a T-shaped HEMT. (From Brech, V. H., Optimization of GaAs based high electron mobility transistors by numerical simulations, PhD Dissertation, 1998.)

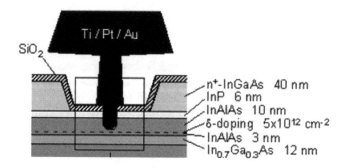

FIGURE 4.5
Schematic cross section of a mushroom-shaped HEMT. (From Brech, V. H., Optimization of GaAs based high electron mobility transistors by numerical simulations, PhD Dissertation, 1998.)

the V_{th} of the AlGaN/GaN HEMTs depends on the design of the epitaxial structure, namely, the Al composition, Si doping concentration and the thickness of the AlGaN barrier. When the threshold voltage is adjusted to be positive, enhancement-mode (E-mode) operation is realized. Compared to the D-mode HEMTs, E-mode devices allow elimination of negative-polarity voltage supply and therefore, reduce the circuit complexity and system cost significantly.

In the generally used AlGaN/GaN heterostructure for HEMTs, where Al composition is in the range of 15%–35% and the AlGaN barrier thickness is around 20 nm, the reduction in the AlGaN thickness by the gate recess results in a decreasing polarization-induced 2DEG density and with the help of the gate metal work function, the threshold voltage can be shifted positively. With a deep enough gate-recess etching, the V_{th} can reach a positive value and E-mode HEMTs are formed. The schematic diagrams of conduction band profile of E-mode HEMT and D-mode HEMT are also shown in Figures 4.6 and 4.7, respectively.

4.6.4 Field Plate Structure of HEMT

The AlGaN/GaN devices have established themselves as microwave power devices. A significant improvement of the device performance has been achieved by adopting the field plate technique [9–11]. With its origins in the context of high-voltage p–n junctions, the main functions of the field plate are to reshape the electric field distribution in the channel and to reduce its peak value on the drain side of the gate edge. The benefit is an increase

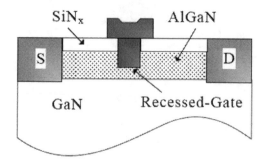

FIGURE 4.6
Schematic diagram of an E-mode Gate Recessed HEMT. (From Ruonan, W., Enhancement/depletion-mode HEMT technology for III-nitride mixed-signal and RF applications, PhD Dissertation, Hong Kong, 2008.)

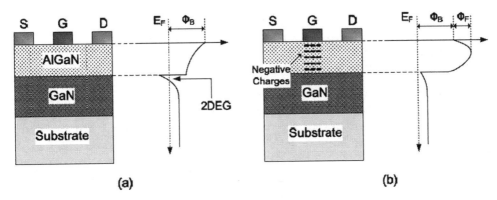

FIGURE 4.7
Schematic diagrams of Conduction band Profile (a) conventional D-mode AlGaN/GaN HEMT and (b) E-mode HEMT. (From Ruonan, W., Enhancement/depletion-mode HEMT technology for III-nitride mixed-signal and RF applications, PhD Dissertation, Hong Kong, 2008.)

of the breakdown voltage and a reduced high-field trapping effect. In HEMTs the polarization charge at the AlGaN/GaN heterointerface is the crucial factor building the channel. In this device, the electric field distribution in the channel is optimized by variation of the geometry of the field plate [10]. The schematic structure of single heterojunction AlGaN/GaN HEMTs with field plates is shown in Figure 4.8.

Two types of field plate structures are available. Normal field-plate is connected to the gate through the common path of the gate and gate feeder in the extrinsic device region and the second one is the intimately connected, i.e., gate and field-plate are intimately connected [11].

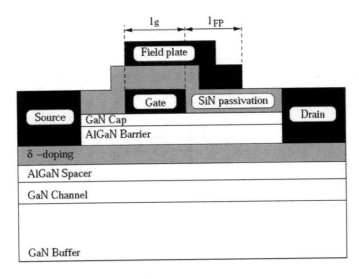

FIGURE 4.8
A schematic structure of single heterojunction AlGaN/GaN HEMTs with field plates. (From Palankovski, V. et al., Field-plate optimization of AlGaN/GaN HEMTs, *IEEE Compound Semiconductor Integrated Circuit Symposium (CSICS)*, At San Antonio, TX, pp. 107–110, 2006.)

4.7 Channel/Barrier Engineering in AlGaN/GaN HEMT

There are numerous methods of designing AlGaN/GaN heterostructure to realize HEMT applications. Particularly for these applications, there are two main figures of merit: 2DEG concentration and mobility. In an AlGaN/GaN heterostructure for HEMT applications, two-dimensional carrier concentration has been found to be controlled by spontaneous and piezoelectric polarizations, surface charges, background concentration and barrier doping etc. whereas the mobility depends on acoustic deformation-potential, piezoelectric, polar optic phonon, alloy disorder, interface roughness, dislocation, remote modulation doping scatterings, [12] etc. Different kind of prevailing AlGaN barrier for HEMT application structures are shown in Figure 4.9.

4.7.1 2DEG and Mobility in AlGaN/GaN Heterostructures

Lu et al. showed that increase in Al molar fraction from 27% to 35% [13] in the AlGaN barrier degraded 2DEG carrier mobility from 1440 $cm^2v^{-1}s^{-1}$ to 1330 $cm^2v^{-1}s^{-1}$ whereas the carrier concentration was found to be increased from $9.80 \times 10^{12} \, cm^{-2}$ to $1.30 \times 10^{13} \, cm^{-2}$. The increase in mobility for those reported heterostructures has been discussed through the screening of polar phonon scattering. However, the decrease in mobility and gradual increase in carrier concentration for $Al_{0.36}Ga_{0.64}N$/GaN heterostructure [13] has been explained through the strain generated lattice defect and strain coherency, respectively. Doping in the barrier layer is another well practiced strategy for increasing the carrier concentration though the impurity scattering attributed by donor ions, decreases the carrier mobility a lot. Similarly the decrease in the mobility due to higher impurity scattering has also been reported when the doping concentration in the $Al_{0.30}Ga_{0.70}N$ barrier was increased. However, compositional grading in doped barrier and in channel layer has been reported as alternative ways to have notable 2DEG carrier concentration with reasonably good mobility. Thus it is evident that high carrier concentration can be achieved at

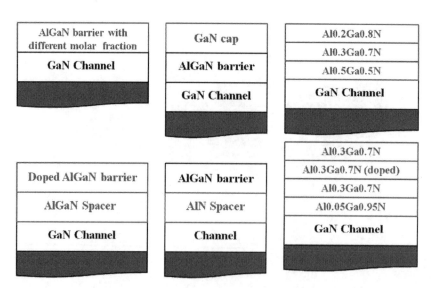

FIGURE 4.9
Different kinds of prevailing AlGaN barrier for HEMT application.

the cost of decrease in mobility due to different scattering factors. As the change in barrier height and its thickness control the confinement as well as the "spill over" probability of the carriers, various physical entities along with the earlier mentioned phenomena are found to be modified. Thus, to control the effect of different polarizations and scatterings in the channel layer, the variation in barrier layer through the change in Al molar fraction is primarily required to have a specific 2DEG with desired mobility. However, for a set of samples having identical substrate, buffer and channel epitaxial layers, the blind variation of Al molar concentration in the barrier might not provide the desired results. At room temperature, the cumulative scattering due to strain-induced acoustic deformation potential (ADP), piezoelectricity (PE) and generation of polar optic phonon (LO) mostly limits the mobility. Theoretically ADP and PE can be minimized by decreasing effective Al molar fraction in barrier as both strain and electro-mechanical coupling coefficients decreases leading to increase in mobility. However, the decrease in effective Al molar fraction decreases the carrier concentration as well. Thus it is required to trade off the value of effective Al molar fraction to get the desired results. As effective Al molar fraction in a graded barrier is the weighted average of molar fractions and thicknesses of all the corresponding layers, it can be modified to achieve any desired carrier concentration along with mobility through proper tuning of these parameters. The decrease in Al molar fractions in all the thin barrier layers thus eventually decreases the 2DEG carrier concentration with a non-linear increment in the confined carrier mobility minimizing the effects of ADP and PE scatterings. Thus a proper estimate of Al molar fractions in the barrier layers may lead to achieve a definite carrier concentration and mobility for an AlGaN/GaN heterostructure for HEMT applications. With this understanding, an alternative method of increasing the piezoelectric polarization has been followed by Palash et al. [14].

4.8 Channel Resistance

HEMT is a suitable candidate due to its high electron mobility, high transconductance, high current carrying capability, low noise, high frequency, high cut-off frequency and high power delivery [15] for different analog VLSI circuits. These devices have not only been identified as the technology of choice for next generation high-power, high-frequency applications but also have shown excellent noise characteristics in Analog VLSI circuits.

GaN is a wide band gap material having high saturation velocity. The property of wide band gap leads to high breakdown voltage and high power level at high temperature. Similarly, the high saturation velocity of the carriers produces high current (I_{max}) and operates at high cut-off frequency (f_t). The high breakdown voltage and high current delivers high output power. Furthermore, the high mobility achieved in the AlGaN/GaN HEMT leads to low R_{on} resistance and low knee voltage (V_{knee}). The low R_{on} resistance leads to high power capability of the device with high efficiency. Therefore, GaN HEMTs emerge as the promising candidates for next generation wireless communication due to its suitable material properties.

Similarly, the AlGaN/GaN-based HEMT produces high 2DEG sheet charge density (n_s) with high mobility of the charge carriers. The high 2DEG density with high saturation velocity produces high current in the device and delivers high output power. Similarly, the low knee voltage of the AlGaN/GaN HEMT also delivers high output power with high power-added efficiency. Superior results have been reported in microwave power

performance of AlGaN/GaN HEMTs, due to the great progress made in material preparation and device fabrication techniques. It is seen from the literature that AlGaN/GaN and
its compound based HEMTs give better electrical characteristics and are usually used as
low noise amplifiers (LNA) in analog VLSI circuits [15].

From the *V~I* characteristics of HEMT, it is observed that for different values of V_{gs}, the
current carrying capability of the device increases with V_{gs}. When $\partial I_d / \partial V_{ds}$ is calculated,
then the peak current occurs at $V_{ds} = \left(V_{gs} - V_{th}\right)$ and is given by

$$I_{d,max} = \mu_n \left(\frac{\varepsilon_s}{\overline{w}}\right)\left(\frac{W}{L}\right)\frac{\left(V_{gs} - V_{th}\right)^2}{2} \qquad (4.23)$$

where $\overline{w}$ is the average depletion layer width in the channel, $\left(\frac{W}{L}\right)$ is the aspect ratio and
$\left(V_{gs} - V_{th}\right)$ is the overdrive voltage or effective voltage.

Further Equation (4.23) can be rearranged and the drain current can be written as in
Equation (4.24). It is observed that the drain current is a linear function V_{ds}.

$$I_d \approx \mu_n \left(\frac{\varepsilon_s}{\overline{w}}\right)\left(\frac{W}{L}\right)\left(V_{gs} - V_{th}\right)V_{ds} \qquad (4.24)$$

The channel from source to drain can be represented by a linear resistor and is written
as [15]

$$R_{on} = \frac{1}{\mu_n C_{dep}\left(\dfrac{W}{L}\right)\left(V_{gs} - V_{th}\right)} \qquad (4.25)$$

Since the MODFET produces a current in response to its gate-source overdrive voltage,
a figure of merit is defined to be how well a device converts a voltage to a current. The
figure of merit is termed as transconductance $\left(g_m\right)$ and is defined as the change in drain
current to the change in gate-source voltage provided the drain-source voltage is kept
constant.

$$g_m = \mu_n C_{dep}\left(\frac{W}{L}\right)\left(V_{gs} - V_{th}\right) = \sqrt{2\mu_n C_{dep}\left(\frac{W}{L}\right)I_d} \qquad (4.26)$$

The g_m also represents the sensitivity of the HEMT and for a high g_m, a small change in V_{gs}
results in a large change in drain current. The g_m, in the saturation region is equal to the
inverse of R_{on} in deep triode region and vice versa, i.e., $R_{on} = 1/g_m$. The transconductance
further can be derived as [15]

$$g_m = \frac{2I_d}{\left(V_{gs} - V_{th}\right)} \qquad (4.27)$$

Equation (4.27) suggests that g_m increases with the overdrive voltage, if W/L is constant
whereas Equation (3.5) implies that g_m decreases with the overdrive voltage if I_d is constant.
The HEMT/MODFET has high transconductance and low R_{on} resistance which makes this
device more suitable for Analog VLSI circuits.

Therefore, the HEMT having high 2DEG density (n_s), high mobility (μ_n), high transconductance (g_m), low R_{on} resistance, low knee voltage (V_{knee}), high cut-off frequency (f_t) and high frequency of Oscillation (f_{max}) is very much suitable for Analog VLSI circuits [16]. Using these techniques, different PhD thesis has been written and few of them are by Lenka [12], Palash [14] and the interested readers can go through that to know more details. They have considered different kind of HEMTs having different structures, different materials to improve the device characteristics such that the HEMT can be used as a suitable candidate in Analog VLSI circuits.

References

1. S. J. Li, "Semiconductor device simulation with equivalent circuit model including quantum effect," PhD Dissertation, Institute of Electrical Engineering, National Central University, China, June 2007.
2. G. L. Snider, I. H. Tan, and E. L. Hu, "Electron states in mesa-etched one-dimensional quantum well wires," *J. Appl. Phys.*, Vol. 68, No. 6, 1990.
3. V. Ziel and E. N. Wu, "High frequency admittance of high-electron mobility transistors (HEMTs)," *Solid State Electron.*, Vol. 26, pp. 753–754, 1983.
4. P. Roblin, S. Kang, A. Ketterson, and H. Morkoc, "Analysis of MODFET microwave characteristics," *IEEE Trans. Electron Devices*, Vol. ED-34, No. 9, pp. 1919–1928, 1987.
5. R. S. Carson, *High Frequency Amplifiers*, 2nd ed. Wiley, New York, 1982.
6. T. Nanjo et al., "Enhancement of drain current by an AlN spacer layer insertion in AlGaN/GaN high-electron-mobility transistors with Si-Ion-Implanted source/drain contacts," *Jpn. J. Appl. Phys.*, Vol. 50, pp. 064101:1–7, 2011.
7. V. H. Brech, "Optimization of GaAs based high electron mobility transistors by numerical simulations," PhD Dissertation, 1998.
8. W. Ruonan, "Enhancement/depletion-mode HEMT technology for III-nitride mixed-signal and RF applications," PhD Dissertation, Hong Kong, 2008.
9. F. Conti and M. Conti, "Surface breakdown in silicon planar diodes equipped with a field plate," *Solid State Electron.*, Vol. 15, pp. 93–105, 1972.
10. V. Palankovski et al., "Field-plate optimization of AlGaN/GaN HEMTs," *IEEE Compound Semiconductor Integrated Circuit Symposium (CSICS)*, At San Antonio, TX, pp. 107–110, 2006.
11. C. Y. Chiang, H. T. Hsu, and E. Y. Chang, "Effect of field plate on the RF performance of AlGaN/GaN HEMT Devices," *Phys. Procedia*, Vol. 25, pp. 86–91, 2012.
12. T. R. Lenka, "Studies on characteristics of III–V compound semiconductor based HEMT/MODFET to use in analog VLSI circuits," PhD Dissertation, Sambalpur University, 2012.
13. J. Lu, Y. Wang, L. Ma, and Z. Yu, "A new small-signal modeling and extraction method in AlGaN/GaN HEMTs," *Solid State Electron.*, Vol. 52, pp. 115–120, 2008.
14. P. Dash, "Modeling, realization and characterization of compositionally Graded AlGaN/GaN heterostructures for electronic applications," PhD Dissertation, IIT, Kharagpur, India, 2012.
15. B. Razavi, *Design of Analog CMOS Integrated Circuits*, Tata McGraw-Hill, New York, 2002.
16. D. Krausse, F. Benkhelifa, R. Reiner, R. Quay, O. Ambacher, "AlGaN/GaN power amplifiers for ISM applications," *Solid-State Electron.*, Vol. 74, pp. 108–113, 2012.

5

AlGaN/GaN HEMTs for High Power Applications

P. Prajoon and Anuja Menokey

CONTENTS

In developing advanced high-power devices, most of the research and development efforts in Solid state devices and IC's are oriented towards High Electron Mobility Transistors (HEMT) and the III-V compound materials. In the past few years, the GaN-based HEMT has becoming an emerging device because of its high-power and high-frequency applications. This chapter demonstrates various GaN-based high-electron mobility transistor structures for high-power applications. The impact of barrier layer in the device structure especially the back-barrier layer. The DC and RF characteristics of the device is explained in detail. Further, versatile method to improve the breakdown voltage of the device is also discussed. A small-signal equivalent circuit for the GaN HEMT is developed and discussed in detail. The chapter concludes with a complete overview of AlGaN/GaN HEMT structures and the various techniques to improve the power performance of the device.

5.1 General Structure of AlGaN/GaN HEMTs

The GaN HEMT is a heterojunction device, like a typical homojunction MESFET. This three-terminal device utilizes the gate to control current flow. However, instead of a thick current carrying channel, HEMT works on the principle of Two-Dimensional Electron Gas (2DEG) formed at the heterojunction interface [1,2]. A simple GaN-based HEMT structure cross-sectional schematic and dimensions are shown in Figure 5.1. It consists minimum of three-layer regions except substrate as shown. The channel is formed at the heterointerface junction of the Barrier/Channel layer in the form of 2DEG [3–5].

The substrates for the GaN devices are sapphire, SiC, AlN, single crystal GaN and ZnO, etc. Rigorously, the sapphire and SiC are the main candidates for the research and fabrication of commercial GaN-based HEMT device, since it renders good temperature stability and acceptable lattice mismatching with GaN material [6–9]. The buffer layer is used to improve the crystal quality of the device by suppressing any effect of lattice mismatch to occur in the channel region, thereby improves the overall performance and inhibit early breakdown of the device. To achieve higher 2DEG density at the heterointerface junction and to attain proper operation of the device, the energy of $Al_xGa_{1-x}N$ barrier layer must be at a higher level than the conduction band of the GaN channel layer. A large conduction band offset between the barrier and the channel layer transfers electrons from the barrier layer to the channel layer. The transferred electrons are confined to a small quantum well region in the channel layer near the heterointerface. Since the electron are confined in third dimension and allows free flow in a two-dimensional sheet, the region is called the 2DEG. The band-diagram showing the 2DEG formation in an AlGaN/GaN heterostructure is given in Figure 5.2. The elements that determine the quality of the 2DEG are growing method, type of substrate, and level of doping of the barrier layer.

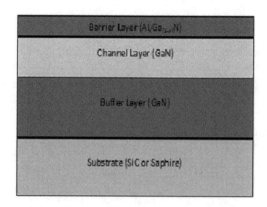

FIGURE 5.1
Basic GaN/AlGaN HEMT structure.

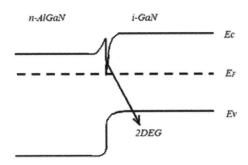

FIGURE 5.2
2DEG formation in an AlGaN/GaN heterostructure.

5.2 Impact of Back Barrier

III-N compound material, especially GaN, has been considered as a promising material for the development of heterostructure devices for high-power switching applications as well as high-power microwave applications. Since GaN exhibits a high breakdown voltage and high-speed characteristics, GaN-based heterostructure field-effect transistors (HFETs) allow reduction of the on-state loss and switching loss in power electronic applications. However, in a conventional AlGaN/GaN heterostructures scaling below 50-nm gate length (L_g) is challenging due to difficulty in maintaining a sufficiently high aspect ratio (L_g/d, where d is the barrier thickness) to suppress short-channel effects (SCEs). Recessed gate structure is one of the best alternatives to overcome this effect; however, it leads to induce damages on the channel as well as creates nonuniformities across the wafer [10–13].

The SCE cannot be ignored in the devices below 30 nm gate length; significant SCEs were noticed in a HEMT device with 30-nm-gate-length and 7 nm barrier thickness [14]. Scaling the top-barrier could improve the aspect ratio and subsequently SCE can be suppressed, but it diminishes the electron charge density and enhances the gate leakage current. Therefore, to get rid of SCE in a GaN HEMT, Thomas Palacios and his team suggested a back-barrier layer below the heterojunction, which sufficiently reduces the SCE problems in a scaled device [11]. The use of InGaN or AlGaN thin layer below the heterojunction raises the conduction band of the GaN buffer with respect to the GaN channel to increase the confinement of the electrons. As proposed in ref. [11,15–17], back-barrier structures have the potential to reduce SCEs by improving the carrier confinement.

Figure 5.3 shows a typical structure of AlGaN/GaN HEMT with AlGaN back barrier. The device performance of AlGaN/GaN high-electron mobility transistors (HEMTs) is analysed and evaluated at different thickness of AlGaN back barrier. It shows that an optimal AlGaN back barrier thickness of 100 nm is necessary for obtaining low leakage current and high I_{on}/I_{off} ratio of approximately 108. Any further variation in the AlGaN thickness leads to deteriorate the device performances because of the generation of more surface defects. The optimal back barrier thickness (t_{bb}) can not only improve the 2DEG confinement, but also prevent electrons penetrating buffer layer. As a result, the trapping effect reduces and shows improvement in dynamic R_{ON}.

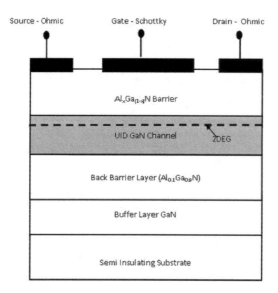

FIGURE 5.3
AlGaN/GaN HEMTs structures with AlGaN back barrier.

Figure 5.4a illustrates the DC I–V characteristics of the HEMTs fabricated on the sample with different $Al_{0.1}Ga_{0.9}N$ back-barrier layer thickness. At minimum 100 nm BBL thickness the devices shows a good pinch-off characteristics with the drain current approximately 500 mA/mm. The relatively higher maximum drain current after introducing the BBL is attributed to the relatively higher electron density. Figure 5.4b present the specific static on-resistance (R_{on_sta}) calculated from the linear region of IV characteristics, it demonstrates a minimum R_{on} at 100 nm BBL thickness.

Figure 5.5 present the on/Off ratio characteristics of the device with different BBL thickness. At 100 nm BBL thickness the leakage current decreases drastically, approximately 10^8 is obtained. It is evident that the incorporation of BBL increases the band offset between

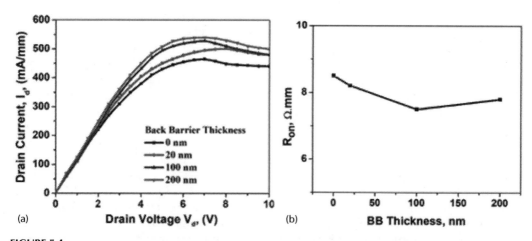

(a)   (b)

FIGURE 5.4
(a) Output characteristics for HEMTs measured at $V_G = 1$ V and BBL thickness between 0 and 200 nm. (b) Static on-resistance of HEMTs at different BBL thickness.

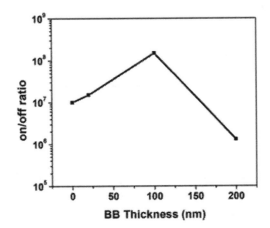

FIGURE 5.5
I_{on}/I_{off} of the device at different BBL thickness.

BBL and GaN channel layer, subsequently increases electron confinement with in the channel and reduces the buffer leakage. Besides, the BBL also alleviate any dislocation and defects from the substrate to extending up to the 2DEG channel, thereby reduces the chances of early breakdown and leakages of the device. However, further increasing the BBL from 100 to 200 nm will deteriorate the device performance.

5.3 DC and RF Characteristics

AlGaN/GaN High Electron Mobility Transistors (HEMTs) are popular in power amplifier systems and offer advantages over another III-V semiconductor heterostructures, such as a large bandgap energy, a low dielectric constant, and a high critical breakdown field. These qualities make GaN a prime candidate for high-power and radiation-hardened applications using a smaller form-factor, different types of semiconductor substrates have been considered for their cost effectiveness and thermal properties. In this book, the DC and RF characteristics of GaN HEMTs grown on Si-substrates will be investigated.

5.3.1 DC Analysis of AlGaN/GaN HEMT

DC characteristics of HEMT is very similar to the operation of MESFET. Figure 5.6 shows the operation of depletion mode HEMT common source configuration. The gate is reverse biased and used as input to the device. The drain is forward biased, and output is taken at the terminal. The depletion region is formed under the gate due to reverse bias. The rectifying effect of depletion region on the channel will affect the electron transport from the source to drain end. Drain voltage varied and gate voltage is fixed, cause current flow from drain to source. MESFET is behave like a resistor at low drain voltage. Depletion width at the drain increases with drain voltage and saturates when the channel could accommodate the maximum velocity of the electrons [14].

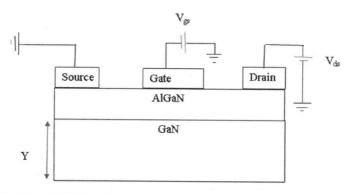

FIGURE 5.6
Cross-sectional view of AlGaN/GaN MESFET.

$$C_{gg} = \frac{dQ}{dV_{gs}} \tag{5.1}$$

$$C_{gd} = \frac{dQ}{dV_{gd}} \tag{5.2}$$

$$C_{gs} = C_{gg} - C_{gd} \tag{5.3}$$

$$g_{ds} = \frac{dI_{ds}}{dV_{gd}} \tag{5.4}$$

$$g_m = \frac{dI_{ds}}{dV_{gs}} \tag{5.5}$$

$$\tau = \frac{dQ}{dI_{ds}} \tag{5.6}$$

The gate length $L_g \gg y$ for long channel devices. The constant low-field mobility and gradual channel approximation the drain current written as [15],

$$I_{ds} = (\text{charge density}) \times (\text{field}) \times (\text{Channel area}) \times (\text{mobility}) \tag{5.7}$$

$$I_{ds} = Z\left[Y - d(x)\right] q \, Nd \, \mu \frac{dV}{dx} \tag{5.8}$$

where Y is the channel depth, Z is the gate width, $d(x)$ is the depletion width and Nd are the doping concentration.

$$d(x) = \left\{ \frac{2\varepsilon_{s[V(x)+V_{bi}} - V_{gs}]}{qN_d} \right\}^{1/2} \tag{5.9}$$

The V_{bi} is the built-in potential and $\varepsilon_s = \varepsilon_0 \times \varepsilon_r$ is the semiconductor dielectric constant.

$$I_{ds} = G_0 \left\{ V_{ds} - \frac{2\left(V_{ds} + V_{bi} - V_{gs}\right)^{3/2} - \left(V_{bi} - V_{gs}\right)^{3/2}}{3V_p^{1/2}} \right\} \tag{5.10}$$

$$G_0 \text{ (the channel conductance)} = \frac{q\mu N_d ZY}{L} \tag{5.11}$$

$V = V_p$ when the depletion width $d(x)$ equals the channel depth Y [15],

$$V_p = \frac{qN_d Y^2}{2\epsilon_s} \tag{5.12}$$

$$I_{ds(sat)} = Z[Y-d]\, q\, N_d V_{sat} \tag{5.13}$$

The intrinsic transconductance g_m,

$$g_m = \frac{I_{ds}}{V_{gs}} \quad (V_{ds} = \text{constant}) \tag{5.14}$$

$$g_m = \frac{\epsilon_s}{d} Z V_{sat} \tag{5.15}$$

where V_{sat} is the saturation velocity.

The transconductance gm give an indication of how good or bad the linearity and the speed of the device and change of the output current I_{ds} with respect to of input voltage (V_{gs}). The intrinsic transconductance g_m is inversely proportional to the gate length and directly proportional to the gate width. The g_{ds} is the output conductance, change with respect to output current I_{ds} and output voltage V_{ds} at a constant input voltage V_{gs} and it is inversely proportional to the gate length.

The output conductance expression is,

$$g_{ds} = \frac{dI_{ds}}{V_{gs}} \quad (V_{gs} \text{ is constant}) \tag{5.16}$$

$$g_{ds} = \frac{\epsilon_s V_{sat} Z}{d} \tag{5.17}$$

The extrinsic transconductance $g_m{}'$ is affected by R_s, and the relation between $g_m{}'$ and R_s can be obtained from Figure 5.7,

1. $\Rightarrow I_{ds} = g_m V_{gs}$
2. $\Rightarrow$ The effect of the source series resistance $I_{ds} = g_m V_{gs}{}'$

$$V_{gs} = V_{gs}{}' + g_m V_{gs}{}' R_s = \left(1 + g_m R_s\right) V_{gs}{}' \tag{5.18}$$

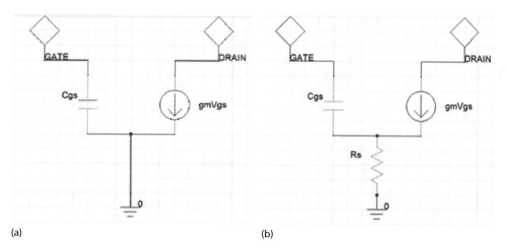

(a)   (b)

FIGURE 5.7
(a) Small-signal equivalent circuit and (b) small-signal equivalent with transconductance circuit with R_s.

$$V_{gs}' = V_{gs} \frac{1}{1 + g_m R_s} \tag{5.19}$$

$$I_{ds} = g_m V_{gs}' = \left\{ \frac{g_m}{1 + g_m R_s} \right\} V_{gs} \tag{5.20}$$

Extrinsic transconductance,

$$g_m' = \frac{g_m}{1 + g_m R_s} \tag{5.21}$$

5.3.1.1 DC Result Analysis of AlGaN/GaN HEMT

The cross-sectional view of AlGaN/GaN HEMT with Schottky contacts and 0.8 μm gate length structure is given in Figure 5.8, The device grown on SiC substrate and consists of a 2 μm thick GaN buffer layer, the thicknesses of the n-Al$_{0.2}$Ga$_{0.8}$N barrier layer, Al$_{0.2}$Ga$_{0.8}$N Spacer layer and GaN channel layer are fixed to 14, 2 and 5 nm, respectively. The Si$_3$N$_4$ passivation layer of 50 nm thickness is deployed in both sides of the gate and Schottky Ni/Au layer is incorporated for source and drain contact. The gate length $L_G = 0.8$ μm. The three sample devices are Al$_{0.3}$Ga$_{0.7}$N (sample A), Al$_{0.2}$Ga$_{0.8}$N (sample B), and Al$_{0.1}$Ga$_{0.9}$N (sample C) is shown in Table 5.1. The sample devices were simulated using TCAD synopsis with variation in the Al mole-concentration.

Figure 5.9 shows I_{ds}–V_{ds} characteristics of the drain current (I_d) mathematical model and simulation with drain-to-source voltage V_{ds} (0–25 V), for the Schottky-based HEMT (Al$_{0.2}$Ga$_{0.8}$N/GaN). Due to high sheet charge density resulting from the strong piezoelectric effect and large conduction band discontinuity, high-drain current is obtained in sample B of 1 A/mm at $V_{gs} = 4$V. The simulated data is compared with modelled date and is validated using experimental data [18,19].

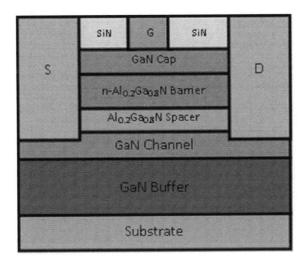

FIGURE 5.8
AlGaN/GaN power HEMT device structure.

TABLE 5.1

Parameters Used to Plot the DC Characteristics Using Drain Current Model

Parameter	Quantity	Sample A	Sample B	Sample C
M	Alloy Composition	$Al_{0.3}Ga_{0.7}N$	$Al_{0.2}Ga_{0.8}N$	$Al_{0.1}Ga_{0.9}N$
n_s	Sheet carrier density (cm^{-2})	1.52E13	1.5E13	1.49E13
L_g	Gate Length (µm)	0.8	0.8	0.8
L_{gd}	Gate to Drain Distance (µm)	1	1	1
V_{th}	Threshold Voltage (V)	−3.0	−2.4	−2.2
W	Gate Width (µm)	75	75	75
V_{sat}	Saturation velocity (V/m)	1.3179E5	1.3179E5	1.3179E5
E_C	Saturation Electric field	190E5	190E5	190E5
μ_0	Low field Mobility (cm^{2}V^{-1})	0.7598	0.7598	0.7598
ΔEc	Conduction band discontinuity b/w AlGaN and GaN (V)	0.1962	0.1953	0.1945
γ_0	Experimental parameter	4.12E−12	4.12E−12	4.12E−12
a(0), a(m)	Lattice Constants	3.17	3.17	3.17
c13(GPa)	Elastic Constant	103.75	103.75	103.75
c33 (GPa)	Elastic Constant	400.2	400.2	400.2
e31(C/m^2)	Piezo Electric Constant of $Al_mGa_{1-m}N$	−0.506	−0.506	−0.506
e33(C/m^2)	Piezo Electric Constant of $Al_mGa_{1-m}N$	0.84	0.84	0.84

The unique material properties, GaN-based HEMTs exhibit very high current density and the depth of the potential well at heterointerface decreases, as the reverse gate voltage is increased. The decrease in the threshold voltage (V_{th}) indicate left shift in the curve w.r.t. increase in the aluminum mole fraction as shown in Figure 5.10. The threshold voltage obtained is −2.4 V, it clearly shows that the device is working in depletion mode, so the device starts conducting in negative gate voltage, and these types of devices are used in high-power transmitters.

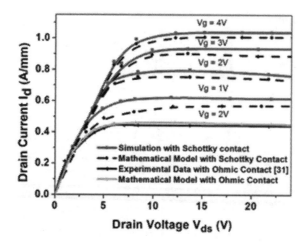

FIGURE 5.9
I_{ds}–V_{ds} characteristics at different gate-to-source biasing (sample B).

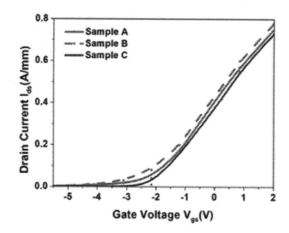

FIGURE 5.10
I_d–V_{gs} characteristics of Schottky S/D contact HEMT.

Figure 5.11 shows the variation of sheet carrier density (n_s) with different gate voltage and the comparison is made for three sample devices sample A, sample B and sample C with 0.8 μm long gate devices. High sheet carrier density is obtained in sample A, which is due to the increased polarization effect with higher mole-concentration of aluminium in the barrier layer. Sheet carrier density is very poor in sample C due to the reduced spontaneous polarization in the barrier and strong influence of interface polarization charge. Higher Al content enhance polarization and the lattice mismatch between $Al_mGa_{1-m}N$ and GaN decreases the actual performance of the devices due to the strain-induced piezoelectric polarization charge.

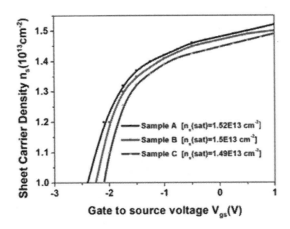

FIGURE 5.11
Variation of n_s with respect to various values of gate-to-source voltage V_{gs}.

5.3.2 RF Analysis of AlGaN/GaN HEMT

5.3.2.1 RF Characteristics of GaN HEMTs

The RF performance of GaN HEMTs used for RF applications includes two important parameters—the unity current-gain cut-off frequency f_T and the unity power-gain cut-off frequency f_{max}. f_T is the frequency at which the magnitude of short-circuit current gain equals unity (or 0 dB), and f_{max} is the frequency at which the unilateral power gain equals unity (or 0 dB). The intrinsic part of the AlGaN/GaN HEMT small-signal model used to find f_T is shown in Figure 5.12.

i_1 is the input current and i_2 is the output current, the equation given by,

$$i_1 = \frac{V_{gs}}{\dfrac{1}{jwC_{gs}} + R_i} + \frac{V_{gs}}{\dfrac{1}{jwC_{gd}} + R_{gd}} \tag{5.22}$$

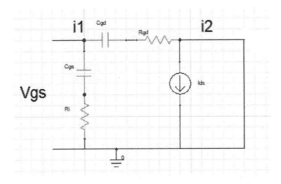

FIGURE 5.12
Intrinsic part of small-signal model of AlGaN/GaN HEMT used to find f_T.

$$i_2 = g_m \, e^{-jwt} \, V_{gs} - \frac{V_{gs}}{\dfrac{1}{jwC_{gd}} + R_{gd}} \tag{5.23}$$

Short-circuit current gain $= \dfrac{i_2}{i_1}$

$$= \frac{\left(1 + jwR_{gd}C_{gd}\right)g_m \, e^{-jwt} - \left(jwC_{gd} - w^2 \, C_{gs}C_{gd}R_i\right)}{jw\left(C_{gs} + C_{gd}\right) - w^2 \, C_{gs}C_{gd}\left(R_i + R_{gd}\right)} \tag{5.24}$$

$$= \frac{g_m}{2\pi f\left(C_{gs} + C_{gd}\right)} \tag{5.25}$$

$wR_{gd}C_{gd} \ll 1$ and $w(C_{gs} + C_{gd}) \gg w^2 \, C_{gs}C_{gd}\left(R_i + R_{gd}\right)$, $wC_{gd} \gg w^2 \, C_{gs}C_{gd}R_i$ and $wC_{gd} \ll g_m$.

Intrinsic current gain cut-off frequency f_T is,

$$= \frac{g_m}{2\pi\left(C_{gs} + C_{gd}\right)} \tag{5.26}$$

both current gain and voltage gain need to be considered for the power-gain cut-off frequency f_{max}. The intrinsic part of small-signal model used to find f_{max} is shown in Figure 5.13. Here, both current gain and voltage gain is considered to determine the power-gain cut-off frequency f_{max}.

The gate resistance R_g and output resistance R_L are included due to their relevance. Voltage gain,

$$A_v = \frac{V_2}{V_1} = \frac{g_m\left(R_{ds} \, / / R_L\right)}{\sqrt{\left(1 + w^2 C_{gs}^{\ 2}\left(R_i + R_g\right)^2\right)}} \tag{5.27}$$

$$\cong \frac{g_m\left(R_{ds} \, / / R_L\right)}{WC_{gs}\left(R_i + R_g\right)} \tag{5.28}$$

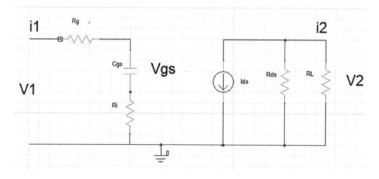

FIGURE 5.13
Intrinsic part of small-signal model used to find f_{max}.

Current gain,

$$\frac{i_2}{i_1} = \frac{g_m}{2\pi f C_{gs}} = \frac{f_T}{f} \tag{5.29}$$

Voltage gain,

$$A_v = \frac{f_T \left(R_{ds} \, / / R_L \right)}{f \left(R_i + R_g \right)} \tag{5.30}$$

$w^2 C_{gs}{}^2 \left(R_i + R_g \right)^2 \gg 1$ for high frequency devices, when $R_{ds} = R_L$ the maximum power gain is obtained at the matched load case and half of the output current flows through the load resistor and the other half through the output resistance of the device.

$$\text{Power gain} = \frac{1}{2} \times \text{current gain} \times \text{voltage gain}.$$

$$= \left(\frac{f_T}{f} \right)^2 \frac{R_{ds}}{\left(R_g + R_i \right)} \tag{5.31}$$

Thus, for $|Gp| = 1$, the power-gain cut-off frequency f_{max} is

$$f_{max} = \frac{f_T}{2} \sqrt{\frac{R_{ds}}{R_g + R_i}} \tag{5.32}$$

The power gain cut-off frequency can be increased by improving f_T, increasing output resistance R_{ds}, minimizing R_g and R_i and the expression of f_T and f_{max} ignore all the extrinsic circuit elements such as R_s, R_d and R_{ds}. The extrinsic elements often limit the intrinsic f_T and f_{max}, when the gate length is very small, considering the extrinsic elements f_T and f_{max} can be written as [20,21]

$$f_T = \frac{g_m}{2\pi f \left(C_{gs} + C_{gd} \right) \dfrac{1 + \left(R_s + R_d \right)}{R_{ds}} + g_m C_{gd} \left(R_s + R_d \right)} \tag{5.33}$$

$$f_{max} = \frac{f_T}{2 \sqrt{\dfrac{\left(R_i + R_g + R_s \right)}{R_{ds}} + 2\pi f_T R_g C_{gd}}} \tag{5.34}$$

5.3.2.2 RF Parameter Extraction Procedure

- The **intrinsic** elements are bias dependent elements, such as g_m, g_d (R_{ds}), C_{gs}, C_{gd}, C_{ds}, R_i and τ [21].
- The **intrinsic** elements are required to be extracted at **hot** bias condition.
- The **extrinsic** elements are independent of the biasing condition, such as L_g, R_g, C_{pg}, L_s, R_s, R_d, C_{pd} and L_d.
- The **extrinsic** elements are extracted using **off** and **cold** bias conditions as shown in Figure 5.14.

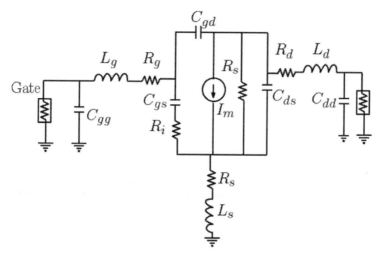

FIGURE 5.14
π small-signal equivalent model for AlGaN/GaN HEMT.

5.3.2.2.1 Cold Bias Extraction

Cold bias extraction contains two parts; the first part is extrinsic parameter extraction, and the second is to subtract the extrinsic part from the whole device model gives intrinsic parameters.

The different operating conditions are needed to extract the extrinsic parameters:

1. OFF condition ($V_{ds} = V_{gs} = 0$)
2. Strong pinch OFF ($V_{ds} = 0$ and $V_{gs} < V_{po}$)

STEP 1: Extracting the Parasitic Inductance and Resistance
The OFF-state Z parameters,

$$Z_{11} = R_s + R_g + .5R_{ch} + j\left[w\left(L_s + L_g\right) - \frac{1}{wC_g} \right] \tag{5.35}$$

$$Z_{12} = Z_{21} = R_s + .5R_{ch} + jwL_s \tag{5.36}$$

$$Z_{22} = R_d + R_s + R_{ch} + jw(L_s + L_d) \tag{5.37}$$

Inductance is extracted from imaginary part of Z parameters,

$$L_s \Rightarrow \mathrm{Im}(Z_{12}) = jwL_s \tag{5.38}$$

$$L_s = \frac{\mathrm{Im}\left(Z_{12}\right)}{W} \tag{5.39}$$

$$L_d \Rightarrow \text{Im}(Z_{22}) = jw(L_s + L_d) \tag{5.40}$$

$$L_d = \frac{\text{Im}(Z_{22})}{W} \tag{5.41}$$

$$L_g \Rightarrow \text{Im}(Z_{11}) = jw(L_s + L_g) \tag{5.42}$$

$$L_g = \frac{\text{Im}(Z_{11})}{w} - L_s \tag{5.43}$$

The resistances can be extracted from the real parts of equations, and under the heavy pinch-off condition where ($V_{DS} = 0$ and V_{GS} much lower than pinch-off voltage V_{po}), the channel is completely off.

$$R_e[Z_{11}(v_{po})] = R_s + R_g \tag{5.44}$$

R_{ch} is extracted using both pinch-off and OFF state,

$$\text{Pinch-off state } R_e[Z_{11}(v_{po})] = R_s + R_g \tag{5.45}$$

$$\text{OFF state } R_e[Z_{11}] = R_s + R_g + .5R_{ch} \tag{5.46}$$

$$R_{ch} = 3 * Z_{11}R_e - R_e[Z_{11}(v_{po})] \tag{5.47}$$

$$R_s \Rightarrow R_e[Z_{12}] = 5R_{ch} + Rs \tag{5.48}$$

$$R_s = R_e[Z_{12}] - 5R_{ch} \tag{5.49}$$

$$R_g \Rightarrow R_g = R_e[Z_{11}] - R_s - 5R_{ch} \tag{5.50}$$

$$R_d \Rightarrow R_d = R_e[Z_{22}] - R_s - R_{ch} \tag{5.51}$$

STEP 2: Extracting the Parasitic Capacitances

The two parasitic capacitances that are required to be extracted, C_{pg} and C_{pd}, which are the pad capacitances connected between gate to ground and drain to ground, respectively. The cold or pinch-off bias condition is used, and the S-parameters measured at this condition, are converted to the respective Z-parameters, the inductances and the resistances computed in the previous section are subtracted.

$$Z_{11} - R_s - R_g - jw(L_s + L_g) \tag{5.52}$$

$$Z_{12} - R_s - jwL_s = Z_{21} - R_s - jwL_s \tag{5.53}$$

$$Z_{22} - R_d - R_s - jw(L_s + L_d) \tag{5.54}$$

The Z parameters are transformed to the respective Y-parameters, and the capacitances can then be extracted,

$$Y_{11} = jw(2C_b + C_{pg}) \tag{5.55}$$

$$Y_{12} = Y_{21} + jwC_b \tag{5.56}$$

$$Y_{22} = jw(C_b + C_{pd} + C_{ds}) \tag{5.57}$$

where C_b is the fringing capacitance due to depleted layer extension at each side of the gate. The extrinsic parameter such as pad parasitic capacitances, inductance and resistance can be extracted using the flow chart as shown in Figure 5.15.

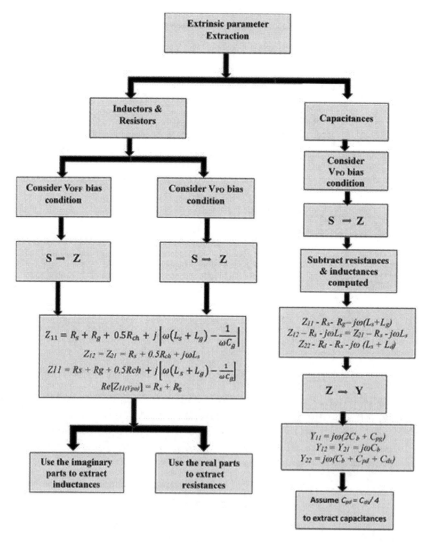

FIGURE 5.15
Flow chart for determining the extrinsic parameters of FETs. (From Lee D. S. et al., *IEEE Electron Device Lett.*, 32, 617–619, 2011.)

$$C_b \Rightarrow C_b = \text{Im}(Y_{12}) \tag{5.58}$$

$$C_{pg} \Rightarrow \text{Im}(Y_{11}) = jw(2C_b + C_{pg}) \tag{5.59}$$

$$C_{pg} = \frac{\text{Im}(Y_{11})}{W} - 2C_b \tag{5.60}$$

$$C_{pg} \Rightarrow \text{Im}(Y_{22}) = jw(C_b + C_{pb} + C_{ds}) \tag{5.61}$$

$$C_{pd} = \frac{\text{Im}(Y_{22})}{5w} - C_b \tag{5.62}$$

5.3.2.2.2 Hot Bias Extraction

The $g_m, g_d, C_{gs}, C_{gd}, C_{ds}, R_i$ and τ are intrinsic elements are required to be extracted at hot biasing conditions and the maximum transconductance value could be determined. The S-parameters of the chosen bias point into Z-parameters and obtained parameters subtract the gate and the drain parasitic series inductances; $Z_{11} - jwL_g$ and $Z_{22} - jwL_d$.

Convert Z-parameters to Y parameters and then the effects of the two parasitic capacitances are eliminated, $Y_{11} - jwC_{pg}$ and $Y_{22} - jwC_{pd}$.

Convert Y-parameters to Z-parameters and source inductances and series resistances are subtracted

$$Z_{11} - R_s - R_g - jwL_s \tag{5.63}$$

$$Z_{12} - R_s - jwL_s = Z_{21} - R_s - jwL_s \tag{5.64}$$

$$Z_{22} - R_d - R_s - jwL_s \tag{5.65}$$

Convert Z-parameters into Y-parameters for the extraction of intrinsic parameter (Table 5.2). The following equations shown in the flow chart Figure 5.16 are then used to determine the intrinsic parameters of AlGaN/GaN device.

TABLE 5.2

RF Extraction Parameters

Intrinsic and Extrinsic Parameters	Physical Description
Gate source-fringe capacitance, C_{gs}	Gate charge modulation by changing V_{GS}
Gate drain-fringe capacitance, C_{gd}	Gate charge modulation by changing V_{DS}
Drain-Source fringe capacitance, C_{ds}	Capacitance between drain and source (e.g., substrate capacitance)
Input resistance, R_i	Lumped representation of distributed channel resistances
Gate-drain resistance, R_{gd}	Complement of R_i, to reflect symmetrical nature of the device
Transconductance, g_m	Drain current gain with respect to the change of gate voltage
Transconductance delay, τ	Time delay between change of gate voltage and drain current
Output resistance, R_{ds}	Variation of drain current by the change of drain voltage
Gate resistance, R_g	Resistance of gate metal strip along the gate current flow
Drain resistance, R_d	Resistance of drain access region and drain ohmic contact
Source resistance, R_s	Resistance of source access region and source ohmic contact
Gate inductance, L_g	Inductance due to the contact of gate
Drain inductance, L_d	Inductance due to the contact of drain
Source inductance, L_s	Inductance due to the contact of source

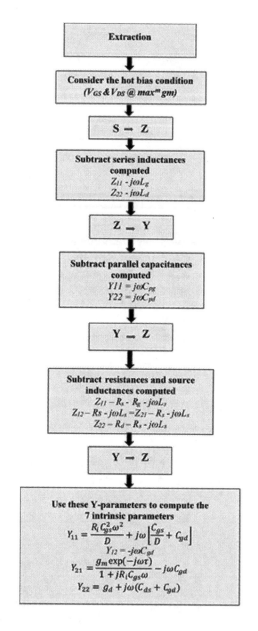

FIGURE 5.16
Flow chart for determining the intrinsic parameters of FETs. (From Lee D. S. et al., *IEEE Electron Device Lett.*, 32, 617–619, 2011.)

$$C_{gd} = -\frac{\text{Im}(Y_{12})}{w} \tag{5.66}$$

$$C_{gs} = \frac{\left[\text{Im}(Y_{11}) - WC_{gd}\right]}{w}\left[1 + \frac{R_e^2(Y_{11})}{\text{Im}(Y_{11} - wC_{gd})^2}\right] \tag{5.67}$$

$$g_{ds} = R_e(Y_{22}) \tag{5.68}$$

$$C_{ds} = \frac{\text{Im}(Y_{22}) - wC_{gd}}{w} \tag{5.69}$$

$$R_i = \frac{R_e(Y_{11})}{R_e^2(Y_{11}) + \left[\text{Im}(Y_{11}) - wC_{gd}\right]^2} \tag{5.70}$$

5.3.2.3 RF Parameter Extraction Using TCAD

HEMTs are used for high-power and high-frequency applications because of the advantages of high electron velocity and high breakdown voltage. High-frequency analysis of the HEMT devices are based on the measurement and evaluation of scattering parameters (S-parameters). The parameters are related to travelling waves that are scattered or reflected when an n-port network is inserted into a transmission line.

The two-port network of HEMT is given in Figure 5.17, where the input port corresponds to gate-source, the output port to drain-source and source is connected to ground. The GaN/AlGaN HEMT various RF parameters such as Cut-off Frequency, Region of Stability, Maximum Frequency of oscillation, Mason's Unilateral gain, Polar plot and Smith Chart can be analyzed with two-port network analysis. The voltage sources are attached to the gate (port 1) and drain (port 2) terminals, and all other terminals are grounded (Figure 5.17).

Two port network S-parameter equations are given below.

$$S_{11} = \frac{B_1}{A_1}(A_2 = 0) \; S_{12} = \frac{B_1}{A_2} \; (A_1 = 0) \tag{5.71}$$

$$S_{21} = \frac{B_2}{A_1}(A_2 = 0) \; S_{22} = \frac{B_2}{A_2}(A_1 = 0) \tag{5.72}$$

S_{11} is the input reflection coefficient, S_{12} the reverse transmission coefficient, S_{21} the forward transmission coefficient, S_{22} the output reflection coefficient and S-parameters are measured as the function of the frequency. RF characterization of device can be calculated based on S-parameter measurement.

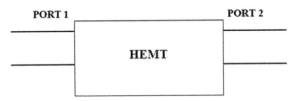

FIGURE 5.17
HEMT 2-port representation.

Maximum Available Gain (MAG): It expresses maximal power gain by conjugate matching at input and output.

Maximum Stable Gain (MSG): It gives power gain of transistor at the stability border, at $K = 1$. K is called stability factor explains the stability of the transistor in frequency range.

Unilateral Power Gain (GU): The transistor exhibits a retroactive capacitance from the output toward input and Unilateral Power Gain considers only the power gain obtained by compensation of these retroaction parameter. unity current gain frequency fT and unity power gain f_{max} are used for the practical application of transistor.

Current Gain Cut-off Frequency f_T: The frequency at which magnitude of the transistor current gain drops to the unity and used for high-speed performance.

Cut-off frequency,

$$f_T = \frac{g_m}{2\pi \left(C_{gs} + C_{gd} \right)} \tag{5.73}$$

where
g_m = Transconductance of the device.
C_{gs} = Gate to source capacitance.
C_{gd} = Gate to drain capacitance.

Maximum Frequency of Oscillation f_{max}: The frequency at which the transistor power gain drops to the unity and estimated by extrapolation of the measured unilateral power gain versus frequency.

5.4 Breakdown Characteristics

The III-Nitride materials especially GaN exhibit unique properties such as wide band gap, high thermal conductivity, high electron velocity and high breakdown voltage. Therefore, GaN-based HEMT delivers a higher output power at RF frequencies. The success of GaN device in high-power application is its ability in controlling the flow of high power while keeping its power dissipation as low as possible. However, achieving a low ohmic contact resistance in a wide band gap GaN is difficult, which ultimately lead to significant power losses in GaN-based HEMTs [22]. In addition to this, the breakdown voltage V_B and on-state resistance R_{on} of the device is directly proportional to each other. Therefore, in developing GaN-based power transistors, there are two key challenges: achievement of a low on-state resistance (including low-resistance Ohmic contacts) and a high breakdown voltage. In view of this, many techniques are proposed to improve the breakdown voltage without affecting the RF characteristics of the device.

5.4.1 Breakdown Voltage Improvement Techniques in AlGaN/GaN HEMT

The breakdown voltage of a HEMT is of two types: On-state and OFF-state breakdown voltages. OFF-state BV is defined as the voltage at which device drain current reaches approximately 1mA at zero gate voltage or below threshold voltage. Similarly, the On-state breakdown voltage is the maximum voltage up to which the device operates safely, beyond which the normal operation of the device fails and drain current increases drastically. In order to tackle breakdown issue of HEMT many investigations are carried out to understand its physical origin. Some of the relevant breakdown mechanism in HEMT devices are

- Impact ionisation mechanism, sudden increase in drain current due to carrier collapse.
- Source-drain breakdown due to short-channel effect.
- High gate electric field, due to the Schottky contact leakage current or through surface related conduction.
- Heterostructure fabrication defects and trap-related problem.

There are different techniques available to improve the breakdown voltage of a transistor; some of the common techniques and their reviews are given.

- Schottky source/drain contact technique.
- High-k passivation technique.
- Field plate engineering technique.

5.4.2 Schottky Source/Drain Contact Technique in HEMT

Since on resistance and BV are in direct relation, the breakdown voltage of the HEMT can be increased by reducing the R_{on}. The R_{on} can be effectively inhibited by incorporating Schottky contact at the source and drain. Binola et al. proposed SSD HEMT; in this type of power HEMT, the On-Resistance (R_{on}) and leakage current is suppressed substantially. Usually in Schottky contact method, a metal layer of Ni/Au is grown over the AlGaN/GaN heterostructure using various techniques [18]. The source Schottky contact can effectively diminish the source carrier injection, since the smooth Schottky metallization produces a more uniform electric-field distribution in GaN buffer.

5.4.3 High-k Passivation Technique

Since GaN is highly sensitive material, the passivation of the top layer is necessary to reduce the surface effects. Severe deterioration in the power performance is observed in GaN HEMT with poor passivation. In conventional GaN-based HEMT, SiO_2 or Si_3N_4 are used as the passivation agent to reduce the surface effects and associated trapping effects. These trapping effects are assumed to be associated with surface states created by dangling bonds, threading dislocations accessible at the surface and ions absorbed from ambient environment. In order to reduce this degradation more effectively, High-k passivation layers are introduced by Binola et al. By increasing the permittivity or by increasing the thickness of the passivation layer, the breakdown voltage increases because of the weakening of the electric field at the drain edge of the gate [23]. Thus by using high-k material and thick passivation layer, breakdown voltage can be increased.

5.4.4 Field Plate Engineering Technique

The semiconductor device performance depends on the physical properties of the semiconductor material used to fabricate. This means that the device performance cannot increase beyond material property boundaries [24]. On the other hand, a noteworthy improvement in the device performance can be obtained by adopting dedicated device structures and fabrication methods. One of the methods in HEMT is called the field plate engineering technique. The FP HEMT devices boosting power performance by 2–4 times compared with conventional NFP devices. The use of FP HEMT renders improvement in breakdown voltage and reduces the parasitic effect or drain current collapse phenomena.

The aim of this section is to provide deep insight of different FP structures and its operation. Figure 5.18 shows a cross-sectional schematic structure of GaN-based field plate HEMT. The metal plate extending from gate towards drain is called the field plate, the purpose of the field plate is to suppress the electric field distribution in the channel and to reduce its peak value on the drain side of the gate edge [25–28]. It is also evident from the literature that the breakdown voltage can be further enhanced by increasing the gate to drain distance L_{gd} and by adjusting the field plate length.

The inclusion of a field plate splits and reshape the electric field distribution at the drain of the gate region, thus the peak electric field reduces and increases the device breakdown. The length of the field-plate dictates the region where field is redistributed. Additionally, the FP reduces the high-field trapping effect, therefore it enhances the current-voltage swings or reduces the drain current collapse effect at high frequencies [29]. The trade-off of the field plate structure includes addition of the gate-drain capacitance at low voltages and extension of the gate-drain depletion length at high voltages, which reduces gain as well as frequency of the device.

5.4.4.1 Gate Field Plate $Al_{0.2}Ga_{0.8}N$/GaN HEMT

The evaluation of the field-plated HEMT is done on a typical GaN structure as shown in Figure 5.18. The numerical simulations are carried out by using Synopsis Sentaurus TCAD simulator [30]. The device structure is composed of Silicon Carbide substrate [5] followed with an undoped GaN Buffer layer, GaN channel, undoped AlGaN spacer, n-doped AlGaN Barrier and GaN cap layer. The mole fraction of the ternary AlGaN material is fixed to 0.26 to obtain a higher polarization effect and to improve the 2DEG sheet carrier density. Considering the physical parameters and the numerical simulations, the analysis of the device structure has been done by means of hydrodynamic simulation including polarization effect, high field saturation effect and gate tunnelling effects, etc. (Figure 5.18).

After having described the device structure let us now move on to device physical parameter that will be simulated to evaluate the effects of the FP geometry on device performance. Figure 5.19 explains the electric field comparison between FP and NFP HEMT device. As shown, an electric field peak is located at the drain edge of the gate contact in NFP device. Since the peak electric field increases at higher drain voltage, the electron tunnelling from the gate to the channel enhances and ultimately leads to increase the gate current and impact ionisation rate, consequently breakdown of the device occurs at a voltage fewer than expected from GaN material. From this it is clear that the breakdown mechanism in HEMT is triggered by the electric field profile at the gate edge. Therefore,

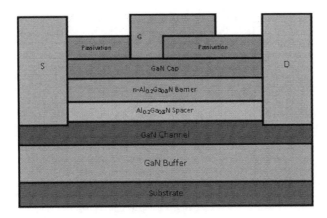

FIGURE 5.18
AlGaN/GaN FP HEMT.

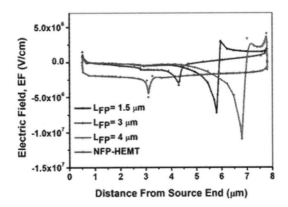

FIGURE 5.19
Electric field variation along the channel in GaN HEMT with different L_{FP}.

in an FP HEMT, the metal plate at the gate edge reduces the peak electric field density and distributes along the field plate length. This makes the AlGaN/GaN HEMT capable of delivering large breakdown voltage and can be used for high-power applications.

5.4.5 Analysis of Field Plate Length on Breakdown Voltage of HEMT

To better empathize field plate operation, it is now possible to start analysing the effects of the field plate geometry on device breakdown. The FP geometry can be varied by playing on two parameters; mainly the field Plate Length (L_{FP}) and Passivation layer thickness (t_{sin}). Various simulation has been carried on the FP HEMT with different field plate length, and the result in terms of breakdown voltage is summarized in Figure 5.20. At first glimpse the breakdown voltage increases for the increase of LFP. However, the breakdown voltage starts falling when the plate length is so comparable with gate to drain distance (L_{gd}). This immature breakdown is caused due to the overlapping of both gate and drain field due to the longer field plate length.

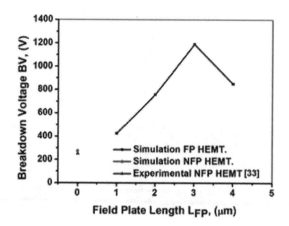

FIGURE 5.20
AlGaN/GaN HEMT breakdown voltage characteristics with field plate length.

5.4.6 Analysis of Gate to Drain Distance on Breakdown Voltage of HEMT

After having described the breakdown characteristics with LFP, let us analyse the impact of gate to drain distance L_{GD} on the BV characteristics of HEMT. Several simulations have been carried out on the FP HEMT with different L_{GD}, and the result in terms of breakdown voltage is summarized in Figure 5.21. Since the concentrated electric field at the gate end diminishes for an increase in L_{GD}, a subsequent increase in BV is observed in the breakdown voltage characteristics of FP HEMT. Further from the analysis, it is found that the fields of drain and field plates are merge together when the length of L_{GD} is comparable with L_{FP}. Since, any increase in device longitudinal dimension reduces the RF high-frequency characteristics of the device, it is desirable to fix an optimum longitudinal dimension in a FP-HEMT [31]. Therefore, the length of the field plate in high-power HEMT is always fixed at a value of about 60% of the length of L_{GD}, at this LFP, the device shows the maximum breakdown voltage. From the simulation it is found that a breakdown voltage of 1660 is obtained for a L_{gd} of 10 μm and LFP of 6 μm.

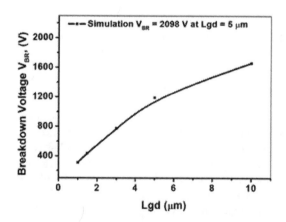

FIGURE 5.21
Breakdown voltage variation with respect to gate to drain distance L_{gd} in field-plated GaN HEMT.

5.4.7 Breakdown Voltage Analysis

With all these analyses it is possible to point out the two reasons behind the early break-down in HEMT devices. First, it is observed that breakdown occurs at the drain-side edge of the gate electrode due to the large concentrated electric field spike under normal operating conditions, via avalanche breakdown. Second, the trap-assisted tunnelling due to higher defect density in the GaN material.

As discussed in the previous section, the breakdown voltage can be increased either by incorporating field plate geometry or by an effective passivation at the top layer. The effect of field plate geometry was already discussed in the last section. When analysing the effect of passivation layer on top of the GaN HEMT, it is found that the BV has a significant improvement when using high relative permittivity dielectric material as passivation layer. Since the electric field gets more weakened as the relative permittivity of the passivation increases, the High-k passivation on GaN HEMT shows a significant improvement in the BV characteristics. In fact, in the insulator, the applied voltage tends to drop uniformly, and the effect will be more significant if the insulator is attached to a semiconductor. In High-k dielectric material the effect of insulator is substantially higher, and the voltage drop along the semiconductor becomes smoother, subsequently increases the BV characteristics of the device. In the case of GaN-based High-k passivated field-plated HEMT structure it is observed that a single gate field plate is enough to suppress the high electric field at the gate edge. Therefore, adapting the FP with High-k passivation on GaN HEMT would increase the breakdown voltage and suppress the parameters responsible for early breakdown in the device.

After having described individual techniques, now, a comparison is carried out to understand the BV behaviour for the different techniques. Figure 5.22 shows the breakdown voltage characteristics of AlGaN/GaN HEMT with different techniques. Since, the BV varies with gate to drain distance L_{gd}, the device is simulated with five different value of L_{gd} from 1 to 5 μm to show the difference in the BV performance. The FP HEMT at 5 μm L_{gd} provides a higher breakdown voltage of 1190 V. similarly, SSD and High-k passivated device shows a breakdown voltage of 320 V and 917 V respectively for the same dimension. In addition, it is clear from the figure that the improvement in breakdown voltage in FP HEMT is more rapid with change in L_{gd} compared with the other two techniques.

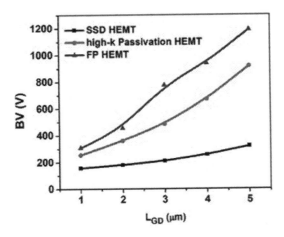

FIGURE 5.22
AlGaN/GaN HEMT breakdown voltage characteristics with different techniques.

5.4.8 Advanced Field Plate Structures

Although, FP-HEMT increases the device breakdown by smoothening the concentrated electric field at the gate edge, it severely reduces the high-frequency small-signal performance. Wu et al. in 2006 demonstrated two different advanced field plate structure in the form of double field plate geometry to further improve the device performance in terms of breakdown and small-signal characteristics [32–34]. The cross-sectional schematic of the two structures are shown in Figure 5.23. In the first technique double-gate field plate geometry is proposed, where basically two field plate terminals are placed in the gate drain access region.

The benefits arising from structure A (Figure 5.23a) is that the breakdown voltage can be further improved compared to the maximum achievable by a single field plate structure. In addition, the effect of added capacitance is decreased, subsequently it is possible to achieve a good RF small-signal performance compared with respect to single field plate geometry. The length of the first field plate L_{FP1} is made to be small and which is in direct contact with the gate metal, the second field plate is placed on top of the first field plate so as to increase the dielectric thickness in between the channel and field plate. Subsequently, reduces the channel capacitance and improves the breakdown voltage as well as small-signal performance.

The numerical simulations of the double field plate structure HEMT is carryout in sentaurus TCAD simulator. The geometry of the FP and passivation is chosen as follows, $t_{SiN1} = 30$ nm, $L_{FP1} = 0.2$ μm, $t_{SiN2} = 90$ nm and $L_{FP2} = 1.4$ μm. After simulation it is found that the electric field profile is shifted to the edges of first and second field plates, respectively. Subsequently, a corresponding improvement in the breakdown voltage is obtained 59.6 V. The double field plates device has reached a power—current gain cut-off frequency product of 50.7, which is larger than 42.2 that represents the best achievable power-current gain cut-off frequency product from a single field plate structure.

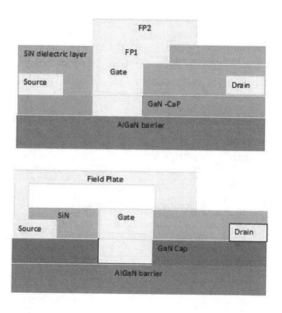

FIGURE 5.23
Advanced FP-HEMT structures. (a) Double field plate structure connected to the gate terminal. (b) Source connected single field plate structure.

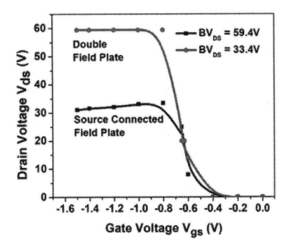

FIGURE 5.24
Simulated off state breakdown measurements at a drain current level of 1mA/mm for a device with a source connected single field plate structure and for a double field plate structure device.

Though the structure reduces the additional capacitance in FP HEMT structure, a significant amount of capacitance is present in double field plate structure. Therefore, a source-connected field plate structure is proposed to completely eliminate the extra capacitance generated by the FP structure, shown in Figure 5.23b. Basically, instead of connecting the field plate terminal to the gate, it is connected to the source, which remains usually grounded during normal operation (Figure 5.24).

The operation of this structure is similar to the double gate structure, except at higher drain current instead of increasing parasitic capacitance C_{GD} an increase in C_{DS} will happen. Generally, the increasing of the drain voltage will lead to start deplete the gate-drain access region in a similar way to what happens when the field plate is connected to the gate.

$$f_t = g_m / \left[2\pi \left(C_{GS} + C_{GD} \right) \right] \tag{5.74}$$

$$f_{max} = f_t \left[4g_o \left(R_s + R_i + R_G \right) + 2 \left(C_{GS} + C_{GD} \right) \left(\left(\frac{C_{GD}}{C_{GS}} \right) + g_m \left(R_s + R_i \right) \right) \right]^{-\frac{1}{2}} \tag{5.75}$$

The main advantage of this structure is however related to the change in parasitic capacitance from C_{GD} to C_{DS}. This capacitance does not have any affect neither in f_t or f_{max} as can be seen in Equations (5.74) and (5.75). Therefore, a GaN HEMT with a source connected field plate can effectively suppress the additional parasitic capacitance. The geometry of the source connected field plate is taken as follows, SiN thickness of 50 nm, $L_{FP} = 0.8\ \mu m$, see Figure 5.23b. The most interesting results are however related to the improvements in terms of small-signal performance. The source connected field plate device has reached a power—current gain cut-off frequency product of 80.5 which is almost twice than 42.2 that represents the best achievable power—current gain cut-off frequency product from a single field plate structure.

5.4.9 Capacitance Modelling of Dual Field-Plate Power AlGaN/GaN HEMT

The field plate AlGaN/GaN HEMT exhibit lower cut-off frequency due to high drain to source and gate-to-drain capacitance, they show improved breakdown voltage, linearity, stability, reliability, efficiency, stability and reliability. Figure 5.25 shows AlGaN/GaN HEMT with gate connected (GFP) and source connected field plate (SFP).

The dual field plate AlGaN/GaN HEMT represent as the series combination of three HEMTs such as intrinsic transistor, gate field transistor and source field transistor. The 2DEG charge density of AlGaN/GaN HEMT devices calculated from the solution of Schrodinger's and Poisson's equations in the quantum well assuming a triangular potential profile, solution of ns considering the two important energy levels are expressed as [35]

$$n_s = D \, V_{th} \left\{ \ln \left[\exp \frac{E_f - E_o}{V_{th}} + 1 \right] + \ln \left[\exp \frac{E_f - E_1}{V_{th}} + 1 \right] \right\} \tag{5.74}$$

$$E_0 = \gamma_0 n_s^{2/3} \tag{5.75}$$

$$E_1 = \gamma_1 n_s^{2/3}$$

$$n_s = \frac{\epsilon}{qd} (V_{g0} - E_f - V_x) \tag{5.76}$$

$$V_{g0} = V_g - V_{off} \tag{5.77}$$

V_{off} is the cutoff voltage, V_x is the channel potential at any point x in the channel.

Subdividing the variation of ns versus the applied gate voltage into different operating regions, the numerical solution for E_f, E_0 and E_1 versus V_g.

The regions are [36],

1. $V_g < V_{off}$ the sub-V_{off} region, where $|E_f|$ is comparable to $|V_{go}|$.
2. $V_g > V_{off}$ and $E_f < E_0$, the moderate 2DEG region.
3. $V_g > V_{off}$ and $E_f > E_0$, the strong 2DEG region.

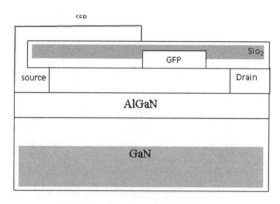

FIGURE 5.25

Cross-sectional view of the dual Field Plate AlGaN/GaN HEMT.

In the sub-V_{off} region $|E_f| > E_0$ and E_1
  $E_f - E_0$ and $= E_f - E_1 \approx E_f$
  $\ln(1 + x) = x$ for $x \ll 1$

$$\frac{qn_s}{c_g v_{th}} e^{\frac{qn_s}{c_g v_{th}}} = \frac{2qD}{c_g} e^{\frac{V_{go}}{V_{th}}} \quad (qn_s \ll c_g v_{th}) \tag{5.78}$$

$$n_{s,sub-off} = 2v_{th}\, D_e^{\frac{V_{go}}{V_{th}}} \quad (E_f \gg E_0, E_1 \text{ and } E_f < 0) \tag{5.79}$$

$$n_{s,above} - V_{off} = \frac{C_g V_{go}}{q} H(V_{go}) \tag{5.80}$$

$$H(V_{g0}) = \frac{V_{go} + v_{th}\left[1 - \ln\left(\beta V_{gon}\right)\right] - \dfrac{\gamma}{3}\left(\dfrac{C_g V_{go}}{q}\right)^{2/3}}{V_{go}\left(1 + \dfrac{V_{th}}{V_{god}}\right) + \dfrac{2\gamma_0}{3}\left(\dfrac{C_g V_{go}}{q}\right)^{2/3}} \tag{5.81}$$

V_{gon} and V_{god} are functions of V_{go} given by the interpolation expression,

$$V_{gox} = \frac{V_{go}\alpha_x}{\sqrt{(V_{go}^2 + \alpha_x^2)}} \tag{5.82}$$

$$\beta = \frac{C_g}{qDV_{th}} \tag{5.83}$$

$$\alpha_n = \frac{e}{\beta} \tag{5.84}$$

$$\alpha_d = \frac{1}{\beta} \tag{5.85}$$

Combine the above expressions,

$$n_{s,unified} = \frac{2V_{th}\left(\dfrac{C_g}{q}\right)\ln\left\{1 + e^{\frac{V_{go}}{2V_{th}}}\right\}}{\dfrac{1}{H\left(V_{go,p}\right)} + \dfrac{C_g}{qD} e^{\frac{-V_{go}}{2V_{th}}}} \tag{5.86}$$

$n_{s,unified}$, 2DEG charge density with the applied voltage is applied to all the regions of operation. Generally, to avoid redundancy in description the model variables are represented with subscript k. Dual field plate HEMT represent as the series combination of 3 transistors T_1, T_2, T_3, respectively as shown in Figure 5.26 [37].

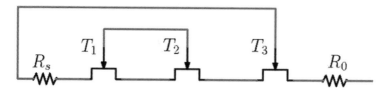

FIGURE 5.26
Model representation of the device. T_1, T_2, and T_3 denote intrinsic device.

$$E_{f,k} = V_{go,k} - \frac{2V_{th} \ln\left(1 + e^{\frac{V_{go,k}}{2V_{th}}}\right)}{\frac{1}{H(V_{go,k})} + \left(\frac{C_{g,k}}{qD}\right)e^{\frac{-V_{go,k}}{2V_{th}}}} \tag{5.87}$$

$$Q_{g,k} = -\int_0^{L_k} WC_{g,k}\left(V_{go,k} - \varphi_{k(x)}\right)dx \tag{5.88}$$

$$Q_g = -WL_kC_{g,k}\left\{V_{go.k} - \frac{1}{2}(\varphi_{s,k} + \varphi_{d,k}) + \frac{\left(\varphi_{d,k} - \varphi_{s,k}\right)^2}{12\left(V_{go,k} - \frac{K_BT}{q} - \frac{1}{2}(\varphi_{s,k} + \varphi_{d,k})\right)^2}\right\} \tag{5.89}$$

$$C_{g,k} = \frac{\epsilon_k}{d_k} \tag{5.90}$$

$C_{g,k}$ is the gate capacitance per unit area, dk is the insulator thickness, W is the channel. Width and L_k is channel length. Gate bias- dependent channel charge per unit length $Q_{ch,k}$ is used to find drain and source charges [38]

$$Q_{ch,k}\left(V_{go,k}, V_{x,k}\right) = WC_{g,k}\left(V_{go,k} - \varphi(x)\right) \tag{5.91}$$

the drain and source charges are defined as

$$Q_{d,k} = \int_0^{L_k} \frac{x}{L_k}Q_{ch,k}\,dx$$

$$= -\frac{1}{2}WL_kC_{g,k}\left\{V_{go.k} - \frac{1}{3}(\varphi_{s,k} + 2\varphi_{d,k}) + \frac{1}{12}\left(\frac{\varphi^2_{ds,k}}{V_{go,k} + \frac{K_BT}{q} - \frac{1}{2}(\varphi_{s,k} + \varphi_{d,k})}\right)\right.$$

$$\left. + \frac{1}{12}\varphi^3_{ds,k}\left(V_{go,k} + \frac{K_BT}{q} - \frac{1}{2}(\varphi_{s,k} + \varphi_{d,k})\right)^2\right. \tag{5.92}$$

$$Q_{s,k} = \int_0^{Lk} \left(1 - \frac{x}{L_k}\right) Q_{ch,k} \, dx$$

(5.93)

$$Q_{s,k} = -\left(Q_{g,k} + Q_{d,k}\right)$$

(5.94)

5.5 Small-Signal Equivalent Circuit

The small-signal analysis of AlGaN/GaN used to determine the electrical properties and the high frequency performance of the device. The quasi-static approach based on analytical expressions are used to determine the bias dependent parameters. The extrinsic resistances are found by fitting the calculated IV characteristics to the measured data. The other bias independent elements are determined using available extraction technique based on fitting the calculated S or Y parameters obtained from circuit analysis to those measured at certain bias levels. The validity of the model is checked through the consistency of the transconductance and cut-off frequency values obtained from DC analysis to those obtained from RF analysis.

Nonlinear characteristics of GaN/AlGaN HEMT modelled by using equivalent circuit with lumped elements. Lumped elements can be divided into two parts such as intrinsic elements and extrinsic elements. Intrinsic elements are bias dependent, which includes C_{gs}, C_{gd}, C_{sd}, R_{gd}, R_{gs}, g_m, g_{ds} and extrinsic elements are bias-independent elements contains R_d, R_g, R_s, L_g, L_s, L_d.

The extraction technique of the small-signal equivalent circuit elements in HEMTs is divide into 3 steps as given below. The equations and extracted parameters are listed in Tables 5.3 and 5.4 respectively.

1. The intrinsic elements (C_{gs}, C_{gd}, g_m and g_{ds}) are determined using DC model.

2. The extrinsic resistors R_s and R_d are extracted through fitting of the I–V characteristics of the device from the DC model with the measured data.

3. The extrinsic elements (C_{gp}, C_{dp}, R_g, L_s and L_d) are determined by fitting the measured Y or S-parameters with obtained from circuit analysis of the equivalent circuit using software. The validity of the model is checked by comparing some device parameters obtained from both DC and RF analysis.

   - The parasitic parameters C_{pg}, C_{pd}, L_g, L_d, L_s, R_g, R_d and R_s are extrinsic parameters and they are bias-independent constant for different biasing.

   - The parasitic capacitances C_{pg} and C_{pd} are that are mainly made up of the sum of capacitances formed between the gate and drain.

   - The capacitances C_{gs} and C_{gd} changes with the depletion region at different gate bias. C_{ds} is the small-signal capacitance between drain and source. This capacitance primarily leads to increase the channel leakage current.

   - The channel resistance R_i, measures how difficult the electrons flow in the channel and are included primarily to improve the match. The g_{ds} which is the inverse value of R_{ds} is the drain to source conductance/resistance called the output conductance of the device and it is again a parameter related to the substrate current.

- g_m indicating how much change in output current from a change of a given input voltage is called the transconductance of the device and can be used to represent the gain of the device.

- The time delay for the output current to be generated with the given input voltage represented as τ and the transconductance takes some time to charge after applying the gate voltage.

TABLE 5.3

Small-Signal Parameters to the Properties of HEMT

Parameter	Comment	Equation
$R_{contact}$	Contact resistance	$\dfrac{1}{Z}\sqrt{\dfrac{\rho_0}{q\mu NdY}}$
R_{SG}	Bulk resistance $R_g = R_{contact} + R_{SG}$ L_{SG} is the distance between drain and gate	$\dfrac{L_{SG}}{q\mu N_d YZ}$
R_{GD}	Bulk resistance, $R_d = R_{contact} + R_{GD}$ L_{SG} is the distance between drain and gate	$\dfrac{L_{GD}}{q\mu N_d YZ}$
R_g	Gate series resistance, m = no of gate strips	$\dfrac{\rho Z}{3m^2 hL}$
C_{gs}	Gate-source capacitance	$\dfrac{\epsilon_s ZL}{d}\left\{\dfrac{X}{2L}-\dfrac{2d}{L+2X}+1\right\}$
C_{gd}	Gate-Drain capacitance	$\dfrac{2\epsilon_s}{1+\dfrac{2X}{L}}$
R_i	Input resistance; μ is the low field mobility	$\dfrac{V_{sat}L}{\mu Ids}$
g_m	Intrinsic transconductance	$\dfrac{\epsilon_s ZV_{sat}}{d}$
τ	Signal delay	$\dfrac{1}{V_{sat}}\left\{\dfrac{X}{2}-\dfrac{2d}{1+\dfrac{2X}{L}}\right\}$
d	Depletion width	$\left\{\dfrac{2\epsilon_s\left(V_{bi}+V_{SG}\right)}{qN_d}\right\}^{1/2}$
X	Depletion extension toward the drain End	$\left\{\dfrac{2\epsilon_s}{qN_d\left(V_{bi}+V_{SG}\right)}\right\}^{1/2}\left(V_{bi}+V_{SG}\right)$
I_{ds}	Saturated drain-Source current	$I_{ds}=Z\left[Y-d\right]qN_dV_{sat}$

Source: Khandelwal, S. and Fjeldly, T.A., *Solid-State Electron.*, 76, 60–66, 2012.

TABLE 5.4

The Extracted Small-Signal Parameter

Parameter	Description	GaN HEMT
C_{pg} (fF)	Parasitic gate capacitance	50.2
C_{pd} (fF)	Parasitic drain capacitance	85.78
L_s (pH)	Source inductance	0.75
L_d (pH)	Drain inductance	69
L_g (pH)	Gate inductance	150
R_s (Ω)	Source resistance	2.97
R_d (Ω)	Drain resistance	5.71
R_g (Ω)	Gate resistance	2.71
C_{gd} (fF)	Gate-drain capacitance	34.2
C_{ds} (fF)	Drain-source capacitance	126.3
C_{gs} (fF)	Gate-source capacitance	152
R_{ds} (Ω)	Output resistance	187
R_i (Ω)	Input resistance	1.6
g_m (mS)	Intrinsic transconductance	66.3
τ (pS)	Signal delay	0.58

Smith chart is used to find the impedance, admittance, reflection coefficient analysis by electrical and electronics engineers specializing in radio frequency (RF) engineering. Smith chart is used to display RF parameters at one or more frequencies. Impedance analysis of the HEMT small-signal circuit in the Figure 5.27 at 45–50 GHz is analysed at different *S*–parameters (Figure 5.28). Figure 5.29 shows how the impedance varies with S11, S12, S21, S22 in AlGaN/GaN HEMT.

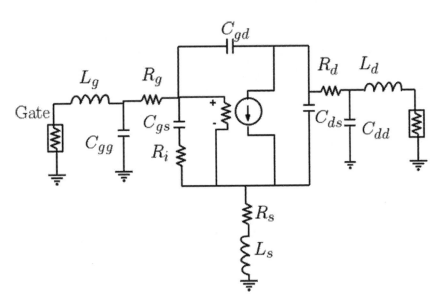

FIGURE 5.27
Small-signal equivalent circuit of AlGaN/GaN HEMT in Ansys at $V_{gs} = -4.8$ V and $V_{ds} = 15$ V.

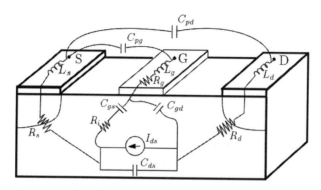

FIGURE 5.28
Small-signal circuit parameters of FET devices. (From Khandelwal, S. and Fjeldly, T.A., *Solid-State Electron.*, 76, 60–66, 2012.)

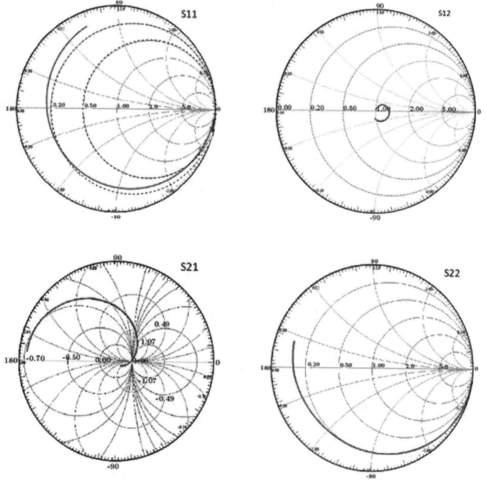

FIGURE 5.29
Impedance analysis using Smith chart at 45–50 GHz.

References

1. Baskaran S., Anbuselvan N., Mohanbabu A., Mohankumar N., Godwinraj D. and Sarkar C. K. (2013) "Modeling of 2DEG sheet carrier density and DC characteristics in spacer based AlGaN/AlN/GaN Hemt devices," *Superlattices and Microstructures*, 64, 470–482.

2. Ambacher O., Foutz B. and Smart J. (2000) "Two-dimensional electron gases induced by spontaneous and piezoelectric polarization in undoped and doped AlGaN/GaN heterostructures," *Journal of Applied Physics*, 87(1), 334–344.

3. Nirmal D., Arivazhagan L., Augustine Fletcher A.S., Ajayan J. and Prajoon P. (2018) "Current collapse modeling in AlGaN/GaN HEMT using small signal equivalent circuit for high power application," *Superlattices and Microstructures (Elsevier)*, 113, 810–820.

4. Rajan S., Xing H., Den Baars S., Mishra U. K. and Jena D. (2004) "AlGaN/GaN polarization-doped field-effect transistor for microwave power applications," *Journal of Applied Physics*, 84, 1591.

5. Ando Y., Okamoto Y., Miyamoto H., Nakayama T., Inoue T. and Kuzuhara M. (2003) "10-W/mm AlGaN-GaN HFET with a field modulatine plate," *IEEE Electron Device Letters*, 24(5), 289–291.

6. Huque M. A., Eliza S. A., Rahman T., Huq H. F. and Islam S. K. (2009) "Temperature dependent analytical model for current–voltage characteristics of AlGaN/GaN power HEMT," *Solid-State Electronics*, 53, 341–348.

7. Chini A., Buttari D., Coffie R., Heikman S., Keller S. and Mishra U. K. (2004) "12 W/mm power density AlGaN/GaN HEMTs on sapphire substrate," *IET Electronics Letters*, 40(1), 73–74.

8. Bouzid-Driad S., Maher H., Defrance N., Hoel V., De Jaeger J.C., Renvoise M. and Frijlink P. (2013) "AlGaN/GaN HEMTs on silicon substrate with 206-GHz FMAX," *IEEE Electron Device Letters*, 34(1), 36–38.

9. Dora Y., Chakraborty A., Mc Carthy L., Keller S., Den Baars S. P. and Mishra U. K. (2006) "High breakdown voltage achieved on AlGaN/GaN HEMTs with integrated slant field plates," *IEEE Electron Device Letters*, 27(9), 713–715.

10. Lee D. S., Gao X., Guo S. and Palacios T. (2011) "InAlN/GaN HEMTs with AlGaN back barriers," *IEEE Electron Device Letters*, 32(5), 617–619.

11. Palacios T., Chakraborty A., Heikman S., Keller S., DenBaars S. P. and Mishra U. K. (2006) "AlGaN/GaN high electron mobility transistors with InGaN back-barriers," *IEEE Electron Device Letters*, 27(1), 13–15.

12. Rennesson S., Damilano B., Vennegues P., Chenot S. and Cordie Y. (2013) "AlGaN/GaN HEMTs with an InGaN back-barrier grown by ammonia-assisted molecular beam epitaxy," *Physica Status Solidi A*, 210(3), 480–483.

13. Medjdoub F., Zegaoui M., Grimbert B., Rolland N. and Rolland P. A. (2011) "Effects of AlGaN back barrier on AlN/GaN-on-silicon high-electron-mobility transistors," *Applied Physics Express*, 4, 124101.

14. Abdel Aziz M., El-Sayed M. and El-Banna M., (1999) "An analytical model for small-signal parameters in HEMTs including the effect of source/drain extrinsic resistances," *Solid-State Electronics*, 43, 891–900.

15. Ladbrook P. H. (1989) *MMIC Design GaAs and HEMTs*. Boston, MA: Artech House.

16. Lee D. S., Chung J. W., Wang H., Gao X., Guo S., Fay P. and Palacios T. (2011) "245-GHz InAlN/GaN HEMTs with oxygen plasma treatment," *IEEE Electron Device Letter*, 32(6), 755–757.

17. Micovic M., Hashimoto P., Ming H. et al. (2004) "GaN double heterojunction field effect transistor for microwave and millimeter-wave power applications," In *IEDM Technical Digest*, pp. 807–810.

18. Jebalin B. K., Rekh A. S., Prajoon P., Godwin Raj D., Kumar N. M. and Nirmal D. (2015) "Unique model of polarization engineered AlGaN/GaN Based HEMTs for high power applications," *Superlattices and Microstructures*, 78, 210–223.

19. Lee H. S., Lee D. S. and Palacios T. (2011) "AlGaN/GaN high-electron-mobility transistors fabricated through a Au-Free technology," *IEEE Electron Device Letters*, 32(5), 623–625.
20. Schwierz F. and Liou J. J. (2003) *Modern Microwave Transistors: Theory, Design, and Performance.* New York: Wiley.
21. Dambrine G., Cappy A., Heliodore F. and Playez E. (1988) "A new method for determining the FET small-signal equivalent circuit," *IEEE Transactions on Microwave Theory and Techniques*, 36(7), 1151–1159.
22. Wong M. H., Keller S., Nidhi S. D. and Dasgupta S. (2013) "N-polar GaN epitaxy and high electron mobility transistors," *Semiconductor Science Technology*, 28, 074009.
23. Jebalin B. K., Rekh A. S., Prajoon P., Mohankumar N. and Nirmal D. (2015) "The influence of High-k passivation layer on breakdown voltage of Schottky AlGaN/GaN HEMTs," *Microelectronics Journal*, 46, 1387–1391.
24. Johnson E. O. (1965) "Physical limitation on frequency and power parameters of transistors," *RCA Review*, 26, 163–176.
25. Asano K., Miyoshi Y., Ishikura K., Nashimoto Y., Kuzuhara M. and Mizuta M. (1998) "Novel high power AlGaAs/GaAs HFET with a field-modulating plate operated at 35 V drain voltage," *Proceedings of the International Electron Devices Meeting (IEDM'98)*, pp. 59–62.
26. Bahl S. R. and del Alamo J. A. (1993) "A new drain injection technique for the measurement of off-state breakdown voltage in FET's," *IEEE Transactions on Electron Devices*, 40(8), 1558–1560.
27. Bhattacharya P. K., Das U., Juang F. Y., Nashimoto Y. and Dhar S. (1986) "Material properties and optical guiding in InGaAs-GaAs strained layer superlattices-a brief review," *Solid-State Electronics*, 29(2), 261–267.
28. Chini A., Lavanga S., Peroni M., Lanzieri C., Cetronio A., Teppati V., Camarchia V., Ghione G. and Verzellesi G. (2006) "Fabrication characetrization and numerical simulation of high breakdown voltage pHEMTs," *Proceedings of the 1st European Microwave Integrated Circuits Conference*, pp. 50–54.
29. Karmalkar S. and Soudabi N. (2006) "A closed-form model of the drain–voltage dependence of the OFF-State channel electric field in a HEMT with a field plate," *IEEE Transactions on Electron Devices*, 53(10), 2430–2437.
30. Lu B. and Palacios T. (2010) "High breakdown (>1500 V) AlGaN/GaN HEMTs by substrate-transfer technology," *IEEE Electron Device Letter*, 31(9), 951–953.
31. Nanjo T., Imai A., Suzuki Y., Abe Y., Oishi T., Suita M., Yagyu E. and Tokuda Y. (2013) "AlGaN channel HEMT with extremely high breakdown voltage," *IEEE Transactions on Electron Devices*, 60(3), 1046–1053.
32. Wu Y.-F., Saxler A., Moore M., Smith R. P., Sheppard S., Chavarkar P. M., Wisleder T., Mishra U. K. and Parikh P. (2004) "30-W/mm GaN HEMTs by field plate optimization," *IEEE Electron Device Letters*, 25(3), 117–119.
33. Wu Y.-F., Moore M., Wisleder T., Chavarkar P. M., Mishra U. K. and Parikh P. (2004) "High-gain microwave GaN HEMTs with source-terminated field-plates," *Proceedings of the International Electron Device Meeting (IEDM'04)*, pp. 1078–1079, December 13–15.
34. Wu Y.-F., Moore M., Wisleder T. and Parikh P. (2006) "40-W/mm double field-plated GaN HEMTs," *Proceedings of the 64th Device Research Conference*, pp. 151–152.
35. Kola S., Golio J. M. and Maracas G. N. (1988) "An analytical expression for fermi level versus sheet carrier concentration for HEMT modelling," *IEEE Electron Device Letter*, 9, 136–138.
36. Khandelwal S. and Fjeldly T. A. (2012) "A physics based compact model of *I–V* and *C–V* characteristics in AlGaN/GaN HEMT devices," *Solid-State Electronic*, 76, 60–66.
37. Ahsan S. A., Ghosh S., Sharma K., Dasgupta A., Khandelwal S. and Chauhan Y. S. (2016) "Capacitance modelling in dual field-plate power GaN HEMT for accurate switching behaviour," *IEEE Transaction Electron Devices*, 63(2): 565–572.
38. Ward, D. E. (1981) "Charge-based modeling of capacitance in MOS transistors," Integrated Circuits Lab, Stanford University, Stanford, CA, Tech. Rep. G201-11, June 1981.

6

AlGaN/GaN HEMT Fabrication and Challenges

Gourab Dutta, Srikanth Kanaga, Nandita DasGupta, and Amitava DasGupta

CONTENTS

Due to their attractive properties, group-III Nitrides (AlN, GaN and InN) and their ternary/quaternary compound based materials are increasingly getting a lot of attention for applications in both optoelectronics and electronics. Due to the wide spectrum of energy bandgap (0.7 eV for InN and 6.2 eV for AlN) these materials find applications from infrared to ultraviolet optoelectronics. GaN-based LEDs have already brought a revolution in the solid-state lighting industry (Khanna 2014, Nakamura and Krames 2013, Nobelprize.org 2014). III-nitrides are also attractive for electronic applications for having properties like wide bandgap, very high electron density at the heterointerface due to polarization, high electron mobility, high critical electric field and high saturation velocity (Mishra *et al.* 2002, 2008, Wu *et al.* 2001). These properties qualify GaN-based HEMTs for applications in high power, high frequency and high temperature applications. Among different GaN-based HEMTs, AlGaN/GaN technology is the most mature and has already been commercialized. This chapter discusses the fabrication process of GaN-based HEMTs primarily emphasizing on AlGaN/GaN HEMTs. Process steps like dry etching of GaN-based materials, ohmic contact metallization, lithography, passivation and some other special techniques like (fluorinated gate HEMT, p-GaN gated HEMT, field plate, MIS-HEMT technology) are separately discussed in detail. Challenges associated with AlGaN/GaN HEMT technology are also discussed.

6.1 HEMT Device Processing

Among different III-Nitrides, AlGaN/GaN heterostructure have shown their potential for high-power and high-frequency applications. This section describes the fabrication processes of typical AlGaN/GaN HEMT. The processing steps include sample cleaning, device isolation, ohmic contact metallization and activation by annealing, gate metal deposition, pad metallization and passivation. Fabrications of different process control monitor (PCM) structures, which are generally included in the HEMT process flow, are also discussed in this section. From the electrical characteristics of these PCM structures, important information regarding the epitaxial layer quality and the individual processes can be extracted. However, before going to the fabrication details of HEMTs, let us first discuss the basic structure of AlGaN/GaN HEMTs which will help us in understanding the process flow of HEMTs.

6.1.1 Basic Structure of AlGaN/GaN HEMTs

AlGaN/GaN heterostructures are typically grown by metal organic chemical vapour deposition (MOCVD) or by molecular beam epitaxy (MBE) on a non-native substrate, viz. (i) c-plane sapphire, (ii) (111) silicon or (iii) SiC (Cordier *et al.* 2003, Ganguly *et al.* 2014, Heikman 2002, Mishra *et al.* 2002, Selvaraj *et al.* 2011). The substrate is the bottom layer of a HEMT structure, and it provides mechanical strength to the HEMT heterostructure. GaN on sapphire technology is quite mature; however, the thermal conductivity of the sapphire substrate is poor resulting significant self-heating effect in AlGaN/GaN/sapphire HEMTs. GaN on SiC would be the best choice in this regard, due to the high thermal conductivity of SiC. But, use of expensive SiC substrates increases the device cost. So, GaN on Si technology has a high potential for fabricating the next generation cost effective power devices. The epitaxial growth process starts with the formation of GaN or AlN nucleation layer on the substrate. After the growth of nucleation layer, a thick (few microns) GaN buffer layer is grown to confine the dislocation defects within it. Then an unintentionally doped GaN channel layer is grown. Finally, the top most AlGaN barrier layer is grown. In some cases, a thin (~2 nm) GaN cap layer is also grown on top of the barrier layer to prevent the unwanted oxidation of the barrier layer.

The difference in electron affinity between the AlGaN barrier layer and GaN channel layer creates a triangular quantum well at AlGaN/GaN heterointerface, which confines the channel electrons. These channel electrons comes from the surface donor states of AlGaN. Unlike AlGaAs/GaAs heterostructure, the two-dimensional electron gas (2DEG) at AlGaN/GaN heterostructure is achieved without any barrier layer doping due to polarization in III-Nitride materials. The difference in polarization between AlGaN and GaN results in a sheet of bound positive charge in AlGaN near the heterointerface. This bound charge controls the 2DEG density, which acts as the channel in GaN-HEMTs. This 2DEG is connected with the outside world through source and drain alloyed ohmic contacts. On application of a drain voltage, a horizontal electric field is created and drain current flows from drain to source terminal. By varying the gate potential, the 2DEG density under the gate and hence the drain current can be modulated. As HEMTs are normally-ON devices, a particular negative gate voltage (threshold voltage) is required for complete pinch-off of the channel or switching the transistor OFF. The regions between source-gate and gate-drain contacts are known as source access region and drain access region respectively. For HEMTs, the access region lengths are expected to be as small as possible to reduce the ON-resistance. However, the length of drain access region (L_{GD}) is limited by the designed breakdown voltage of the HEMT as it was found that breakdown voltage is almost linearly proportional to L_{GD} (Herbecq *et al.* 2016). Exposed AlGaN surface is passivated by depositing a suitable dielectric layer primarily to reduce the current collapse.

6.1.2 Fabrication Process of HEMT

Fabrication process of an AlGaN/GaN HEMT starts with the sample cleaning. Here, "sample" may be a part of the AlGaN/GaN wafer diced as required or the complete AlGaN/GaN wafer. The complete fabrication process is schematically shown in Figure 6.1. Samples are cleaned with different organic solvents, e.g., trichloroethylene (TCE), acetone and isopropyl alcohol (IPA). Samples are first cleaned with hot TCE followed by cleaning using hot acetone and finally rinsing in IPA. Ultrasonic agitation is preferred during these cleaning. After cleaning with these organic solvents, samples are rinsed with de-ionized water and dried using nitrogen blow.

Device isolation is an essential step to isolate individual devices from each other and different techniques like mesa isolation by dry etching or Ar/Kr/F ion-implantation can be used. In mesa isolation, the epitaxial layer is etched sufficiently deep into the unintentionally doped GaN-buffer layer. As a result, there exists no continuous channel between two adjacent devices, i.e. devices are electrically isolated from each other. Chlorine-based dry plasma etching using the Inductively Coupled Plasma Reactive Ion Etching (ICP-RIE) system is widely used for etching the III-nitrides. Mechanisms of Cl-based dry etching are described in detail in Section 6.4. During plasma etching, the active regions of devices are protected by mask. Different masking materials can be used, for example: Si_3N_4, SiO_2, Ni, photoresist, etc. Among them, use of photoresist as an etch mask simplifies the overall process as it is easier to be removed compared to other hard masks. The etch rate and etch profile depends strongly on different etching parameters of ICP-RIE (see Section 6.4). Typical etch depth used for mesa isolation is between 100 nm and 200 nm. A schematic 3D view of a mesa is shown in Figure 6.2. For device isolation using multi-energy implantation, heavy/negatively charged ions (e.g., Ar, Kr, F, O, etc.) are implanted into the AlGaN/GaN heterostructure (while protecting the active regions by suitable mask), which creates defects in the implanted regions and isolates adjacent devices by depleting the 2DEG at the heterojunction (Arulkumaran *et al.* 2014, Hua *et al.* 2015a, 2015b, Shiu *et al.* 2007). Isolation by implantation results in planner device structure.

After mesa isolation and removal of the masking material, source/drain ohmic contact lithography is done using the optical lithography. Prior to metallization, samples are treated with dilute HCl or low power Cl-plasma to remove the native surface oxide. Ti-based ohmic contact (e.g., Ti/Al/Mo/Au or Ti/Al/Ni/Au) is primarily used for GaN-HEMTs. These metal stacks are deposited by electron beam evaporation system at a very high vacuum and then patterned using lift-off process. Contact activation is then done by rapid thermal annealing (RTA) at a temperature range of 800°C–850°C to form the alloyed ohmic contact. Mechanism of ohmic contact formation and the role of each metal in ohmic contact formation are discussed in Section 6.3.

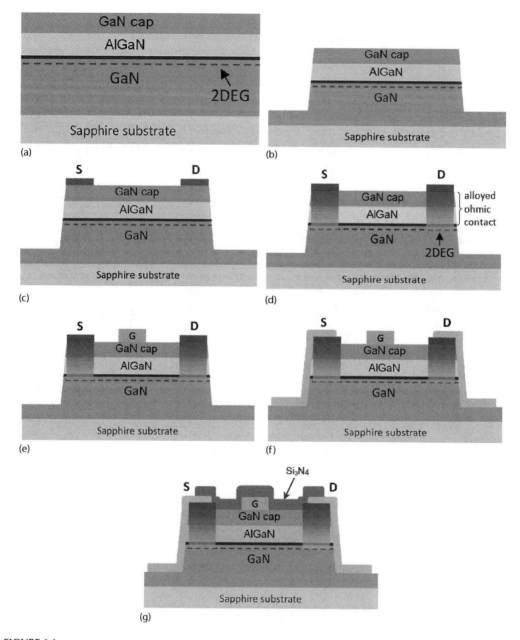

FIGURE 6.1
Fabrication steps of AlGaN/GaN HEMTs where passivation is the last process. (a) Starting wafer, (b) mesa isolation by dry etching, (c) source-drain ohmic contact formation, (d) ohmic contact activation by annealing, (e) gate contact formation, (f) contact pad metallization and (g) passivation with deposited Si_3N_4 layer and its patterning.

Then, gate contact lithography is done by using either optical or electron beam lithography. For fabricating gate structures with sub-micron geometry, electron beam lithography is used. Ni/Au or Pt/Au metal stack is quite popular as a gate metal. Gate metals are also patterned by lift-off process. "T"-shaped gate structures can also be created using the electron beam lithography system. More details about the lift-off process and about the

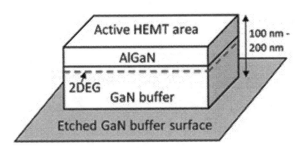

FIGURE 6.2
3D-schematic view of AlGaN/GaN mesa structure.

formation of T-gate structure can be found in Section 6.2. In passivation first process flow (contrary to this current process where passivation is done at last), gate windows have to be opened in the passivation layer. Two consecutive lithography processes is generally performed to get a "T" or "Γ" shape gate structure. The first lithography is for opening the gate foot print in the passivation layer and the second one is for patterning the deposited gate metal. It is well understood that the second lithography pattern is larger than the first one.

After gate lithography, contact pad metallization is done using a bilayer metal scheme. In this process contact pads are created which are required for electrical characterization and wire bonding. Finally, the fabricated devices are passivated by depositing a suitable dielectric material. Si_3N_4 is the common choice as a passivation layer. Contacts pads are then opened by dry/wet etching of Si_3N_4 passivation layer.

For passivation first process, the process flow is slightly different as shown in Figure 6.3. Here, the Si_3N_4 passivation layer is deposited after ohmic contact activation. Then gate pattern is opened in the Si_3N_4 layer by lithography and dry etching and "T" shaped gate contact is formed as mention earlier. Then after opening the source and drain ohmic contacts, contact pad metallization is performed. In some cases, an additional thick Si_3N_4 layer is deposited over the pre-passivated devices to further reduce the current collapse. Details of current collapse are discussed later.

It is mentioned earlier that conventional AlGaN/GaN HEMTs are normally-ON devices, i.e. have negative threshold voltage (V_{Th}), which depends on the Al mole fraction, AlGaN layer thickness and work function of the gate metal. So, one possible way to engineer the V_{Th} of HEMTs is by altering the thickness of the barrier layer. Reduction of barrier layer thickness, leads to positive shift in V_{Th} due to reduced 2DEG density. However, this achievement comes with the penalty of increased access region resistance which negatively impacts the device performance. Recessed-gate technique fits well in this regard where the barrier layer just under the gate is thinned down to modulate the V_{Th} without altering the access region resistance. Schematic cross-sectional view of recessed-gate AlGaN/GaN HEMT is shown in Figure 6.4. For fabricating these devices, after gate contact lithography the barrier layer is partially etched by using controlled dry etching. Then the same gate lithography can also be used to pattern the gate metal for realizing the self-aligned recessed-gate structure. Other parts of the fabrication process are same as that of mentioned earlier.

Breakdown voltage of HEMTs can be increased by employing the gate- and/or source-connected field-plate structure. Field plate structures are the extension of gate or source contact layer separated from the AlGaN surface by a dielectric layer and extended beyond the primary gate contact towards drain. Working principle of field plates is discussed in

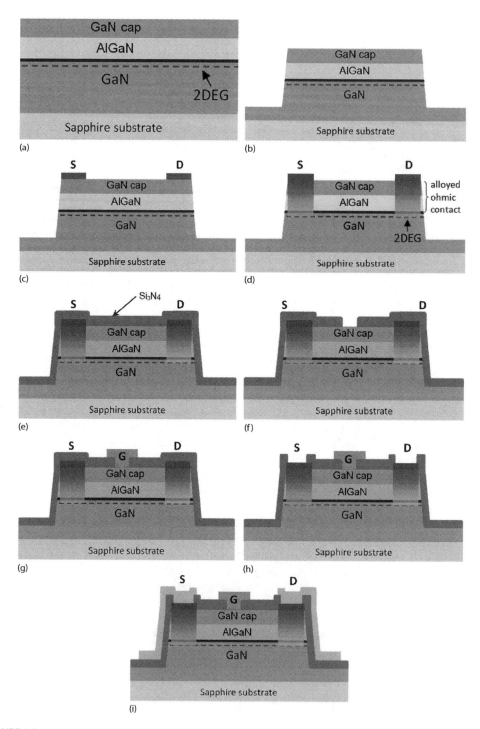

FIGURE 6.3
HEMT fabrication process flow (passivation First process). (a) starting wafer/sample, (b) mesa isolation by dry etching, (c) ohmic contact formation, (d) ohmic contact activation, (e) deposition of Si_3N_4 passivation layer, (f) gate lithography and gate pattern opening in the deposited Si_3N_4, (g) T–shape gate contact formation by using second lithography, (h) dry etching of Si_3N_4 layer to open contact pads, (i) contact pad metallization and patterning.

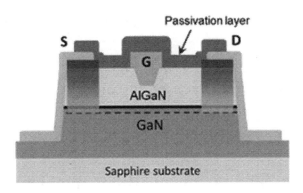

FIGURE 6.4
Schematic cross section of recess-gate AlGaN/GaN HEMT.

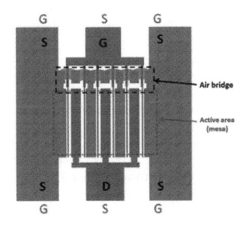

FIGURE 6.5
Schematic illustration of multi-fingered HEMT. Source-contact air-bridge structure is pointed out in the figure.

Section 6.6. For fabrication of HEMT structure with field plates, additional lithography and metallization processes are required.

Another important aspect of GaN-HEMT is to achieve a high drive current, which can be achieved by increasing the gate width. Multi-finger FET structure is widely used to increase the drive current without increasing the gate resistance. Schematic top view of a multi-finger HEMT is shown in Figure 6.5. Air bridge technique can be used to connect the source pads without shorting it with other electrodes.

6.1.3 Process Control Monitor Structures

Now let us briefly discuss some of the important PCM structures and their usefulness. These PCM structures are generally fabricated along with HEMTs on the same wafer.

1. **Mesa isolation structure** is used to check the buffer layer quality of GaN HEMT wafers. This structure consists of square shape mesas separated by a distance of 5–10 μm as shown in Figure 6.6. The structure is fabricated by mesa isolation followed by ohmic contact metallization. The leakage current between these two mesa-ohmic contacts is measured to examine the quality of the buffer and a very

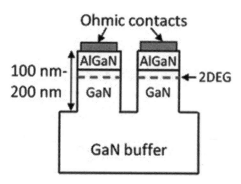

FIGURE 6.6
Mesa isolation structure to measure the buffer leakage current.

low leakage current suggests a good quality of the buffer layer. Inferior buffer quality creates a parasitic conduction path between drain and source, and the HEMT suffers from high OFF state leakage, low control of gate and reduced three terminal breakdown voltage.

2. **Transfer length method (TLM)** structures are used to measure the ohmic contact resistance. To fabricate linear TLM structures, at first a mesa is formed by using dry etching followed by alloyed ohmic contact formation. These identical ohmic contact pads are separated from each other with an increasing gap (d) as shown in Figure 6.7. Resistance between consecutive pads are measured and contact resistance can be found from the resistance versus d plot (Duffy *et al.* 2017).

3. **FATFET** structures are used to extract the channel mobility of HEMTs (Dutta *et al.* 2016b, Katz *et al.* 2003). Schematic top view of FATFET structures is shown in Figure 6.8. Fabrication process is same as that of conventional HEMTs. The name FATFET originates from the large gate length of the structure. Much larger gate length compared to access region lengths are used to neglect the source and drain access region resistances.

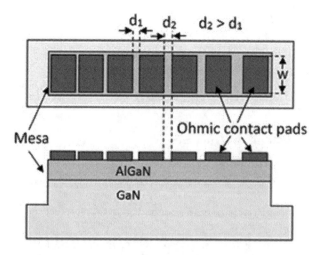

FIGURE 6.7
Schematic top view and cross-sectional view of TLM structure.

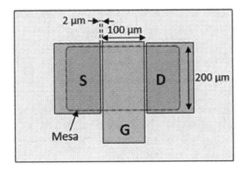

FIGURE 6.8
Simple FATFET structure.

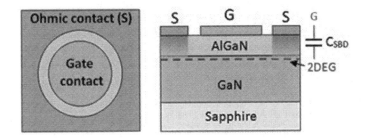

FIGURE 6.9
Schematic top view and cross-sectional view of AlGaN/GaN Schottky barrier diode (SBD).

4. **Schottky barrier diode (SBD)** structures are used to measure the gate capacitance of HEMT. SBDs are fabricated by first ohmic contact metallization followed by circular Schottky gate metallization to form a concentric structure as shown in Figure 6.9. From measured capacitance-voltage (C–V) characteristics, we can extract the barrier layer thickness, 2DEG density and V_{Th} of the heterostructure. The thickness of the AlGaN Barrier layer can be extracted from the capacitance value at zero gate bias by knowing only the dielectric constant of the barrier. 2DEG density at any particular gate bias can be extracted by integrating the C–V characteristics. V_{Th} of the HEMTs can also be extracted from this structure. V_{Th} is normally defined as the gate voltage corresponding to the capacitance of $C_{min} + 0.1(C_{max}-C_{min})$, where C_{max} and C_{min} are the maximum and minimum capacitance respectively (Dutta *et al.* 2014, Tapajna *et al.* 2009).

6.1.4 Challenges Associated with AlGaN/GaN HEMTs

The challenges associated with AlGaN/GaN HEMT technology are fabrication related and device related. For cost competitiveness of GaN-based HEMTs compared to Si devices, large diameter GaN on Si wafers should be process compatible with existing CMOS process line (Marcon *et al.* 2013). Large lattice and thermal expansion coefficient mismatch between GaN and Si makes the growth process very challenging for large area wafers. Controlling of epitaxial layer uniformity and dislocation density in the grown layers are among other challenges associated with GaN-on-Silicon technology. Besides growth, the HEMT process also has to be CMOS compatible. As mention earlier, Au-containing metallization schemes are generally used in GaN-HEMT technology for making contact. However, Au is a rapidly diffusing contaminant for Si which deteriorates the minority carrier life-time. Development of Au-free stable ohmic contact having low contact resistance is also challenging.

Device related challenges include high gate leakage current (I_G), inverse piezoelectric effect, normally-ON operation and trapping. AlGaN/GaN HEMTs suffer from high I_G due to the strong polarization induced electric field in the barrier layer and Schottky gate contact. High I_G increases the OFF-state power dissipation, can act as a noise source and is also a concern for device reliability. Besides, Schottky gate contact also limits the forward gate swing which limits the performance of AlGaN/GaN HEMTs. Due to lattice mismatch between AlGaN and GaN, the top thin AlGaN layer is under compressive stress which induces additional 2DEG due to piezoelectric polarization. Similarly, stress can also be induced in III-nitride layer when operating under high electric field, which is known as inverse piezoelectric effect. During high voltage operation of AlGaN/GaN HEMTs, stress induced in the AlGaN barrier layer in addition with the built-in stress can be significantly high and may create permanent defects at the gate edge of the HEMT (del Alamo and Joh 2009). This is one of the serious concerns associated with AlGaN/GaN HEMTs. Polarization induced 2DEG channel also make GaN-based devices a normally-ON type. However, for fail-safe operation and compact design of the circuit, normally-OFF devices are preferred. Realization of high-performance and reliable normally-OFF GaN-HEMTs are still challenging. The presence of large concentration of traps in different regions of the device is another serious concern in GaN-based HEMTs (Jia *et al.* 2018, Joh and del Alamo 2011, Zheng *et al.* 2016). Trapping, which can occur at the exposed ungated AlGaN surface, AlGaN barrier layer and at bulk GaN can dramatically deteriorate the device performance in terms of output power density and maximum operating frequency. Complete understanding of different trapping phenomenon in GaN-HEMT is still ongoing and mitigating the effect of trapping is still a challenge.

6.2 Lithography

In 1796, Alois Senefelder successfully transferred a shape on a stone onto paper using appropriate chemicals. The word lithography originated from this process. In Greek, lithos refers to stone and grapho means to write. Today, lithography technique is widely used in the semiconductor industry to transfer micron and sub-micron dimension patterns on a mask onto a desired substrate. The micron range pattern transfer is mostly achieved with the help of photolithography, whereas the sub-micron range patterns are realized via high resolution lithography techniques such as Deep UV (DUV), Extreme UV (EUV) and Electron Beam Lithography (EBL).

Fabrication of GaN based HEMTs involves at least three stages of lithography. In the first lithography, active regions are formed in the wafer to isolate the devices from each other. The area for the source and drain ohmic contact formation is then defined by the second lithography. As the dimensions for the mesa isolation and ohmic contact are in several microns, they can be easily realized with the help of photolithography. The third and critical lithography step is to realize the sub-micron Schottky gate contact. This can be achieved with the help of DUV, EUV or EBL.

6.2.1 Photolithography

As the name suggests photolithography is a technique in which the pattern transfer takes place with the help of near ultraviolet (UV) radiation. In this process, the sample is coated with photoresist, which is an organic polymer (Ando *et al.* 1994) sensitive to UV radiation. Usually, photoresists are of two types, viz. positive and negative photoresist (Thompson *et al.* 1983). When a positive photoresist is exposed to light, the polymer chains break and the exposed region can be removed with the help of a developer. Negative resist behaves in the opposite way, i.e. the polymer chains linkup in regions exposed to light and the unexposed region gets washed away in the developer. As the photoresist is sensitive to light, care should be taken while spin coating the photoresist so that it is not exposed to light. Usually the samples are spin coated in a clean room with yellow light ($\lambda > 500$ nm). The thickness of the resist is inversely related to the spin speed. After spin coating the photoresist on the samples, the samples are pre-baked at 80°C–100°C for 20 minutes in a convection oven to remove any residual solvents from the resist. Pre-baking also ensures proper adhesion of the resist layer to the wafer. In the next step, the desired patterns on the photomask are transferred on to the photoresist coated substrate on exposure to UV radiation. A photomask is usually a flat glass plate with an 80–100 nm thick patterned chromium layer. The chromium layer is opaque whereas the glass plate where chromium layer is not present is transparent to UV radiation. Based on the contact between the photomask and the wafer, photolithography can be classified into three types, viz. (i) contact lithography, (ii) proximity lithography and (iii) projection lithography (Thompson *et al.* 1983). In contact lithography process, the photomask makes a physical contact (hard contact) with the wafer. Due to physical contact with the wafer, these photomasks degrade fast. Hence this problem is overcome in the proximity lithography technique where the photomask is raised about 10–20 μm above the wafer. Though in this case the wear on the photomask is less, the image resolution is poorer compared to contact lithography. In case of projection lithography, the UV radiation passing through the transparent regions of the photomask is steered via imaging optics on to the wafer (Ito and Okazaki 2000). Figure 6.10 compares the three

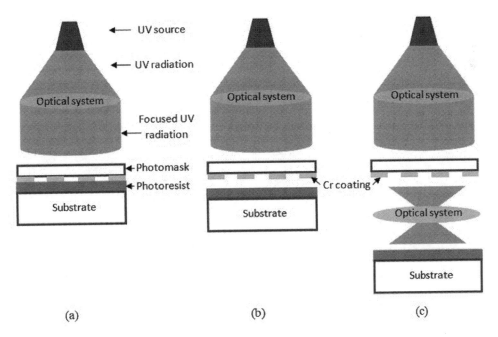

FIGURE 6.10
Schematic diagram comparing the different lithography techniques (a) contact lithography (b) proximity lithography and (c) projection lithography.

lithography techniques. Projection lithography is considered to be more reliable than contact and proximity lithography. After the photolithography process, the exposed regions of the photoresist are then removed by rinsing the wafer in a developer solution. Aqueous alkaline solutions are generally used to develop the positive photoresist whereas the negative resists are developed with organic solutions.

In case of the ohmic contact formation during the fabrication of GaN-based HEMTs, the photolithography process is followed by metallization. After metallization the samples are immersed in acetone for few minutes with mild agitation. This dissolves the underlying resist structure. Along with the resist layer the undesired metal layer is also lifted off. This process is termed as "lift-off" (Madou 2011). The lift-off process is demonstrated in Figure 6.11. An undercut profile in the photoresist is necessary to facilitate a clean lift-off. The undercut profile can be achieved by different techniques. In case of single layer photoresist, the wafer is immersed in chlorobenzene solution for one minute after pre-baking. The top few nanometers of the resist layer, influenced by the chlorobenzene solution, develop slowly whereas the bulk resist layer develops faster compared to the top layer, thereby facilitating an undercut profile. The undercut profile can also be achieved with the help of bilayer resists, in which the top layer develops slowly compared to the bottom layer.

As photolithography process involves light, the resolution of the patterns transferred is affected by diffraction of light. As the pattern size approaches to the wavelength of the light, the diffraction is more pronounced. The image resolution in the photolithography is governed by (Madou 2011)

$$R = K\sqrt{\lambda\left(d + \frac{t}{2}\right)}$$ (6.1)

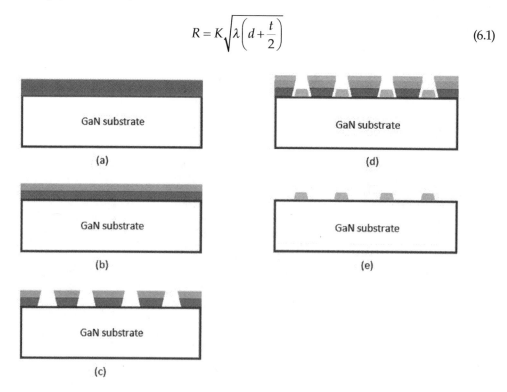

FIGURE 6.11
Cross-sectional view of the sample under process during ohmic contact formation (a) sample coated with positive resist (b) after chlorobenzene dip (c) after UV exposure and developing (d) after metallization (e) after lift-off.

where d is the distance between the photoresist and the mask, t is the thickness of the photo-resist, λ is the wavelength of the UV light and K is a constant with a value ~1.5. With near UV radiation of $\lambda = 400$ nm and a photoresist thickness of 1 µm, the minimum resolution that can be achieved is ~1 µm. As wavelength is one of the deciding factors in obtaining high-resolution patterns, techniques such as DUV and EUV has emerged. The difference between DUV and EUV lithography is the wavelength of the light source. DUV has a wavelength of ~248 nm with KrF excimer laser and 193 nm with ArF excimer laser (Ito and Okazaki 2000) whereas EUV's wavelength is ~13.5 nm (Harriott 1999, Madou 2011). Further high-resolution patterns which are limited by diffraction of the light in optical lithography can be achieved with the help of EBL.

6.2.2 E-Beam Lithography

Optical lithography uses light to transfer the patterns. In contrast, EBL uses focused electron beams to perform the same task. Due to shorter wavelength of electrons, EBL offers high resolution patterning compared to optical lithography. An EBL system consists of an electron gun which acts as the source of the electrons. As the electron is a charged particle, its motion can be guided and focused with the help of electric or magnetic fields. Usually, in EBL systems this task is accomplished with the help of magnetic fields. This focused e-beam of size ~1 nm is then used to expose the electron-beam resist. Figure 6.12 shows the basic schematic diagram of an EBL system. E-beam resists are usually of two types, viz. positive and negative resist. Poly methyl methacrylate (PMMA) is widely used as the positive resist whereas Hydrogen Silsesquioxane (HSQ) is a negative tone resist. The samples coated with e-beam resist are usually placed on a laser controlled x/y stage inside a vacuum chamber. For writing the desired patterns on the samples, EBL generally adopts two writing strategies, viz. (i) raster scan and (ii) vector scan (Helbert 2001). In raster scan, the beam is swept across the entire write field while blanking the beam in those portions

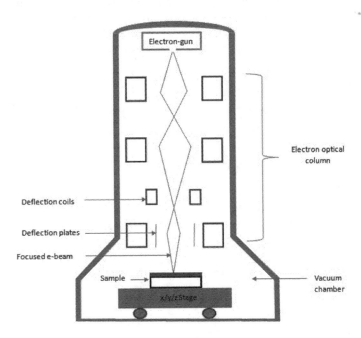

FIGURE 6.12
Schematic diagram showing the basic parts of electron beam lithography system.

of the mask where the patterns are not present. This is a time consuming process because irrespective of sparse or dense patterns the beam scans the entire write field. A vector scan writes only in the areas where the patterns are located, thus saving time in writing sparse patterns. After exposure to electron beam, the resist changes its chemical property and hence can be developed in a developer solution. A ratio of one part of methyl isobutyl ketone (MIBK) in three parts of isopropyl alcohol (IPA) is commonly used as developer to achieve high resolution patterns. Unlike photolithography, in EBL we do not need a physical mask plate for pattern transfer. Here the pattern transfer is done via a soft copy of the mask file designed with the help of CAD software.

Gate length scaling in GaN-based HEMT devices improves the device speed and makes them strong contenders for microwave electronics. However, smaller gate lengths increase gate resistance. To reduce gate resistance, GaN HEMTs with T-shape (or mushroom gate) gate technology are widely preferred for RF applications. Both rectangular and T-shaped gate can be realized with the help of EBL. Rectangular gate can be realized using a single layer resist whereas for T-shape gate we need two resist layers. In this bilayer process (Matsumura *et al.* 1981, Ocola *et al.* 2003, Todokoro 1980), first PMMA is spin coated on top of the substrate followed by highly sensitive copolymer MAA (methacrylic acid). After exposure to the electron beam, the MAA layer develops at a faster rate compared to PMMA as shown in Figure 6.13, thus facilitating a T-gate after metallization and liftoff process. Figure 6.14 shows the cross-section SEM image of T-shaped 150 nm gate length HEMT device (Piotrowicz *et al.* 2018). As we know, GaN is a high bandgap material with semi-insulating nature. Therefore, when the GaN substrate coated with the e-beam resist is exposed to the electrons, the electrons cannot find a conductive path and hence they start charging the resist. In order to avoid this, a thin layer (5–10 nm) of gold is deposited on top of the resist to facilitate a conductive path for the electrons. After the lithography the metal layer is first etched out and then the e-beam resist is developed.

As diffraction defines the resolution of patterns in photolithography, similarly in EBL the resolution is limited mainly by the electron scattering mechanism in the resist and the substrate. As a result of this scattering phenomenon, undesired areas in the resist get exposed. This is termed as proximity effect in EBL (Cabrini and Kawata 2012). When electrons travel

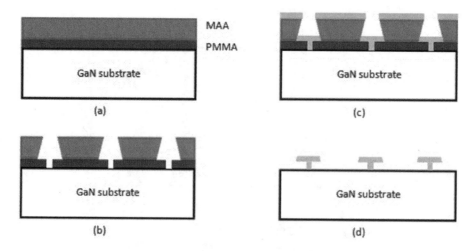

FIGURE 6.13
Cross sectional view of sample under process during T-gate formation (a) GaN substrate coated with PMMA and MMA (b) after exposure to e-beam and developing (c) metallization (d) lift-off of the undesired metal layer.

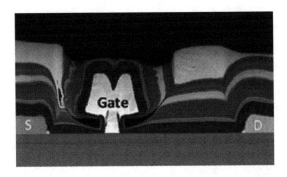

FIGURE 6.14
Cross-section SEM image of HEMT device having T-shaped gate of length 150 nm. (From Piotrowicz, S. *et al.*, *Int. J. Microw. Wirel. Technol.*, 10, 39–46, 2018. Copyright 2018 Cambridge University Press and the European Microwave Association.)

through the matter they are affected by scattering. There are two types of scattering mechanisms, viz. elastic and inelastic scattering. Elastic scattering is a phenomenon in which an electron passing through the matter do not lose energy but simply undergo deviation in its path, whereas electron-electron interaction is considered to be inelastic which slows down the electrons. When the electron beam penetrates through the resist coated substrate, it may interact with an atom in the substrate or resist and change its direction and transfer a part of its energy to the interacting atom. This is called forward scattering, and it is inelastic in nature. The energy transferred by the primary electron may be sufficient enough to ionize the atom resulting in the secondary electron formation. Secondary electrons play an important role in the resist exposure. As a result of forward scattering there would be a small deviation in the pattern transferred as compared to that in the mask file. Another scattering effect that is more prominent in EBL is the backscattering of the electrons. Most of the electrons which pass through the resist will undergo various scattering in the bulk substrate. When an electron collides with heavier nucleus, the electron retains its energy but changes its direction. These backscattered electrons expose wider areas of the resist as they have large scattering angle compared to the forward scattering.

Though very high resolution patterns can be achieved with EBL, it is a time consuming process. Hence EBL is mostly preferred for writing high resolution sparse patterns and also to write the mask used for optical lithography.

6.3 Metallization and Annealing

Metallization is one of the important and critical process steps in fabrication of GaN-HEMTs. At least two metallization are required for connecting these devices with the external world, one for ohmic source/drain contact and the other one for the Schottky gate contact. In this section both these metallization processes will be discussed.

6.3.1 Ohmic Contact

For optimal performance of GaN-based HEMTs, low resistance and thermally stable ohmic contacts are essential. The ohmic contact can significantly affect the following device performances: maximum drain current, ON-resistance, extrinsic transconductance, power dissipation at ohmic contacts, and hence the unity gain cut-off frequency. So, formation of good ohmic contact with low contact resistance is essential for GaN-based HEMTs. In general, realization of low resistance ohmic contact is difficult for wide bandgap semiconductors. From our basic understanding of metal-semiconductor contacts, we can argue that metal with low work function would be the primary choice for ohmic contact formation. From our earlier discussion, we have seen that typically grown III-nitride layers are unintentionally doped (n-type, ~10^{16} cm^{-3}) and for achieving low contact resistance the metal layer should be in contact with the 2DEG channel. So, mere use of low work function metals cannot serve the purpose of achieving of good ohmic contact. Generally, multi-layer metals are used for making the source/drain contact to III-Nitride HEMTs which forms alloyed-ohmic contact with the 2DEG after annealing. So, for GaN-HEMT devices the activation of ohmic contact by high temperature annealing is essential which makes annealing an integral part of ohmic contact formation. However, as a result of high temperature thermal annealing, diffusion of metals and out-flow of their liquid phase creates difficulty in realizing good surface morphology and good line edge which in turn effect the further processing and can reduce the source-drain spacing. Apart from low contact resistance, thermal stability of the ohmic contact is another important concern which also has to be dealt seriously.

In general, a Ti/Al based metal stack is used to realize the ohmic contact on III-Nitride layers. During annealing, for Ti/Al-based contacts metals react with each other and with the III-nitride forming several intermediate materials, which helps in realizing good ohmic contact. For example, in case of Ti/Al-based ohmic contact, formation of phases like TiN, AlN and Al$_3$Ti can be observed. Different sophisticated techniques (e.g., XPS, EDS, TEM, EELS) were used to analyze the intermediate states (Liu *et al.* 1998). Titanium (Ti) reacts with N in the III-Nitride layer to form TiN. This TiN has a low work function (<4 eV), which assist in ohmic contact formation by reducing the barrier height (Chaturvedi *et al.* 2006). In addition, created N-vacancies in the nitride layer become heavily doped n-type. This heavily doped interface under the ohmic contact further improves the ohmic behavior by enhancing the tunneling. Al reacts with Ti and forms Al$_3$Ti compound, which prevents the oxidation of underneath Ti layer. Al also reacts with N from the semiconductor and forms AlN, which also creates N-vacancies (Daele *et al.* 2005) and further improves the ohmic contact as mentioned before. However, both Ti and Al have high affinity to oxygen and can get oxidized even at room temperature. So, small presence of oxygen during metal deposition and/or annealing can significantly impede the ohmic contact. Therefore, use of Ti/Al ohmic contact is not very reliable. One possible way to improve the reliability of this ohmic contact is by capping the Ti/Al stack with Au-layer which is very difficult to oxidize. However, a barrier is required between Ti/Al and Au to reduce/prevent the intermixing of Ti/Al and Au layers. Different metals (Ni, Ti, Mo etc.) have been investigated as the barrier layer.

Now, let us discuss the Ti/Al/Mo/Au based ohmic contact formation mechanism on AlGaN/GaN heterostructure using a simplified model (Figure 6.15). During annealing, Ti and Al diffuse in and react with AlGaN, and form TiN and AlN. This creates a highly doped n-type region underneath the ohmic contact due to the formation of nitrogen vacancies. This highly doped n-region increases the tunneling current. Low work function of TiN also improves the ohmic behavior of the contact. On the contrary, AlN has a wide bandgap and expected to have a higher barrier height which in turn restricts the conduction by thermionic emission. However, formation of highly doped

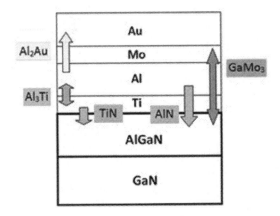

FIGURE 6.15
Schematic illustration of Ti/Al/Mo/Au ohmic contact formation mechanism on AlGaN/GaN heterostructure.

nitride layer below the thin AlN is believed to be the reason for overall low ohmic contact resistance. Besides, Al and Ti also react among themselves and form Al_3Ti. This intermetallic alloy Al_3Ti, which has higher resistance to oxidation, makes the contact thermally more stable and reliable. The resistivity of this intermetallic alloy is also lower than that of Ti. So, individual thicknesses of Ti and Al layer and their (Ti and Al) ratio can significantly affect the ohmic contact resistance and a careful optimization is required. Regarding the ratio of Ti/Al, 1:5 is typically used. Though the Al and Au layer is separated by the metal barrier layer (Mo or Ni or Ti), intermixing happens between them during annealing, which produces Al_2Au phase as determined from XRD analysis. Intermetallic Al_2Au is believed to be the reason for rough surface morphology in case of Ti/Al/Ti/Au and Ti/Al/Ni/Au based ohmic contacts. However, in case of Ti/Al/Mo/ Au the smooth surface morphology after annealing is possibly due to the formation of $GaMo_3$ alone with Al_2Au (Chaturvedi *et al.* 2006). Figure 6.16 shows the surface morphology of Ti/Al/Mo/Au and Ti/Al/Ti/Au-based ohmic contact. Surface morphology of Ti/Al/Ni/Au (30/140/40/100 nm) ohmic contact is also shown for different

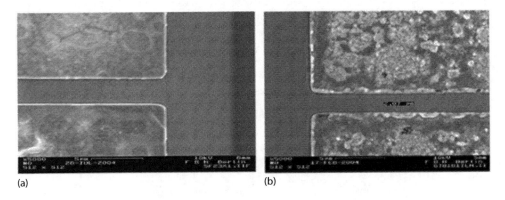

(a)   (b)

FIGURE 6.16
SEM image of ohmic contact on AlGaN/GaN heterostructure using (a) Ti/Al/Mo/Au and (b) Ti/Al/Ti/Au metal scheme. (Reprinted with permission from Chaturvedi, N. *et al.*, Mechanism of ohmic contact formation in AlGaN/GaN high electron mobility transistors, *Semi. Sci. Tech.*, 21, 175–179, 2006. Copyright 2006, Institute of Physics.)

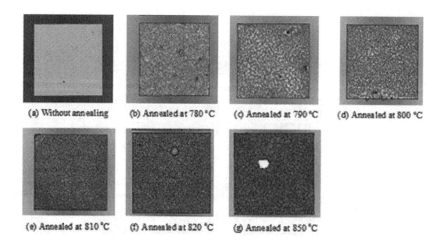

FIGURE 6.17
Surface morphology of Ti/Al/Ni/Au ohmic contact on AlGaN/GaN heterostructure annealed at different temperatures. (a) Without annealing (b) Annealed at 780°C (c) Annealed at 790°C (d) Annealed at 800°C (e) Annealed at 810°C (f) Annealed at 820°C (g) Annealed at 850°C.

post-annealing conditions in Figure 6.17. So, from the above discussion, it is clear that surface morphology of ohmic contact is a strong function of metallization scheme and annealing schedule. It is worth to mention here that these individual metals layers are mostly deposited by electron beam evaporation system at low/room temperature and without breaking the chamber vacuum in order to pattern the metal stack by lift-off process and to avoid the atmospheric exposure (primarily oxygen) to the intermediate layers respectively. Surface treatments before the ohmic metal deposition can significantly improve the contact resistance by removing the native oxide layer. Dilute HCl and/or HF surface treatment prior to metal deposition is widely used for GaN-HEMT fabrication. For contact activation by thermal annealing computer controlled rapid thermal annealing (RTA) systems are generally preferred in order to tightly control the annealing temperature and its duration. Formation of contact through the AlGaN layer can further be explained by "spike mechanism" (Wang *et al.* 2007), which postulates that during annealing TiN islands penetrate through the AlGaN layer and creates a direct contact with the 2DEG. This penetration of TiN happens preferably through the treading dislocation present in the AlGaN layer which forms spikes of TiN from metal contact to 2DEG (Figure 6.18). Contact resistances of AlGaN/GaN HEMTs for some of the reported metallization schemes and annealing schedules are listed in Table 6.1.

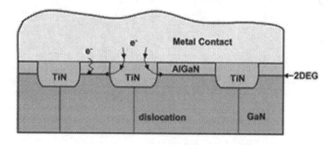

FIGURE 6.18
A schematic of spike mechanism for ohmic contact formation. (From Wang, L. *et al.*, *J. Appl. Phys.*, 101, 013702, 2007. With permission.)

TABLE 6.1

Different Metallization Schemes and Respective Contact Resistances for AlGaN/GaN HEMTs

Metal Stack	Annealing Conditions	Contact Resistance (Ω-mm)	References
Ti/Al/Ni/Au	850°C, 30 sec	0.6	Tang *et al.* (2013a, 2013b)
Ti/Al/Ti/Au	850°C, 60 sec	0.7	Husna *et al.* (2012)
Ti/Al/Mo/Au	830°C, 15 sec	0.42	Chaturvedi *et al.* (2012)
Ti/Al/W	600°C, 60 sec	0.65	Hove *et al.* (2012)
Si/Ti/Al/Mo/Au	780°C, 60 sec	0.45	Choi *et al.* (2014)

Another popular technique to reduce the contact resistance is by pre-metal-deposition plasma etching of the AlGaN-barrier layer (Arulkumaran *et al.* 2010, Buttari *et al.* 2002, Lee *et al.* 2011). This technique is known as recessed ohmic contact formation (Figure 6.19). Improvements in contact resistance in this process happens primarily because of the removal of surface oxide layer, enhanced tunneling through the thinner AlGaN barrier layer and formation of surface donor states due to additional *N*-vacancies created by ion bombardment during plasma etching. However, precise control of AlGaN layer thickness is quite critical for achieving the low contact resistance as over-etched AlGaN layer can cause the depletion of the 2DEG layer at AlGaN/GaN interface, hence degrading the contact resistance. Reduction of contact resistance from 0.45 Ω-mm for conventional unetched AlGaN-barrier HEMT to 0.27 Ω-mm for 7-nm recessed etched HEMT was reported (Buttari *et al.* 2002). Further, etching of AlGaN layer showed increasing ohmic contact resistance. Recessed ohmic contact formed by the complete etching of AlGaN barrier layer and low temperature annealing (500°C) of Ti/Al/Ti/Au metal stack has shown quite high contact resistance (3.2 Ω-mm) (Lao *et al.* 2009). Improvement in ohmic contact resistance was also observed by BCl$_3$ plasma treatment prior to metal deposition even without the recess etching of AlGaN layer. BCl$_3$ plasma treated AlGaN/GaN samples in an ICP-RIE system at an ICP power of 100 W and RF power of 0 W (to prevent recess etching) showed reduction in contact resistance from 0.34 Ω-mm to 0.17 Ω-mm. This improvement is due to removal of native surface oxide and formation of surface donor states as confirmed from XPS analysis (Fujishima *et al.* 2013).

For deeply downscaled GaN-based HEMTs, it is essential to reduce the contact resistance even further. Reduction of ohmic contact resistance which is the dominant part of parasitic resistance can significantly improve the RF performance of the device. Another technique called "Regrown ohmic contact" has successfully been demonstrated (Joglekar *et al.* 2016, Song *et al.* 2016). In this technique, the source and drain contact windows are first etched using Cl-based deep plasma etching till few nm to tens of nm inside the

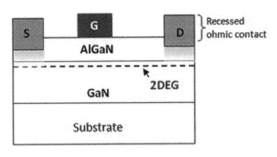

FIGURE 6.19
Cross-sectional view of AlGaN/GaN HEMT with recessed ohmic contact.

GaN-buffer layer while protecting the remaining active area of the device by deposited SiO$_2$ hard mask. Then heavily doped n-GaN (~10^{20} cm^{-3} Si doped) is deposited either by MOCVD or MBE. After deposition of n^+ GaN layer, the polycrystalline GaN deposited on the SiO$_2$ is lifted-off/removed by using buffered hydrofluoric acid/dry etching. A non-alloyed ohmic contact metal (Ti/Au or Mo/Au) is then deposited on the regrown ohmic contact regions (Figure 6.20). Very low contact resistance (0.1 Ω-mm) can be achieved by using this technique. For regrowth, MBE is preferred compared to that of MOCVD as very heavily Si-doped (>10^{19} cm^{-3}) GaN causes surface roughness and formation of cracks.

Contact resistances (R_C) are typically estimated from the Linear Transfer Length Method (LTLM) (Schroder 2006). Details of TLM structure is already discussed in Section 6.1. Resistances between consecutive pads are first measured from the linear current-voltage characteristics between two pads (Figure 6.21a). From the Y-axis intercept (R_Y) of resistance vs. pad spacing plot (Figure 6.21b) contact resistance (R_C) can extracted by $R_C = 0.5\ R_Y\ W$, where W is the width of the TLM pad. These contact resistances were found to be a strong function of the annealing temperature and annealing duration. Another variety of TLM is circular TLM or CTLM structures where resistance is measured between inner circular ohmic contacts and conducting outer region contacts separated by an increasing gap. Mesa structures are not required in case of CTLM.

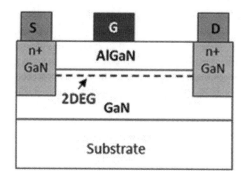

FIGURE 6.20
Schematic illustration of AlGaN/GaN HEMT with regrown ohmic contact.

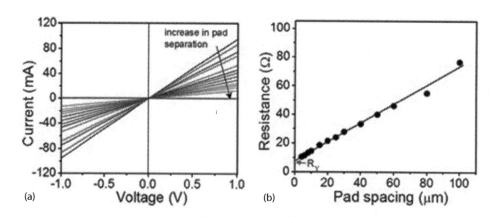

FIGURE 6.21
(a) I–V characteristics of typical LTLM structure fabricated on a GaN-heterostructure. (b) Corresponding extracted resistance vs. pad spacing plot of the LTLM structure.

To make AlGaN/GaN HEMTs commercially viable, significant effort has been made to fabricate the GaN HEMTs on standard sized Si wafers and to integrate the GaN-HEMT process technology with that of standard Si-CMOS process line. In addition to different challenges associated with the growth and with the technology integration, implementation of Au-free ohmic contacts has become essential as Au contaminates the Si-process line. For ohmic contact formation, different Au-free metallization schemes, e.g., Ti/Al/W and Ti/Al/Ni/Ta/Cu/Ta with annealing temperature around/above 800°C have been reported (Alomari 2009, Lee *et al.* 2011). For realization of CMOS-process compatible Au-free Ti/Al/TiN-based ohmic contact with low annealing temperature (550°C) has also been reported (Firrincieli 2014). It is important to mention here that for realizing low-resistance gold free contacts, the barrier layer from the source-drain contact regions is partially/totally etched before depositing the ohmic metals. The obtained contact resistances are in the range of 0.46–0.62 Ω.mm which is comparable to that of conventional ohmic contacts having Au as one of the layer.

6.3.2 Schottky Gate Contact

For realizing a good Schottky contact, metals with high work function is intended. Ni has been widely used as a Schottky gate contact for AlGaN/GaN HEMTs due to its high work function and good adhesion with GaN-based materials. Though single layered Ni was reported as a gate metal, Ni/Au stack is more popular in order to prevent the oxidation of Ni. Gate dimensions from microns to hundreds of nanometer can be successfully realized using this metallization scheme. T-gate structure with Ni/Au gate metals has already been reported. This T-gate structure reduces the gate foot-print while maintaining gate resistance which is essential for improving the RF performance of HEMTs. Schottky contacts with large barrier height can significantly suppress the gate leakage current hence improving the device performance and reliability. Barrier height of a Schottky contact can be extracted from the gate current vs. forward-gate voltage characteristics, capacitance-voltage characteristics and x-ray photoelectron spectroscopy (XPS) characteristics. It is found that Ni-based gate contacts are quite stable up to 500°C annealing temperature. Pt, Cu and Ti based Schottky contacts are also investigated for GaN- HEMTs. Pt has higher work function compared to Ni which helps in reducing the gate leakage current. Similar to that of ohmic contact, gold-free gate contact is also essential for manufacturing the GaN-HEMT in CMOS process line. Different gold-free gate contacts, e.g. TiN, TaN, ITO, have been also been investigated (Ahn *et al.* 2013, Li *et al.* 2017, Pei *et al.* 2009). Among these, TiN exhibits many advantages: large work function (~5 eV), good thermal stability and process compatibility to dry etching required for CMOS processing.

6.4 Dry Etching

Etching is one of important process steps in fabrication of semiconductor devices. In simple words "etching" means removal of materials from the substrate. Depending on the medium of etching environment, this process can be classified into two main categories: "wet etching" and "dry etching." In wet etching, samples are immersed into an etching solution (primarily acid/base) where a chemical reaction happens between the etching solution and the substrate and a volatile/soluble product is formed. The volatile/soluble product is then removed from the substrate by diffusion and the etching continues. Wet etching is a simple and inexpensive technique. In wet etching, physical agitation, temperature and concentration of etching solution have significant effect on the etch rate. Contrary to this, in dry etching, samples are exposed to plasma (partially ionized gas) which contains the etching species. As dry etching happens in the presence of plasma, it is also known as plasma etching. Plasma etching system is relatively more complex and expensive. However, certain advantages in case of plasma etching as compared to that of wet etching can surely justify the additional cost and complicity.

In a typical etching process (both wet and dry), depending on the fabrication requirements, parts of the substrate are exposed to the etching solution/plasma while the remaining regions are protected with a suitable mask material. Selection of a mask material depends on the fabrication process flow, temperature withstandability and etch rate of the mask material for a particular etching environment. It is expected that the etch rate of the mask material should be much smaller than that of the substrate material. A figure of merit, "selectivity" can be introduced in this regard which can be defined as the ratio of etch rate of the substrate to that of the mask material. Typically, a selectivity of 30–50 is considered reasonable for an etching process (DasGupta 2017). Another important figure of merit associated with etching process is the directionality of etching. When a sample is subjected to etching, the unprotected region (not covered with mask) comes in the direct contact with etchant and ideally only vertical etching is expected to progress resulting in a perfectly vertical sidewall as shown in Figure 6.22a. This type of etching is called anisotropic etching. However, in reality, the etching can progress in both vertical and horizontal direction creating a slanted sidewall profile and an undercut as shown in Figure 6.22b. Directionality can be defined as $(1-R_h/R_v)$, where R_h is the horizontal etch rate and R_v is the vertical etch rate. For perfect anisotropic etching R_h should be close to zero, i.e., directionality closes to 1. However, if $R_v \approx R_h$, then the directionality is nearly zero and the etch profile is called isotropic etch profile.

In wet etching, selectivity is quite good as the etchant solution only reacts with the substrate. However, as wet etching process is chemically dominated, a limited amount of directionality can only be achieved. If the wet process is diffusion limited (transport of fresh etchant to the etch surface or transport of reaction product from the surface), then the etch profile will be of isotropic in nature. By contrary, if the process is reaction rate limited then the etch profile depends on the crystallographic orientation of the substrate and limited directionality can be achieved. In early days, wet etching was used extensively in semiconductor processing. However, in modern semiconductor manufacturing, dry etching is the most popular etching technique. The main advantage of dry etching is that the etch profile can be easily tailored by varying different process parameters of the plasma. Besides, dry etching can be easily automated and scaled-up, and it does not use corrosive wet chemicals like that of wet etching.

Dry etching is carried out in presence of plasma, which is a partially ionized gas with low degree of ionization and having equal number of positive ions and negative charges (i.e., electrons). In plasma, electrons can acquire very high energy (equivalent to electron

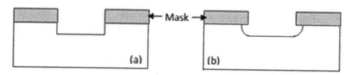

FIGURE 6.22
(a) Anisotropic and (b) isotropic etch profile.

temperature of 10,000 K), which allows high-temperature reactions to take place at much lower temperature and generates free radicals.

Due to inert chemical nature and high bond energies of group-III nitrides, it is difficult to use wet chemical etching to effectively etch these materials. As a result, dry etching has become the dominant etching technique for group-III nitrides. For processing of GaN-based HEMTs, dry etching is primarily used for mesa isolation and realization of recess-gate structure. However, depending on the fabrication process, dry etching has also been used for opening the source/drain and gate contact window in the passivation layer. Requirements of etch characteristics are also different for different cases. For example, during mesa etching, high etch rate, smooth sidewall, anisotropic etch profile and equal etch rate of different III-Nitrides is desired. However, for gate recess etching, slow etch rate is essential to ensure precise etch depth control, repeatability and low plasma damage.

Plasma etching can proceed in three different ways: physical sputtering, chemical etching and a combination of both physical and chemical etching. In physical sputtering, accelerating high energy ions present in the plasma transfer the momentum to the substrate surface while ejecting materials from the substrate. As a result, the etch profile becomes anisotropic. However, sputter etching suffers from poor selectivity, rough surface morphology and non-stoichiometric etch surface. Due to high bond energies of group-III Nitrides the physical etch rate is low. The etch rate can be increased by increasing the ion energy but that creates more surface damage. On the other hand, chemical etch mechanisms rely on the formation of volatile etch products due to the absorption of reactive species present in the plasma on the surface and removal of the volatile products from the surface. In chemical etch mechanism, the etch rates are almost similar in both vertical and horizontal directions, which results in an isotropic etch profile and loss of critical dimensions. By using a balanced combination of physical and chemical component of dry etching intended etch profile at high resolution and with minimum surface damage can be obtained.

Dry etching is usually carried out in Reactive Ion Etching (RIE) systems which utilize both physical and chemical components of dry etching mechanism to realize both anisotropic etch profile and high etch rate. Figure 6.23 shows a simple schematic diagram of a RIE system which consists of two parallel plate electrodes housed in a low pressure (typically 1 mtorr–1 torr) chamber. A high frequency voltage is applied between these two electrodes which ionizes the gas and produces glowing plasma. The colour of the plasma glow depends on the chamber gas. An RF generator (typically operating at 13.56 MHz) is generally used to supply the voltage (Plummer *et al.* 2012). When plasma strikes, a voltage

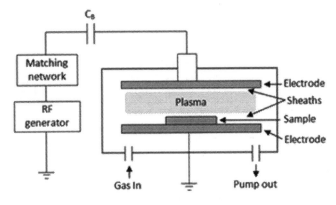

FIGURE 6.23
Schematic illustration of Reactive Ion Etching (RIE) system.

difference occurs between the plasma and the electrodes as a result of difference in mobility of electrons and ions. As electrons diffuse faster leaving the positive charges in the plasma, a positive plasma potential is created. The regions adjacent to both the electrodes are depleted of charges (sheath regions) and the potential drop between the plasma and electrode occurs in this regions. One of the electrodes in RIE system is connected to ground while the other is connected to the RF power supply through a blocking capacitor which prevents the discharge of the electrode through the power supply. Heavy ions cannot respond to the RF supply frequency but it can respond to the DC electric field created due to the sheath potential. As a result, ion bombardment to the substrate surface will be normal, which provides directional etching in case of dry etching.

Etch rate of RIE can be increased by increasing the RF power. However, high RF power also results in high energy plasma which creates surface damage due to enhanced physical etching. Etch rate can also be increased by increasing the plasma density. So, for achieving high etch rate while limiting the surface damage, decoupling of plasma density from plasma energy is essential. This can be achieved by using Inductively Coupled Plasma RIE (ICP-RIE) system which offers a plasma density, which is orders of magnitude higher, compared to RIE by using additional RF power source (ICP power). By applying a RF power to the ICP-coil which encircles the ICP-chamber the ICP-plasma is created. Alternating electric field in the ICP-coil induces a strong magnetic field which helps in creating the high density plasma. On the other hand, the RF power applied between the two electrodes determines the plasma energy as in the case of RIE system. Thus, energy and plasma density can be effectively decoupled in case of ICP-RIE, and plasma induced damage can be effectively controlled, while achieving a high etch rate. So, III-Nitrides can be etched at a high rate with controlled surface roughness by using ICP-RIE.

For mesa etching, different plasma chemistries (e.g., iodine-, bromine-, methane- and chlorine-based) have been investigated (Pearton *et al.* 2000). Among these, chlorine-based binary etch chemistries are the most popular and mostly used due to the volatility of group-III chlorides. Many Cl-based gas mixtures have been investigated, such as Cl_2/Ar, Cl_2/BCl_3, $Cl_2/BCl_3/Ar$, etc. ICP-etching of III-Nitrides using Cl_2/BCl_3 and Cl_2/Ar are increasingly becoming more popular because of its excellent anisotropy and smooth surface morphology (Rawal *et al.* 2012a, 2012b, 2014). In both gas mixtures Cl_2 is the primary etching gas. In many etching processes, Ni, Al, Cr and SiO_2 have been used as a hard mask, which requires additional process steps like deposition, patterning and removal of mask adding complexity to the overall process. Use of photoresist as a etch mask significantly simplifies the process and has been used successfully as discussed in the following paragraph. We will discuss a case study of GaN etching using an ICP-RIE system to show the dependence of different etch parameters like: RF power, ICP power, chamber pressure, gas flow rate and their ratio, etc. on the etch characteristics.

Increase in RF power increases the DC bias (plasma potential) which enhances both physical and chemical etching components. At high RF power, increased ion bombardment modifies the reactivity of the substrate surface and enhances the chemical etch reaction between the neural etching species and the substrate material. Increased ion impact also enhances the physical sputtering process and helps in removing the volatile etch product. As a result GaN etch rate increases linearly with increasing RF power for both Cl_2/Ar and Cl_2/BCl_3 based dry etching. For a constant RF power, with increase in ICP-power, DC bias reduces and the etch rate increases. Increase in ICP-power enhances the ion density in the plasma which results in reduction in DC bias. Higher etch rate at higher ICP power is due to the enhanced concentration of reactive chlorine radical and higher ion flux on the sample surface. A higher etch rate was observed for Cl_2/Ar gas chemistry compared to that of Cl_2/BCl_3 for the same process parameters as a result of higher dissociation efficiency of Cl_2 in presence of Ar. Etch rate decreases monotonically with increase in chamber pressure. This is due to the fact that at higher pressure the mean free path of ions decreases as a result of increased collisions, which reduces the energy of bombarding ions and hence the etch rate. Moreover, anisotropy of etching also gets hampered due to the broadening of angle distribution in which the ions hits the substrate. It is important to mention here that both RF power and chamber pressure can affect the mask selectivity by influencing the physical sputtering process. Mask selectivity decreases with increasing RF power and decreasing process pressure. Percentage of Cl_2 in Cl_2/Ar mixer has marginal effect on the etch rate up to 60% Cl_2 concentration. Further increase in Cl_2 in mixer gas reduces the etch rate due to lack of Ar which helps in removing the etched byproduct from the substrate surface. However, increase in Cl_2/BCl_3 ratio significantly enhances the each rate due to increase in Cl^+ density in the gas mixer. However, at high Cl_2/BCl_3 ratio visible erosion in the mesa sidewall was observed possibly due to high Cl-ion scattering at the mesa sidewall.

From the above discussion it is clear that a careful optimization is required for realizing a mesa with clear sharp edges, vertical and smooth mesa sidewalls. SEM image of GaN mesa sidewall using RF power of 75 W, ICP power of 500 W, chamber pressure of 10 mTorr and gas flow rate of 25/10 sccm is shown in Figure 6.24 for both Cl_2/Ar and Cl_2/BCl_3 chemistry (Rawal 2014).

Another application where dry etching can be used in HEMT fabrication is for realizing recessed-gate structure. A possible way to engineer the V_{Th} of GaN-HEMTs is by tailoring the thickness of the barrier layer. However, this will also affect the access region resistance and hence performance of HEMTs. So, in order to only engineer the V_{Th} of devices without altering the access region resistance, the recessed gate technology can be used where the barrier layer just under the gate was partially etched. Besides threshold voltage engineering, an improvement in breakdown voltage, transconductance and device linearity is also reported for devices employing gate recess structure. For gate recess etching, a very low etch rate with minimum surface damage is essential. Different dry etching recipes have also been investigated for realizing recess gate structure. Among them, Cl_2-based dry etching is the most popular. In gate recess etching, actual depth control as well as the damage associated with dry etching is a critical concern. By using only Cl_2-based dry etching, a dead time (delay in initiation of etching) was observed for AlGaN/GaN heterostructure which is explained as a result of low etch rate of native oxide layer present on the substrate surface. A pretreatment of BCl_3 plasma was found effective in etching native oxide layer and minimizing the dead time.

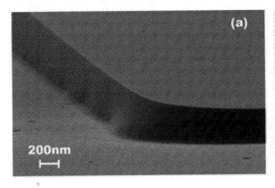
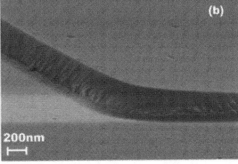

FIGURE 6.24
SEM image of GaN mesa sidewall etched by (a) Cl_2/Ar and (b) Cl_2/BCl_3 plasma. (From Rawal, D. S. *et al., J. Vac. Sci. Technol. A*, 32, 2014.)

Presence of BCl_3 also reduces the etch selectivity between GaN and AlGaN. An etch rate of ~8.7 nm/min was reported at an input power of 15 W and a chamber pressure of 10 mT at Cl_2 flow rate of 10 sccm after 60 sec of BCl_3 exposure (Buttari *et al.* 2003). A higher etch rate of ~18 nm/min is also reported using a gas mixer of $Cl_2/BCl_3/Ar$ plasma at 30 W RF and 100 W ICP power using SiN_x as etch mask (Anderson *et al.* 2010). A new gas chemistry using only BCl_3 as process gas and photoresist as etch mask has also been reported to realize self-aligned recess-gate structure (Dutta *et al.* 2016a). Using RF power of 20 W and at chamber pressure 10 mT, an intended low etch rate ~1.5 nm/min was reported and this rate is almost independent of Al mole fraction. Another important feature of this recipe is minimal damage of the photoresist mask used. So this technique can be used for fabricating self-aligned recessed gate HEMT devices using a single gate lithography step for gate recessing and gate metal patterning. Use of this technique simplifies the device processing considerably and helps in avoiding the lithographic misalignment. A simple process flow for fabrication of self-aligned recess-gate HEMT is shown in Figure 6.25.

Another technique which is becoming increasingly popular is called the "Digital etch technique" (Burnham *et al.* 2010, Yamanaka *et al.* 2015) where the thin top layer AlGaN/GaN heterostructure is first oxidized by exposing it to O_2 plasma followed by etching the thin oxide layer by using dry etching (using BCl_3) or by wet chemical etching (using diluted HCl). This cycle is repeated for certain no. of times depending on the required etch depth. From repeatability of etch depth and smoothness of etch surface point of view, this particular technique is quite promising.

Another application where dry etching can be used for HEMT fabrication is for opening the source/drain and/or gate contact window in the passivation layer (Tadjer *et al.* 2010). Depending on the fabrication process of AlGaN/GaN HEMT, a passivation layer is deposited (insitu/exsitu) before the formation of source/drain and/or gate contact. In these cases, the passivation layer has to be etched in order to open the ohmic (source and drain) and Schottky (gate) contact windows. The parameters for dry etching are dependent on the passivation material and required etch profile structure. Silicon nitride (Si_3N_4) deposited by different methods has been widely accepted as a passivation layer for GaN-based HEMTs.

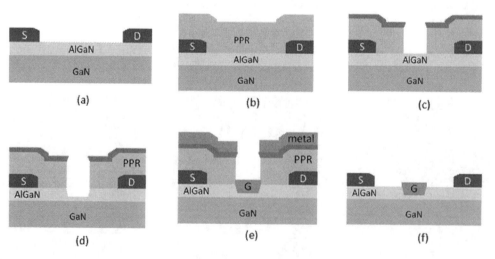

FIGURE 6.25

Process steps for formation of self-aligned recessed-gate HEMT. (a) AlGaN/GaN heterostructure after source/drain ohmic contact, (b) Positive photoresist (PPR) coating, (c) soaking in chlorobenzene (which hardens the top PPR layer) and gate pattern developing, (d) gate recess etching, (e) gate metal deposition and (f) gate metal patterning by lift-off.

For opening contact windows in Si_3N_4 passivation layer, F-based dry etching (using CF_4 or SF_6 as etchant gas) is typically used. However, careful measures should be taken to avoid the negative fluorine ion incorporation in the barrier layer which can deplete the 2DEG under the gate region as discussed later in Section 6.6. It was found that addition of O_2 along with CF_4 etch chemistry can reduce the fluorine-induced 2DEG degradation, though on the other hand it degrades the RF performance (Hahn *et al.* 2012).

6.5 Passivation

Passivation of GaN-based HEMT is one of the critical steps in the development of reliable AlGaN/GaN HEMT technology. Without proper passivation, frequency dependent current slump (Mishra *et al.* 2007, Nakajima *et al.* 2007) and knee-walkout (Green *et al.* 2003) significantly affects the performance of AlGaN/GaN HEMTs. This frequency dependent current slump or RF dispersion is commonly known as current collapse and this phenomenon is related to surface traps as well as bulk traps in the device (Roff *et al.* 2009). The bulk trapping can be significantly reduced by improving the growth technology and with specially designed epi-layer. However, the surface trapping can be effectively reduced by passivating the surface with proper dielectric material and by surface treatment immediately prior to the dielectric deposition. This passivation layer also helps to protect the device from different chemical and mechanical damage.

Unique properties of GaN-based materials also lead to different challenges in GaN HEMT operation. High critical electric field of GaN allows very high-voltage operation, which leads to a high degree of charge injection and trapping. As a wide bandgap material, GaN, AlN or their ternary compound (AlGaN) also have wide variety of deep-level traps with emission-time ranging from microseconds to seconds or even more (Joh and del Alamo 2011, Mase *et al.* 2017). So, the charge transfer process in these traps is too slow to follow the applied high frequency signal. Electrons trapped in these defects act as a negative charge, which reduces the carrier density (2DEG) at the channel and increases the knee voltage. This results in reduction in the drain current and increase in the ON-resistance which limits the output power of a device. This phenomenon, which is popularly known as Current Collapse, can significantly deteriorate the device performance. Current collapse phenomenon is schematically shown in Figure 6.26. The possible areas in AlGaN/GaN

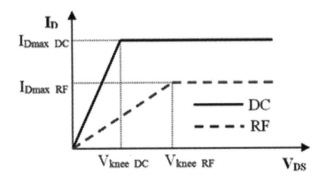

FIGURE 6.26
Schematic representation of current collapse from HEMT output characteristics.

HEMTs where trapping can happen are ungated AlGaN surface, AlGaN barrier layer, GaN buffer layer and AlGaN/GaN interface. These deep level defects can originate from exposure and damage of the surface during device processing, threading dislocation due to lattice mismatch, point defects, impurity addition during growth, high Al mole fraction, and relaxation of the AlGaN barrier layer (Chaturvedi 2007). Growth-related defects can be significantly reduced by careful optimization of growth process and surface trappings can be mitigated by careful device processing. In the following section we will concentrate on the surface trapping in AlGaN/GaN HEMTs, its effect, and its reduction by surface passivation.

When AlGaN/GaN HEMT is biased in the OFF-state (at a gate voltage more negative than the threshold voltage for depletion mode HEMTs) while retaining the applied positive drain voltage, a large electric field exists at the drain side of the gate-edge. Negative gate voltage in the gate terminal allows the supply of large number of electrons which can be driven out to surface states in the presence of applied electric field. Once the electrons are captured/trapped in the surface states, they behave like a fixed negative charge until they are emitted from the surface states in the absence of electric field. However, as the emission process is slower, the trapped charges cannot follow the rapid change in gate potential. In this situation, density of carrier in the channel is not only influenced by the gate potential but also by the trapped surface charges. This phenomenon is shown in Figure 6.27. As the trapped charges behave like a fixed negative charge, the dynamic or AC drain current would be lower than that of static or DC condition. Presence of negative trapped charges in the gate-drain access region of HEMT can also be observed by using Kelvin-probe microscopy after OFF-state biasing stress (Sabuktagin *et al.* 2005). This phenomenon can also be described as an effect of virtual gate which results in a slump of AC output current.

Surface trapping can be mitigated by passivating the exposed AlGaN surface with proper dielectric layer. Proper passivation reduces the population of trapping defects present at the exposed HEMT surface hence mitigating the surface trapping and prevents the virtual gate formation. This leads to negligible/small reduction in the number of channel carrier and reduced current collapse. Different dielectrics (e.g., Si_3N_4, AlN, SiO_2, Al_2O_3 etc.) or their suitable combination (SiO_2/Si_3N_4, AlN/Si_3N_4) deposited by different techniques have been investigated as a passivation layer (Chou *et al.* 2014, Javorka *et al.* 2003, Koehler *et al.* 2013,

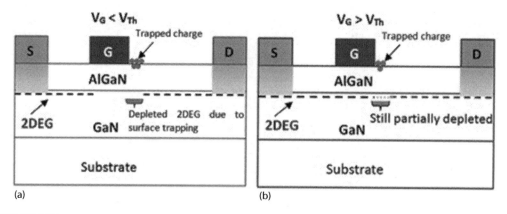

FIGURE 6.27
(a) Charge trapping at the gate edge (towards drain side) of an unpassivated AlGaN/GaN HEMT during OFF-state biasing ($V_G < V_{Th}$). (b) Partial recovery of channel charge at the gate edge when the device is just switched from OFF-state to ON-state.

Mattalah *et al.* 2012, Romero *et al.* 2008, Tang *et al.* 2013a, 2013b). Surface treatment prior to dielectric deposition also plays an important role in determining the overall performance and stability of passivated HEMTs (Lee *et al.* 2014, Romero *et al.* 2008, 2012). Among different dielectrics, Si_3N_4 is widely used for passivation of GaN-based HEMTs. Surface trap density as low as 10^{11} cm^{-2} eV^{-1} can be achieved at Si_3N_4/GaN interface by N_2 plasma treatment prior to dielectric deposition (Hashizume *et al.* 2003). The effectiveness of Si_3N_4 as a passivation layer is possibly due to the formation of controlled nitrogen-vacancy-related defects at III-Nitride surface during its deposition, which leads to reduced surface states generation (Yatabe *et al.* 2016). However, passivation may also affect the device characteristics in adverse ways. For example, in many cases, reduction in breakdown voltage of HEMT as well as increase in the surface leakage current was observed. Dielectric material and its quality, deposition technique and pre-deposition surface treatment were found to have a significant effect in determining the breakdown voltage and surface leakage current (Chaturvedi 2007, Liu *et al.* 2011).

SiN$_x$ deposited by PECVD and LPCVD has been widely used as a passivation layer for GaN-based HEMTs. Deposition condition and method of SiN$_x$ can significantly affect the device performance. For example, AlGaN/GaN HEMTs passivated with PECVD SiN$_x$ deposited at different conditions have showed different levels of current collapse and surface leakage current. HEMTs passivated with standard SiN$_x$ (refractive index or RI of 1.88–2) and high refractive index SiN$_x$ (RI of 2.9) both deposited at 300°C showed much lower current collapse and surface leakage current compared to that of low temperature (40°C) deposited PECVD SiN$_x$ (RI 1.79) (Tan *et al.* 2006). On the other hand, compared to PECVD SiN$_x$, AlGaN/GaN HEMTs passivated with LPCVD SiN$_x$ (deposited at 780°C) showed reduction in current collapse, surface leakage current and improved RF power performance (Wang *et al.* 2015). Performance improvement in case of LPCVD SiN$_x$ is due to the reduced oxygen contamination at AlGaN surface as confirmed from TEM-EDS mapping.

Current collapse in AlGaN/GaN HEMTs can further be reduced by using the field plate structure. Field plate is a lateral extension of gate metal that sits on the surface of passivation layer in a direction towards drain contact. In addition to lowering the effect of trapping, field plate also helps in enhancing the breakdown voltage of HEMT. More discussion on field plate can be found in Section 6.6.

Significant reduction in current collapse and dynamic ON-resistance can also be achieved by using thin AlN passivation layer deposited by plasma enhanced atomic layer deposition (Huang *et al.* 2012). High-density positive polarization-charge at mono-crystalline like AlN/III-Nitride interface was claimed to be the reason for improved RF performance (Huang *et al.* 2013). This positive charge can effectively compensate the trapped negative charge, eliminating the effect of virtual gate.

Pulsed I–V is generally used to observe the current collapse in AlGaN/GaN HEMTs. This technique is different compared to DC I–V characteristics. Gate and drain bias waveforms during DC and pulsed I_D–V_{DS} measurements (for a particular V_{GS}) are shown in Figure 6.28. In pulsed I–V characteristics, the devices are normally switched-ON only for a small duration for taking measurements and are in OFF-state for the remaining part of the time-period. Switching-ON the HEMTs for a small interval can minimize the self-heating effect. Besides, by making the measurement timescale comparable to the time constant of traps causing this current collapse, we can obtain the degraded characteristics that would have happened during RF operation. The OFF-state biasing voltages are known as quiescent bias points (V_{GS0}, V_{DS0}) which cause trapping in the AlGaN/GaN HEMT. Deep level traps have large time constant, and emission from these traps cannot happen during

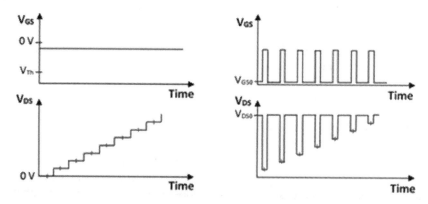

FIGURE 6.28

Waveforms of applied V_{GS} and V_{DS} biases for (a) DC and (b) pulsed $I_D–V_{DS}$ characteristics measurements. Small red mark on the V_{DS} waveform indicates the time where measurements were taken during each bias step.

the very small measurement time (ON-state). So, degraded I–V characteristics can be captured by using pulsed measurement. Unpassivated AlGaN/GaN HEMTs show degraded $I_D – V_{DS}$ characteristics when measured in pulsed condition (Shen *et al.* 2003). Reduction in output current and increase in ON-resistance can be seen in case of pulsed characteristics. This pulsed I–V characteristics strongly depend on the pulse-width during measurement, as small pulse-width can even capture the effect of traps with very small time-constant (Shen *et al.* 2003) as well as on the biasing condition during stress. Current collapse can be significantly improved by proper surface passivation. Deposition condition of the passivating layer can strongly influence the device characteristics and hence the current collapse. DC and pulsed output characteristics of AlGaN/GaN HEMTs passivated with Si_3N_4 deposited by PECVD and Cat-CVD are shown Figure 6.29. Lower current slump during pulsed measurement was observed in case of Cat-CVD Si_3N_4 passivated devices (Oku *et al.* 2008). Magnitude of current collapse is typically quantified as the ratio of dynamic ON-resistance to static or DC ON-resistance, where ON-resistance is calculated from the respective slope of $I_D – V_{DS}$ characteristics in the linear region. Current collapse can also be defined as the ratio of slump in dynamic output current to that of static DC current.

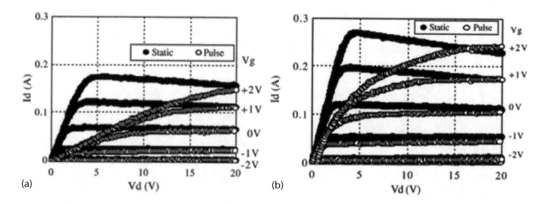

FIGURE 6.29

DC (static) and pulsed characteristics of AlGaN/GaN HEMTs passivated with Si_3N_4 film deposited by (a) PECVD and (b) Cat-CVD. (Reprinted from *Thin Solid Films*, 516, Oku, T. *et al.*, AlGaN/GaN HEMTs passivated by Cat-CVD SiN Film, 545–547, 2008, Copyright 2007, with permission from Elsevier.)

6.6 Special Techniques in GaN-HEMT Processing

In this section we will discuss some more techniques related to GaN-HEMT processing resulting in improved performance of the device. It is important to discuss these popular techniques which are closely associated with AlGaN/GaN HEMT technology. Fluorinated-gate and p-(Al)GaN gate HEMT technologies are discussed first, which helps in realizing enhancement mode (E-mode) or normally-OFF devices. Field-plated HEMTs are discussed next, which shows enhanced breakdown voltage compared to conventional HEMTs. Finally, AlGaN/GaN MIS-HEMT technology is discussed which can significantly reduce the gate leakage current of HEMTs and increases the forward gate bias swing. For the sake of completeness, other promising III-nitride heterostructures (other than AlGaN/GaN) are also identified and their relative advantages are also discussed.

6.6.1 Fluorinated Gate HEMTs

Fluorinated gate HEMT or fluorine plasma treated HEMT is a technique used for realization of normally-OFF HEMTs (Cai 2006, Medjdoub *et al.* 2008). Conventional GaN-HEMTs are normally-ON devices due to the presence of the 2DEG in the channel even at zero gate bias. A negative gate bias has to be applied to deplete the channel or switching OFF the HEMT. However, from circuit application point of view, normally-OFF devices would be preferred. In fluorinated gate HEMT devices, AlGaN-barrier layer just under the gate is treated with fluorine plasma, which incorporates negative fluorine ions in the barrier layer. This plasma treatment is done before the gate contact metallization typically in a RIE system using CF_4-plasma. Plasma damage due to F-ion incorporation can be recovered by post-gate annealing at 400°C (Cai 2006). These negative fluorine ions deplete the 2DEG under the gate region and can eventually shift the threshold voltage from negative to positive. The extent of the positive shift in the threshold voltage depends on the RF power and duration of plasma treatment. So, threshold voltage of the devices can be tailored by varying these parameters. Gate leakage current in fluorinated devices also reduces due to the increased barrier height between gate metal and AlGaN. Both fluorination and gate recess technique can be simultaneously used for realizing devices with high positive voltage. Figure 6.30 illustrates the schematic diagram of Fluorine plasma treated HEMT. Though this fluorine plasma treatment is a useful and effective technique to modulate the

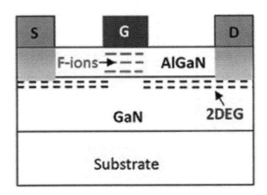

FIGURE 6.30
Simplified representation of fluorinated gate HEMT.

threshold voltage of GaN-based HEMT, thermal stability of threshold voltage of fluorinated gate HEMTs is a concern (Lorenz *et al.* 2009).

6.6.2 *p*-GaN or *p*-AlGaN Gated HEMT

In this technique, a heavily doped (generally with Mg) *p*-type III-Nitride (GaN, AlGaN) cap layer of thickness ~100 nm (Uemoto *et al.* 2007) is grown in-situ on conventional HEMT structure and carefully etched from the source and drain access regions as shown in Figure 6.31. The *p*-GaN or *p*-AlGaN layer present in the gate region of AlGaN/GaN heterostructure depletes the 2DEG channel just under the gate. This results in a normally-OFF operation of GaN-based HEMT. This technique is considered as one of the most promising techniques for realization of normally-OFF GaN-HEMTs. However, incorporation of *p*-GaN/*p*-AlGaN cap layer increases the growth complexity and precise etching of only *p*-(Al)GaN cap is also critical. Besides, etching related damage is also a concern. In this regard, another technique has been proposed where instead of etching the *p*-III-Nitride layer from the access regions, the devices are treated with hydrogen plasma after gate metallization which compensates for the holes in *p*-GaN present in the access region (Hao *et al.* 2016).

6.6.3 GaN-HEMTs with Field-Plates

GaN HEMTs have already shown their potential for applications in both high-power and high-frequency devices. High breakdown voltage and low ON-resistance are the key requirements of power switching devices. Similarly, reduction in current collapse is one of the essential criteria for high-frequency devices. Field-plated structures can be used to effectively increase the breakdown voltage and reduce the current collapse in AlGaN/GaN HEMTs. Use of field plates can improve the device performance by re-distributing the electric field in the gate-drain region and suppressing the DC to RF dispersion (current collapse). Field plates can be of different types and the most commonly used types are gate-connected and source-connected field plates (Mishra and Singh, 2008). In gate connected field plate (Figure 6.32) the extension of the gate metal on a dielectric layer can modulate the channel beyond the primary gate which reduces the peak electric field by spreading it into two peaks (as shown in Figure 6.33). Distribution of electric field in dual

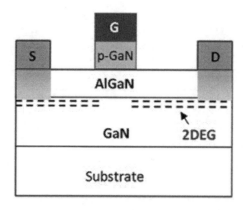

FIGURE 6.31
Schematic cross-sectional view of *p*-GaN gated normally-OFF HEMT.

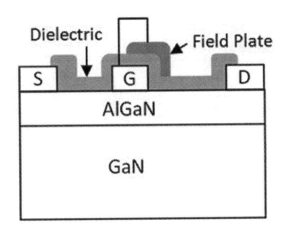

FIGURE 6.32
Schematic illustration of AlGaN/GaN HEMT with gate connected field plate.

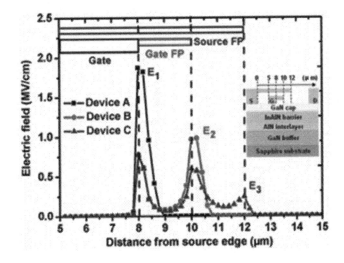

FIGURE 6.33
Simulated distribution of electric field along the channel for devices without field plate (Device A), with gate connected field plate (Device B) and with gate and source dual field plates (Device C). Devices are biased at OFF-state ($V_{GS} = -8$ V and $V_{DS} = 100$ V). The inset shows the schematic cross sectional view of dual field plated HEMT structure. (Reprinted with permission from Li, W. *et al.*, Impact of dual field plates on drain current degradation in InAlN/AlN/GaN HEMTs, *Semi. Sci. Tech.*, 31, 125003, 2016. Copyright 2016, Institute of Physics.)

field plated (gate and source connected) HEMT is also shown in Figure 6.33. Though the breakdown voltage of the HEMT can be increased by using the gate connected field plate, the primary drawback of this technique is the enhanced gate-drain capacitance, which impacts the high frequency operation by reducing the gain. Electric field in the gate-drain access region can also be modulated by using source connected field plate; however, this method comes with the penalty of enhanced gate-source and drain-source capacitances. So, there is a trade-off between the high-frequency gain and breakdown voltage which can be optimized based on the requirements, by varying different design parameters like horizontal extension of a FP beyond the gate, separation between field plate and AlGaN surface, and shape of the field plate.

6.6.4 AlGaN/GaN MIS-HEMT

In spite of its many advantages, GaN-based HEMTs suffer from high gate leakage current (I_G) due to the Schottky gate contact and polarization induced strong electric field in the barrier layer. High I_G increases the OFF-state power dissipation, limits positive gate voltage swing, introduces reliability issues and acts as an additional noise source. Hence, suppression of high-gate leakage current is essential to improve the device performance and its reliability. High I_G in AlGaN/GaN HEMT can be mitigated by inserting an insulating layer between the AlGaN barrier layer and the gate metal. This structure is commonly known as Metal Insulator Semiconductor HEMT or MIS-HEMT (Figure 6.34). Compared to HEMTs, fabrication process of MIS-HEMTs has one additional step: gate dielectric deposition. Different gate dielectrics (e.g. Al_2O_3, HfO_2, Si_3N_4, etc.) deposited by different methods have been investigated for forming the MIS-HEMT structure. MIS-HEMTs showed significant reduction in I_G and increase in forward gate voltage swing. However, threshold voltage (V_{Th}) instability of MIS-HEMTs is a serious concern which arises because of the high density interface traps at gate oxide/III-Nitride interface possibly due to the presence of Ga-O bonds (Hua *et al.* 2015a, 2015b). From this point of view, nitride based gate dielectrics (e.g., Si_3N_4, AlN, etc.) are gaining a huge popularity. Stable device performance of AlGaN/GaN MIS-HEMTs was demonstrated using Si_3N_4 as gate dielectric (Dutta *et al.* 2016b, Hua *et al.* 2015a, 2015b).

A common observation for GaN-based MIS-HEMTs is the negative shift in V_{Th} of MIS-HEMTs compared to that of HEMTs. In other words, threshold voltage of MIS-HEMTs is typically more negative than that of HEMTs. This is due to the presence of positive interfacial charge density at dielectric/III-nitride interface whose magnitude is higher than the negative polarization charge density at the barrier layer surface. If by some means the resultant interfacial charge density at the dielectric/III-Nitride interface can be made negative, a positive shift in V_{Th} can be expected, which can be effectively used in realizing the E-mode MIS-HEMTs. E-mode MIS-HEMTs are intended for designing fail-safe, compact, and energy efficient power converter systems. By using interface charge engineering, positive shift in V_{Th} hence the enhancement mode MIS-HEMTs have been demonstrated by using reactive ion sputtered Al_2O_3 as gate dielectric (Dutta *et al.* 2014, 2016a). Gate recess and fluorination techniques can also be included with MIS-HEMT technology for realizing enhancement mode MIS devices.

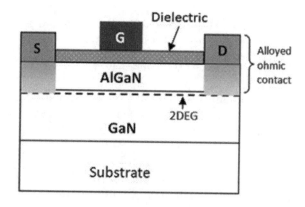

FIGURE 6.34

Schematic cross-sectional view of AlGaN/GaN MIS-HEMT. Presence of gate dielectric layer improves the device performance.

6.6.5 Other III-Nitride HEMTs

AlGaN/GaN heterostructure is most commonly used for fabrication of HEMT devices due to its existing mature technology. However, AlGaN/GaN HEMT devices suffers from reliability issue due to the built-in strain in the thin AlGaN barrier layer as a result of lattice mismatch between AlGaN and GaN, and additional strain due to inverse piezoelectric effect during device reverse bias stress. Under high voltage operation, the total strain can create permanent defects at the gate edge of drain side and degrade the device performance. AlInN/GaN and AlN/GaN heterostructures are also emerging due to some advantages compared to conventional AlGaN/GaN devices (Medjdoub *et al.* 2011, Sun *et al.* 2010). Unlike AlGaN/GaN, AlInN with an Al mole fraction of ~83% is lattice matched with the underneath GaN layer. So, lattice matched $Al_{0.83}In_{0.17}N$/GaN heterostructure is free from any initial stress and expected to have a higher reliability. Though piezoelectric polarization is absent in AlInN barrier layer, its higher spontaneous polarization creates higher 2DEG density at AlInN/GaN heterointerface compared to conventional AlGaN/GaN system (Sun *et al.* 2010). Further, lattice matched AlInN/GaN has higher conduction band offset compared to AlGaN/GaN, which improves the carrier confinement in the channel. Also, AlInN barrier layer thickness can be made much thinner (typically 7–10 nm) without sacrificing the 2DEG density. Thinner barrier layer provides better gate control and improves the device transconductance and cut-off frequency. Among different III-Nitride HEMTs, AlN/GaN offers the highest 2DEG density and hence very high channel conductance (Zimmermann *et al.* 2009). Besides, very thin (typically 3–6 nm) AlN barrier layer also helps in realizing high transconductance.

References

Ahn, W. *et al.*, Various Schottky contacts of AlGaN/GaN Schottky Barrier Diodes (SBDs), *ECS Transactions*, vol. 53, no. 2, p. 171, 2013.

Alomari, M. *et al.*, Au free ohmic contacts for high temperature InAlN/GaN HEMT's, *ECS Transactions*, vol. 25, no. 12, pp. 33–36, 2009.

Anderson, T. J. *et al.*, Characterization of recessed-gate AlGaN/GaN HEMTs as a function of etch depth, *Journal of Electronic Materials*, vol. 39, no. 5, pp. 478–481, 2010.

Ando, S. *et al.*, *Polymers for Microelectronics: Resists and Dielectrics*, American Chemical Society, Washington, DC, 1994.

Arulkumaran, S. *et al.*, Low specific on-resistance AlGaN/AlN/GaN high electron mobility transistors on high resistivity silicon substrate, *Electrochemical and Solid-State Letters*, vol. 13, no. 5, pp. H169–H172, 2010.

Arulkumaran, S. *et al.*, High-frequency microwave noise characteristics of InAlN/GaN high-electron mobility Transistors on Si (111) substrate, *IEEE Electron Device Letters*, vol. 35, no. 10, pp. 992–994, 2014.

Burnham, S. D. *et al.*, Gate-recessed normally-off GaN-on Si HEMT using a new O_2-BCl_3 digital etching technique, *Physica Status Solidi C*, vol. 7, no. 7–8, pp. 2010–2012, 2010.

Buttari, D. *et al.*, Systematic characterization of Cl_2 reactive ion etching for improved ohmics in AlGaN/GaN HEMTs, *IEEE Electron Device Letters*, vol. 23, no. 2, pp. 76–78, 2002.

Buttari, D. *et al.*, Origin of etch delay time in Cl_2 dry etching of AlGaN/GaN structures, *Applied Physics Letters*, vol. 83, no. 23, pp. 4779–4781, 2003.

Cabrini, S. and S. Kawata, *Nanofabrication Handbook*, CRC Press, 2012.

Cai, Y. *et al.*, Control of threshold voltage of AlGaN/GaN HEMTs by fluoride-based plasma treatment: From depletion mode to enhancement mode, *IEEE Transactions on Electron Devices*, vol. 53, no. 9, pp. 2207–2215, 2006.

Chaturvedi, N. *et al.*, Mechanism of ohmic contact formation in AlGaN/GaN high electron mobility transistors, *Semiconductor Science and Technology*, vol. 21, pp. 175–179, 2006.

Chaturvedi, N., Development and study of AlGaN/GaN microwave transistors for high power operation, Doctoral Thesis, Technical University Berlin, Germany, 2007.

Choi, W. *et al.*, High-voltage and low-leakage-current gate recessed normally-off GaN MIS-HEMTs with dual gate insulator employing PEALD-SiN$_x$/RF-sputtered HfO$_2$, *IEEE Electron Device Letters*, vol. 35, no. 2, pp. 175–177, 2014.

Choi, W. *et al.*, Improvement of V_{Th} instability in normally-off GaN MIS-HEMTs employing PEALD-SiN$_x$ as an interfacial layer, *IEEE Electron Device Letters*, vol. 35, no. 1, pp. 30–32, 2014.

Chou, B. Y. *et al.*, Al$_2$O$_3$-passivated AlGaN/GaN HEMTs by using nonvacuum ultrasonic spray pyrolysis deposition technique, *IEEE Electron Device Letters*, vol. 35, no. 9, pp. 903–905, 2014.

Cordier, Y. *et al.*, MBE growth of AlGaN/GaN HEMTS on resistive Si (1 1 1) substrate with RF small signal and power performances, *Journal of Crystal Growth*, vol. 251, pp. 811–815, 2003.

DasGupta, N., Etching in *Semiconductor Manufacturing Handbook*, Edited by H. Geng, McGraw-Hill Publication, 2017.

del Alamo, J. A. and J. Joh, GaN HEMT reliability, *Microelectronics Reliability*, vol. 49, pp. 1200–1206, 2009.

Duffy, S. J. *et al.*, Low source/drain contact resistance for AlGaN/GaN HEMTs with high Al concentration and Si-HP [111] substrate, *ECS Journal of Solid State Science and Technology*, vol. 6, no. 11, pp. S3040–S3043, 2017.

Dutta, G. *et al.*, Effect of sputtered-Al$_2$O$_3$ layer thickness on the threshold voltage of III-nitride MIS-HEMTs, *IEEE Transactions on Electron Devices*, vol. 63, no. 4, pp. 1450–1458, 2016a.

Dutta, G. *et al.*, Low-temperature ICP-CVD SiN$_x$ as gate dielectric for GaN-based MIS-HEMTs, *IEEE Transactions on Electron Devices*, vol. 63, no. 12, pp. 4693–4701, 2016b.

Dutta, G. *et al.*, Positive shift in threshold voltage for reactive-ion-sputtered Al$_2$O$_3$/AlInN/GaN MIS-HEMT, *IEEE Electron Device Letters*, vol. 35, no. 11, pp. 1085–1087, 2014.

Firrincieli, A. *et al.*, Au-free low temperature ohmic contacts for AlGaN/GaN power devices on 200 mm Si substrates, *Japanese Journal of Applied Physics*, vol. 53, p. 04EF01, 2014.

Fujishima, T. *et al.*, Formation of low resistance ohmic contacts in GaN-based high electron mobility transistors with BCl$_3$ surface plasma treatment, *Applied Physical Letters*, vol. 103, p. 083508, 2013.

Ganguly, S. *et al.*, Plasma MBE growth conditions of AlGaN/GaN high-electron-mobility transistors on silicon and their device characteristics with epitaxially regrown ohmic contacts. *Applied Physics Express*, vol. 7, no. 10, p. 105501, 2014.

Green, B. M. *et al.*, Microwave power limits of AlGaN/GaN HEMTs under pulsed-bias conditions, *IEEE Transactions on Electron Devices*, vol. 51, no. 2, pp. 618–623, 2003.

Hahn, H. *et al.*, Oxygen addition to fluorine based SiN etch process: Impact on the electrical properties of AlGaN/GaN 2DEG and transistor characteristics, *Solid State Electronics*, vol. 67, pp. 90–93, 2012.

Hao, R. *et al.*, Normally-off p-GaN/AlGaN/GaN high electron mobility transistors using hydrogen plasma treatment, *Applied Physics Letters*, vol. 109, no. 15, p. 152106, 2016.

Harriott, L. R., Next generation lithography, *Materials Today*, vol. 2, no. 2, pp. 9–12, 1999.

Hashizume, T. *et al.*, Surface passivation of GaN and GaN/AlGaN heterostructures by dielectric films and its application to insulated-gate heterostructure transistors, *Journal of Vacuum Science & Technology B*, vol. 21, no. 4, pp. 1828–1838, 2003.

Heikman, S. J., MOCVD growth technologies for applications in AlGaN/GaN high electron mobility transistors, PhD Dissertation, University of California Santa Barbara, 2002.

Helbert, J. N., *Handbook of VLSI Microlithography Second Edition Principles, Technology and Applications*, Noyes Publication, 2001.

Herbecq, N. *et al.*, Above 2000 V breakdown voltage at 600 K GaN-on-silicon high electron mobility transistors, *Physica Status Solidi A*, vol. 213, no. 4, pp. 873–877, 2016.

Hove, M. *et al.*, CMOS process-compatible high-power low leakage AlGaN/GaN MISHEMT on silicon, *IEEE Electron Device Letters*, vol. 33, no. 5, pp. 667–669, 2012.

Hua, M. *et al.*, Characterization of leakage and reliability of SiN$_x$ gate dielectric by low-pressure chemical vapor deposition for GaN based MIS-HEMTs, *IEEE Transactions on Electron Devices*, vol. 62, no. 10, pp. 3215–3222, 2015b.

Hua, M. *et al.*, GaN-based metal-insulator-semiconductor high-electron-mobility transistors using low-pressure chemical vapor deposition SiN_x as gate dielectric, *IEEE Electron Device Letters*, vol. 36, no. 5, pp. 448–450, 2015a.

Huang, S. *et al.*, Effective passivation of AlGaN/GaN HEMTs by ALD-grown AlN thin film, *IEEE Electron Device Letters*, vol. 33, no. 4, pp. 516–518, 2012.

Huang, S. *et al.*, Mechanism of PEALD-grown AlN passivation for AlGaN/GaN HEMTs: Compensation of interface traps by polarization charges, *IEEE Electron Device Letters*, vol. 34, no. 2, pp. 193–195, 2013.

Husna, F. *et al.*, High-temperature performance of AlGaN/GaN MOSHEMT with SiO_2 gate insulator fabricated on Si (111) substrate, *IEEE Transactions on Electron Devices*, vol. 59, no. 9, pp. 2424–2429, 2012.

Ito, T. and S. Okazaki, Pushing the limits of lithography, *Nature*, vol. 406, pp. 1027–1031, 2000.

Javorka, P. *et al.*, Influence of SiO_2 and Si_3N_4 passivation on AlGaN/GaN/Si HEMT performance, *Electronics Letters*, vol. 39, no. 15, pp. 1155–1157, 2003.

Jia, Y. *et al.*, Characterization of buffer-related current collapse by buffer potential simulation in AlGaN/GaN HEMTs, *IEEE Transactions on Electron Devices*, vol. 65, no. 8, pp. 3169–3175, 2018.

Joglekar, S. *et al.*, Impact of recess etching and surface treatments on ohmic contacts regrown by molecular-beam epitaxy for AlGaN/GaN high electron mobility transistors, *Applied Physics Letters*, vol. 109, p. 041602, 2016.

Joh, J. and J. A. del Alamo, A current-transient methodology for trap analysis for GaN high electron mobility transistors, *IEEE Transactions on Electron Devices*, vol. 58, no. 1, pp. 132–140, 2011.

Katz, O. *et al.*, Electron mobility in an AlGaN/GaN two-dimensional electron gas I-Carrier concentration dependent mobility, *IEEE Transactions on Electron Devices*, vol. 50, no. 10, pp. 2002–2008, 2003.

Khanna, V. K., *Fundamentals of Solid State Lighting: LEDs, OLEDs, and Their Application in Illumination and Displays*, CRC Press, 2014.

Koehler, A. D. *et al.*, Atomic layer epitaxy AlN for enhanced AlGaN/GaN HEMT passivation, *IEEE Electron Device Letters*, vol. 34, no. 9, pp. 1115–1117, 2013.

Lao, W. S. *et al.*, Formation of Ohmic contacts in AlGaN/GaN HEMT structures at 500°C by Ohmic contact recess etching," *Microelectronics Reliability*, vol. 49, no. 5, pp. 558–561, 2009.

Lee, H. S. *et al.*, AlGaN/GaN high-electron-mobility transistors fabricated through a Au-Free Technology, *IEEE Electron Device Letters*, vol. 32, no. 5, pp. 623–625, 2011.

Lee, N. H. *et al.*, The effect of SF_6 plasma and in-situ N_2 plasma treatment on gate leakage, sub-threshold slope, and current collapse in AlGaN/GaN HEMTs, *CS MANTECH Conference*, Denever, CO, pp. 245–248, 2014.

Li, Y. *et al.*, AlGaN/GaN high electron mobility transistors on Si with sputtered TiN gate, *Physica Status Solidi A*, vol. 214, no. 3, p. 1600555, 2017.

Liu, Q. Z. *et al.*, A review of the Metal-GaN contact technology, *Solid-State Electronics*, vol. 42, no. 5, pp. 677–691, 1998.

Liu, Z. H. *et al.*, Reduced surface leakage current and trapping effects in AlGaN/GaN high electron mobility transistors on silicon with SiN/Al_2O_3 passivation, *Applied Physics Letters*, vol. 98, pp. 113506 (1–3), 2011.

Lorenz, A. *et al.*, Influence of thermal annealing steps on the current collapse of fluorine treated enhancement mode SiN/AlGaN/GaN HEMTs, *Physica Status Solidi (C)*, vol. 6, no. S2, pp. S996–S998, Jun. 2009.

Madou, M. J., *Fundamentals of Microfabrication and Nanotechnology—vol - 2—Manufacturing Techniques for Microfabrication and Nanotechnology*, CRC Press, 2011.

Marcon, D. *et al.*, Manufacturing challenges of GaN-on-Si HEMTs in a 200 mm CMOS fab, *IEEE Transactions on Semiconductor Manufacturing*, vol. 26, no. 3, pp. 361–367, 2013.

Mase, S. *et al.*, Analysis of carrier trapping and emission in AlGaN/GaN HEMT with bias-controllable field plate, *Physics Status Solidi A*, vol. 214, no. 8, pp. 1600840(1–5), 2017.

Matsumura, M. *et al.*, Submicrometre lift-off line with T-shaped cross-sectional form, *Electronics Letters*, vol. 17, no. 12, 1981.

Mattalah, M. *et al.*, Analysis of the SiO_2/Si_3N_4 passivation bilayer thickness on the rectifier behavior of AlGaN/GaN HEMTs on (111) silicon substrate, *Physica Status Solidi C*, vol. 9, no. 3–4, pp. 1083–1087, 2012.

Medjdoub, F. *et al.*, Effect of fluoride plasma treatment on InAlN/GaN HEMTs, *Electronics Letters*, vol. 44, no. 11, pp. 696–698, 2008.

Medjdoub, F. *et al.*, High-performance low-leakage-current AlN/GaN HEMTs grown on silicon substrate, *IEEE Electron Device Letters*, vol. 32, no. 7, pp. 874–876, 2011.

Mishra, U. K. and J. Singh, Field effect transistors in *Semiconductor Device Physics and Design*, Springer, 2008.

Mishra, U. K. *et al.*, GaN-based RF power devices and amplifiers, *Proceedings of the IEEE*, vol. 96, no. 2, pp. 287–305, 2008.

Mishra, U. K., P. Parikh, and Y. F. Wu, AlGaN/GaN HEMTs-an overview of device operation and applications, *Proceedings of the IEEE*, vol. 90, no. 6, pp. 1022–1031, 2002.

Nakajima, A. *et al.*, Analysis of drain lag and current slump in GaN-based HEMTs and MESFETs, in *IEEE International Symposium on Signals, Systems and Electronics (ISSSE'07)*, pp. 193–196, 2007.

Nakamura, S. and M. R. Krames, History of gallium–nitride-based light-emitting diodes for illumination, *Proceedings of the IEEE*, vol. 101, no. 10, pp. 2211–2220, 2013.

"New light to illuminate the world," *Nobleprize.org*, 2014, Available online at: https://www.nobelprize.org/nobel_prizes/physics/laureates/2014/press.html (accessed on July 15, 2018).

Ocola, L. E. *et al.*, Bilayer process for T-gates and Γ-gates using 100-kV e-beam lithography, *Microelectronic Engineering*, vol. 67–68, pp. 104–108, 2003.

Oku, T. *et al.*, AlGaN/GaN HEMTs passivated by Cat-CVD SiN Film, *Thin Solid Films*, vol. 516, pp. 545–547, 2008.

Pearton, S. J. *et al.*, A review of dry etching of GaN and related materials, *MRS Internet Journal of Nitride Semiconductor Research*, vol. 5, no. 11, 2000.

Pei, Y. *et al.*, AlGaN/GaN HEMT with a transparent gate electrode, *IEEE Electron Device Letters*, vol. 30, no. 5, p. 439, 2009.

Piotrowicz, S. *et al.*, InAlN/GaN with AlGaN back-barrier HEMT technology on SiC for Ka-band applications, *International Journal of Microwave and Wireless Technologies*, vol. 10, no. 1, pp. 39–46, 2018.

Plummer, J. D., M. D. Deal and P. B. Griffin, Etching in *Silicon VLSI Technology*, Pearson Publication, New Delhi, India, 2012.

Rawal, D. S. *et al.*, Effect of BCl_3 concentration and process pressure on the GaN mesa sidewalls in BCl_3/Cl_2 based inductively coupled plasma etching, *Vacuum*, vol. 86, pp. 1844–1849, 2012a.

Rawal, D. S. *et al.*, Comparative study of GaN mesa etch characteristics in Cl2 based inductively coupled plasma with Ar and BCl3 as additive gases, *Journal of Vacuum Science and Technology A*, vol. 32, no. 3, pp. 031301-1–031301-10, 2014.

Roff, C. *et al.*, Analysis of DC–RF dispersion in AlGaN/GaN HFETs using RF Waveform Engineering, *IEEE Transactions on Electron Devices*, vol. 56, no. 1, pp. 13–19, 2009.

Romero, M. F. *et al.*, Effects of N_2 plasma pretreatment on the SiN passivation of AlGaN/GaN HEMT, *IEEE Electron Device Letters*, vol. 29, no. 3, pp. 209–211, 2008.

Romero, M. F. *et al.*, Impact of N_2 plasma power discharge on AlGaN/GaN HEMT performance, *IEEE Transactions on Electron Devices*, vol. 59, no. 2, pp. 374–379, 2012.

Sabuktagin, S. *et al.*, Surface charging and current collapse in an AlGaN/GaN heterostructure field effect transistor, *Applied Physics Letters*, vol. 86, p. 083506(1–3), 2005.

Schroder, D. K., *Semiconductor Material and Device Characterization*, John Wiley & Sons, 2006.

Selvaraj, S. L. *et al.*, Enhanced mobility for MOCVD grown AlGaN/GaN HEMTs on Si substrate, *IEEE Device Research Conference (DRC)*, pp. 221–222, 2011.

Shen, L. *et al.*, Temperature dependence of the current-voltage characteristics of AlGaN/GaN HEMT, *IEEE Device Research Conference*, pp. 63–64, 2003.

Shiu, J. Y. *et al.*, Oxygen ion implantation isolation planar process for AlGaN/GaN HEMTs, *IEEE Electron Device Letters*, vol. 28, no. 6, pp. 476–478, 2007.

Song, B. *et al.*, Ultralow-leakage AlGaN/GaN high electron mobility transistors on Si with non-alloyed regrown ohmic contacts, *IEEE Electron Device Letters*, vol. 37, no. 1, p. 16, 2016.

Sun, H. *et al.*, Ultrahigh-speed AlInN/GaN high electron mobility transistors grown on (111) high-resistivity silicon with F_T 143 GHz, *Applied Physics Express*, vol. 3, no. 9, p. 094101, 2010.

Tadjer, M.J. *et al.*, Electrical and optical characterization of AlGaN/GaN HEMTs with In Situ and Ex Situ deposited SiN_x layers, *Journal of Electronic Materials*, vol. 39, no. 11, pp. 2452–2458, 2010.

Tan, W. S. *et al.*, Surface leakage currents in SiNx passivated AlGaN/GaN HFETs, *IEEE Electron Device Letters*, vol. 27, no. 1, pp. 1–3, 2006.

Tang, Z. *et al.*, 600-V normally off SiN_x/AlGaN/GaN MIS-HEMT with large gate swing and low current collapse, *IEEE Electron Device Letters*, vol. 34, no. 11, pp. 1373–1375, 2013a.

Tang, Z. *et al.*, Atomic layer epitaxy AlN for enhanced AlGaN/GaN HEMT passivation, *IEEE Electron Device Letters*, vol. 34, no. 3, pp. 366–368, 2013b.

Tapajna, M. *et al.*, Thermally induced voltage shift in capacitance—Voltage characteristics and its relation to oxide/semiconductor interface states in $Ni/Al_2O_3/InAlN/GaN$ heterostructures, *Semiconductor Science Technology*, vol. 24, no. 3, pp. 035008-1–035008-5, 2009.

Thompson, L. F *et al.*, Introduction to microlithography—Theory, materials and processing, American Chemical Society, ACS symposium series 219, 1983.

Todokoro, Y., Double layer resist films for submicrometer electron beam lithography, *IEEE Journal of Solid State Circuits*, vol. SC-15, no. 4, 1980.

Uemoto, Y. *et al.*, Gate injection transistor (GIT)—A normally-off AlGaN/GaN power transistor using conductivity modulation, *IEEE Transactions on Electron Devices*, vol. 54, no. 12, pp. 3393–3399, 2007.

Van Daele, B. *et al.*, The role of Al on Ohmic contact formation on n-type GaN and AlGaN/GaN, *Applied Physics Letters*, vol. 87, no. 6, pp. 061905 (1–3), 2005.

Wang, L. *et al.*, Differences in the reaction kinetics and contact formation mechanisms of annealed Ti/Al/Mo/Au Ohmic contacts on n-GaN and AlGaN/GaN epilayers, *Journal of Applied Physics*, vol. 101, p. 013702, 2007.

Wang, X. *et al.*, Comparative study of AlGaN/GaN HEMTs with LPCVD- and PECVD-SiN_x passivation, *CS MANTECH Conference*, Scottsdale, AZ, pp. 173–176, 2015.

Wu, Y. F. *et al.*, Very-high power density AlGaN/GaN HEMTs, *IEEE Transactions on Electron Devices*, vol. 48, no. 3, pp. 586–590, 2001.

Yamanaka, R. *et al.*, Normally-off AlGaN/GaN high-electron-mobility transistor using digital etching technique, *Japanese Journal of Applied Physics*, vol. 54, p. 06FG04, 2015.

Yatabe, Z. *et al.*, Insulated gate and surface passivation structures for GaN-based power transistors, *Journal of Physics D: Applied Physics*, vol. 49, no. 39, 2016.

Zheng, X. *et al.*, Identifying the spatial position and properties of traps in GaN HEMTs using current transient spectroscopy, *Microelectronics Reliability*, vol. 63, pp. 46–51, 2016.

Zimmermann, T. *et al.*, AlN/GaN insulated-gate HEMTs with 2.3 A/mm output current and 480 mS/mm transconductance, *IEEE Electron Device Letters*, vol. 30, no. 3, p. 209, 2009.

7

Analytical Modeling of High Electron Mobility Transistors

N. B. Balamurugan

CONTENTS

7.1 Introduction

In semiconductor industry, device modeling deals with developing set of equations which describe the electrical behavior of any device which is under consideration. The modeling of device is broadly classified as physical model and compact model. Physics-based device modeling is found to be accurate, but it is not fast enough for circuit simulators such as SPICE-like advanced level tools. Hence circuit simulators normally use compact models that consist of more empirical fitting parameters and they do not directly model the underlying physics of the device. Analytical model for HEMT device is not available so far, but a lot of papers in the literature have reported simulation results alone. Therefore, in order to understand the operation and behavior of the AlInSb/Insb High Electron Mobility Transistor (HEMT) device modeling is very much essential.

7.2 AlInSb/InSb Device Structure

Figure 7.1 represents the cross-sectional view of single gate AlInSb/InSb HEMT with $L = 60$ nm as the length of the gate metal M. The gate metal used in this device is gold material which is having a work function of 5.3 eV. The contact resistance of the source and drain contacts is minimized with the help of highly doped capacitive layer. The δ-doping layer is the donor of charge carriers, which provides electrons to the channel. The work functions of gate metal M is $\Phi = 5.3$ eV.

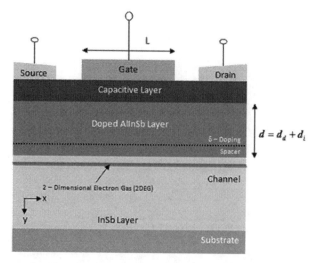

FIGURE 7.1
Cross-sectional view of single gate AlInSb/InSb HEMT.

7.3 Analytical Modeling of Channel Potential for AlInSb/InSb HEMT

The flat band voltage of the metal gate is given by $\phi_{FB} = \phi_M - \phi_S$. The 2-D Poisson equation for single gate HEMT is rewritten from the Poisson equation available from the book authored by Colinge (2007),

$$\frac{d^2\phi(x,y)}{dx^2} + \frac{d^2\phi(x,y)}{dy^2} = \frac{-qN_d}{\varepsilon_a} \text{ for } 0 \le x \le L \tag{7.1}$$

where q is the electronic charge, ε_a is permittivity, $\phi(x,y)$ is the electric potential and N_d Doping concentration.

The 2-D Poisson equation is solved using the parabolic approximation method. The parabolic approach is applied to estimate the potential distribution $\phi(x,y)$ over the 2-D space (along the device length and device depth) and the solution for the potential is given as,

$$\phi(x,y) = \phi_l(x) + C_1(x)(y-d) + C_2(x)(y-d)^2 \tag{7.2}$$

where $\phi_l(x)$ is the channel potential, $d = d_d + d_i$ is the thickness of AlInSb layer with thickness of doped AlInSb layer (d_d) and thickness of spacer layer (d_i). $C_1(x)$, $C_2(x)$ are arbitrary coefficients. The Poisson equation is solved using the following boundary conditions,

1. The channel potential at $y = 0$ is

$$\phi(x,0) = V_g = V_{gs} - V_{fb} \tag{7.3}$$

2. The electric field at $y = d$ is

$$\left. \frac{d\phi(x,y)}{dy} \right|_{y=d} = -E_{int} \tag{7.4}$$

3. The potential at the source end is

$$\phi(0,d) = V_{bi} = \phi_l(0) \tag{7.5}$$

4. The potential at the drain end is

$$\phi(L,d) = V_{bi} + V_{ds} = \phi_l(L) \tag{7.6}$$

where V_{gs} is the gate-source voltage, V_{fb} is the flat-band voltage, V_g is the gate voltage, V_{bi} is the built in voltage, E_{int} is the electric field at the interface, L is the channel length and d is the thickness of AlInSb layer.

The constants in Equation (7.2) can be found from boundary condition (7.3) to (7.6) and on substituting their value in Equation (7.2), we get

$$\phi(x,y) = \phi_l(x) - E_{int}(y-d) + \left[\frac{V_g - \phi_l(x)}{d^2} - \frac{E_{int}}{d}\right](y-d)^2 \tag{7.7}$$

The channel potential $\phi_l(x)$ is obtained by substituting (7.7) in (7.1), we get

$$\frac{d^2\phi_l(x)}{dx^2} + V_g - \frac{\phi_l(x)}{\lambda^2} = \frac{-qN_d}{\varepsilon_a} + \frac{2E_{int}}{d} \tag{7.8}$$

where λ is the characteristic length given by $\lambda = \sqrt{\dfrac{d^2}{2}}$.

Introducing new variable $\eta(x)$ in Equation (7.8)

$$\eta(x) = \phi_l(x) - V_g - \left[\frac{qN_d}{\varepsilon_a} - \frac{2E_{int}}{d}\right]\lambda^2 \tag{7.9}$$

Substitute $\phi_l(x)$ in Equation (7.8) we get,

$$\frac{d^2\eta(x)}{dx^2} - \frac{\eta(x)}{\lambda^2} = 0 \tag{7.10}$$

Using (7.9) $\eta(x=0) = \eta_1$, $\eta(x=L) = \eta_2$ and substitute boundary condition (7.5), (7.4) and $d = d_d + d_i$

$$\eta_1 = V_{bi} - V_g - \left[\frac{qN_d}{\varepsilon_a} - \frac{2E_{int}}{d}\right]\lambda^2 \tag{7.11}$$

$$\eta_2 = \eta_1 + V_{ds} \tag{7.12}$$

On solving Equation (7.10) for $\eta(x)$ and substituting in (7.9), we obtain the channel potential $\phi_l(x)$ as

$$\phi_l(x) = \frac{\eta_1 \sinh\left(\dfrac{L-x}{\lambda}\right) + \eta_2 \sinh\left(\dfrac{x}{\lambda}\right)}{\sinh\dfrac{L}{\lambda}} + V_g + \left[\frac{qN_d}{\varepsilon_a} - \frac{2E_{int}}{d}\right]\lambda^2 \tag{7.13}$$

where L is the channel length, λ is the characteristic length, N_d is the doping concentration, d is the thickness and E_{int} is the electric field at the interface.

The electric field pattern along the channel determines the electron transport velocity through the channel. The electric field component along the channel in the x direction is given by,

$$E(x) = \frac{d\phi(x,y)}{dx}\bigg|_{y=d} = \frac{d\phi_l(x)}{dx} \tag{7.14}$$

The surface potential along the x direction is an important factor. On differentiating Equation (7.13), the channel potential $\phi_l(x)$ with respect to x direction, we get,

$$E(x) = \frac{1}{\sinh\dfrac{L}{\lambda}}\left(\frac{1}{\lambda}\eta_2 \cosh\left(\frac{x}{\lambda}\right) - \frac{1}{\lambda}\cosh\left(\frac{L-x}{\lambda}\right)\right) \tag{7.15}$$

Figure 7.2 depicts the variation of channel potential with the normalized channel position of single gate HEMT with different V_{ds} values. As the drain voltage increases, the channel potential at the drain end increases. The potential drop in source side channel beneath the metal gate, the potential begins to drop more in channel under metal and drain junction region, then reduction in DIBL. To verify the accuracy of the results, the results are compared with the TCAD Sentaurus device simulator. The symbol represents the simulated model and solid line represents analytical model results.

Figure 7.3 shows the variation of channel potential with the normalized channel position of single gate HEMT with different V_{gs} values. Also the figure shows the model and simulated data under different bias conditions. As the more negative voltage increases, the channel potential decreases are compared with the MOSFET results (Sze 2004).

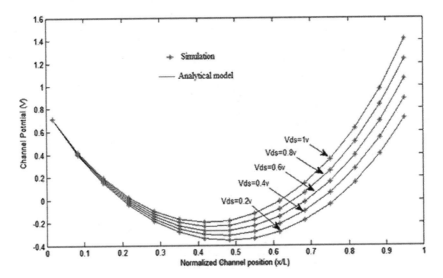

FIGURE 7.2
Variation of channel potential versus normalized channel position with different V_{ds} and $N_d = 1 * 10^{23}$ m^{-3} and $L = 60$ nm.

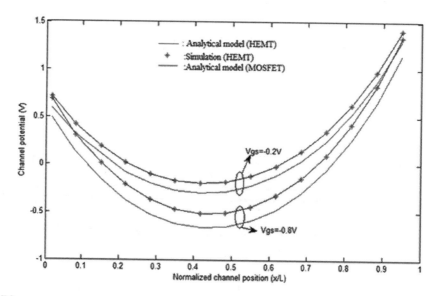

FIGURE 7.3

Variation of channel potential verses normalized channel position with different V_{gs} and $N_d = 1 * 10^{23}$ m^{-3} and $L = 60$ nm. Our results are compared with the MOSFET results.

The HEMT can reduce the DIBL effect than MOSFET. The model predictions are very good approximation for all regions of the device operation (for weak [low V_{gs}] as well as strong [high V_{gs}] inversion). The analytical results agree very well with simulation data's under different values of V_{gs}.

Figure 7.4 shows the variation of channel potential with the normalized channel position of single gate HEMT with different T values. It shows the analytical and simulated model of channel potential has exactly matched. The channel potential decreases as the temperature increases. At the room temperature, i.e., 77 K, it has good potential then others. It is useful for cryogenic application. To verify the accuracy of the results, the results are compared with the TCAD Sentaurus device simulator. The symbol represents the simulated model and solid line represents analytical model results.

Figure 7.5 depicts the variation of electric field with the normalized channel position of single gate HEMT with different V_{ds} values. As the drain voltage increases, the electric field at the drain end decreases. The high electric field at the source end has low velocity and at the drain end has maximum drift velocity. It has increased the speed of the device. This is due to the impact of InSb material that is used in the proposed device. Indium antimonide has the highest electron mobility and saturation velocity of any known semiconductor, and is therefore a promising material for ultra-high-speed devices. AlInSb/InSb high-electron mobility transistor (HEMT) devices are considered to be very promising candidates for high-speed and high-power applications.

These devices offer an advantage such as high breakdown voltage, high charge density and good electron mobility. Figure 7.6 shows the variation of electric field with the normalized channel position of single gate HEMT with different V_{gs} values. As the gate voltage increases, the electric field at the drain end decreases with respect to the average velocity

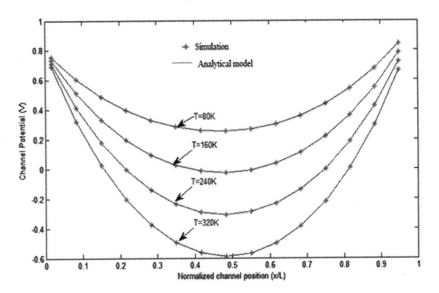

FIGURE 7.4
Variation of channel potential verses normalized channel position with different temperatureand $N_d = 1 * 10^{23}$ m^{-3} and $L = 60$ nm.

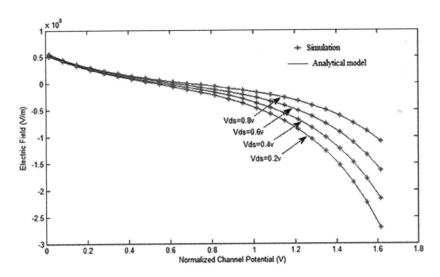

FIGURE 7.5
Variation of electric field verses normalized channel position with different V_{ds} and $N_d = 1 * 10^{23}$ m^{-3} and $L = 60$ nm.

and it is clearly evident that the short-channel effects have been reduced reasonably. The symbol represents the simulated model and solid line represents analytical model results. It clearly shows that calculated values of analytical model matches with the simulated model very well.

Figure 7.7 shows the variation of electric field with the normalized channel position of single gate HEMT with different temperature values. The impact of InSb compound

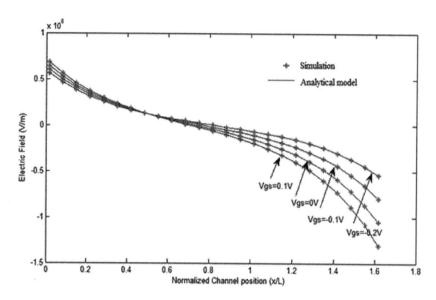

FIGURE 7.6
Variation of channel potential verses normalized channel position with different V_{gs} and $N_d = 1 * 10^{23}$ m^{-3} and $L = 60$ nm.

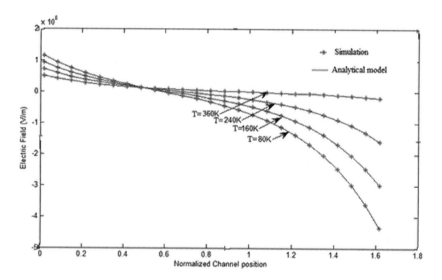

FIGURE 7.7
Variation of electric field verses normalized channel position with different T and $N_d = 1 * 10^{23}$ m^{-3} and $L = 60$ nm.

semiconductor material has been observed from the graph that the electric field in the channel becomes more uniform.

InSb material has the lightest effective mass and narrowest band gap along with extremely high electron mobility and saturation velocity in the channel region in quantum well heterostructure. Hence it has improved the carrier transport efficiency of the device and speed of the device. To verify the accuracy of the results, the results are compared with the TCAD Sentaurus device simulator.

7.4 Analytical Modeling Threshold Voltage and Subthreshold Current for AlInSb/InSb HEMT

Drain-induced barrier lowering (DIBL) is the most prevalent short-channel effect. The concept of DIBL effect is described by the channel potential where the sub-threshold leakage current often occurs at the minimum channel potential. The device threshold voltage plays a major role in the control of DIBL effect. The threshold voltage is the minimum potential required to turn on the device. Here gives a measure of drain-induced barrier lowering. Hence the position of minimum channel potential is evaluated by setting,

$$\frac{d\phi_l(x)}{dx} = 0 \tag{7.16}$$

And the minimum channel potential is given by,

$$\phi_l(x_{min}) = \frac{\eta_1 \sinh\left(\dfrac{L - x_{min}}{\lambda}\right)}{\sinh\left(\dfrac{L}{\lambda}\right)} + V_g + \left[\frac{qN_d}{\varepsilon_a} - \frac{2E_{int}}{d}\right]\lambda^2 \tag{7.17}$$

where q is the electronic charge, is permittivity and is the position of minimum channel potential.

In short-channel devices, the sub-threshold drain current, which occurs at the position of minimum channel potential increases rapidly, and hence it is vital to determine the magnitude of this current. The drain current in the sub threshold regime is function of V_{gs} as well as V_{ds} (Sze 2004) and is given by after making proper substitutions,

$$I_{ds} = \frac{W\mu\varepsilon_a}{\sqrt{2}L}\left(\frac{KTn_i}{N_d q}\right)^2 \sqrt{\left(\frac{N_d q^2}{KT\varepsilon_a}\right)}\left[1 - \exp\left(\frac{-qV_{ds}}{KT}\right)\right]\exp\left(\frac{q\phi_{l\,min}}{KT}\right)\sqrt{\frac{\phi_{l\,min}}{KT}} \tag{7.18}$$

where $\phi_{l\,min}$ is the minimum channel potential, μ is the mobility of the carriers, which has been extracted from device simulation. The value extracted is 30,000 cm^2/V-s and W is the width of the channel which has been taken to be 20 μm in our model.

Figure 7.8 shows the normalized channel position verses threshold voltage for various values of thickness (d), i.e., $d = 20$ nm, $d = 25$ nm, and $d = 30$ nm. As normalized channel position increases, the threshold voltage also increases. The model correctly predicts the threshold voltage roll-off at shorter channel lengths and validated through the simulations for $d = 20$ nm, $d = 25$ nm, and $d = 30$ nm. In the figure the symbol represents the simulated model and solid line represents analytical model results. To verify the accuracy of the results, the results are compared with the TCAD Sentaurus device simulator. The result of the analytical model matches well with the simulation results.

Figure 7.9 shows the drain current verses drain to source voltage with various values of V_{gs} i.e., $V_{gs} = -2$ V, $V_{gs} = -1$ V, $V_{gs} = 0$ V, $V_{gs} = 1$ V. In saturation region, as V_{gs} increases the drain current also increases. The saturation region predicts the current more accurate than other regions as the current is almost constant in this region. It can provide good continuity of a charge control models and increases the electron velocity in the channel. This is due to the fact that thelength of the charge control region has reduced while keeping the total gate length a constant.

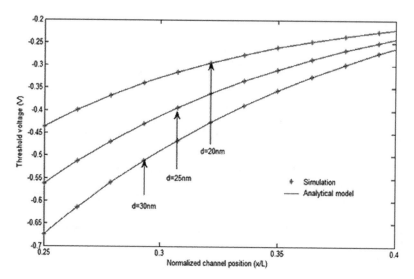

FIGURE 7.8
Variation of threshold voltage verses normalized channel position with thickness $d = 20$ nm, $d = 25$ nm, $d = 30$ nm.

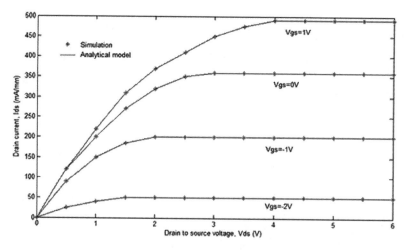

FIGURE 7.9
Variation of drain current verses drain to source voltage with various values of V_{gs}, like $V_{gs} = -2$ V, $V_{gs} = 1$ V, $V_{gs} = 0$ V, $V_{gs} = 1$ V.

7.5 AlInSb/InSb HEMT Analytical Modeling of Charge Density and Fermilevel

7.5.1 Two-Dimensional Electron Gas

A two-dimensional electron gas (2DEG) is a sheet of electrons free to move in two dimensions, but tightly confined in the third. This tight confinement leads to quantized energy levels for motion in that direction, which can then be ignored for most problems. Thus the electrons appear to be a 2D sheet embedded in a 3D world. The analogous construct of holes is called a two-dimensional hole gas (2DHG), and such systems have many useful and interesting properties. Most 2DEGs are found in transistor-like structures made from semiconductors. The most commonly encountered 2DEG is the layer of electrons found in MOSFETs. When the transistor is in inversion mode, the electrons underneath the gate oxide are confined to the semiconductor-oxide interface, and thus occupy well defined energy levels. Nearly always, only the lowest level is occupied, and so the motion of the electrons perpendicular to the interface can be ignored. However, the electron is free to move parallel to the interface, and so is quasi two-dimensional.

7.5.2 Device Structure for AlInSb/InSb HEMT

The cross-sectional view of AlInSb/InSb HEMT device for two different experimentally determined position of the 2DEG channel in the device structure as shown in Figure 7.10. The layer sequence, from bottom to top, is InSb-undoped, UID-AlInSb, n doped-AlInSb, gate metal, with 2DEG channel formed in the interface between InSb-undoped and UID-AlInSb.

A doped AlInSb barrier layer saves the carriers from Coulomb scattering caused by ionized impurities and provides carriers a channel at a zero gate voltage. The device thus works in a depletion mode with carriers confined in the InSb Quantum well. The device structure is also known as a modulation-doped field-effective transistor (MODFET) as suggested by Lee et al. (1983). The quantum well formed of ternary compound with mole fraction of $Al_{0.3}In_{0.7}Sb/InSb$ HEMT is used. It has been demonstrated that when the mole fraction of Al increases to 30%, the electron wave function is perfectly confined within the well.

7.5.3 Development of 2DEG Charge Density Model

The difficulty in modeling the 2DEG charge density arises from the complicated variation of E_f with n_s in the quantum well (Kola et al. 1988). This relationship is given by

$$n_s = Dv_{th}\left\{\ln\left[\exp\left(\frac{E_f - E_0}{V_{th}}\right) + 1\right] + \ln\left[\exp\left(\frac{E_f - E_1}{V_{th}}\right) + 1\right]\right\} \tag{7.19}$$

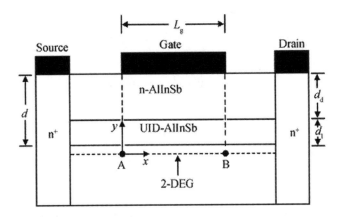

FIGURE 7.10
Cross-sectional view of AlInSb/InSb HEMTs with gate length L_g, d_i spacer layer thickness and d_d n-AlInSb layer thickness.

where $E_0 = \gamma_0 n_s^{2/3}$ and $E_1 = \gamma_1 n_s^{2/3}$ are the levels of the two lowest sub bands, E_f is the Fermi level expressed in volts, and $V_{th} = kT / q$ is the thermal voltage.

$$n_s = Dv_{th}\left\{\ln\left[\exp\left(\frac{E_f - E_0}{V_{th}}\right) + 1\right]\right\} \tag{7.20}$$

Assuming that the AlInSb layer is completely ionized, we can write

$$n_s = \frac{\varepsilon}{qd\left(v_{g0} - E_f\right)} \tag{7.21}$$

where $v_{g0} = v_g - v_{off}$, and v_{off} are the cutoff voltage. The variation of n_s with respect to v_g is a complicated transcendental function. Hence, we propose few assumptions in order to simplify these equations, which allow us to develop a precise analytical expression for n_s versus v_g for the typical operating range of gate voltages and temperatures and drain current of the AlInSb/InSbHEMT device.

It is interesting to note that the relative positions of the sub band levels when compared with E_f in this range. The charge density increases linearly with the energy bands. Upper-level E_1 is larger than E_f for the complete range of gate voltages. Hence contribution to the charge density from the second sub band can be safely ignored. Here E_0 have two distinct regions namely Region I and Region II. In Region I, E_0 is larger than E_f corresponding to $V_{g0} < 0.9$ V. In Region II where E_0 is lower than E_f corresponding to $V_{g0} > 0.9$ V. We consider the full-scale range of gate voltages except very close to cutoff. The Fermi level equation is derived as follows for the Regions I and II.

The second term can be neglected in Equation (7.20) for the calculation of Fermi's level (Kola et al. 1988). This is written in a simplified form as,

$$\frac{n_s}{Dv_{th}} = e^{\frac{E_f - E_0}{v_{th}}} \tag{7.22}$$

Taking ln on both sides

$$\ln\left[\frac{n_s}{Dv_{th}}\right] = \frac{E_f - E_0}{v_{th}}\ln e \tag{7.23}$$

By solving we get the equation for Fermi's level as follows,

$$E_f = \gamma_0 n_s^{2/3} + v_{th}\ln\left[\frac{n_s}{Dv_{th}}\right] \tag{7.24}$$

7.5.4 Charge Density Model in Region I

By considering the full-scale range of gate voltages, except very close to cutoff, E_f will be much smaller than V_{g0}. The model for n_s is separately developed in regions I and II. These models are then combined to give a unified model covering both regions.

The Fermi level in Region I is given by

$$E_f^I = \gamma_0 n_s^{2/3} + v_{th}\ln\left[\frac{n_s}{Dv_{th}}\right] \tag{7.25}$$

It is important to point out here that, for vertically scaled devices with $d < 5$ nm, the 2DEG distance from the interface, i.e., δ_d, will become comparable with d. In such cases, d should be replaced by $d_{eff} = d + \delta_d$ in the model expressions.

7.5.5 Fermi Level and Sheet Carrier Density for Region I

The Fermi level formulation for Region I is obtained by substituting the charge density equation n_s to the general Fermi level equation as follows,

$$\gamma_0\left[\frac{\varepsilon}{qd}\left(v_{g0} - E_f\right)\right]^{2/3} + v_{th}\ln\left[\frac{\varepsilon}{Dv_{th}qd}\left(v_{g0} - E_f\right)\right] = E_f^I \tag{7.26}$$

Also substituting for $C_g = \varepsilon/d$, and $\beta = C_g/(qDv_{th})$ and solving we get Fermi's level in Region 1 as,

$$E_f^I = v_{g0}\frac{\gamma_0\left(\dfrac{c_g v_{g0}}{q}\right)^{\frac{2}{3}} + v_{th}\ln\left(\beta v_{g0}\right)}{v_{g0} + v_{th} + \dfrac{2}{3}\gamma_0\left(\dfrac{c_g v_{g0}}{q}\right)^{\frac{2}{3}}} \tag{7.27}$$

Using Fermi level in region I we can derive the sheet carrier density equation in region I. Assuming that the AlInSb layer is completely ionized, we can write,

$$n_s = \frac{\varepsilon}{qd}\left(v_{g0} - E_f\right) \tag{7.28}$$

Substituting Fermi level in region I in the above equation

$$n_s = \frac{\varepsilon}{qd}\left[v_{g0} - v_{g0}\frac{\gamma_0\left(\dfrac{c_g v_{g0}}{q}\right)^{\frac{2}{3}} + v_{th}\ln(\beta v_{g0})}{v_{g0} + v_{th} + \dfrac{2}{3}\gamma_0\left(c_g v_{g0}/q\right)^{2/3}}\right] \tag{7.29}$$

$$n_s = \frac{\varepsilon v_{g0}}{qd} \left[1 - \frac{\gamma_0 \left(\dfrac{c_g v_{g0}}{q} \right)^{\frac{2}{3}} + v_{th} \ln(\beta v_{g0})}{v_{g0} + v_{th} + \dfrac{2}{3} \gamma_0 \left(\dfrac{c_g v_{g0}}{q} \right)^{\frac{2}{3}}} \right] \tag{7.30}$$

$$n_s = \frac{c_{gv_{g0}}}{q} \left[1 - \frac{\gamma_0 \left(\dfrac{c_g v_{g0}}{q} \right)^{\frac{2}{3}} + v_{th} \ln(\beta v_{g0})}{v_{g0} + v_{th} + \dfrac{2}{3} \gamma_0 \left(c_{g v_{g0}} / q \right)^{2/3}} \right] \tag{7.31}$$

We develop an analytical model for n_s versus gate voltage in Region I, the sheet carrier density in Region I is given by,

$$n_s^{I} = \frac{c_{gv_{g0}}}{q} \left[\frac{\left[v_{g0} + v_{th}(1 - \ln(\beta v_{g0})) - \dfrac{\gamma_0}{3} \left(\dfrac{c_{gv_{g0}}}{q} \right)^{\frac{2}{3}} \right]}{v_{g0} + v_{th} + 2\dfrac{\gamma_0}{3} \left(\dfrac{c_{gv_{g0}}}{q} \right)^{\frac{2}{3}}} \right] \tag{7.32}$$

7.5.6 Charge Density Model in Region II

The relationship approximate expression for n_s^{II} and it is given by,

$$n_s^{II} \approx D \left(E_f^{II} - E_0 \right) \tag{7.33}$$

$$\frac{n_s^{II}}{D} = E_f^{II} - E_0 \tag{7.34}$$

$$\frac{n_s^{II}}{D} + E_0 = E_f^{II} \tag{7.35}$$

Since AlInSb layer is completely ionized, we can substitute n_s in the above equation.

7.5.7 Fermi Level and Sheet Carrier Density for Region II

The Fermi level formulation for Region II is obtained by substituting the charge density equation to the general Fermi level.

$$\frac{1}{D} \left[\frac{\varepsilon}{qd} \left(v_{g0} - E_f \right) \right] + \gamma_0 n_s^{2/3} = E_f^{II} \tag{7.36}$$

$$\frac{c_g}{qD}\left(v_{g0}-E_f\right)+\gamma_0\left[\frac{\varepsilon}{qd}\left(v_{g0}-E_f\right)\right]^{\frac{2}{3}}=E_f^{\mathrm{II}} \tag{7.37}$$

By solving the above equation, we get the Fermi level in region II.

$$E_f^{\mathrm{II}}=\frac{v_{g0}\beta v_{g0}v_{th}+\dfrac{\gamma_0}{3}\left(\dfrac{c_{gv_{g0}}}{q}\right)^{\frac{2}{3}}}{v_{g0}\left(1+\beta v_{th}\right)+\dfrac{2}{3}\gamma_0\left(\dfrac{c_{gv_{g0}}}{q}\right)^{\frac{2}{3}}} \tag{7.38}$$

SinceAlInSb layer is completely ionized, we can use n_s equation for substitution of Fermi level in Region II. Substituting Fermi's level E_f of region II

$$n_s=\frac{\varepsilon}{qd}\left(v_{g0}-\frac{v_{g0}\beta v_{g0}v_{th}+\dfrac{\gamma_0}{3}\left(\dfrac{c_{gv_{g0}}}{q}\right)^{\frac{2}{3}}}{v_{g0}\left(1+\beta v_{th}\right)+\dfrac{2}{3}\gamma_0\left(\dfrac{c_{gv_{g0}}}{q}\right)^{\frac{2}{3}}}\right) \tag{7.39}$$

$$n_s=\frac{\varepsilon v_{g0}}{qd}\left(1-\frac{\beta v_{g0\,v_{th}}+\dfrac{\gamma_0}{3}\left(\dfrac{c_{gv_{g0}}}{q}\right)^{\frac{2}{3}}}{v_{g0}\left(1+\beta v_{th}\right)+\dfrac{2}{3}\gamma_0\left(\dfrac{c_gv_{g0}}{q}\right)^{\frac{2}{3}}}\right) \tag{7.40}$$

The sheet carrier density in region II is given by,

$$n_s^{\mathrm{II}}=\frac{c_{gv_{g0}}}{q}\frac{\left[v_{g0}-\dfrac{\gamma_0}{3}\left(\dfrac{c_{gv_{g0}}}{q}\right)^{\frac{2}{3}}\right]}{v_{g0}\left(1+\beta v_{th}\right)+\dfrac{2}{3}\gamma_0\left(\dfrac{c_gv_{g0}}{q}\right)^{\frac{2}{3}}} \tag{7.41}$$

7.5.8 Unified Charge Density Model

In order to find a unifying expression for E_f and n_s applicable for the full-scale range of gate voltages, we observe that the respective expressions for Regions I and II differ only in one term to the numerator and one term to a denominator, both with a factor of V_{th}. Hence, both the regional models are expected to provide a fairly reasonable approximation for full scale within the voltage range, except near cutoff. To improve accuracy, we propose to use a unified form that combines the two regional models. Specifically, the expression for n_s is given by,

$$n_s = \frac{c_g v_{g0}}{q} \frac{v_{g0} + v_{th}\left(1 - \ln\left(\beta v_{g0}\right)\right) - \frac{\gamma_0}{3}\left(\frac{c_g v_{g0}}{q}\right)^{\frac{2}{3}}}{v_{g0}\left(1 + \frac{v_{th}}{v_{god}}\right) + \frac{2}{3}\gamma_0\left(\frac{c_g v_{g0}}{q}\right)^{\frac{2}{3}}} \qquad (7.42)$$

Here, V_{gon} and V_{god} are functions of V_g given by the interpolation expression.

$$V_{gox} = \frac{V_{g0}\alpha_x}{\sqrt{V_{g0}^2 + \alpha_x^2}} \qquad (7.43)$$

where $\alpha_n = \frac{e}{\beta}$ and $\alpha_d = \frac{1}{\beta}$, we observe that unified model has the proper limitation for charge density of Region I and Region II. A similar unified expression can be also easily established for Fermi level by applying unified charge density in the fundamental charge density equation. Figure 7.11 plot shows the Fermi level of first sub band and second sub band with the gate voltage variation. The charge density increases linearly with the energy bands. Upper-level E_1 is larger than E_f for the complete range of gate voltages. Hence contribution to the charge density from the second sub band can be safely ignored.

Here E_0 have two distinct regions namely Region I and Region II. In Region I, E_0 is larger than E_f corresponding to $V_{g0} < 0.9$ V. In Region II where E_0 is lower than E_f corresponding to $V_{g0} > 0.9$ V. We consider the full-scale range of gate voltages except very close to cutoff.

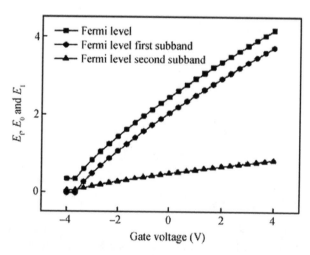

FIGURE 7.11
Numerical calculations of E_f, E_0, and E_1 versus gate voltage at $T = 298$ K.

Figure 7.12 plot shows the relative positions of Fermi levels with the first sub band and second sub band with the gate voltage. The Fermi level of first sub band will not increase with charge density, but the conduction pattern of the triangular quantum well of second sub band will cause the charge density to increase linearly with applied gate voltage. Figures 7.13 and 7.14 plot shows the comparison of charge density model for Region I with thermal voltage for different values of gate voltage.

From this plot, we can say that, first sub band energy level increases with respect to gate voltage, which causes the energy level less than 0.9 V in Region I and greater than 0.9 V in Region II. The observations are shown in Figures 7.13 and 7.14 such that the charge density for electron decreases while accounting for the thermal voltage and increases by leaving to the thermal voltage respectively. The model results are validated with experimental data.

Figure 7.15 plot shows the comparison of a unified model for charge density of AlInSb/InSb with the gate voltage. It is applicable for the full-scale range of gate voltage, we observe that the respective expressions for Regions I and II differ only in one term to the

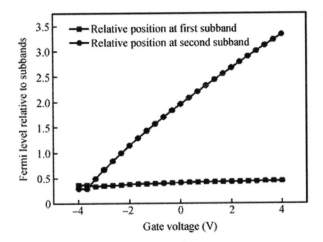

FIGURE 7.12
Numerical calculations of $E_f - E_0$ and $E_f - E_1$ versus gate voltage at $T = 298$ K.

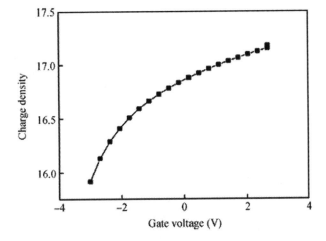

FIGURE 7.13
Charge density for Region I with thermal voltage versus gate voltage.

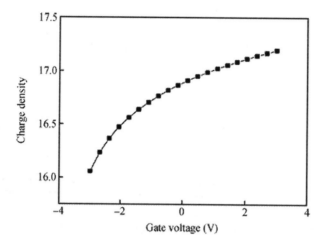

FIGURE 7.14
Charge density for Region I without thermal voltage versus gate voltage.

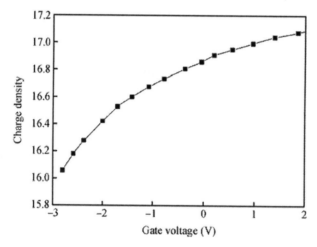

FIGURE 7.15
Unified model for charge density versus gate voltage.

numerator and one term to a denominator both with a factor of thermal voltage. Hence both the regional models are expected to provide a fairly reasonable approximation for full-scale voltage range, except near cut off. For achieving accuracy, we relate the unified two regional models expression. Here the function of gate voltage is to conduct electron for on and off states. Electrons in Region I perform quite well with respected to Region II, electrons in Region II models performs relatively poorly in Region I, with a rapidly increasing in full-scale gate voltage.

Figure 7.16 plot shows the comparison of charge density with different temperature versus the gate voltage ($T = 200$ K, 300 K, 400 K, 500 K). From the graph, we conclude that the InSb operates at high power at elevated temperatures. In this Figure 7.16, the unified model for charge density is compared with experiment data for different temperatures up to 500 K. It is apparent from the results that the model agrees quite well with the experimental data throughout the temperature range.

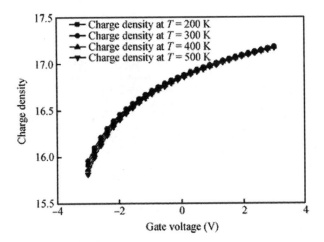

FIGURE 7.16
Charge density at various temperatures (T = 200 K, 300 K, 400 K, 500 K) versus gate voltage.

The increase in error is found to be almost linear with increasing temperature for bias point gate voltage is 0.2 V.

7.6 AlInSb/InSb HEMT Scale Length Modeling with Effective Conducting Path Effect

Figure 7.17 represents the cross-sectional view of single gate AlInSb/InSb HEMT with L = 60 nm as the length of the gate metal M. The T-shape structure of the gate as suggested by Du et al. (2011) helps to minimize the gate resistance. The highly doped capacitive layer helps in minimizing the contact resistance of the source and drain contacts. The δ-doping layer is the donor of charge carriers, which provides electrons to the channel.

As electrons occupy the state which is of lowest energy, they drain into the potential well (Khandelwal et al. 2011) and form the confined 2DEG in the channel. The work functions of gate metal M is $\phi_M = 5.3$ eV. The flat band voltage of the metal gate is given by $\phi_{FB} = \phi_M - \phi_S$.

7.6.1 Analytical Model

The 2-D Poisson equation for single gate HEMT, as stated by Colinge (2007), is given in Equation (7.44) as

$$\frac{d^2\phi(x,y)}{dx^2} + \frac{d^2\phi(x,y)}{dy^2} = \frac{-qN_d}{\varepsilon_a} \; for \;\; 0 \leq x \leq L; 0 \leq y \leq d \tag{7.44}$$

where q is the electronic charge, ε_a is permittivity, $\phi(x,y)$ is the electric potential, and N_d is the Doping concentration of doped AlInSb layer.

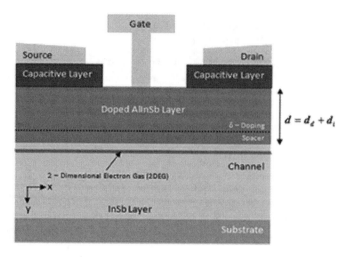

FIGURE 7.17
Cross-sectional view of single gate AlInSb/InSb HEMT.

The 2-D Poisson equation is solved using the parabolic approximation method. The parabolic approach is applied to estimate the potential distribution $\phi(x,y)$ over the 2-D space (along the device length and device depth) and the solution for the potential (Sze 2004) is given in Equation (7.45) as,

$$\varphi(x,y) = \varphi_l(x) + C_1(x)(y-d) + C_2(x)(y-d)^2 \qquad (7.45)$$

where $\phi_l(x)$ is the channel potential and $d = d_i + d_d$ thickness of AlInSb layer which is the combined depth of doped AlInSb layer and spacer layer from the surface. $C_1(x)$ and $C_2(x)$ are the arbitrary coefficients. The boundary conditions required to solve the Poisson Equation (7.44) are,

1. The channel potential at $y = 0$ is,

$$\phi(x,0) = V_g \qquad (7.46)$$

   where $V_g = V_{gs} - V_{fb}$, V_{gs} is the gate-source voltage and V_{fb} is the flat band voltage.
2. The electric field at $y = d$ is

$$\left.\frac{d\phi(x,y)}{dy}\right|_{y=d} = -E_{int} \qquad (7.47)$$

   where

$$E_{int} = \frac{\varepsilon_{ox}}{\varepsilon_{si}} \frac{(\phi_l(x) - V_g)}{t_{ox}} \qquad (7.48)$$

3. The potential at the source end is

$$\Phi(0,d) = V_{bi} = \phi_l(0) \tag{7.49}$$

4. The potential at the drain end is

$$\phi(L,d) = V_{bi} + V_{ds} = \phi_l(L) \tag{7.50}$$

where V_{ds} drain to source voltage, V_{bi} built in voltage and E_{int} electric field in interface. The arbitrary coefficients in Equation (7.45) are obtained from boundary condition (7.47) to (7.50) as follows,

$$C_1(x) = -E_{int} \tag{7.51}$$

$$C_2(x) = \frac{V_g - \phi_l(x)}{d^2} - \frac{E_{int}}{d} \tag{7.52}$$

On substituting the above coefficients in Equation (7.45), we get,

$$\phi(x, y) = \phi_l(x) - E_{int}(y-d) + \left[\frac{V_g - \phi_l(x)}{d^2} - \frac{E_{int}}{d} \right](y-d)^2 \tag{7.53}$$

Considering ECPE in equation the most leakage path $y = d_{eff}$ is at the position between surface $(y = d)$ and channel center of $(y = 0)$. The potential in the effective conducting path can be,

$$\phi_{d_{eff}}(x) = \phi_l(x) - E_{int}(d_{eff} - d) + \left[\frac{V_g - \phi_l(x)}{d^2} - \frac{E_{int}}{d} \right](d_{eff} - d)^2 \tag{7.54}$$

On rearranging Equation (7.54), we get

$$\phi_{d_{eff}}(x) + E_{int}k + \frac{E_{int}k^2}{d} = \phi_l(x) + \left(\frac{V_g - \phi_l(x)}{d^2} \right)k^2 \tag{7.55}$$

$$\phi_{d_{eff}}(x) + E_{int}k + E_{int}\frac{k^2}{d} - V_g\frac{k^2}{d} = \phi_l(x)\left(1 - \frac{k^2}{d^2} \right) \tag{7.56}$$

where $k = d_{eff} - d$ and the remaining constant terms are grouped and equated as,

$$A = kE_{int} + \frac{k^2}{d}E_{int} - V_g\frac{k^2}{d^2} \tag{7.57}$$

From Equation (2.56) the channel potential is obtained as,

$$\phi_I(x) = \frac{\phi_{d_{eff}}(x) + A}{1 + \dfrac{\varepsilon_{ox}}{\left(1 - \dfrac{k^2}{d^2}\right)}} \tag{7.58}$$

Substituting Equation (2.58) in Equation (2.53), we obtain

$$\phi(x,y) = \frac{\phi_{d_{eff}}(x) + A}{1 - \dfrac{k^2}{d^2}} + \left(\frac{v_g(1 - \dfrac{k^2}{d^2}) - \phi_{d_{eff}}(x) - A}{d^2\left(1 - \dfrac{k^2}{d^2}\right)} - \frac{E_{int}}{d}\right)(y - d)^2 - E_{int}(y - d) \tag{7.59}$$

A simple scaling equation for single gate HEMT is obtained by substituting Equation (7.59) into Equation (7.46) and setting $y = d_{eff}$. On differentiating $\phi(x,y)$ with respect to x and y we get,

$$2\left\{\frac{V_g\left(1 - \dfrac{k^2}{d^2}\right) - \phi_{d_{eff}}(x) - A}{d^2 - k^2} - \frac{E_{int}}{d}\right\} + \frac{d^2\phi_{d_{eff}}(x)}{dx^2}\left[\frac{1}{1 - \dfrac{k^2}{d^2}} - \frac{(y-d)^2}{d^2\left(1 - \dfrac{k^2}{d^2}\right)}\right] = \frac{-qN_d}{\varepsilon_a} \tag{7.60}$$

$$\frac{d^2\phi_{d_{eff}}(x)}{dx^2}\left(\frac{2yd - y^2}{2d^2}\right) - \phi_{d_{eff}}(x) = (d^2 - k^2)\left[\frac{E_{int}}{d^2} + \frac{V_g}{d^2} - \frac{qN_d}{\varepsilon_a}\right] + A \tag{7.61}$$

On rearranging, we obtain the final scaling equation as,

$$\frac{d^2\phi_{d_{eff}}(x)}{dx^2} - \frac{\phi_{d_{eff}(x)}}{\dfrac{2d_{eff}d - d_{eff}^2}{2d^2}} = \frac{A + B(d^2 - k^2)}{\dfrac{2d_{eff}d - d_{eff}^2}{2d^2}} \tag{7.62}$$

where

$$B = \left(\frac{E_{int} + V_g}{d^2}\right) - \left(\frac{qN_d}{\varepsilon_a}\right)$$

The common denominator part of the scaling equation $\frac{2d_{eff}d - d_{eff}^2}{2d^2}$ is considered as λ^2, where λ is the natural length as suggested by Chiang (2005) and is given as,

$$\lambda = \sqrt{\frac{2d_{eff}d - d_{eff}^2}{2d^2}} \tag{7.63}$$

In Equation (7.63), the ECPE-related natural length scale is derived. It is clearly observed that λ is dependent on the depth of the effective conducting path of d_{eff}, which corresponds to the depth of the doped layer and this parameter can be used to adjust the scaling design for different doping levels. This generalized scaling equation which is on a base of ECPE can be utilized further in the derivation of generalized scaling factor.

7.6.2 Derivation of Subthreshold Conduction

Once $\phi_{d_{eff}}(x)$ is determined, the potential of $\phi(x,y)$ can be obtained by using Equation (7.54). The natural length λ is introduced to describe the potential distribution. The scaling Equation (7.63) is a simple second-order 1D differential equation, and can be uniquely solved by Equations (7.44) and (7.45). The potential in the effective conducting path is obtained as,

$$\phi_{d_{eff}}(x) = \frac{1}{\text{Sinh}\left(\frac{L_{eff}}{\lambda}\right)}\left[\left(V_{bi}+V_{ds}+D\right)\text{Sinh}\left(\frac{x}{\lambda}\right)+\left(V_{bi}+D\right)\text{Sinh}\left(\frac{L_{eff}-x}{\lambda}\right)\right]-D \qquad (7.64)$$

where

$$D = A + B(d^2 - k^2) \qquad (7.65)$$

The minimum potential region in the effective conducting path is the point from which the HEMT device is in ON state. The subthreshold leakage current that occurs due to the high mobility in the channel has to be minimized in order to obtain a low subthreshold swing (Balamurugan et al. 2008). This leakage current depends primarily on the minimum channel potential in the effective conducting path. The minimum of the potential will occur at x_{min} by setting

$$\frac{d\phi_{eff}(x)}{dx} = 0 \qquad (7.66)$$

Solving Equation (7.66) by using Equation (7.45) gives us,

$$x_{min} = \frac{1}{2}\lambda\left[\frac{L - M\exp\left(\frac{L_{eff}}{\lambda}\right)}{M\exp\left(-\frac{L_{eff}}{\lambda}\right) - L}\right] \qquad (7.67)$$

where

$$L = V_{bi} + V_{ds} + D \qquad (7.68)$$

$$M = V_{bi} + D \qquad (7.69)$$

we define the minimum potential as,

$$\phi_{d_{eff}, min} \approx \sqrt{LM} \exp\left(\frac{-L_{eff}}{2\lambda}\right) - D \tag{7.70}$$

and the minimum channel position is obtained as,

$$x_{min} \approx \frac{L_{eff}}{2} + \frac{\lambda}{2} \ln\left(\frac{M}{L}\right) \tag{7.71}$$

Here if we set $V_{ds} \ll 1$ volt then $L = M$ in Equation (7.71), which ignores the drain-induced barrier lowering (DIBL) effects. Since punch through current at subthreshold region depends on the potential difference between the minimum potential in the effective conducting path and that at the source end, hence from the exponential term of Equation (7.72), the scaling factor with ECPE can be expressed as,

$$\alpha = \frac{L_{eff}}{2\lambda} \tag{7.72}$$

Scaling factor (α) is a constant which is taken out as a common factor from the Equation (7.72), this parameter is directly related to the effective gate length, for a minimum effective gate length (L_{eff}) the lowest scaling factor is achieved. Any increase in the scaling factor corresponds to the decrease in natural length which thereby reduces the drain-induced barrier lowering (DIBL) effect considerably. Once α is determined, the thickness of AlInSb layer can be defined as,

$$d = \frac{\varepsilon_{ox}(\phi_l(x) - V_g)}{\varepsilon_{si} E_{int}} \tag{7.73}$$

where V_g is the gate to source voltage, $V_g = V_{gs} - V_{fb}$ and V_{fb} is the flat band voltage, $V_{fb} = \phi_M + X + \frac{E_g}{2}$, ϕ_M is the material work function = 5.3 eV, X is the electron affinity and E_g is the energy of band gap.

Figure 7.18 shows the variation of minimum gate length with thickness of AlInSb layer for various depth of effective conducting path. Natural length λ is dependent on d_{eff} and this parameter can be used to adjust the scaling design for different doping levels. It is also clear from the figure that as the thickness of AlInSb layer increases, the effective gate length also increases. By varying d_{eff} over a range of values for the highly doped AlInSb layer with $N_d = 10^{23} m^{-3}$, the minimum effective gate length is achieved for $d_{eff} = 0.9*d$, which corresponds to a higher doping level.

Figure 7.19 depicts the dependence of subthreshold swing on scaling factor. It is clearly observed that as the natural length λ decreases, ideal subthreshold swing of 70 mV/dec is obtained and hence leakage current is also minimized. This ideal subthreshold swing, however, makes the scaling factor to go beyond the value of 4 and effectively suppresses the short-channel effects (SCEs). Hence the proposed model can be prominently used in SPICE simulation. The analytical results of single gate AlInSb/InSb HEMT is compared with TCAD simulation results, and a close match is observed with analytical and simulation results.

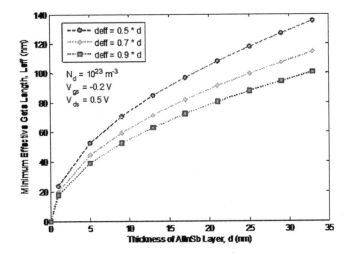

FIGURE 7.18
Variation of the minimum gate length with thickness of AlInSb layer for various depth of effective conducting path.

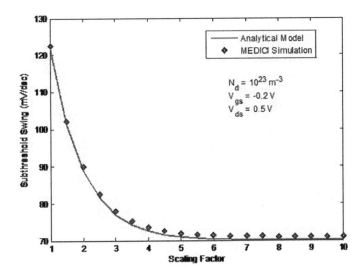

FIGURE 7.19
Dependence of subthreshold swing on scaling factor and the analytical results are compared with simulation results.

Figure 7.20 shows the dependence of subthreshold swing on effective gate length. The above figure also demonstrates the influence of ECPE on subthreshold swing. With $N_d = 10^{23} \text{m}^{-3}$ the calculated analytical results of the proposed work match with those simulated by TCAD. The results are also compared with Yan model, which deviate from the simulation results because of the ignorance of ECPE and unrealistic assumption of surface conduction mode in MOSFET. But single gate AlInSb/InSb HEMT with ECPE provides an ideal subthreshold swing.

Figure 7.21 shows the channel length verses threshold voltage roll-off for various values of AlInSb layer thickness (d), i.e., $d = 20$ nm, $d = 25$ nm, and $d = 30$ nm. As channel length increases, the threshold voltage also increases. For $d = 30$ nm, the threshold voltage roll-off

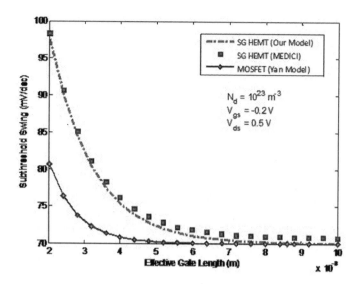

FIGURE 7.20
Dependence of subthreshold swing on effective gate length and the analytical results are compared with simulation results and MOSFET results.

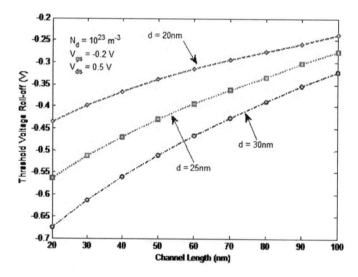

FIGURE 7.21
Variation of threshold voltage roll-off with channel length for various values of thickness of AlInSb layer.

is minimum. It is obvious that due to small threshold voltage roll-off, this proposed work can be used in designing the single gate HEMT for small-device technology.

7.7 Gate Engineered AlInSb/InSb HEMT

AlInSb/InSb high electron mobility transistors (HEMTs) has vital role in high-current, high-voltage, high-power and high-frequency operations. The materials used in this HEMT device possess the advantage of high-current drivability, thermal stability and high-breakdown fields. Indium Antimonide is a dark grey material with zincblende crystal structure has the advantage of lowest band gap 0.1 eV at room temperature and 0.23 eV at 80 K and highest lattice constant 0.64 nm than any other quantum well formed of III and V group material system. The largest ambient-temperature electron mobility of undopedInSb quantum well is found to be approximately as high as 78,000 cm²/V-s than any other material. Hence quantum well formed by sandwiching layers of AlInSb/InSb help to construct fastest transistors. It is a trend in the compound semiconductor industry to continuously develop devices, which are extremely small, fast and consumes less power.

7.7.1 Device Structure of AlInSb/InSb TMG HEMT

In the proposed TMG HEMT structure, three materials of different work functions are used. Figure 7.22 shows the schematic view of the TMG AlInSb/InSb HEMT with three gate materials of different work functions Φ_{M1}, Φ_{M2}, and Φ_{M3}. The gate materials are $\Phi_{M1} = 4.8\,\text{eV}$ (Au), $\Phi_{M2} = 4.4\,\text{eV}$ (Ti) and $\Phi_{M3} = 4.1\,\text{eV}$ (Al), which are chosen in such a way that $\varphi_{M1} > \varphi_{M2} > \varphi_{M3}$. As a result, the threshold voltage will be in the order of $V_{th1} > V_{th2} > V_{th3}$. The gate region M_1 near the source side is the first screen gate region, the gate region M_2 in the middle is the control gate region and the gate region M_3 near the drain side is the second screen gate region.

Basically, the structure consists of a semi-insulating GaAs substrate, on which is first grown a undoped InSb layer; then an delta doped AlInSb layer followed by the spacer, then a doped AlInSb layer and the capacitive layer. The capacitive layer is highly doped in order to minimize the contact resistance of the source and drain contacts. The transistor is finally realized by depositing three different gate materials to form a Schottky barrier and serve as the gate, and by providing two ohmic contacts to serve as the source and the drain. The transistors may be operated in two modes, normally on mode and normally off mode. This is determined by the thickness of the AlInSb layers. When these layers are thick enough, charge is supplied by the layer to fill up the surface states at the interface between it and the

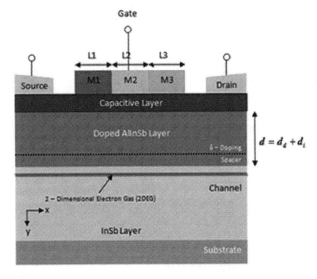

FIGURE 7.22
Schematic view of TMG AlInSb/InSb HEMT, L_1, L_2, and L_3 are the lengths of gate materials M_1, M_2, and M_3.

gate metal and also to the InSb layer for the alignment of the Fermi levels. The transistor is normally on. If, on the other hand, the barrier layer is made thin, then the charge available in the layer is not enough to cause alignment of the Fermi levels and the InSb layer is required to supply additional charge. The layer being depleted, the transistor is normally off. Both these transistors are required for different functional circuits.

In high-electron-mobility-transistors (HEMTs) the heterojunction between two semiconducting materials confine electrons to a triangular quantum well and form two-dimensional electron gas (2DEG). The two-dimensional electron gas is nothing but electrons which are free to move in two dimensions, but tightly confined in the third dimension. Thus the electrons appear to be a two dimensional sheet of charge. The mobility of electrons confined to the heterojunction of HEMTs is higher than that of MOSFETs. Here the junction between the n-doped AlInSb and undopedInSb layer is the source of 2DEG. The model which includes the polarization effect would be valuable but it is beyond scope of this work. However, the model in its current form would be useful for predicting the performance enhancement of the DMG HEMT and the device structure optimization as the currents in the sub-threshold regime are small.

7.7.2 Analytical Modeling of Channel Potential for AlInSb/InSb TMG HEMT

The proposed model analyses the channel potential, electric field distribution and threshold voltage which are obtained by solving the 2-D Poisson equation using parabolic approximation method. The solution to Poisson equation can be obtained by different methods such as superposition method, Fourier series expansion, Green function, Newton Raphson method, parabolic approximation method. Since the other methods are slightly difficult and complex, we use the parabolic approximation method in order to make the model simple and accurate. The channel region is divided into three parts, since three materials have been used in the gate region. The source and the drain regions are uniformly doped. The 2-D potential distribution (x, y) can be obtained by solving the 2-D Poisson's equation which is rewritten from the book authored by Jean-Pierre Colinge [14], and is given as

$$\frac{d^2\phi(x,y)}{dx^2} + \frac{d^2\phi(x,y)}{dy^2} = -\frac{qN_d}{\varepsilon_a} \quad \text{for} \quad 0 \le x \le L \tag{7.74}$$

where N_d is the doping density of the AlInSb layer, q is the electric charge, ε_a is the dielectric constant and $L = L_1 + L_2 + L_3$ is the effective channel length. The 2-D Poisson equation is solved using the parabolic approximation method. The parabolic approach is applied to estimate the potential distribution $\phi(x, y)$ over the 2-D space (along the device length and device depth) and the solution for the potential (Sze 2004) is given as

$$\phi(x,y) = \phi_l(x) + C_1(x)(y-d) + C_2(x)(y-d)^2 \tag{7.75}$$

ϕ_l is the channel potential, $d = d_d + d_i$ is the thickness of the AlInSb layer with d_s as the thickness of the doped AlInSb and d_i as the thickness of the spacer layer. $C_1(x)$ and $C_2(x)$ are arbitrary coefficients. Equation (7.75) must satisfy the following boundary conditions in order to get the solution for Poisson's equation.

The flat band voltages of the three metal gates are different and are given by

$$V_{FB1} = \phi_{M1} - \phi_S \tag{7.76}$$

$$V_{FB2} = \phi_{M2} - \phi_S \tag{7.77}$$

$$V_{FB3} = \phi_{M3} - \phi_S \tag{7.78}$$

where

$$\phi_S = \chi_a + \frac{E_g}{2} - \phi_F \tag{7.79}$$

χ_a is the electron affinity, E_g is the band gap at $T = 300$ K and ϕ_F is the Fermi potential given by, $\phi_F = \frac{kT}{q} . \ln\left(\frac{N_d}{n_i}\right)$, n_i being the intrinsic carrier concentration.

In the TMG HEMT structure, the gate has been divided into three parts; hence the potential under the different gate region M_1, M_2, and M_3 is given as follows. The potential under the gate region M_1 near the source side (first screen gate region) is

$$\phi_1(x,y) = \phi_{l1}(x) + C_{11}(x)(y-d) + C_{21}(x)(y-d)^2 \quad \text{for } 0 \le x \le L_1 \tag{7.80}$$

The potential under the gate region M_2 in the middle (control gate region) is

$$\phi_2(x,y) = \phi_{l2}(x) + C_{12}(x)(y-d) + C_{22}(x)(y-d)^2 \quad \text{for } L_1 \le x \le L_1 + L_2 \tag{7.81}$$

The potential under the gate region M_3 near the drain side (second screen gate region) is

$$\phi_3(x,y) = \phi_{l3}(x) + C_{13}(x)(y-d) + C_{23}(x)(y-d)^2 \quad \text{for } L_1 + L_2 \le x \le L_1 + L_2 + L_3 \tag{7.82}$$

The Poisson's equation is solved separately under the three gate regions (M_1, M_2 and M_3) using the following boundary conditions:

1. Channel potential at the interface of the three dissimilar metals is continuous

$$\phi_1(L_1, d) = \phi_2(L_1, d) \quad \text{at } x = L_1 \tag{7.83}$$

$$\phi_2(L_1 + L_2, d) = \phi_3(L_1 + L_2, d) \tag{7.84}$$

2. Electric field at the interface of the three dissimilar metals is continuous

$$\left. \frac{d\phi_1(x,y)}{dx} = \frac{d\phi_2(x,y)}{dx} \right|_{x = L_1} \tag{7.85}$$

$$\left. \frac{d\phi_2(x,y)}{dx} = \frac{d\phi_3(x,y)}{dx} \right|_{x = L_1 + L_2} \tag{7.86}$$

3.
$$\phi_1(x,0) = V_{g1} \tag{7.87}$$

$$\phi_2(x,0) = V_{g2} \tag{7.88}$$

$$\phi_3(x,0) = V_{g3} \tag{7.89}$$

where $V_{g1} = V_{gs} - V_{FB1}$, $V_{g2} = V_{gs} - V_{FB2}$, $V_{g3} = V_{gs} - V_{FB3}$ and V_{gs} is the gate to source voltage.

$$\left. \frac{d\phi_1(x,y)}{dy} \right|_{y=d} = -E_{int1} \tag{7.90}$$

$$\left. \frac{d\phi_2(x,y)}{dy} \right|_{y=d} = -E_{int2} \tag{7.91}$$

$$\left. \frac{d\phi_3(x,y)}{dy} \right|_{y=d} = -E_{int3} \tag{7.92}$$

where E_{int1}, E_{int2}, E_{int3} are the fields at the hetero interface under M_1, M_2, and M_3 respectively.

4. The potential at the source end is

$$\phi_1(0,d) = V_{bi} \tag{7.93}$$

where V_{bi} is the built-in voltage.

5. The potential at the drain end is

$$\phi_2(L_1 + L_2 + L_3, d) = V_{bi} + V_{ds} = \phi_{12}(L_1 + L_2 + L_3) \tag{7.94}$$

where V_{ds} is the drain to source voltage of the device.

The constants in (7.78), (7.79) and (7.80) can be found from boundary conditions (7.81) and (7.82) and on substituting their values in (7.78), (7.79) and (7.80), we get

$$\phi_j(x,y) = \phi_{lj}(x) - E_{int}(y-d) + \left[\frac{V_{gj} - \phi_{lj}(x)}{d^2} - \frac{E_{intj}}{d} \right](y-d)^2 \tag{7.95}$$

where $j = 1,2,3$ for regions under M_1, M_2 and M_3 respectively.

The channel potential $\phi_{lj}(x)$ is obtained by substituting (7.93) in Equation (7.74). On substitution, we get

$$\frac{d^2\phi_{lj}(x,y)}{dx^2} + \frac{V_{gj} - \phi_{lj}(x)}{\lambda^2} = -\frac{qN_d}{\varepsilon_a} + \frac{2E_{intj}}{d} \tag{7.96}$$

where λ is the characteristic length and it is given by

$$\lambda = \sqrt{\frac{d^2}{2}} \qquad (7.97)$$

Introducing a new variable

$$\eta_j(x) = \phi_{lj}(x) - V_{gj} - \left[\frac{qN_d}{\varepsilon_a} - \frac{2E_{intj}}{d}\right]\lambda^2 \qquad (7.98)$$

and substituting (7.96) in (7.94), we get

$$\frac{d^2\phi(x,y)}{dx^2} + \frac{\eta_j(x)}{\lambda^2} = 0 \qquad (7.99)$$

Using (3.23), $\eta_j(x=0) = \eta_{11}$, $\eta_j(x = L_1 + L_2) = \eta_{22}$, $\eta_j(x = L_1 + L_2 + L_3) = \eta_{33}$ and the boundary conditions (7.80) and (7.81) we get

$$\eta_{11} = V_{bi} - V_{g1} - \left[\frac{qN_d}{\varepsilon_a} - \frac{2E_{int1}}{d}\right]\lambda^2 \qquad (7.100)$$

$$\eta_{22} = (\eta_{11} + V_{ds}) - (V_{fb1} - V_{fb2}) - (b_2 - b_1)^2 \qquad (7.101)$$

$$\eta_{33} = (\eta_{11} + V_{ds}) - (V_{fb1} + V_{fb2} - V_{fb3}) - (b_2 + b_1 - b_3)^2 \qquad (7.102)$$

where $b_1 = \frac{qN_d}{\varepsilon_a} - \frac{2E_{int1}}{d}$, $b_2 = \frac{qN_d}{\varepsilon_a} - \frac{2E_{int2}}{d}$ and $b_3 = \frac{qN_d}{\varepsilon_a} - \frac{2E_{int3}}{d}$

Solving (7.97) for $\eta_{j(x)}$ and substituting in Equation (7.96), we obtain the channel potential $\phi_{lj}(x)$ as

$$\phi_{l1}(x) = \frac{\eta_{11}\sinh\left[\dfrac{L_1 - x}{\lambda}\right] + \eta_{12}\sinh\left[\dfrac{x}{\lambda}\right]}{\sinh\left[\dfrac{L_1}{\lambda}\right]} + V_{g1} + \left[\frac{qN_d}{\varepsilon_a} - \frac{2E_{int1}}{d}\right]\lambda^2 \qquad (7.103)$$

$$\phi_{l2}(x) = \frac{\eta_{22}\sinh\left[\dfrac{x - L_2}{\lambda}\right] + \eta_{21}\sinh\left[\dfrac{L_1 + L_2 - x}{\lambda}\right] + \eta_{23}\sinh\left[\dfrac{L_1 + L_2 + L_3 - x}{\lambda}\right]}{\sinh\left[\dfrac{L_2}{\lambda}\right]} + V_{g2} + b_2\lambda^2 \qquad (7.104)$$

$$\phi_{l3}(x) = \frac{\eta_{33}\sinh\left[\dfrac{x - L_3}{\lambda}\right] + \eta_{32}\sinh\left[\dfrac{L_1 + L_2 + L_3 - x}{\lambda}\right]}{\sinh\left[\dfrac{L_3}{\lambda}\right]} + V_{g3} + b_3\lambda^2 \qquad (7.105)$$

Let

$$\alpha_i = \frac{L_i}{\lambda}, \quad \text{where} \quad i=1,2,3 \tag{7.106}$$

Using boundary conditions (7.81) to (7.83) and Equation (7.98) to (7.104), we obtain the values of $\eta_{12}, \eta_{21}, \eta_{32}, \eta_{23}, \eta_{31}, \eta_{13}$ as

$$\eta_{12} = \frac{\eta_{11}\sinh\alpha_2 + \eta_{22}\sinh\alpha_1 + (V_{g2}-V_{g1})\sinh\alpha_1\cosh\alpha_2 + (b_2-b_1)^2\sinh\alpha_1\cosh\alpha_2}{\cosh\alpha_1\sinh\alpha_2 + \cosh\alpha_2\sinh\alpha_1} \tag{7.107}$$

$$\eta_{21} = \frac{\eta_{22}\sinh\alpha_1 + \eta_{11}\sinh\alpha_2 + (V_{g1}-V_{g2})\sinh\alpha_2\cosh\alpha_1 + (b_1-b_2)^2\sinh\alpha_2\cosh\alpha_1}{\cosh\alpha_2\sinh\alpha_1 + \cosh\alpha_1\sinh\alpha_2} \tag{7.108}$$

$$\eta_{32} = \frac{\eta_{33}\sinh\alpha_2 + \eta_{22}\sinh\alpha_3 + (V_{g2}-V_{g3})\sinh\alpha_3\cosh\alpha_2 + (b_2-b_3)^2\sinh\alpha_3\cosh\alpha_2}{\cosh\alpha_3\sinh\alpha_2 + \cosh\alpha_2\sinh\alpha_3} \tag{7.109}$$

$$\eta_{23} = \frac{\eta_{22}\sinh\alpha_3 + \eta_{11}\sinh\alpha_2 + (V_{g3}-V_{g2})\sinh\alpha_2\cosh\alpha_3 + (b_3-b_2)^2\sinh\alpha_2\cosh\alpha_3}{\cosh\alpha_2\sinh\alpha_3 + \cosh\alpha_3\sinh\alpha_2} \tag{7.110}$$

$$\eta_{31} = \frac{\eta_{33}\sinh\alpha_1 + \eta_{11}\sinh\alpha_3 + (V_{g1}-V_{g3})\sinh\alpha_3\cosh\alpha_1 + (b_1-b_3)^2\sinh\alpha_3\cosh\alpha_1}{\cosh\alpha_3\sinh\alpha_1 + \cosh\alpha_1\sinh\alpha_3} \tag{7.111}$$

$$\eta_{13} = \frac{\eta_{11}\sinh\alpha_3 + \eta_{33}\sinh\alpha_1 + (V_{g3}-V_{g1})\sinh\alpha_1\cosh\alpha_3 + (b_3-b_1)^2\sinh\alpha_1\cosh\alpha_3}{\cosh\alpha_1\sinh\alpha_3 + \cosh\alpha_3\sinh\alpha_1} \tag{7.112}$$

7.7.3 Analytical Modeling Threshold Voltage and Subthreshold Current for AlInSb/InSb TMG HEMT

The surface electric field component along the channel in the x-direction is an important parameter as the electron transport velocity through the channel is directly related to the electric field along the channel, whose x-component under M_1 is given by

$$E_1(x) = \frac{d}{dx}\Phi_1(x,y)\bigg|_{y=d} = \frac{d\phi_{l1}(x)}{dx} \tag{7.113}$$

$$E_1(x) = \frac{\eta_{11}\cosh\left[\dfrac{L_1-x}{\lambda}\right]\left[-\dfrac{1}{\lambda}\right] + \eta_{12}\cosh\left[\dfrac{x}{\lambda}\right]\left[\dfrac{1}{\lambda}\right]}{\sinh\alpha_1} \tag{7.114}$$

and x-component under M_2 is

$$E_2(x) = \frac{d}{dx}\Phi_2(x,y)\bigg|_{y=d} = \frac{d\phi_{l2}(x)}{dx} \tag{7.115}$$

$$E_2(x) = \frac{\eta_{22}\cosh\left[\dfrac{x-L_2}{\lambda}\right]\left[\dfrac{1}{\lambda}\right] + \eta_{21}\cosh\left[\dfrac{L_1+L_2-x}{\lambda}\right]\left[-\dfrac{1}{\lambda}\right] + \eta_{23}\cosh\left[\dfrac{L_1+L_2+L_3-x}{\lambda}\right]\left[-\dfrac{1}{\lambda}\right]}{\sinh\alpha_2}$$

$$\tag{7.116}$$

and x-component under M_3 is

$$E_3(x) = \frac{d}{dx} \Phi_3(x,y) \bigg|_{y=d} = \frac{d\phi_{l3}(x)}{dx} \tag{7.117}$$

$$E_3(x) = \frac{\eta_{33} \cosh\left[\dfrac{x - L_3}{\lambda}\right]\left[\dfrac{1}{\lambda}\right] + \eta_{32} \sinh\left[\dfrac{L_1 + L_2 + L_3 - x}{\lambda}\right]\left[-\dfrac{1}{\lambda}\right]}{\sinh \alpha_3} \tag{7.118}$$

7.7.3.1 Threshold Voltage Model

The most common short-channel effect is drain-induced barrier lowering (DIBL). The sub-threshold leakage current often occurs at the point of minimum channel potential $\phi_{l\,min}$. It gives a measure of the DIBL. In TMG structures, because of the different work functions of M_1, M_2 and M_3, the minimum channel potential is determined by the metal gate with the higher work function. Hence, the position of the minimum channel potential x_{min} lies under M_1 and is given by

$$\frac{d\phi_{l1}(x)}{dx} = 0 \tag{7.119}$$

and the minimum channel potential is

$$\Phi_{l1}(x_{min}) = \phi_{l\,min} = \frac{\eta_{11} \sinh\left[\dfrac{L_1 - x}{\lambda}\right] + \eta_{12} \sinh\left[\dfrac{x}{\lambda}\right]}{\sinh\left[\dfrac{L_1}{\lambda}\right]} + V_{g1} + \left[\frac{qN_d}{\varepsilon_a} - \frac{2E_{int1}}{d}\right]\lambda^2 \tag{7.120}$$

The threshold voltage is taken to be that value of gate to source voltage, Vgs, at which the minimum channel potential $\phi_{l\,min}$ equals $2\phi_F$. Hence, using $\phi_{l\,min} = 2\phi_F$ and $V_{gs} = V_{th}$ in Equation (7.118), we get

$$V_{th} = 2\phi_F + \frac{\eta_{11} \sinh\left[\dfrac{L_1 - x_{min}}{\lambda}\right] + \eta_{12} \sinh\left[\dfrac{x_{min}}{\lambda}\right]}{\sinh\left[\dfrac{L_1}{\lambda}\right]} + V_{FB1} + \left[\frac{qN_d}{\varepsilon_a} - \frac{2E_{int1}}{d}\right]\lambda^2 \tag{7.121}$$

7.7.3.2 Sub-threshold Drain Current Model

In short-channel devices, the sub-threshold drain current, which occurs at the position of minimum channel potential increases rapidly, and hence, it is vital to determine the magnitude of this current. The drain current in the sub-threshold regime is a function of V_{gs} as well as V_{ds} and is given by

$$I_{ds} = \frac{W\mu\varepsilon_a}{\sqrt{2}(L_1 + L_2)}\left(\frac{kTn_i}{N_d q}\right)^2 \sqrt{\left(\frac{N_d q^2}{kT\varepsilon_a}\right)}\left[1 - \exp\left(\frac{-qV_{ds}}{kT}\right)\right]\exp\left(\frac{q\varphi_{l\,min}}{kT}\right)\sqrt{\frac{q\phi_{l\,min}}{kT}} \tag{7.122}$$

In the sub-threshold region, the electron concentration in the channel is controlled not only by gate bias but also by the nearness of the source and drain depletion regions. As the drain voltage increases, the gate slowly loses its control over the channel. Hence, a larger swing in gate voltage is needed to obtain better control in very short-channel devices. The sub-threshold slope, S, an important parameter in the sub-threshold regime, represents the efficiency of coupling between the gate potential and the channel potential, and is defined as the gate voltage swing required to reduce the sub-threshold current, I_{ds}, by one decade and is given by

$$S = \frac{\partial V_{gs}}{\partial (\log I_{ds})} \tag{7.123}$$

It is a key parameter to design high-gain amplifiers and VLSI chips with low standby power dissipation.

Figure 7.23 shows the variation of channel potential with the normalized channel position for TMG AlInSb/InSb. It can be inferred that when compared with DMG, TMG has greater channel potential and this leads to greater ON current. Figure 7.24 depicts the plot of channel potential variation of TMG AlInSb/InSb HEMT for different V_{ds}. It is observed that due to the difference in work functions of three metal regions, the drain side is screened from the source side thereby reducing short-channel effect, it is observed that as distance from channel increases, the channel potential also increases.

Figure 7.25 shows channel potential variation with normalized channel position for different combinations of gate length L_1, L_2, and L_3 of M_1, M_2, and M_3 respectively, keeping the sum $L = L_1 + L_2 + L_3$ constant at 120 nm for TMG and DMG. In this graph, The channel potential is with respect to the normalized channel position. It can be inferred from the graph that in TMG HEMT, the electric field at the drain side is considerably low. This suppresses the short-channel effect in the device.

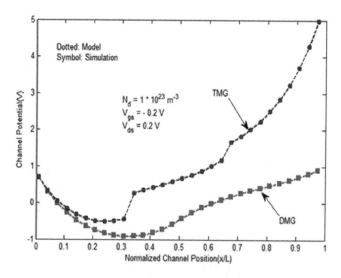

FIGURE 7.23

Variation of channel potential with the normalized channel position for TMG AlInSb/InSb HEMT.

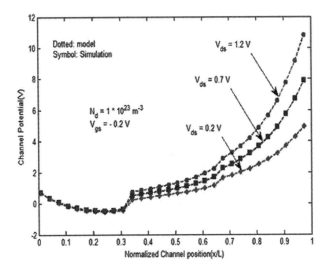

FIGURE 7.24
Variation of channel potential of TMG AlInSb/InSb HEMT for different V_{ds}.

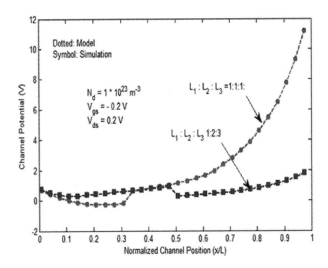

FIGURE 7.25
Variation of channel potential of TMG AlInSb/InSb HEMT for different gate material length.

Figure 7.26 shows the variations of channel potential along the channel position for different values of temperature. It is observed that increase in temperature leads to decrease in surface potential in the channel. This clearly shows that as temperature decreases to cryogenic temperature, there is a greater channel potential and leads to increased performance. Hence, HEMT can be used for cryogenic applications.

Figure 7.27 shows the plot of variation of channel potential for different values of gate to source voltages. While we plot for various V_{gs}, it is observed that for more negative V_{gs}, there is an increase in the channel potential. The TCAD simulations are compared with the analytical results and a good match between them is observed.

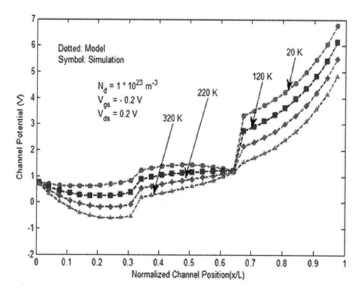

FIGURE 7.26
Variation of channel potential of TMG AlInSb/InSb HEMT for different temperature.

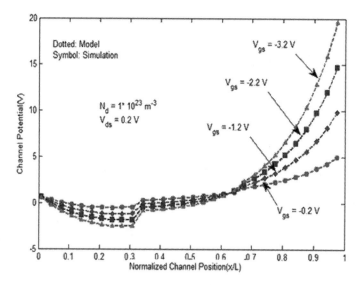

FIGURE 7.27
Variation of channel potential of TMG AlInSb/InSb HEMT for different V_{gs}.

Figure 7.28 shows the electric field variation along the channel position for TMG and DMG HEMT. It is inferred that the electric field at the drain end is considerably low for TMG HEMT than DMG HEMT. This is due to the screening effect, where excess drain voltage is absorbed by the M_3 region. Hence M_1 region is screened from the drain potential variations, which reduces the peak electric field towards the drain end. This ensures reduced short-channel effects and increased lifetime of the device.

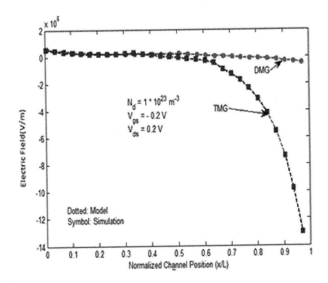

FIGURE 7.28
Variation of electric field of TMG AlInSb/InSb HEMT.

7.8 Summary

The effectiveness of the triple material gate concept to the AlInSb/InSb HEMTs has been examined for the first time by developing a 2-D analytical model. Expressions for the channel potential, electric field, and sub-threshold drain current have been obtained, proving the superiority of TMG AlInSb/InSb HEMT in suppressing the SCEs over the conventional DMG structure. The results obtained from the model agree well with the simulation results. It is apparent from the results that the TMG architecture exhibits improved gate controllability over the channel. In addition to this, the step function in the channel potential profile induces the screening of the drain potential variations, thereby suppressing the SCEs. Moreover, the enhanced electric field near the source leads to a more uniform average drift velocity of electrons in the channel, resulting in improved carrier transport efficiency. Results also emphasize that the device optimization in terms of the metal gate length ratios, work function differences, and barrier layer thicknesses, leads to improvement in the device performance. Thus, the introduction of the TMG structures over their single gate counterparts offers a new way of improving the short-channel behavior of AlInSb/InSb HEMTs.

References

Balamurugan, NB, Sankaranarayanan, K and Suguna, M (2008). A new scaling theory for the effective conducting path effect of dual material surrounding gate nanoscale MOSFETs. *Journal of Semiconductor Technology and Science*, 8(1), 92–97.

Chiang, TK (2005). A scaling theory for fully-depleted, surrounding-gate MOSFET's: Including effective conducting path effect. *Microelectronic Engineering*, 77(2), 175–183.

Colinge, JP (2007). *FinFETs and Other Multi-Gate Transistors*, Springer, New York.

Du, YD, Cao, HZ, Yan, W, Han, WH, Liu, Y, Dong, XZ, Zhang, YB et al. (2011). T-shaped gate AlGaN/GaN HEMTs fabricated by femtosecond laser lithography without ablation. *Applied Physics A*, 106(3), 575–579.

Khandelwal, S, Goyal, N and Fjedly, TA (2011). A physics-based analytical model for 2DEG charge density in AlGaN/GaN HEMT devices. *IEEE Transactions on Electron Devices*, 58(10), 3622–3625.

Kola, S, Golio, JM and Maracas, GN (1988). An analytical expression for Fermi level versus sheet carrier concentration for HEMT modeling. *IEEE Electron Device Letters*, 9(3), 136–138.

Lee, K, Shur, M, Drummond, TJ and Morkoc H (1983). Current–voltage and capacitance–voltage characteristics of modulation doped field transistors. *IEEE Transactions on Electron Devices*, 30(3), 207–212.

Sze, SM (2004). *Physics of Semiconductor Devices*, John Wiley & Sons, New York.

8

Polarization Effects in AlGaN/GaN HEMTs

Palash Das, T. R. Lenka, Satya Sopan Mahato, and A. K. Panda

CONTENTS

8.1 Polarization in AlGaN/GaN HEMT

Piezoelectric and spontaneous polarization effects are two of the most interesting parameters in GaN-based research. Locally strain generated fields in conjunction with threading dislocations can create piezoelectric-induced sheet-charge density and electric fields at a free surface or a heterostructure interface near the dislocation. An additional important structural property of group III nitrides with wurtzite structure, which strongly affects the orientation of the spontaneous and piezoelectric polarization, is the polarity of the crystals. The polarization effect improves device performance and induces in many device applications; like it increases barrier height, reduces gate leakage current, enhances current handling capability. This misfit strain as discussed in the following section causes the piezoelectric polarization (Figure 8.1).

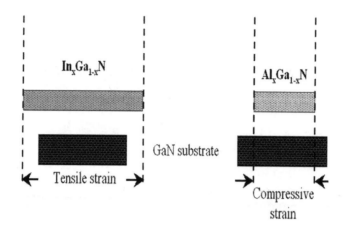

FIGURE 8.1
Lattice constant difference causes different strain: different piezoelectric effect.

8.2 Piezoelectric Polarization

8.2.1 Misfit Strain and Critical Thickness

A pseudomorphic epitaxial layer of a crystal can be grown on a substrate of slightly different lattice constants. The thickness of the layer is considered sufficiently small. The structure of a lattice mismatched layer is shown in Figure 2.3, where the lattice constant of the epilayer is bigger than that of the substrate. If the thickness h of the layer is smaller than a certain thickness, known as critical thickness h_c, the misfit between the layer and the substrate is accommodated by a tetragonal compression of the layer (Figure 8.2).

The homogeneous strain in the layer is known as misfit strain. The in-plane lattice constant of the layer becomes the same as that of the substrate. If the lattice constant of the epilayer is smaller than that of the substrate, the strain is tensile. Such homogeneously strained layers are known as pseudomorphic strained layers. When $h > h_c$, misfit strain begins to relax by the introduction of misfit dislocations as shown in Figure 2.3b. Strain has an intense effect on many properties of the epilayers [1,2]. Lattice mismatch is measured by the misfit parameter f_m which is defined as

$$f_m = \frac{a_l - a_{sub}}{a_{sub}} \tag{8.1}$$

where a_l and a_{sub} are the lattice constants of the epilayer and the substrate. If the epilayer consists of an alloy, such as $Al_xGa_{1-x}N$ grown on GaN, its lattice constants depend of the composition x. Assuming that Vegard's law is valid, the lattice constant $a_l(x)$ of the alloy epilayer is given by

$$a_l(x) = a_{GaN} + (a_{AlN} - a_{GaN})x \tag{8.2}$$

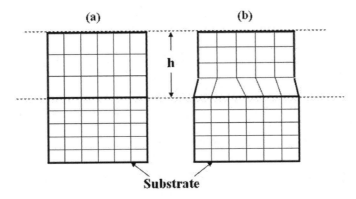

FIGURE 8.2
Structure of an epilayer under biaxial compression: (a) pseudomorphic and (b) relaxed with misfit dislocations.

The misfit parameter is also a function of the alloy composition x, i.e., $f_m = f_m(x)$. It can be calculated using the values of lattice constants. The in-plane strain components in the x and y directions are given by

$$\varepsilon_{xx} = \varepsilon_{yy} = -f_m(x) \tag{8.3}$$

The above equations can be formulated using the convention that the strains ε_{xx} and ε_{yy} are negative if the layer is under compression and positive if it is under tension. There are two approaches for deriving expression for h_c, both known as equilibrium theories of critical thickness. The first approach is based on the principle of energy minimization and was first expressed by Frank and Van der Merwe in 1949 and the second, known as the force balance theory, was given by Matthews and Blakeslee in the 1960s. Two different numerical values of h_c were obtained for the same layer calculated using the two theories. However, if we consider equal values of dislocation energy, then the values of h_c by both the methods should be identical.

A dislocation half loop nucleates at the surface and expands until it touches the interface. Its threading arms then expand and deposit misfit dislocation at the interface. The threading arm of an existing threading dislocation can also move and deposit the misfit dislocation. The force G acting on the threading arm is given by [3].

$$G = 2\mu \frac{1-v}{1+v} h \left| f_m b_1 \right| - 2E_D^{\infty} \tag{8.4}$$

where E_D^{∞} is the dislocation tension or its energy per unit length

$$E_D^{\infty} = \frac{b^2 \mu}{4\pi(1-v)} (1-v^2) \ln \frac{\rho_c h}{q} \tag{8.5}$$

where b is the Burgers vector, $b_1 = b\cos\beta$, q is the core cutoff parameter and is assumed to be equal to b, μ is the shear modulus of elasticity, n is the Poisson's ratio, and ρ_c is the core energy parameter. β is the angle between the dislocation line and its Burgers vector and has a value 60° for the 60° dislocations. Only 60° dislocation can glide at the interface. When $h < h_c$, the line tension, i.e., the force which opposes the motion of the threading dislocation is larger than the driving force due to misfit strain given by the first term and G is negative. In that case, the misfit dislocations cannot be generated, and the layer remains pseudomorphic. With increasing h, the driving force increases faster than the opposing force given by the dislocation tension. At $h = h_c$, G = 0. For $h > h_c$, G is positive and the threading dislocation moves and deposits the misfit dislocation. The thickness h_c is the maximum thickness up to which the layer can remain pseudomorphic. Equating G to 0, we obtain the expression for the critical thickness

$$h_c = \frac{b^2(1 - v\cos^2\beta)}{8\pi f_m(1+v)b_1}\ln\frac{p_c h_c}{q} \tag{8.6}$$

The theory does not take into account the energy needed for the nucleation and motion of the dislocations. The only material parameter involved in Equation (8.6) is Poisson's ratio v, which has a value approximately equal to 0.3 for most semiconductors. Therefore theoretical h_c versus f_m curve is approximately the same for most lattice mismatched epilayers. Experimental values of h_c are generally larger than those calculated using Equation (8.6). The discrepancy between the theoretical and the experimental values of h_c is believed to be due to the fact that the energy needed for the propagation of dislocations is not available at the growth temperatures. In high quality strained epilayers, the number of existing threading dislocations is very small, and nucleation of dislocation loops is also required for the generation of misfit dislocations. If there is low-defect density in strained layer, that is the layers are of high quality, nucleation is homogeneous and needs large energy [3]. This also limits the generation of misfit dislocations. There are effects of interactions between dislocations on the critical thickness and strain relaxation. For $h > h_c$, the equilibrium theory requires that misfit dislocations to be introduced to relieve the misfit strain. Two perpendicular arrays of misfit dislocations are created under the ideal equilibrium condition. The distribution of dislocations in each array is periodic. In the presence of the dislocations, the strain is given by

$$\varepsilon_{xx} = \varepsilon_{yy} = -\left(f_m(x) - \frac{b_1}{p}\right) \tag{8.7}$$

where p is the distance between two neighboring dislocations. In practice, the distribution is not periodic and an average value of p is used. The relation between the thickness h and strain ε_{xx} is obtained by replacing f_m by the strain ε_{xx} in Equation (8.6).

$$f_m(x) - \frac{b_1}{p} = \frac{b^2(1 - v\cos^2\beta)}{8\pi h(1+v)b_1}\ln\frac{p_c h_c}{q} \tag{8.8}$$

The value of p and therefore the number of misfit dislocations in a layer of thickness h can be calculated using Equation (8.8). The number of dislocations observed in a thick layer (i.e., a layer with $h > h_c$) is generally smaller than predicted by Equation (8.8). In many cases of mismatch between the epilayer and the substrate is very large, e.g., GaAs grown on Si and GaN grown on sapphire substrates. In such cases the layers are grown by using the so-called two-step method. If the mismatch is ≥4% and layer thickness is not too small, the strain is almost completely relaxed by generation of misfit dislocations during the growth and/or annealing of the layer at the higher temperatures. The layer can be regarded as "lattice matched" to the substrate.

On cooling the layers to room temperature, stresses and strains are generated due to the difference in the values of coefficients of thermal expansion α_{th}. The misfit parameter due to thermal strain is given by [4]

$$f_{(m,th)} = (\alpha_{th,layer} - \alpha_{th,substrate})(T_g - RT) \tag{8.9}$$

where T_g is the growth temperature. All results of the previous section can be used for thermal strain if $f_{(m,th)}$ is used instead of $f_m(x)$. Generally, the experimental values of thermal strain do not agree with the values calculated using this equation. If the thickness of the layer is not sufficiently large or if the growth temperature is not sufficiently high, relaxation of strain at the growth temperature is not complete and the lattice mismatch strain persists on cooling the layer to lower temperatures. Also, additional misfit dislocations may be generated during the cooling process, relieving some of the thermal strain. Both factors will make the observed strain different from the strain calculated using Equation (8.9). It is difficult to take these factors into account and calculate the thermal strain more accurately.

The experimental thickness of GaN on AlN is around 2 monolayers (ML). The behavior of AlN epilayers grown on relaxed GaN is somewhat different. The change in lattice parameter due to strain relaxation was observed after 3 ML, a value similar to that for GaN grown on AlN. Akasaki and Amano [5] investigated the critical thickness of $In_xGa_{1-x}N$ and $Al_xGa_{1-x}N$ layers. In both cases the layers were grown on relaxed GaN layers. The concentration x of Al and In was in the range $0.05 < x < 0.2$. The values of critical thickness were in the range 300–700 nm for $Al_xGa_{1-x}N/GaN$ and 400 nm for $In_xGa_{1-x}N/GaN$. There was no significant change of the critical thickness with the composition x of the epilayers.

8.3 Spontaneous Polarization

8.3.1 Polarity

Noncentrosymmetric compound crystals exhibit two different sequences of atomic layering in the two opposing directions parallel to certain crystallographic axes. As a consequence, crystallographic polarity along these axes can be observed. For binary A–B compounds with wurtzite structure, the sequence of the atomic layers of the constituents A and B is reversed along the [0001] and [0001] directions. The corresponding (0001) and (0001) faces are the A-face and B-face, respectively. In the case of heteroepitaxial growth of thin films of a noncentrosymmetric compound, the polarity of the material cannot be predicted in a straightforward way and must be determined by experiment. This is the case for group III nitrides with wurtzite structure and GaN-based heterostructures with the most common growth direction normal to the {0001} basal plane, where the atoms are arranged in bilayers. These bilayers consist of two closely spaced hexagonal layers, one formed by cations and the other formed by anions, leading to polar faces. Thus, in the case

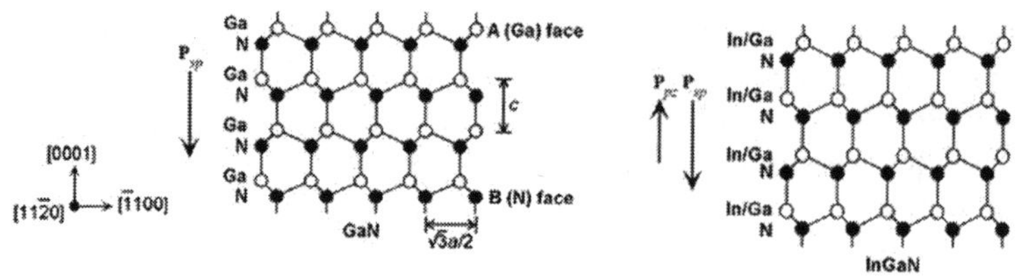

FIGURE 8.3
Piezoelectric study.

of GaN, a basal surface should be either Ga- or N-face. Ga-face means Ga atoms are placed on the top position of the {0001} bilayer, corresponding to the [0001] polarity (Figure 8.3) (by convention, the [0001] direction is given by a vector pointing from a Ga atom to a nearest-neighbour N atom).

8.3.2 Spontaneous Polarization

The investigation related to pyroelectric properties of alloys show that the spontaneous polarization of relaxed alloys due to varying cation–anion bond length for a given composition depends linearly on the average molar fraction parameter. The following calculated relation is observed for the spontaneous polarization $P_{SP}ABN$ (in cm^{-2}) of relaxed ABN alloys as a function of x. The bowing parameters can be experimentally obtained.

$$P_{SP}ABN(x) = P_{SP}ANx + P_{SP}BN (1 - x) + bx (1 - x)$$

$$P_{SP}AlGaN(x) = -0.090x - 0.034 (1 - x) + 0.021x (1 - x)$$

$$P_{SP}InGaN(x) = -0.042x - 0.034 (1 - x) + 0.037x (1 - x)$$

$$P_{SP}AlInN(x) = -0.090x - 0.042 (1 - x) + 0.070x (1 - x)$$

8.4 Polarization-Induced Effects on AlGaN/GaN HEMT

The polarization-based engineering of Heterojunction Field Effect Transistor (HFET) structure plays important roles toward the benefit of the device. For an example the incorporation of indium into the buffer layer in nitride HFET can improve the performance of the device a lot.

If we take a glance on the crystal asymmetric structure of hexagonal GaN, the piezoelectricity and polarization effects are obvious for it. Spontaneous and Piezoelectric polarization generated fields give rise to large internal electrostatic field which reduces the overlap of electron hole wave functions. In an optical device this poor overlap results in long radiative life time with low internal quantum efficiency due to competing non-radiative recombination channel at elevated temperatures. Cubic GaN having symmetric crystal structure does not have the polarization effects.

8.5 Polarization Effects on 2DEG of AlGaN/GaN HEMT

The self consistent solution of Schrodinger and Poisson's equation along with the total charge depletion approximation [7] leads to a converging solution to the 2DEG charge density. However, due to the transcendental nature [1] of the above-mentioned equations, some approximations in different cases are taken into consideration to obtain final solution. Firstly, the well-known polarization model [2] of AlGaN/GaN heterostructure is used to determine the polarization-dependent charge at the AlGaN/GaN interface. Here only the Ga face calculation has been considered. The values of polarization generated charges are the main factor for determining carrier concentration at the interface.

$$P_{PZ,Al_xGa_{1-x}N} = 2\left(\frac{a_{GaN} - a_{Al_xGa_{1-x}N}}{a_{Al_xGa_{1-x}N}}\right)\left(e_{31,Al_xGa_{1-x}N} - e_{33,Al_xGa_{1-x}N} \times \frac{C_{13,Al_xGa_{1-x}N}}{C_{33,Al_xGa_{1-x}N}}\right) \quad (8.10)$$

Here P_{PZ}, a, e_{31}, e_{33}, C_{13}, C_{33} and x are piezoelectric polarization generated charge, in-plane lattice constant, elastic constant, piezoelectric constant and aluminum (Al) molar fraction in AlGaN layer respectively and the suffix stands for the corresponding material.

The piezoelectric and elastic constants of $Al_xGa_{1-x}N$ for different Al percentages (x) are calculated with Vegard's law.

The spontaneous polarization for different Al percentages is calculated as [3]:

$$P_{SP,Al_xGa_{1-x}N} = -0.09x - 0.034(1-x) + 0.021x(1-x) \quad (8.11)$$

Here the composite AlGaN barrier over GaN is primarily considered as the HEMT structure, hence it is necessary to incorporate the composite barrier polarization effects on total polarization driven charge concentration, which is formulated as:

$$\sigma_{tot} = P_{SP,tot} - P_{SP,GaN} + P_{PZ,tot} \quad (8.12)$$

Here σ_{tot} is the total polarization driven charge, $P_{SP,tot}$ and $P_{PZ,tot}$ are the total spontaneous and piezoelectric charges which has been introduced here as follows:

$$P_{SP,tot} = \sum_i \frac{d_i}{d_{tot}} \times P_{SP,(Al_xGa_{1-x}N)i}(i = 1,2,3\ldots \text{for all the barrier layers}) \quad (8.13)$$

$$P_{PZ,tot} = \sum_i \frac{d_i}{d_{tot}} \times P_{PZ,(Al_xGa_{1-x}N)i}(i = 1,2,3\ldots \text{for all the barrier layers}) \quad (8.14)$$

where d_i is the thickness of ith barrier layer and d_{tot} is the total barrier thickness. $P_{PZ,(Al_xGa_{1-x}N)i}$ and $P_{SP,(Al_xGa_{1-x}N)i}$ represents the spontaneous and piezoelectric polarization charge due to ith AlGaN layer with Al molar fraction of x.

Under the complete charge depletion, the 2DEG carrier density is governed by the following equation [7]:

$$n_s = \frac{\varepsilon(x)}{qd_{tot}} \times (V_{GS} - V_{th} - E_F) \quad (8.15)$$

where x is the Al molar fraction in the AlGaN barrier, $\varepsilon(x)$ is the dielectric constant of $Al_xGa_{1-x}N$, d_{tot} is the AlGaN layer thickness, V_{GS} is the applied gate bias and V_{th} is the threshold voltage defined as follows [7]:

$$V_{th} = \varnothing_b(x) - \Delta E_C(x) - \frac{qN_Dd_d^2}{2\varepsilon(x)} - \frac{q\sigma(x)d_{tot}}{\varepsilon(x)} \tag{8.16}$$

where $\varnothing_b(x)$ is the Schottky barrier height, $\Delta E_C(x)$ is conduction band discontinuity between AlGaN and GaN and N_D is the doping concentration of doped AlGaN layer of thickness d_d and $\sigma(x)$ is the polarization driven charge at the AlGaN/GaN interface.

It can be observed in Equation (8.16), that there are some "unit area capacitive" terms, e.g., $\varepsilon(x)/d_{tot}$. The individual AlGaN barrier layers grown one on another has been considered as individual materials with different dielectric constants electrically connected in series. Hence the unit area capacitive terms in conventional single barrier AlGaN/GaN heterostructure has been modified to an equivalent capacitance and the series capacitance effect can be considered for all the AlGaN layers with different dielectric constants and layer widths. Hence the following modified version of Equation (8.17) is newly considered to calculate the total 2DEG carrier density:

$$n_s = \frac{C \parallel}{q}(V_{GS} - V_{th} - E_F) \tag{8.17}$$

Here $C \parallel$ and V_{th} is defined with the above-mentioned equivalent unit area capacitive terms as follows:

$$\frac{1}{C \parallel} = \sum_i \frac{d_i}{\epsilon_i} \, (i = 1,2,3 \ldots \text{for all the barrier layers}) \tag{8.18}$$

$$V_{th} = \varnothing_b - \Delta E_c - \frac{qN_dd_d}{2C \parallel_d} - \frac{q\sigma_{tot}}{C \parallel} \tag{8.19}$$

$$\frac{1}{C \parallel_d} = \sum_i \frac{d_i}{\epsilon_i} (i = 1,2,3 \ldots \text{for the doped layers only}) \tag{8.20}$$

The position of Fermi energy level (E_F) inside the triangular potential well formed at AlGaN/GaN interface is related to the 2DEG density at AlGaN/GaN interface by the following equation, which is obtained [4] by the Fermi Dirac statistics:

$$n_s = Dk_BT ln\left(\left(1 + exp\left(\frac{E_F - E_0}{k_BT}\right)\right) \times \left(1 + exp\left(\frac{E_F - E_1}{k_BT}\right)\right)\right) \tag{8.21}$$

Here, D is the two-dimensional density of states, E_0 and E_1 are the allowed energy levels and can be obtained by the following equations [5]:

$$E_i = \gamma_i n_s^{2/3} (for \ i = 1 \ and \ 2) \tag{8.22}$$

where [1]

$$\gamma_1 = 2.123 \times 10^{-12} Vm^{4/3} \ and \ \gamma_2 = 3.734 \times 10^{-12} Vm^{4/3}$$

Combining Equations (2.12) and (2.13) and separating n_S and E_F we get:

$$exp\left(\frac{n_s}{Dk_BT}\right)-1=exp\left(\frac{E_F-\gamma_0 n_S^{\frac{2}{3}}}{k_BT}\right)+exp\left(\frac{E_F-\gamma_1 n_S^{\frac{2}{3}}}{k_BT}\right)+exp\left(\frac{2E_F-\gamma_0 n_S^{\frac{2}{3}}-\gamma_1 n_S^{\frac{2}{3}}}{k_BT}\right) \quad (8.23)$$

Thus the equation can be simplified in the following form:

$$\frac{A}{B}+\frac{A}{C}+\frac{A^2}{BC}=D \quad (8.24)$$

where A, B, C, and D are defined as follows:

$$A=exp\left(\frac{E_F}{k_BT}\right),B=exp\left(\frac{\gamma_0 n_S^{\frac{2}{3}}}{k_BT}\right),C=exp\left(\frac{\gamma_1 n_S^{\frac{2}{3}}}{k_BT}\right),D=exp\left(\frac{n_S}{Dk_BT}\right)-1 \quad (8.25)$$

After solving the Equation (8.15) and relating E_F and n_S within the n_S range of 2×10^{12} cm^{-2} to 2×10^{13} cm^{-2} which are normally found in AlGaN/GaN HEMT heterostructures, the E_F versus n_S curve is plotted as shown in Figure 8.4. The fitting of the curve shows dependence of n_S on as follows:

$$E_F = 1.2101 \times 10^{-13} \times n_S^{0.7452} \text{(at 300 K)} \quad (8.26)$$

Equation (8.17) can again be simplified as the following straight-line equation with less than 2.5% error in the n_S range of 5.3×10^{12} cm^{-2} to 2×10^{13} cm^{-2} as shown in Figure 8.4.

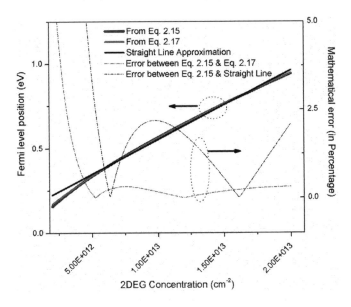

FIGURE 8.4
Fermi energy level versus 2DEG carrier concentration with mathematical error.

$$E_F = 0.1429 + 4.096 \times 10^{-18} \times n_S \, (at \, 300K) \tag{8.27}$$

Now Equation (8.8), if incorporated with Equation (8.18), can be simplified as:

$$n_S = \frac{V_{GS} - V_{th} - 0.1429}{4.096 \times 10^{-18} + \dfrac{q}{C \,\|}} \, (at \, 300K) \tag{8.28}$$

The effects of temperature on 2DEG carrier concentration on temperature has also been calculated with Equation (8.15) and the same has been fitted in the range of 2×10^{12} to 2×10^{13} cm^{-2} carrier concentration range for different temperatures. The temperature (T) dependent carrier concentration is as follows

$$n_S = \frac{V_{GS} - V_{th} - a(T)}{b(T) + \dfrac{q}{C \,\|}} \tag{8.29}$$

The temperature-dependent terms are related with the following best-fitted equations with less than 3% error within the range of 300–500 K

$$a(T) = 0.179 - 2.174 \times 10^{-4} T$$

$$b(T) = (1.182 \times 10^{-3} T + 3.954) \times 10^{-18} \tag{8.30}$$

To determine the threshold voltage V_{th} it is necessary to calculate the Schottky barrier height and it can be calculated as [6] (for Ni gate contact):

$$\varnothing_b(x) = 0.7841 + 1.8559x \tag{8.31}$$

The 2DEG has been found to be growth dependent and its value may vary [45] with the growth conditions and even with different epitaxial growth equipments [7] for the similar Al molar fraction and the AlGaN barrier thickness. The grown surface is one of the main dependable factors of different growth environments. Hence, it is hereby considered that the surface energy level pinning happens differently for different growth environments. A new parameter named Surface Factor (SF) has been introduced in this work and considered to be linearly modifying the surface pinning. Then the modified expression for threshold voltage can be written as

$$V_{th} = \varnothing_b + SF - \Delta E_c - \frac{qN_d d_d}{2C \,\|_d} - \frac{q\sigma_{tot}}{C \,\|} \tag{8.32}$$

The conduction band offset has been calculated with the help of band alignment factor between GaN and AlGaN and the bandgap of AlGaN barrier. The band alignment factor that is the ratio between ΔE_c and ΔE_G is considered as 70% and AlGaN Bandgap has been calculated [7] as:

$$E_{g, Al_xGa_{1-x}N} = 6.2x + 3.43(1-x) - 0.7x(1-x) \tag{8.33}$$

The calculations give rise the predicted values of E_F, ground state energy level (E_0) and first excited state energy level (E_1). These energy states provide the information about the carrier confinement, which has been discussed in the next parts.

8.5.1 Polarization in III-Nitrides

Polarization is a physical effect that dominates device behavior and may also determine defect density in the grown film due to the polar nature of the GaN and AlGaN. The lack of GaN substrates necessitates heteroepitaxy on compatible substrates, commonly sapphire (Al_2O_3) and silicon carbide (SiC). The epitaxial layers may be either grown entirely by MBE (Molecular Beam Epitaxy) or MOCVD (metal-organic chemical vapor deposition) technique. Heteroepitaxy on such severely lattice-mismatched substrates makes the nucleation layer one of the most critical aspects of the growth. With sapphire as a substrate, the nucleation layer consists of GaN or AlN deposited at a low temperature (typically 600°C), which is then heated up to the growth temperature of the main layer. The GaN and AlGaN layers are typically grown at 1000°C at growth rates of ~1 µm/hr. Nucleation on SiC is typically performed using AlN grown at 900°C.

Due to the epitaxial growth of $Al_xGa_{1-x}N$ or $In_xGa_{1-x}N$ layer over GaN leads to the formation of polarization charges. The polarization in GaN is described by Morkoc et al. in detail [7]. There are two types of polarizations seen in III-Nitrides.

- Spontaneous polarization
- Piezoelectric polarization

The wurtzite crystal structures of III-Nitrides are tetrahedrally coordinated and lack the inversion symmetry in the unit cell. The schematics of the crystal structure of wurtzite Ga-face and N-face GaN are shown in Figure 8.5. Currently all high-quality material is grown with Ga-face polarity. Due to the non-centro symmetric structure and large ionicity associated with the covalent metal nitrogen bond, a spontaneous polarization occurs along the hexagonal c-axis.

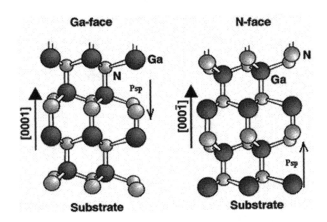

FIGURE 8.5
Schematics of the crystal structure of wurtzite Ga-face and N-face GaN. The spontaneous polarization (P_{sp}) direction is also shown (From Dash, P., Modeling, realization and characterization of compositionally Graded AlGaN/GaN heterostructures for electronic applications, PhD Dissertation, IIT, Kharagpur, 2012.)

Group III-Nitride materials have also large piezopolarization (PE) coefficients. Misfit strain and thermal strain, which are caused by the thermal expansion coefficient difference between the substrate and epitaxial layer, cause the strain-related piezoelectric effect in heterostructures. The effect of strain-induced polarization on band structures can have a significant effect when there are heterointerfaces in a structure. Heterojunction devices such as HFETs are based on the control of free carriers and band structures at heterointerface.

In AlGaN/GaN based HFETs, the polarization is due to piezoelectric effects and the difference in spontaneous polarization between AlGaN and GaN layers. The effect of strain-induced polarization is lowered when the strain is reduced by misfit dislocations [7]. Polarization has an important effect on III-Nitride electronic devices, because piezoelectric field is superimposed to the band energies. If the modulation charge is the basis of operation for the device, polarization charge and the strain management of the thin film layers can result in increasing the sheet carrier density (n_s) at the interface, which forms a two-dimensional electron gas [7]. The PE polarization is negative for tensile and positive for compressive strained layers, while the SP polarization for GaN and AlN is always negative [7]. The polarization induced sheet charge density and directions for the spontaneous (SP) and piezoelectric (PE) polarization in Ga- and N-face AlGaN/GaN heterostructures are shown in Figure 8.6.

AlGaN and GaN possess polarized wurtzite crystal structures, having dipoles across the crystal in the [0001] direction as shown in Figure 8.5. In the absence of external fields, this macroscopic polarization includes spontaneous (pyroelectric) and strain-induced (piezoelectric) contributions [7]. The primary effect of polarization is an interface charge due to abrupt divergence in the polarization at the AlGaN/GaN heterointerface. Here it is assumed that GaN bulk is fully relaxed and, therefore, its polarization vector contains only the spontaneous component, $P_{sp}(\text{GaN})$. But for AlGaN layer, in addition to the spontaneous component $P_{sp}(\text{AlGaN})$, the piezopolarization component is also present, which is generated by strain due to Al content in $Al_xGa_{1-x}N$ layer. GaN and AlN values and linear interpolation are adopted in the computation of all mole fraction dependent piezoelectric and mechanical constants of AlGaN. The strain is computed using the in-plane lattice constant (a) of GaN and AlGaN and is given by,

$$\varepsilon = \left(1 - r\right)\left(\frac{a - a_0}{a_0}\right) \tag{8.34}$$

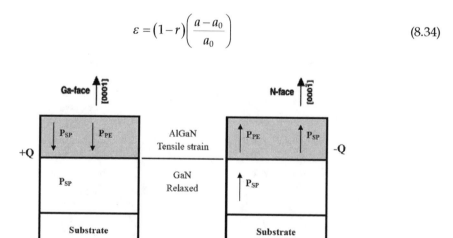

FIGURE 8.6

Polarization-induced sheet charge density and directions for the spontaneous (SP) and piezoelectric (PE) polarization in Ga- and N-face AlGaN/GaN heterostructures (From Dash, P., Modeling, realization and characterization of compositionally Graded AlGaN/GaN heterostructures for electronic applications, PhD Dissertation, IIT, Kharagpur, 2012.)

where (a) is the in-plane lattice constant and (r) is the amount of strain relaxation. The piezopolarization component of $Al_xGa_{1-x}N$ is given by,

$$P_{PE}(AlGaN) = 2\varepsilon\left[e_{31} - e_{33}\left(\frac{c_{13}}{c_{33}}\right)\right] \tag{8.35}$$

In relaxed material there exist a built-in or spontaneous polarization and this polarization is dependent on the Al mole fraction (x) with the following relation.

$$P_{SP}(Al_xGa_{1-x}N) = -0.052x - 0.29 \text{ C/m}^2 \tag{8.36}$$

and

$$P_{SP}(GaN) = -0.029 \text{ C/m}^2 \tag{8.37}$$

This polarization points toward the substrate for Ga-face material and points toward the surface in N-face material. The polarization in the material can be changed by placing it under strain. This change in polarization is commonly called the piezoelectric polarization and is given by:

$$P_{PE}(Al_xGa_{1-x}N) = 2(\frac{a(0)-a(x)}{a(x)})\left[e_{31}(x) - e_{33}(x)\frac{c_{13}(x)}{c_{33}(x)}\right] \text{C/m}^2 \tag{8.38}$$

where $a(0)$ is the lattice constant under strain and $a(x)$ is the lattice constant of the relaxed material. The constants $c_{13}(x)$ and $c_{33}(x)$ are the elastic deformation constants and $e_{31}(x)$ and $e_{33}(x)$ are piezoelectric constants of AlGaN layer. The constants used to calculate the polarization in III-Nitride layers such as GaN, AlN and InN are tabulated in Table 8.1.

The total polarization in a given layer is simply the sum of the spontaneous and piezoelectric polarization. Therefore, the total polarization vectors for GaN and AlGaN are given by

$$P_{Tot}(GaN) = P_{SP}(GaN) + 0$$
$$P_{Tot}(AlGaN) = P_{SP}(AlGaN) + P_{PE}(AlGaN) \tag{8.39}$$

TABLE 8.1

The Constants Used to Calculate the Polarization in III-Nitride Layers

Constants	GaN	AlN	InN
P_{sp} [C/cm²]	-2.9×10^{-6}	-8.1×10^{-6}	-3.2×10^{-6}
e_{31} [C/cm²]	-0.49×10^{-4}	-0.6×10^{-4}	-0.57×10^{-4}
e_{33} [C/cm²]	0.73×10^{-4}	1.46×10^{-4}	0.97×10^{-4}
c_{13} [GPa]	103	108	92
c_{33} [GPa]	405	373	224
a_0 (A°)	3.189	3.112	3.54

At a heterojunction there is usually a change in the polarization on each side. This abrupt change in polarization causes a bound sheet charge. In general, the bound sheet charge is the polarization of the bottom layer minus the polarization of the top layer, i.e., $\sigma = P_{Bottom} - P_{Top}$. The heterointerface charges can be computed for the AlGaN surface and the AlGaN/GaN interface by

$$\left| \sigma_{surf} \right| = \left| 0 - P_{Tot} \left(Al_x Ga_{1-x} N \right) \right| \tag{8.40}$$

$$\left| \sigma(x) \right| = \left| \sigma_{AlGaN/GaN}(x) \right|$$

$$= \left| P_{Tot,Top} \left(Al_x Ga_{1-x} N \right) - P_{Tot,Bottom} \left(GaN \right) \right| \tag{8.41}$$

$$= \left| 2 \left(\frac{a(0) - a(x)}{a(x)} \right) \left\{ e_{31}(x) - e_{33}(x) \frac{C_{13}(x)}{C_{33}(x)} \right\} + P_{SP}(x) - P_{SP}(0) \right|$$

In the above expression, it has been assumed that GaN layer is fully relaxed. Thus, $P_{pz}(GaN)$ is almost zero. Hence, by increasing the Al-content of the AlGaN donor layer, the piezoelectric and spontaneous polarization of AlGaN will increase. The tensile strain caused by the growth of $Al_x Ga_{1-x} N$ on GaN results in a piezoelectric polarization, P_{pz}, that adds to the net spontaneous polarization, Psp, is given by

$$P_{Tot} \left(Al_x Ga_{1-x} N \right) = P_{PZ} + P_{SP}$$

$$= \left| \left(3.2x - 1.9x^2 \right) \times 10^{-6} x - 5.2 \times 10^{-6} x \right| Ccm^- \tag{8.42}$$

where P_{Tot} is the net total polarization.

The net positive charge at the AlGaN/GaN interface caused by the sum of the net spontaneous polarization and piezoelectric polarization between AlGaN and GaN is shown in Figure 8.7.

It shows that the polarization situation for both Ga-face and N-face material for an AlGaN/GaN HEMT structure. In a Ga-face structure, a positive bound charge is created at the deeper interface, which causes the formation of a 2DEG at the lower interface. Similarly,

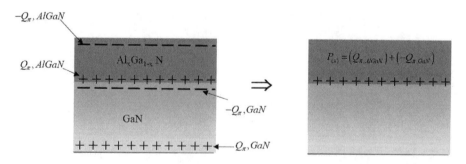

FIGURE 8.7
The net positive charge at the AlGaN/GaN interface caused by the sum of the net spontaneous polarization and piezoelectric polarization between AlGaN and GaN (From Dash, P., Modeling, realization and characterization of compositionally Graded AlGaN/GaN heterostructures for electronic applications, PhD Dissertation, IIT, Kharagpur, 2012.)

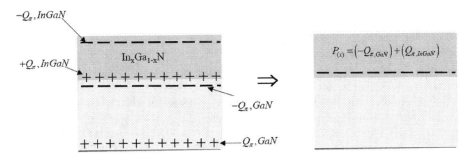

FIGURE 8.8
The net negative charge at the InGaN/GaN interface caused by the compressive strain resulting from growth of In$_x$Ga$_{1-x}$N on GaN (From Dash, P., Modeling, realization and characterization of compositionally Graded AlGaN/GaN heterostructures for electronic applications, PhD Dissertation, IIT, Kharagpur, 2012.)

in the N-face material, the positive bound charge is present at the upper interface and the 2DEG is also formed there as shown in Figure 8.7.

Similarly, the compressive strain caused by the growth of In$_x$Ga$_{1-x}$N on GaN causes a net negative piezoelectric polarization charge at the In$_x$Ga$_{1-x}$N/GaN interface as shown in Figure 8.8. The magnitude of the charge is given by

$$P_{Tot}\left(\text{In}_x\text{Ga}_{1-x}\text{N}\right) = P_{PZ} + P_{SP}$$

$$= \left|\left(14.1x + 4.9x^2\right) \times 10^{-6} - \left(0.3 \times 10^{-6}\right)x\right| Ccm^{-2} \tag{8.43}$$

With the high positive polarization induced sheet charge density at the AlGaN/GaN interface for Ga-face material, the maximum possible sheet carrier concentration found at the 2DEG formed at the heterointerface of the unintentionally doped structure is given by

$$n_s\left(x\right) = \frac{+\sigma\left(x\right)}{e} - \left(\frac{\varepsilon_0\varepsilon\left(x\right)}{de^2}\right)\left[e\Phi_b\left(x\right) + E_F\left(x\right) - \Delta E_c\left(x\right)\right] \tag{8.44}$$

where d is the width of the AlGaN donor layer, $\left(e\phi_{bn}\right)$ is the Schottky barrier of a gate contact, E_F is the Fermi energy level with respect to the GaN conduction band energy, and ΔE_c is the conduction band offset at the AlGaN/GaN interface.

References

1. S. N. Mohammad and H. Morkoc, "Progress and prospects of group-III nitride semiconductors," *Prog. Quant. Electron.*, Vol. 20, No. 5–6, pp. 361–525, 1996.
2. D. Jena, PhD Dissertation, "Polarization induced electron populations in III–V nitride semiconductors Transport, growth, and device applications," University of California, Santa Barbara, 2003.
3. O. Ambacher, J. Smart, J. R. Shealy, N. G. Weimann, K. Chu, M. Murphy, W. J. Schaff, and L. F. Eastman, "Two-dimensional electron gases induced by spontaneous and piezoelectric polarization charges in N- and Ga-face AlGaN/GaN heterostructures," *J. Appl. Phys.*, Vol. 85, No. 6, 1999.

4. O. Ambacher, B. Foutz, J. Smart, J. R. Shealy, N. G. Weimann, K. Chu, M. Murphy, A. J. Sierakowski, W. J. Schaff, and L. F. Eastman, "Two-dimensional electron gases induced by spontaneous and piezoelectric polarization in undoped and doped AlGaN/GaN heterostructures," *J. Appl. Phys.*, Vol. 87, No. 1, 2000.
5. O. Ambacher *et al.*, "Pyroelectric properties of Al(In)GaN/GaN hetero and quantum well structures," *J. Phys. Condens. Matter*, Vol. 14, pp. 3399–3434, 2002.
6. T. R. Lenka, "Studies on characteristics of III–V compound semicondctor based HEMT/MODFET to use in analog VLSI circuits," PhD Dissertation, Sambalpur University, 2012.
7. P. Dash, "Modeling, realization and characterization of compositionally Graded AlGaN/GaN heterostructures for electronic applications," PhD Dissertation, IIT, Kharagpur, 2012.

9

Current Collapse in AlGaN/GaN HEMTs

Sneha Kabra and Mridula Gupta

CONTENTS

9.1 Introduction

Group III-V nitride compound semiconductors such as Gallium Nitride (GaN) and Aluminum Gallium Nitride (AlGaN) exhibit distinctive combination of properties like high breakdown field, large energy band gap, good thermal conductivity, high mobility and high saturation velocity. Due to these properties, GaN-based devices outperform existing Si- and SiC-based devices for power electronic applications. There has been a rapid progress in the design and development of GaN-based devices and circuits since the demonstration of first GaN-based transistor [1]. Since early 1990s, Ga-N has been considered as a very interesting and highly promising material for both optical and high power microwave applications. Owing to the high defect density present in native GaN material, early studies of GaN and related compounds showed uncertainty on whether devices based on this material could ever be used for practical applications. Nevertheless, with better understanding of effect of these defects on device performance along with improvement in the epitaxial growth technology, GaN-based technology has been successfully in use for high-power applications. Trapping related reliability issues still remain a major obstacle for AlGaN/GaN High Electron mobility Transistors (HEMT), thus preventing them from being adopted widely for high-voltage and high-temperature applications. It is essential to analyze and study trapping effects in AlGaN/GaN HEMTs because of two important reasons. First, they limit DC and RF performance of the device. Secondly, they also play an important role in reliability of the device in high-power applications. The device performance severely degrades when it operates at high voltage as the electrons get trapped in various locations in the device. It has also been seen that the trapping effects rise after device degradation. Performance and thus reliability of the device thus greatly reduces by increased trapping [2–5].

Traps cause transient and recoverable reduction in drain current at high drain voltage. This phenomenon is called as current collapse, which is also referred as current slump, current compression, current instability, knee walk out [6], kink in output current voltage [7] and RF dispersion. The excess charge associated with the trapped carriers leads to the formation of depletion region in the conducting channel. It leads to partial pinch off of the device and a severe degradation of drain current characteristics. Another important problem attributed to presence of traps in AlGaN/GaN HEMT is gate leakage current [8]. Some of the other effects of traps include threshold voltage shift [9], light sensitivity, transconductance frequency dispersion, increase in ON resistance of the device, drain lag transients, gate lag transients and limited microwave power output. In AlGaN/GaN HEMT, electrically active traps are present in bulk of AlGaN as well as at the hetero interface. The traps present on the surface and in the bulk of heterostructure change the density of two-dimensional electron gas (2DEG) formed in the channel of HEMT. This adversely affects performance of the device [4].

There has been great technological advancements in the growth, fabrication and characterization of AlGaN/GaN HEMTs [10–12], various trapping-based parasitic effects still exist and limit the device performance [13]. Various analytical models have also been developed to investigate short-channel effects in GaN MESFET and AlGaN HEMTs [14–17].

Degradation in drain current is mainly caused by vertical electric field across the AlGaN barrier [5,18]. In AlGaN/GaN HEMT, the trapping effects have been reported to exist in AlGaN barrier layer, AlGaN/GaN interface, GaN buffer and at the device surface [4,18–22] as shown in Figure 9.1. Hence, on the basis of their location, traps can be classified as:

1. *Surface traps* [20,23–26]: Surface traps are present on the surface of the device which, and they mainly occur due to poor passivation of the device surface. They can also be caused due to trapping of electrons at the interface between the semiconductor and the passivation layer. Injection of electrons from the gate into the passivation layer also results in surface trapping.

2. *Interface traps*: Interface traps are present on AlGaN/GaN heterostructure Interface [27].

3. *Bulk traps* [26,28–30]: Bulk traps are present within the heterostructure, which could be in AlGaN barrier layer or in GaN buffer layer. They are caused due to high electric field.

This chapter provides a detailed experimental and theoretical study on the properties and origin of various types of traps present in AlGaN/GaN HEMTs. The first part of this chapter (Section 9.2) describes origin and properties of interface and near interface charge. The next

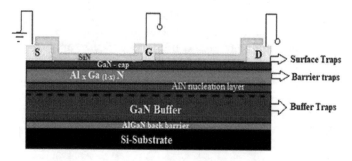

FIGURE 9.1
Pictorial representation of different traps present in AlGaN/GaN HEMTs.

section (Section 9.3) of this chapter deals with properties and influence of surface traps. In this section, we have discussed the concept of virtual gate in detail to describe current dispersion due to surface trapping states. Section 9.4 presents the properties and influence of bulk traps on the device behavior. In this section, phenomenon of current collapse is discussed in detail. Subsequently, in Section 9.5, effect of temperature on traps in AlGaN/GaN HEMT is described. Finally, Section 9.6 reviews the impact of traps on behavior of AlGaN/GaN HEMT. Various methodologies that are used for trap measurement and characterization have been described in detail in this section. A brief overview of behavior of traps from DC to radio-frequency operation mode is also presented in this section. It is based on pulsed current-voltage measurements, low-frequency dispersion measurements, and RF-drain current measurements, which may help in device optimization for improved performance.

9.2 Origin and Properties of Interface and Near Interfacial Traps

AlGaN/GaN HEMTS are either fabricated homoepitaxially or heteroepitaxially. In homoepitaxially grown HEMTs, high-resistive free-standing GaN-on sapphire templates and free-standing GaN substrates are used. On the other hand, in heteroepitaxially grown HEMTs, foreign substrates such as Si, sapphire and SiC are used along with insulating and stress mitigating buffer layers. Although it is best to grow AlGaN/GaN HEMTs on GaN substrate, but, due to the lack of large free-standing GaN substrates and high cost, GaN HEMTs are usually grown on foreign substrates. This leads to crystalline imperfections which in turn generate trap centers within the band-gap of a semiconductor that assist the creation of free electrons and holes in the conduction band and valance band, respectively.

The important properties of the substrates used for the growth of AlGaN/GaN HEMTs are listed in Table 9.1. It summarizes the lattice-mismatch and thermal-mismatch between GaN material with respect to different substrates. The large lattice- and thermal-mismatch between GaN and other substrate materials cause a large number of defects, which adversely affect the device performance leading to various reliability issues. Threading defects are usually oriented parallel to the c-axis of the material. Their density varies between 10^8 and 10^{11} cm^{-2} for GaN layers grown on a sapphire or SiC substrate. Dislocations and point defects which are present in AlGaN/GaN HEMTs due to lattice-mismatch can

TABLE 9.1

Important Properties of the Substrates Used for the Growth of AlGaN/GaN HEMTs

Substrate Properties	Si (111)	Sapphire	6H-SiC	GaN (0001)
Lattice constant (Å)	3.846	4.758	3.081	3.189
Lattice-mismatch (%)	17	14	3.5	–
Thermal expansion coefficient (10^{-6} K^{-1})	2.6	7.5	4.5	5.6
Thermal-mismatch (%)	116	36.4	24	–
Thermal conductivity (W cm^{-1} K^{-1})	1.5	0.35	4.5	1.3
Wafer size	2″ to 12″	2″ to 8″	2″ to 6″	2″
Price	Low	Medium	Very high	Extremely high

become charged and act as centers of Coulomb scattering. Lattice-mismatch also results in a high density of dislocations in the GaN epitaxial layers, which further leads to charge trapping [31,32].

For switching and high-power amplifier applications, when a high drain to source bias is applied along with the gate bias, it drives the device into the sub-threshold regime. As a result, the device characteristics become significantly different from the static characteristics for a period of time ranging from microseconds to hours.

The near interface and interface traps are formed at the heterointerfaces in semiconductor heterojunctions [33]. Studies have shown that the majority of electrically active traps are present at the heterointerface and the near the region of heterointerface. However traps have also been observed in the region near the metal/AlGaN interface [34]. The source of these traps is found to be due to epitaxial growth, crystalline imperfections, and due to various device fabrication processes. A substantial amount of tensile strain is stored in the heterointerface due to the lattice-mismatch between GaN and AlGaN as shown in Figure 9.2. This strain contributes to the formation of high-density two-dimensional electron gas due to the piezoelectric polarization effects depicted in Figure 9.3. It also causes dislocation at the heterointerface. This leads to interface traps, near interface traps and crystallographic defects in the bulk of AlGaN layer.

These traps also cause gate leakage current which leads to performance instability in AlGaN/GaN HEMT [5,8]. Similarly, for AlGaN/GaN Schottky diode, reverse bias stress induces a gradual increase in the leakage current [35]. Capacitance-voltage characteristics get significantly modified when the device is subjected to stress. It is caused due to the generation of donor traps in GaN-buffer, close to AlGaN/GaN interface by the injection of accelerated electrons towards the GaN buffer or by applying high electric field. It has also been found that depending on device processing technique, nitrogen deficiency is expected to be created in bulk of AlGaN barrier layer [36] leading to trapping effects.

Frequency dependent capacitance and conductance analysis conducted by Stoklas et al. show the density of interface traps to be $2 \times 10^{12} \text{cm}^{-2}\text{eV}^{-1}$ to $3 \times 10^{12} \text{cm}^{-2}\text{eV}^{-1}$ at the heterointerface [37,38].

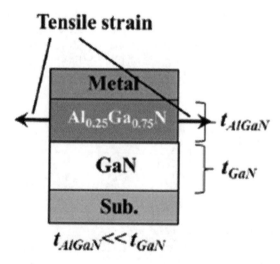

FIGURE 9.2
Tensile Strain at the interface of AlGaN/GaN.

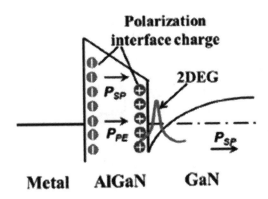

FIGURE 9.3
Formation of Polarization induced interface traps.

Researchers have shown that following mechanisms play significant function in assessing the performance degradation of AlGaN/GaN HEMT:

1. Generation of donor like traps in the AlGaN layer, which can be measured by Deep Level Transient Spectroscopy [39].

2. Oxidation effects, which cause the electrochemical degradation of the device surface [40].

3. Results reported by Marcon et al. and Meneghini et al. [41,42] show that the gate-drain diode of a AlGaN/GaN HEMT can degrade at low and moderate stress voltage levels. They have also reported that the reverse-bias degradation mechanism is time dependent. It was demonstrated that degradation of HEMTs exposed to negative gate voltage was due to defect generation and percolation process. If the device is exposed for sufficiently long stress times, it leads to generation of permanent leakage paths.

Therefore it is important to evaluate distribution of near-interface and interface traps.

There are two types of traps which exist in a device:acceptor and donor like traps. Figure 9.4 pictorially depicts the characteristics of the acceptor-like traps and donor-like traps present in device. The donor-like traps are located in the lower half of the energy band gap. They are neutral when occupied by an electron and positively charged when empty. In contrast, acceptor-like traps are located in the upper half of the energy band gap. They are negatively charged when occupied by an electron and neutral when empty.

The process of trapping and de-trapping of electrons and holes follow the Shockley Read Hall theory [43,44]. The defects present in AlGaN/GaN HEMTs are also categorized with respect to their energy level. The traps having an energy level close to the valence band or conduction band are called shallow level traps. The traps having an energy level within the forbidden band gap are called deep level traps.

The interactions between the free-carriers and the generation/recombination mechanisms for a deep level transition to or from a energy band has been described by Morkoc [45].

The overall trap occupancy rate in conduction band is given by Equation 9.1.

$$\frac{\partial n_T}{\partial t} = (c_n(N_T - n_T) - e_n n_T)$$

(9.1)

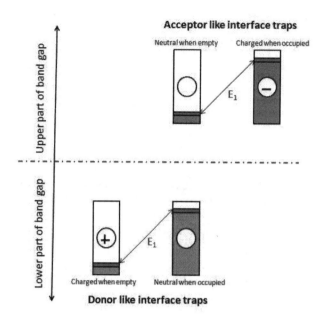

FIGURE 9.4
Donor like interface traps (green) and acceptor like interface traps (red).

where c_n is the electron capture rate, N_T is the number of defect states, n_T is the number of defect states filled with electrons and e_n is the electron emission rate.

It has been observed that for AlGaN/GaN hetero-interfaces both near interface traps and interface traps contribute to extra capacitance by trapping mobile carriers. The capacitance effect causes the screening of gate-to-channel control. The electric field under the gate cannot control the Fermi level in the semiconductor effectively. As a consequence, the sub threshold control is reduced, as the channel is not properly depleted. Since sufficient amount of mobile charge carriers cannot be formed at the surface hence the on state current of the device is also degraded.

Semra et al. [46] have investigated location, distribution and physical characteristics of interface traps which exist at AlGaN/GaN interface using frequency dependent measurement of conductance and capacitance. It has been observed that conductance changes due to filling and emptying of trap states. Similarly, capacitance also gets modified due to charge storage in trapped states. An electrical model has also been developed with two levels of traps at AlGaN/GaN interface having activation energy and capture cross sections of traps as $\Delta E_{a1} = 0.201$ eV, $\sigma_{n1} = 4.992 \times 10^{-18}$ cm^2, $\Delta E_{a2} = 0.053$ eV, $\sigma_{n2} = 2.585 \times 10^{-21}$ cm^2, and one trap level located in AlGaN barrier with activation energy and capture cross section as $\Delta E_{a3} = 0.121$ eV, $\sigma_{n3} = 1.256 \times 10^{-20}$ cm^2.

The equivalent circuit used to evaluate interface traps is shown in Figure 9.5.

It consists of barrier capacitance corresponding to the depleted AlGaN layer (C_b), channel capacitance, (C_s), parallel conductance, (Gp/ω), Interfacial state capacitance Cit represents capacitance due to charged traps and Rit represents losses due to recombination/generation through interface states, $D_{it} =$ Surface trap density.

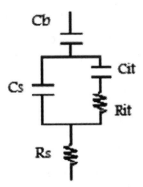

FIGURE 9.5
Equivalent circuit model for measuring interface traps. (From Semra, L. et al., *Surf. Interface Anal.*, 42, 799–802, 2010.)

Parallel capacitance and parallel conductance for single level of trap can be expressed as [46] by Equations 9.2 and 9.3, respectively.

$$C_P = C_s + \frac{C_{it}}{\omega\tau\tan(\omega\tau)}$$

(9.2)

$$\frac{G_P}{\omega} = \frac{q\omega\tau D_{it}}{1+(\omega\tau)^2}$$

(9.3)

For continuum trap levels, Parallel capacitance and parallel conductance can be evaluated by the following Equations 9.4 and 9.5:

$$C_P = C_s + \frac{C_{it}}{\omega\tau\tan(\omega\tau)}$$

(9.4)

$$\frac{G_P}{\omega} = \frac{qD_{it}}{2\omega\tau}\ln\left(1+(\omega\tau)^2\right)$$

(9.5)

where

$$C_{it} = qD_{it}$$

(9.6)

Time constant, τ, is given by Equation 9.7:

$$\tau = R_{it}C_{it}$$

(9.7)

Figure 9.6a and b shows variation of parallel capacitance and parallel conductance with frequency.

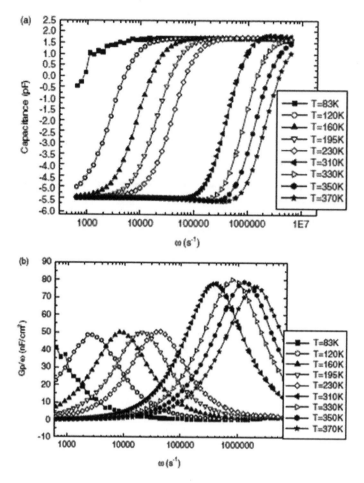

FIGURE 9.6
(a) Variation of parallel capacitance with frequency for different temperatures (From Semra, L. et al., *Surf. Interface Anal.*, 42, 799–802, 2010.); (b) Variation of parallel conductance with frequency for different temperatures. (From Semra, L. et al., *Surf. Interface Anal.*, 42, 799–802, 2010.)

9.3 Properties and Influence of Surface Traps

Surface traps are caused due to the presence of surface states. Tanaka et al. proposed the mechanism of formation of charged surface states and the subsequent origin of electrons in the 2DEG at the AlGaN/GaN interface [30]. In order to preserve overall charge neutrality in AlGaN/GaN HEMT structure, 2DEG at the AlGaN/GaN hetero interface should be balanced by positive charge in the AlGaN barrier and charge at the surface. This positive charge required to maintain charge neutrality can in principle be provided by holes attracted at the AlGaN surface by the negative polarization charge and by ionized donor like surface traps. It has been demonstrated that long hours of ON-state electrical stress also leads to the generation of interface and surface related traps due to hot electron effects [47,48]. At high-drain voltages, electrons can modulate surface charge, as hot electrons in the channel may overcome the potential barrier in AlGaN layer and potential barrier at the interface and get trapped at the surface. This phenomenon has also been observed in AlGaAs/GaAs HEMTS [49] and GaN MESFETs [16].

Researchers have proposed two main mechanisms which limit HEMT lifetime, when they are exposed to low-power high-electric field stress. First is the formation of a pit or a crack in the crystal due to mechanical stress caused by the inverse piezoelectric effect [2,5,50–53]. This occurs when the device voltage exceeds a critical voltage. The changes in the surface morphology of electrically stressed device leads to surface pitting in the vicinity of the gate terminal [54]. Second is the time-dependent degradation mechanism proposed by Marcon et al. [55]. They have shown that the time-dependent degradation effects, result from the formation of percolation paths that are created in the AlGaN barrier.

Wang et al. have demonstrated that the hot-electron effects make a negligible contribution to the negative differential conductance [56] of an AlGaN/GaN dual channel HEMT [57]. However, self heating causes a reduction in carrier mobility due to the enhanced phonon scattering which limit the output-power-density at microwave frequencies. Thus trapping and hot electron induced effects severely degrade GaN HEMT performance. A detailed study of these effects has been reported by Meneghini et al. using combined electrical and optical measurements [58].

Jia et al. [59] have developed an analytical model to predict surface electric field distribution around the drain side gate edge. This model can be effectively used to understand various surface trapping mechanisms resulting from high electric field around the drain-side gate edge. Density of 2DEG, gate leakage current and drain current collapse are dependent on gate metals [60–63]. Takamasa et al. have studied and compared the impact of various gate metals namely, titanium nitride, nickle and tungsten on drain current collapse in AlGaN/GaN HEMT [64]. As shown in Figure 9.7, TiN metal mitigate the collapse

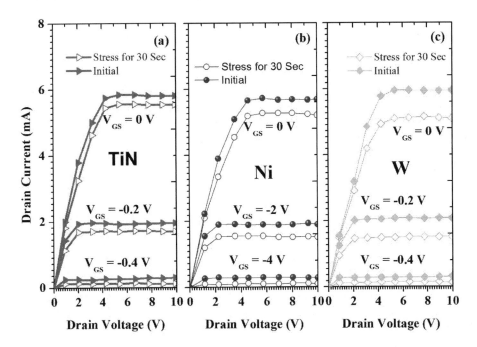

FIGURE 9.7
Impact of different gate metals on the drain current collapse in 25 μm AlGaN/GaN HEMT with (a) TiN gate (b) Ni gate (c) W gate. (From Kawanago, T. et al., *IEEE Trans. Electron Devices*, 61, 785–792, 2014.)

of drain current after application of stress. Whereas, a reduction in drain current can be observed in conventional Tungsten and Ni gate. It has also been demonstrated that TiN metal gates with higher nitrogen concentrations show less reduction in drain current with an increase in electrical stress at the drain.

The frequency dependent conductance analysis done by Stoklas et al. [37] shows that two kinds of traps exist in AlGaN/GaN HFET—slow traps and fast traps. The density of slow traps was found to be 2.5×10^{11} cm^{-2}eV^{-1} to 4×10^{11} cm^{-2}eV^{-1}. The slow traps have been ascribed to surface states which are responsible for Fermi level pinning. The density of fast traps was found to be 2×10^{12} cm^{-2}eV^{-1} to 3×10^{12} cm^{-2}eV^{-1}. The fast traps have been ascribed to bulk defects and their density was found to be independent of surface passivation.

Vetury et al. [20] described the current dispersion due to surface states using the concept of "virtual gate." On direct measurement of surface potential along the gate drain access region, it was observed that the lateral extension of the depletion region was not consistent with the ionized positive donor density. This was attributed to the presence of negative charge on the surface, therefore making the surface potential more negative. This leads to the fast extension of the gate depletion region. Therefore now, along with the metal gate, an additional virtual gate, connected in series exists on the surface of the device, between drain and source terminal, which act as electron trap. The net amount of trapped charges in gate drain access region control the potential, V_{Vg} of the virtual gate. Trapping of electrons on the surface of the device, in donor like states decreases the quantity of total positive charge present on the surface which leads to the formation of virtual gate. The drain bias was applied after applying gate signal, which is large enough to drive the device into both saturation and cut off. After that, the envelope of output drain current, called trapping transient was measured. It shows decay of maximum drain current which is caused due to formation of virtual gate. It was observed that the electrons trapped by the surface states still remained even when the gate bias was removed. Cardwell et al. have determined the length of virtual gate to be several hundred nanometers using Kelvin probe force microscopy [65].

In order to overcome the current degradation due to virtual gate, it was proposed to passivate the surface of the device using Silicon Nitride (SiN). Current collapse in passivated device is much smaller as compared to unpassivated device. This also proves that surface states lead to the formation of virtual gate and reduce the drain current and increases ON resistance of the device. The virtual gate is completely driven by the trapping and detrapping of electrons by the surface states. Hence virtual gate causes delay in the response of the device. This delayed response of the drain current to the applied gate voltage, when gate voltage is changed from OFF-state to ON-state is also called the gate-lag effect.

Another method to reduce surface trapping effect is by using appropriate surface treatment [66,67]. It has been demonstrated that surface traps can also be controlled by well-designed field plate along with dielectric encapsulation [68].

Meneghesso et al. [69] described the influence of surface sates on the pulsed drain-current characteristics of AlGaN/GaN HEMT using 2-D simulations. N$^-_{pol}$ is negative fixed charge density and N$_{TD}$ is the density of donor like traps. It can be observed that the net negative charge distribution is maximum at $t = 0^-$ because at that time, the channel under the gate is pinched off. At $t = 0+$, when $V_{GS} = 0$ V is applied, the channel under the gate opens immediately but, it does not changes simultaneously in access regions due to the lag with which surface states respond to the applied gate voltage. This is the reason for drain current collapse in gate turn-on pulsing mode. The net negative charge is distributed when holes are emitted from surface states during induced transient. It has been seen that in gate drain access region, the net negative charge increase with time causing reduction in the drain current.

Koley et al. also confirmed the presence of surface traps using scanning Kelvin probe microscopy [70]. The other mechanisms which lead to formation of surface traps include dangling-bonds or impurities at the surface and surface point-defects formed during the device growth process.

9.4 Properties and Influence of Bulk Traps

Bulk traps are the traps that are present in GaN buffer or AlGaN barrier or anywhere within the device heterostructure. The presence of bulk traps in the AlGaN barrier layer has been shown by Yang et al. [27] using electron tunneling spectroscopy. When a high drain-to-source voltage is applied to the device, a high electric field is generated. This may cause injection of electrons from 2DEG channel into the traps within the heterostructure, in GaN buffer region.

It has been reported that bulk traps can also arise because of charge tunneling from the gate into the semiconductor layer.

Further, Jungwoo Joh et al. [71] have described two major trapping processes, T_{P1} (AlGaN barrier traps) and T_{P2} (Bulk traps) having time constant as 3 s and 0.1 ms respectively at 30°C in AlGaN/GaN HEMT. It was found that time constant for trapping process, T_{P1}, is thermally activated and arises due to self heating, while the time constant for T_{P2} was not sensitive to temperature. This is because, in trapping process T_{P1}, the electrons have to overcome an energy barrier before they get trapped, and they also show thermal broadening. T_{P1} is due to electron injection from the gate and trapping at the surface region close to the gate or inside the AlGaN barrier. Whereas, T_{P2} occurs only when the current flows through the channel. Hence, it is related to the trapping of electrons in the channel through a tunneling process.

It has been observed that the trap level in the AlGaN layer cause the current collapse at a trap level of 0.6e V from the conduction band edge. This happens due to nitrogen vacancies [72]. The gate current injects electrons into AlGaN barrier layer and the surface in the ON-state. A few of these electrons get trapped in these regions. Traps in the channel or in the buffer, capture some of the electrons.

Even though the trapping at the surface is supposed to be an important cause for current collapse, electron trapping in the GaN buffer layer and in the AlGaN barrier layer can also lead to drain current collapse [73]. The characteristics indicate a reduction in drain current for high V_{DS}. The fall in drain current upon application of high drain voltage is termed as Current Collapse.

This happens because at high V_{DS}, electrons get injected into the GaN buffer layer. These electrons get trapped in the buffer layer. The trapped charges deplete the two-dimensional electron gas, and therefore result in a drain current dispersion for subsequent traces. The drain current collapse caused due to bulk traps is also known as "drain-lag." Measurements of drain-lag as a delayed response of drain current to V_{DS} change is also one of the methods to analyse current collapse due to bulk trapping effects. Tirado et al. [19] used drain-lag turn-ON technique to analyze current collapse and differentiate between surface traps and bulk traps. The drain current obtained using this technique suggested that the bulk related traps exhibited a slow emission rate while surface related traps exhibit comparatively faster emission rate.

Meneghesso et al. observed a positive shift in threshold voltage and a decrease in transconductance as a result of drain current collapse in GaN MESFET [69]. They characterized the traps responsible for current collapse by means of photo-transient measurements.

The current collapse was attributed to electron trapping in the regions under the gate and in gate to drain access region. Using photoionization spectroscopy, traps were also identified in unintentionally doped GaN buffer layer and at the GaN buffer/channel interface.

The origin of bulk traps is related to material growth conditions of the AlGaN/GaN heterostructure and threading dislocations in the GaN layer due to the large lattice-mismatch between non-native substrates and GaN. It has been reported that the threading dislocations in GaN epitaxial layers give rise to a huge number of deep level states which are spread across the energy bandgap of GaN. It has been observed that the location of bulk traps may be affected by the gate electrode because of the strain modulation in the AlGaN layer.

Compensation charges, which occur due to the background doping effects present in unintentionally doped GaN buffer, are also one of the major source of bulk traps present in buffer layer. These deep acceptors also act as traps resulting in drain current collapse under high voltage operation. The amount of current collapse is found to be directly related to the resistivity of the buffer layer. Drain current collapse has also been attributed to hot electron injection in GaN buffer. When the hot electrons are captured by trapping centers in AlGaN/GaN buffer, the channel gets depleted and the drain current reduces.

Kaushik et al. have proposed that the injection of hot-electrons into the bulk traps can lead to "kink effect" in dc characteristics of the device [74,75]. It is an undesirable and abrupt change in drain current at a certain drain voltage. It is dependent on device operating point and temperature. They measured drain current with drain voltage at different base plate temperatures. The measurement duration at each point was 1 sec. Figure 9.8 presents variation of measured and modeled steady-state drain current with

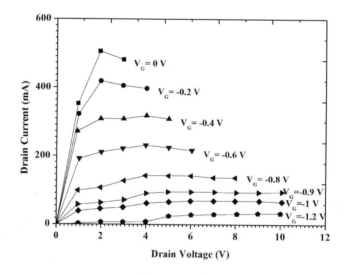

FIGURE 9.8

Variation of measured and modeled steady-state drain current with drain voltage. Solid lines show the modeled drain current and measured current is depicted by symbols. Inset shows the variation of channel temperature with drain voltage, extracted from thermal resistance model using measured base plate temperature. (From Kaushik, J.K. et al., *IEEE Trans. Electron Devices*, 60, 3351–3357, 2013.)

drain voltage. A kink in I_{DS} curves can be observed, which is maximum near the pinch off voltage and decreases as the gate voltage reduces. It can also be observed that for higher drain currents, kink shifts towards lower drain voltage values. The increase in junction temperature with drain voltage is considered to be because of self-heating of AlGaN/GaN HEMT under electrical stress. It has been suggested that kink effect in steady state current voltage characteristics may be associated with combined effect of high electric field and temperature. Kink effect has also been attributed to the accumulation generated by impact ionization.

Recently, researchers have used drain current transient measurements to investigate the properties of traps, which are responsible for current collapse in Iron (Fe) doped AlGaN/GaN HEMT. It was observed that current collapse results from two traps, E1 and E2. Trap activation energy of E1 was found to be 0.82 eV and its cross section was found out to be $8.5 \times 10^{-13} cm^2$. Trap activation energy of E2 was found to be 0.63 eV and its cross section was found out to be $2.3 \times 10^{-14} cm^2$.

The most common methods used to probe and characterize bulk traps include deep-level transient spectroscopy and photoionization spectroscopy. Soft switched pulsed I–V measurement technique is used to distinguish between surface and buffer-related current collapse in AlGaN/GaN HEMTs. It has been demonstrated that current collapse related to buffer shows a nonmonotonic bell-shaped behavior with the applied gate bias. The hot electron trapping in the buffer is suggested to be the foremost mechanism of current collapse. It was also observed that surface traps induced current collapse increases with the negative gate bias because of the enhanced gate injection.

9.5 Effect of Temperature on Traps in AlGaN/GaN HEMT

Currently, there is great interest in studying wide band gap materials for high power and high temperature applications. These include severe environment electronics applications like jet engine control systems, mineral exploration, automotive and space technologies. For example, a typical military temperature range is from −55°C to 125°C. Some commercial applications may require even more stringent temperature conditions. Owing to their wide band gap, high intrinsic temperature, and stoichiometric melting point of around 2573 K [73] III-V nitrides are attractive materials for high-temperature, high-power electronics operation.

Kenichiro et al. [30] investigated the transient response of the AlGaN/GaN HEMT at temperatures up to 125°C. It was observed that current collapse increases at elevated temperatures. Activation energy for the emission of deep level electrons was found to be $0.78 \pm 0.02°eV$ and the activation energy for capture of deep level electrons was found to be $0.73 \pm 0.02°eV$. Figure 9.9 depicts leakage current, J flowing through the source and drain electrodes as a function of the inverse of the temperature for AlGaN/GaN HFET with gate length 2 μm and gate width 100 μm.

Kabra et al. have reported that for GaN MESFET, gate leakage current crossing the metal semiconductor rectifying contact increases with an increase in temperature as shown in Figure 9.10 [76]. With an increase in temperature, electrons are injected more easily into the channel from the metal gate leading to the premature device failure. The inset of Figure 9.10 shows the gate diode characteristics. It can also be observed from Figure 9.11 that, as temperature increases, transconductance decreases because current degrades due to presence of traps at elevated temperature.

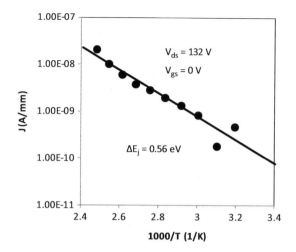

FIGURE 9.9
Leakage current, *J*, flowing through the source and drain electrodes as a function of the inverse of the temperature for AlGaN/GaN HFET with gate length 2 μm and gate width 100 μm. (From Tanaka, K. et al., *Jpn. J. Appl. Phys.*, 52, 04CF07, 2013.)

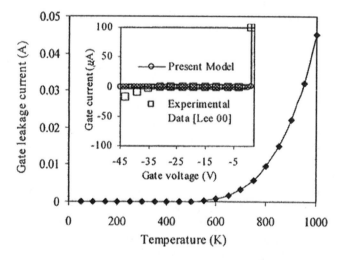

FIGURE 9.10
Variation of gate leakage current with temperature for 0.25 μm GaN MESFET. (From Kabra, S. et al., *Solid-State Electron.*, 52, 25–30, 2008.)

Recently, Zhang et al. [77] have studied the temperature dependence of current collapse induced by surface traps and bulk traps present in buffer layer of carbon-doped AlGaN/GaN HEMT. Figure 9.12 shows the Arrhenius plot of two trap levels *E*1 and *E*2. Trap level *E*1 has an activation energy of 0.08 eV and is related to the surface trapping. Trap level *E*2 has an activation energy of 0.22 eV and is located in the GaN buffer layer. Lower activation energy of trap level *E*1 shows that detrapping mechanism of surface traps induced current collapse has a weak dependence on temperature.

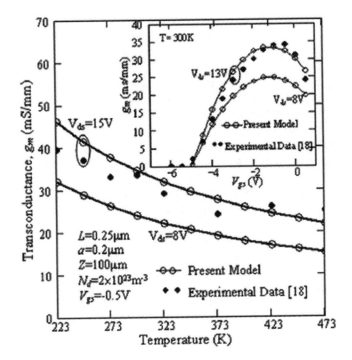

FIGURE 9.11
Transconductance as a function of temperature for a 0.25 25 µm GaN MESFET at V_{ds} = 13 V, 8 V and V_{gs} = −0.5 V. Inset: variation of transconductance with gate bias at 300 K for V_{ds} = 13 and 8 V. (From Kabra, S. et al., *Solid-State Electron.*, 52, 25–30, 2008.)

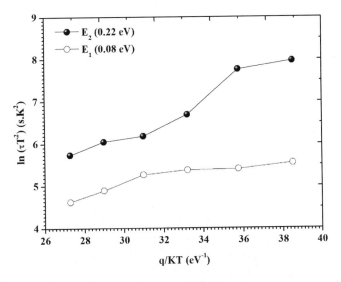

FIGURE 9.12
Arrhenius plot of two trap levels E1 and E2 having activation energies 0.08 and 0.22 eV respectively for carbon–doped AlGaN/GaN HEMT with gate length 1.5 µm. (From Zhang, C. et al., *IEEE Trans. Electron Devices*, 62, 2475–2480, 2015.)

9.6 Impact of Traps in AlGaN/GaN HEMT

Hot electron-induced effects and trapping phenomema are the two most significant mechanisms that strongly affect the reliability and performance of AlGaN/GaN HEMT. Various trap related effects are depicted in Figure 9.13. In order to analyse the impact of traps, it is important to determine the location of traps, their activation energies, their effect on electric field, threshold voltage shift and ON resistance of the device.

If the traps directly below the gate dominate, a shift in dynamic threshold is observed. But, if the traps in the drain access region dominate, they deplete the 2DEG in this region and thus increase the dynamic on resistance of the device.

Eliana et al. very recently proposed a double gated edge termination structure which substantially improves the breakdown voltage and prolongs the device lifetime of AlGaN/GaNSchottky barrier diode.

Current dispersion in DC and RF characteristics of AlGaN/GaN HEMT due to trapping effects has been reported by many researchers and it has been observed that traps also limit the microwave power output of the device.

The current collapse due to surface and buffer traps leads to a decrease in the maximum current and an increase in the knee voltage (V_{knee}) of the device. As a result, the maximum output power (P_{max}) also reduces. P_{max} can be expressed by Equation 9.8,

$$P_{max} = (\Delta V_D \times \Delta I_D)/8$$

(9.8)

where ΔV_D is the maximum voltage swing, which is the difference between the breakdown voltage and the knee voltage. ΔI_D is the maximum current swing.

It has been proposed that dynamic ON resistance of the device can be decreased by adding a gate stack under the gate. Another method to reduce dynamic ON resistance is by encapsulating the device with benzocyclobutane.

Anand et al. [78] have shown that surface leakage currents caused due to surface traps drastically reduce by using benzocyclobutane encapsulation. It has been observed that current collapse reduces by 55% and gate leakage current also decreases by using bilayer gate dielectric insulation [79].

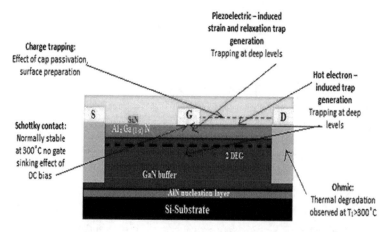

FIGURE 9.13
Impact of traps in AlGaN/GaN HEMT.

The various methods which have been employed to study and characterize traps are:

1. *Pulsed I–V Characterization* [29]: This technique is based on pulsed I_D–V_D measurements. It gives a preliminary and quick characterization of various trapping mechanisms which may affect performance of the device. As large gate-drain voltage swing is applied, a decrease in the drain current and increase of the knee voltage is obtained. The increase in knee voltage and decrease in drain current causes current collapse [47].

2. *Deep-Level Transient Spectroscopy* [80,81]: It is based on the analyzing dependence of drain current transients on the channel temperature. This technique can provide very precise information on the activation energy and cross section of the trap levels that limit the performance of AlGaN/GaN HEMTs. Figure 9.14 shows DLTS spectra as a function of rate window measured in the pinch-off region for AlGaN/GaN Schottky barrier diode. A_1 is the dominant trap and trap A_2 also appears. The Arrhenius plot in the inset of Figure 9.14 shows that the activation energy for trap A_1 is 1.02eV and apparent capture cross section area for trap A_1 was determined to be $2 \times 10^{-12}\,cm^2$. The trap density of A_1 was found to be $2 \times 10^{14}\,cm^{-3}$.

3. *Photo Ionization Spectroscopy* [47]: It is an optical technique that probes the characteristics of defects. The spectral dependence yields detailed information on lattice coupling, location of the defects in the device structure and trap depth.

4. *Electron Tunneling Spectroscopy (ETS)* [27,82–87]: ETS studies suggest the existence of electron traps both at the interface of heterostrucure and in the bulk of AlGaN. As shown in Figure 9.15, the majority of electrically active traps are located within 0.5 eV band below the conduction band edge of the AlGaN barrier.

5. *Gate Lag Measurements*: In this technique, a transient voltage step is applied to the gate terminal and a fixed voltage is applied to drain terminal. The device is operated under this situation from the initial pinch off to an open channel condition. The drain–current transient (i.e., gate lag) versus time is recorded and analyzed.

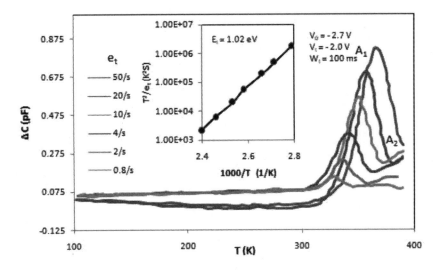

FIGURE 9.14
DLTS spectra measured as a function of rate window for the unpassivated Schottky barrier diode. Inset shows Arrhenius plot for trap A1. (From Fang, Z.-Q. et al., *Appl. Phys. Lett.*, 87, 182115, 2005.)

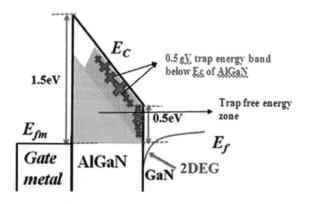

FIGURE 9.15
The energy distribution deduced by ETS analysis of the traps in the AlGaN/GaNHEMT using circular gate electrode with a radius of 150 µm. (From Yang, J. et al., *Appl. Phys. Lett.*, 103, 223507, 2013.)

6. *Drain Lag Measurements:* In this technique, a transient voltage step is applied to the drain terminal and a fixed voltage is applied to gate terminal. The device is operated under this condition from the initial pinch off to an open channel state.

7. *Non Linear Drain Current Transient Spectroscopy (DCTS)* [88]: This method involves extraction of time constants of AlGaN/GaN HEMT operating under large-signal RF conditions to study various trapping mechanisms and related dispersion effects. The increase of load line excursions creates a strong impact on detrapping behavior of the device.

8. *Capacitance Deep-Level Transient Spectroscopy:* This method is used when the gate area of the device is big enough for standard capacitance measurements. It is based on the analysis of the gate capacitance transient induced by a abrupt voltage variation for a wide range of frequencies.

9. *Low-Frequency Noise Measurement:* This technique is used for characterization of material defects and identification of trap locations. Current spectral density (SID) of aged devices is one to three orders higher than current spectral density of unstressed device [89].

10. *Output Conductance and Transconductance Frequency Dispersion Spectroscopy:* Low-frequency dispersion of transconductance and output conductance is attributed to trapping phenomena [90]. Figure 9.16 shows positive extrinsic output conductance dispersion for all bias conditions. Dispersion is less than 33% for V_{DS} values below 10 V whereas a large increase is observed for $V_{DS} > 13$ V. Figure 9.17 depicts that transconductance frequency dispersion is not as significant as that of the output conductance and is less than 8% for the applied bias. It can be seen that for drain to gate voltage, V_{DG} below 3 V, the device shows negative frequency dispersion for the entire frequency range. When V_{DG} is increased, a positive dispersion occurs. This behavior is attributed to surface states present in the ungated regions of the device.

The surface morphology studies performed on electrically stressed AlGaN/GaN HEMT show that when voltage below a critical voltage is applied, a groove is developed in the GaN cap layer along the gate edges, at the drain side of the gate which further extends across the entire length of the device [51,91]. A very faint and discontinuous groove is also

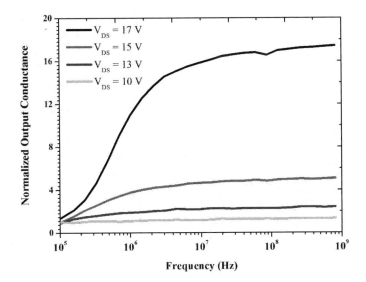

FIGURE 9.16
Frequency dispersion of the extrinsic output conductance measured for 0.25 μmAlInN/GaN HEMT. (From Nsele, S.D. et al., *IEEE Trans. Electron Devices*, 60, 1372–1378, 2013.)

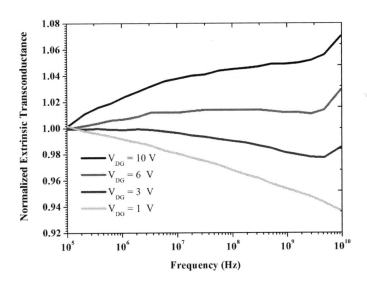

FIGURE 9.17
Frequency dispersion of the extrinsic transconductance. (From Nsele, S.D. et al., *IE EE Trans. Electron Devices*, 60, 1372–1378, 2013.)

noticed on the source side. Whereas, in unstressed devices the surface is smooth under the gate and no groove was observed. The groove on the source side becomes more evident for devices stressed at higher voltages. For stress voltages beyond the critical voltage, localized pits are formed along the groove line. Their density and size was found to get increased with an increase in drain to gate stress voltage. Figure 9.18 depicts the progressive structural damage with varying voltage stress, (a) unstressed device, (b) $V_{DGstress} = 15\ V < V_{crit}$

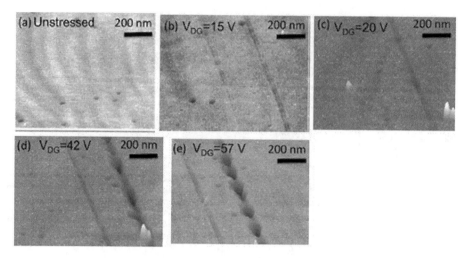

FIGURE 9.18
Progressive structural damage with electrical stress (a) unstressed device, (b) $V_{DGstess} = 15\ V < V_{crit}$, (c) $V_{DGstess} = 20\ V = V_{crit}$, (d) $V_{DGstess} = 42\ V$ and (e) $V_{DGstess} = 57\ V$. (From Makaram, P. et al., *Appl. Phys. Lett.*, 96, 233509, 2010.)

(c) $V_{DGstress} = 20\ V = V_{crit}$, (d) $V_{DGstress} = 42\ V$, and (e) $V_{DGstress} = 57\ V$ for 0.25 µm Gate length AlGaN/GaN HEMT grown on SiC Substrate.

For telecommunications and radar applications, AlGaN/GaN device is fed by modulated signals and the nonlinear dynamics of the dispersion effects play a major role in the modification of the large-signal RF performances [92,93].

Benvegnù et al. have evaluated the impact of DC and RF excitation on trapping mechanisms by means of the drain current transient measurements [88]. Yu Syan et al. [94] have proposed square gate design to reduce the effects of traps. Device with square gate layout improves current collapse and gate lag effects as compared to the conventional two finger layout.

References

1. Khan, M.A. et al., Short-channel GaN/AlGaN doped channel heterostructure field effect transistors with 36.1 cutoff freq. *Electronics Letters*, 1996. **32**(4): p. 357.
2. Joh, J. and J.A. del Alamo. Mechanisms for electrical degradation of GaN high-electron mobility transistors. In *Electron Devices Meeting, 2006. IEDM'06. International*. 2006. IEEE.
3. Pavlidis, D., P. Valizadeh, and S. Hsu. AlGaN/GaN high electron mobility transistor (HEMT) reliability. In *Gallium Arsenide and Other Semiconductor Application Symposium, 2005. EGAAS 2005. European*. 2005. IEEE.
4. Meneghesso, G. et al., Reliability of GaN high-electron-mobility transistors: State of the art and perspectives. *IEEE Transactions on Device and Materials Reliability*, 2008. **8**(2): pp. 332–343.
5. del Alamo, J.A. and J. Joh, GaN HEMT reliability. *Microelectronics Reliability*, 2009. **49**(9–11): pp. 1200–1206.
6. Doo, S.J. et al., Effective suppression of IV knee walk-out in AlGaN/GaN HEMTs for pulsed-IV pulsed-RF with a large signal network analyzer. *IEEE Microwave and Wireless Components Letters*, 2006. **16**(12): pp. 681–683.

7. Lin, C.-H. et al., Transient pulsed analysis on GaN HEMTs at cryogenic temperatures. *IEEE Electron Device Letters*, 2005. **26**(10): pp. 710–712.
8. Kim, H. et al., Gate current leakage and breakdown mechanism in unpassivated AlGaN/GaN high electron mobility transistors by post-gate annealing. *Applied Physics Letters*, 2005. **86**(14): p. 143505.
9. Meneghesso, G. et al. Diagnosis of trapping phenomena in GaN MESFETs. In *Electron Devices Meeting, 2000. IEDM'00. Technical Digest. International*. 2000. IEEE.
10. Atmaca, G. et al., Negative differential resistance observation and a new fitting model for electron drift velocity in GaN-Based hetero structures. *IEEE Transactions on Electron Devices*, 2018. **65**(3): pp. 950–956.
11. Rennesson, S. et al., Optimization of $Al_{0.29}$ $Ga_{0.71}$ N/GaN High electron mobility hetero structures for high-power/frequency performances. *IEEE Transactions on Electron Devices*, 2013. **60**(10): pp. 3105–3111.
12. Soltani, A. et al., Power performance of AlGaN/GaN high-electron-mobility transistors on (110) silicon substrate at 40 GHz. *IEEE Electron Device Letters*, 2013. **34**(4): pp. 490–492.
13. Paine, B.M. et al. GaN HEMT lifetesting–characterizing diverse mechanisms. In *Proceedings of the International Conference on Compound Semiconductor Manufacturing Technology* (Invited paper). 2015.
14. Kabra, S. et al., A semi empirical approach for submicron GaN MESFET using an accurate velocity field relationship for high power applications. *Microelectronics Journal*, 2006. **37**(7): pp. 620–626.
15. Kumar, S.P. et al., Threshold voltage model for small geometry AlGaN/GaN HEMTs based on analytical solution of 3-D Poisson's equation. *Microelectronics Journal*, 2007. **38**(10–11): pp. 1013–1020.
16. Kabra, S. et al., Two-dimensional subthreshold analysis of sub-micron GaN MESFET. *Microelectronics Journal*, 2007. **38**(4–5): pp. 547–555.
17. Kumar, S.P. et al., An analysis for AlGaN/GaN modulation doped field effect transistor using accurate velocity-field dependence for high power microwave frequency applications. *Microelectronics Journal*, 2006. **37**(11): pp. 1339–1346.
18. Sarua, A. et al., Piezoelectric strain in Al Ga N/Ga N heterostructure field-effect transistors under bias. *Applied Physics Letters*, 2006. **88**(10): p. 103502.
19. Tirado, J.M., J.L. Sanchez-Rojas, and J.I. Izpura, Trapping effects in the transient response of AlGaN/GaN HEMT devices. *IEEE Transactions on Electron Devices*, 2007. **54**(3): pp. 410–417.
20. Vetury, R. et al., The impact of surface states on the DC and RF characteristics of AlGaN/GaN HFETs. *IEEE Transactions on Electron Devices*, 2001. **48**(3): pp. 560–566.
21. Binari, S.C., P. Klein, and T.E. Kazior, Trapping effects in GaN and SiC microwave FETs. *Proceedings of the IEEE*, 2002. **90**(6): pp. 1048–1058.
22. Anand, M.J. et al., Distribution of trap energy level in AlGaN/GaN high-electron-mobility transistors on Si under ON-state stress. *Applied Physics Express*, 2015. **8**(10): p. 104101.
23. Meneghesso, G. et al., Trapping phenomena in AlGaN/GaN HEMTs: A study based on pulsed and transient measurements. *Semiconductor Science and Technology*, 2013. **28**(7): p. 074021.
24. Zhang, N.-Q. et al. Effects of surface traps on breakdown voltage and switching speed of GaN power switching HEMTs. In *Electron Devices Meeting, 2001. IEDM'01. Technical Digest. International*. 2001. IEEE.
25. Mitrofanov, O. and M. Manfra, Mechanisms of gate lag in GaN/AlGaN/GaN high electron mobility transistors. *Superlattices and Microstructures*, 2003. **34**(1–2): pp. 33–53.
26. Faqir, M. et al., Mechanisms of RF current collapse in AlGaN–GaN high electron mobility transistors. *IEEE Transactions on Device and Materials Reliability*, 2008. **8**(2): pp. 240–247.
27. Yang, J. et al., Electron tunneling spectroscopy study of electrically active traps in AlGaN/GaN high electron mobility transistors. *Applied Physics Letters*, 2013. **103**(22): p. 223507.
28. Anand, M. et al., Effect of OFF-state stress induced electric field on trapping in AlGaN/GaN high electron mobility transistors on Si (111). *Applied Physics Letters*, 2015. **106**(8): p. 083508.

29. Binari, S.C. et al., Trapping effects and microwave power performance in AlGaN/GaN HEMTs. *IEEE Transactions on Electron Devices*, 2001. **48**(3): pp. 465–471.

30. Tanaka, K. et al., Effects of deep trapping states at high temperatures on transient performance of AlGaN/GaN heterostructure field-effect transistors. *Japanese Journal of Applied Physics*, 2013. **52**(4S): p. 04CF07.

31. Marino, F.A. et al., Effects of threading dislocations on AlGaN/GaN high-electron mobility transistors. *IEEE Transactions on Electron Devices*, 2010. **57**(1): pp. 353–360.

32. Joshi, R. et al., Analysis of dislocation scattering on electron mobility in GaN high electron mobility transistors. *Journal of Applied Physics*, 2003. **93**(12): pp. 10046–10052.

33. Gassoumi, M. et al., Investigation of traps in AlGaN/GaN HEMTs by current transient spectroscopy. *Materials Science and Engineering: C*, 2006. **26**(2–3): pp. 383–386.

34. Yang, J. et al., A study of electrically active traps in AlGaN/GaN high electron mobility transistor. *Applied Physics Letters*, 2013. **103**(17): p. 173520.

35. Meneghini, M. et al. A novel degradation mechanism of AlGaN/GaN/Silicon heterostructures related to the generation of interface traps. In *Electron Devices Meeting (IEDM), 2012 IEEE International*. 2012. IEEE.

36. Hashizume, T. and H. Hasegawa, Effects of nitrogen deficiency on electronic properties of AlGaN surfaces subjected to thermal and plasma processes. *Applied Surface Science*, 2004. **234**(1–4): pp. 387–394.

37. Stoklas, R. et al., Investigation of trapping effects in AlGaN/GaN/Si field-effect transistors by frequency dependent capacitance and conductance analysis. *Applied Physics Letters*, 2008. **93**(12): p. 124103.

38. Miller, E. et al., Trap characterization by gate-drain conductance and capacitance dispersion studies of an AlGaN/GaN heterostructure field-effect transistor. *Journal of Applied Physics*, 2000. **87**(11): pp. 8070–8073.

39. Chini, A. et al., Evaluation of GaN HEMT degradation by means of pulsed I–V, leakage and DLTS measurements. *Electronics Letters*, 2009. **45**(8): pp. 426–427.

40. Gao, F. et al., Role of oxygen in the OFF-state degradation of AlGaN/GaN high electron mobility transistors. *Applied Physics Letters*, 2011. **99**(22): p. 223506.

41. Marcon, D. et al. A comprehensive reliability investigation of the voltage-, temperature-and device geometry-dependence of the gate degradation on state-of-the-art GaN-on-Si HEMTs. In *Electron Devices Meeting (IEDM), 2010 IEEE International*. 2010. IEEE.

42. Meneghini, M. et al., Time-dependent degradation of AlGaN/GaN high electron mobility transistors under reverse bias. *Applied Physics Letters*, 2012. **100**(3): p. 033505.

43. Hall, R.N., Electron-hole recombination in germanium. *Physical Review*, 1952. **87**(2): p. 387.

44. Shockley, W. and W. Read Jr, Statistics of the recombinations of holes and electrons. *Physical Review*, 1952. **87**(5): p. 835.

45. Morko, H., *Handbook of Nitride Semiconductors and Devices*. 2008. Berlin, Germany: Springer.

46. Semra, L., A. Telia, and A. Soltani, Trap characterization in AlGaN/GaN HEMT by analyzing frequency dispersion in capacitance and conductance. *Surface and Interface Analysis*, 2010. **42**(6–7): pp. 799–802.

47. Klein, P. and S. Binari, Photoionization spectroscopy of deep defects responsible for current collapse in nitride-based field effect transistors. *Journal of Physics: Condensed Matter*, 2003. **15**(44): p. R1641.

48. Sozza, A. et al. Evidence of traps creation in GaN/AlGaN/GaN HEMTs after a 3000 hour on-state and off-state hot-electron stress. In *Electron Devices Meeting, 2005. IEDM Technical Digest. IEEE International*. 2005. IEEE.

49. Meneghesso, G. et al., Evidence of interface trap creation by hot-electrons in AlGaAs/GaAs high electron mobility transistors. *Applied Physics Letters*, 1996. **69**(10): pp. 1411–1413.

50. Joh, J. and J.A. Del Alamo, Critical voltage for electrical degradation of GaN high-electron mobility transistors. *IEEE Electron Device Letters*, 2008. **29**(4): pp. 287–289.

51. Makaram, P. et al., Evolution of structural defects associated with electrical degradation in AlGaN/GaN high electron mobility transistors. *Applied Physics Letters*, 2010. **96**(23): p. 233509.

52. Demirtas, S. and J.A. del Alamo. Critical voltage for electrical reliability of GaN high electron mobility transistors on Si substrate. In *Reliability of Compound Semiconductors Digest (ROCS), 2009.* 2009. IEEE.

53. Jimenez, J. and U. Chowdhury. X-band GaN FET reliability. In *Reliability Physics Symposium, 2008. IRPS 2008. IEEE International.* 2008. IEEE.

54. Hodge, M.D. et al., Analysis of time dependent electric field degradation in AlGaN/GaN HEMTs. *IEEE Transactions on Electron Devices*, 2014. **61**(9): pp. 3145–3151.

55. Marcon, D. et al., Reliability analysis of permanent degradations on AlGaN/GaN HEMTs. *IEEE Transactions on Electron Devices*, 2013. **60**(10): pp. 3132–3141.

56. Braga, N. et al., Simulation of hot electron and quantum effects in AlGaN/GaN heterostructure field effect transistors. *Journal of Applied Physics*, 2004. **95**(11): pp. 6409–6413.

57. Wang, X.-D. et al., The study of self-heating and hot-electron effects for AlGaN/GaN double-channel HEMTs. *IEEE Transactions on Electron Devices*, 2012. **59**(5): pp. 1393–1401.

58. Meneghini, M. et al., Investigation of trapping and hot-electron effects in GaN HEMTs by means of a combined electrooptical method. *IEEE Transactions on Electron Devices*, 2011. **58**(9): pp. 2996–3003.

59. Si, J. et al., Electric field distribution around drain-side gate edge in AlGaN/GaN HEMTs: Analytical approach. *IEEE Transactions on Electron Devices*, 2013. **60**(10): pp. 3223–3229.

60. Yu, L. et al., Ni and Ti Schottky barriers on n-AlGaN grown on SiC substrates. *Applied Physics Letters*, 1998. **73**(2): pp. 238–240.

61. Ao, J.-P. et al., Copper gate AlGaN/GaN HEMT with low gate leakage current. *IEEE Electron Device Letters*, 2003. **24**(8): pp. 500–502.

62. Hoon Shin, J. et al., Gate metal induced reduction of surface donor states of AlGaN/GaN hetero structure on Si-substrate investigated by electro reflectance spectroscopy. *Applied Physics Letters*, 2012. **100**(11): p. 111908.

63. Esposto, M. et al. Comparison of Cu-gate and Ni/Au-gate GaN HEMTs large signal characteristics. In *Solid State Device Research Conference, 2009. ESSDERC'09. Proceedings of the European.* 2009. IEEE.

64. Kawanago, T. et al., Gate technology contributions to collapse of drain current in AlGaN/GaN Schottky HEMT. *IEEE Transactions on Electron Devices*, 2014. **61**(3): pp. 785–792.

65. Cardwell, D. et al., Nm-scale measurements of fast surface potential transients in an AlGaN/GaN high electron mobility transistor. *Applied Physics Letters*, 2012. **100**(19): p. 193507.

66. Arulkumaran, S., G. Ng, and S. Vicknesh, Enhanced breakdown voltage with high johnson's figure-of-merit in 0.3-μm T-gate AlGaN/GaN HEMTs on silicon by $(NH_4)_2 S_x$ treatment. *IEEE Electron Device Letters*, 2013. **34**(11): pp. 1364–1366.

67. Hu, C.-Y. and T. Hashizume, Non-localized trapping effects in AlGaN/GaN heterojunction field-effect transistors subjected to on-state bias stress. *Journal of Applied Physics*, 2012. **111**(8): p. 084504.

68. Xing, H. et al., High breakdown voltage AlGaN-GaN HEMTs achieved by multiple field plates. *IEEE Electron Device Letters*, 2004. **25**(4): pp. 161–163.

69. Meneghesso, G. et al., Surface-related drain current dispersion effects in AlGaN-GaN HEMTs. *IEEE Transactions on Electron Devices*, 2004. **51**(10): pp. 1554–1561.

70. Koley, G. et al., Slow transients observed in AlGaN/GaN HFETs: Effects of SiN_x passivation and UV illumination. *IEEE Transactions on Electron Devices*, 2003. **50**(4): pp. 886–893.

71. Joh, J. and J.A. Del Alamo, A current-transient methodology for trap analysis for GaN high electron mobility transistors. *IEEE Transactions on Electron Devices*, 2011. **58**(1): pp. 132–140.

72. Katsuno, T. et al., Improvement of current collapse by surface treatment and passivation layer in p-GaN gate GaN high-electron-mobility transistors. *Japanese Journal of Applied Physics*, 2013. **52**(4S): p. 04CF08.

73. Binari, S. et al., Fabrication and characterization of GaN FETs. *Solid-State Electronics*, 1997. **41**(10): pp. 1549–1554.

74. Kaushik, J.K. et al., On the origin of kink effect in current–voltage characteristics of AlGaN/GaN high electron mobility transistors. *IEEE Transactions on Electron Devices*, 2013. **60**(10): pp. 3351–3357.

75. Mazzanti, A. et al., Physics-based explanation of kink dynamics in AlGaAs/GaAs HFETs. *IEEE Electron Device Letters*, 2002. **23**(7): pp. 383–385.

76. Kabra, S. et al., Temperature dependent analytical model of sub-micron GaN MESFETs for microwave frequency applications. *Solid-State Electronics*, 2008. **52**(1): pp. 25–30.

77. Zhang, C. et al., Temperature dependence of the surface-and buffer-induced current collapse in GaN high-electron mobility transistors on Si substrate. *IEEE Transactions on Electron Devices*, 2015. **62**(8): pp. 2475–2480.

78. Anand, M.J. et al., Low k-dielectric benzocyclobutane encapsulated AlGaN/GaN HEMTs with Improved off-state breakdown voltage. *Japanese Journal of Applied Physics*, 2015. **54**(3): p. 036504.

79. Anand, M. et al., Reduction of current collapse in AlGaN/GaN MISHEMT with bilayer SiN/Al$_2$O$_3$ dielectric gate stack. *Physica Status Solidi* (c), 2013. **10**(11): pp. 1421–1425.

80. Okino, T. et al., Drain current dlts of algan-gan mis-hemts. *IEEE Electron Device Letters*, 2004. **25**(8): pp. 523–525.

81. Fang, Z.-Q. et al., Traps in AlGaN/GaN/SiC heterostructures studied by deep level transient spectroscopy. *Applied Physics Letters*, 2005. **87**(18): p. 182115.

82. Lye, W.-K. et al., Quantitative inelastic tunneling spectroscopy in the silicon metal-oxide-semiconductor system. *Applied Physics Letters*, 1997. **71**(17): p. 2523–2525.

83. Wang, M., W. He, and T. Ma, Electron tunneling spectroscopy study of traps in high-k gate dielectrics: Determination of physical locations and energy levels of traps. *Applied Physics Letters*, 2005. **86**(19): p. 192113.

84. He, W. and T. Ma, Inelastic electron tunneling spectroscopy study of ultrathin HfO 2 and HfAlO. *Applied Physics Letters*, 2003. **83**(13): pp. 2605–2607.

85. Liu, Z. and T. Ma, Determination of energy and spatial distributions of traps in ultrathin dielectrics by use of inelastic electron tunneling spectroscopy. *Applied Physics Letters*, 2010. **97**(17): p. 172102.

86. Reiner, J.W. et al., Inelastic electron tunneling spectroscopy study of thin gate dielectrics. *Advanced Materials*, 2010. **22**(26–27): pp. 2962–2968.

87. Liu, Z. et al., Effect of H on interface properties of Al2O3/In0. 53Ga0. 47As. *Applied Physics Letters*, 2011. **99**(22): p. 222104.

88. Benvegnù, A. et al., Characterization of defects in AlGaN/GaN HEMTs based on nonlinear microwave current transient spectroscopy. *IEEE Transactions on Electron Devices*, 2017. **64**(5): pp. 2135–2141.

89. Tartarin, J.-G. et al. I-DLTS, electrical lag and low frequency noise measurements of trapping effects in AlGaN/GaN HEMT for reliability studies. In *Microwave Integrated Circuits Conference (EuMIC), 2011 European*. 2011. IEEE.

90. Nsele, S.D. et al., Broadband frequency dispersion small-signal modeling of the output conductance and transconductance in AlInN/GaN HEMTs. *IEEE Transactions on Electron Devices*, 2013. **60**(4): pp. 1372–1378.

91. Chowdhury, U. et al., TEM observation of crack-and pit-shaped defects in electrically degraded GaN HEMTs. *IEEE Electron Device Letters*, 2008. **29**(10): p. 1098–1100.

92. Quéré, R. et al. Low frequency parasitic effects in RF transistors and their impact on power amplifier performances. In *Wireless and Microwave Technology Conference (WAMICON), 2012 IEEE 13th Annual*. 2012. IEEE.

93. Delprato, J. et al., Measured and simulated impact of irregular radar pulse trains on the pulse-to-pulse stability of microwave power amplifiers. *IEEE Transactions on Microwave Theory and Techniques*, 2014. **62**(12): pp. 3538–3548.

94. Lin, Y.-S. et al., Square-gate AlGaN/GaN HEMTs with improved trap-related characteristics. *IEEE Transactions on Electron Devices*, 2009. **56**(12): pp. 3207–3211.

10

AlGaN/GaN HEMT Modeling and Simulation

Binit Syamal and Atanu Kundu

CONTENTS

10.1 Introduction to the Compound Semiconductor

Silicon (Si), germanium (Ge), tin (Sn), and diamond (C) are the Group-IV elemental semiconductor materials. The band gaps of the elementary semiconductors are Si:1.12eV, Ge: 0.67 eV, Sn: 0.08 eV, C: 5.5 eV. Among these materials, silicon is the most popular semiconductor material in VLSI industry for manufacturing of discrete devices and integrated circuits due to its several inherent advantages, like more reliable, many ways easier to use and lower cost [1]. Silicon is prepared from raw materials S and or SiO_2 or silica or quartz. This is available in earth crust and foundries, extract pure Si (99.99) from the silica through several process. Silicon to silicon-dioxide formation is very essential as it is required as insulating material to prevent leakages, device isolation and thin gate oxide formation for MOS/CMOS architecture. Therefore, silicon is designated as the God's gift to the world of microelectronics. Si to SiO_2 formation is quite easy and less expensive compared to Ge. It is difficult to grow oxide on Ge, whereas, SiO_2 grows naturally on Si. Doping is Si is also a very easy process, and crystal property is very sensitive to dopant mount change. However, though silicon has important properties and advantages, it also has several limitations, like indirect band-gap material, not suitable for high-frequency applications (system operating at 40 GHz and above), difficult to design optical devices, not appropriate for high-voltage/temperature electronics applications where operable temperature goes above 200°C and not suitable for cryogenic electronics required for space instruments that operate at 4.2 K and below.

10.2 Homojunction/Heterojunction

The semiconductor junctions are classified into two types: homojunction and heterojunction. When a semiconductor junction is formed of similar crystalline semiconductors having equal band gaps, it is called homojunction; whereas any junction that is formed of different band gaps semiconductors is called heterojunction. To form a heterojunction, the compound semiconductor material is preferred for its various advantages. In the compound semiconductor materials, band gap engineering is possible and suitable for high-frequency applications due to many compound semiconductors having larger carrier mobility. As shown in Table 10.3, the materials majorly have direct band gap semiconductors, therefore, it is also appropriate for optoelectronic devices fabrication.

Different materials of the periodic table form the compound semiconductor. Compound semiconductors are classified as binary and alloy semiconductors. Binary compound semiconductors materials are mostly direct band-gap semiconductor materials and covers wide range of band gap from 0.17 to 6.36 eV. The binary compound semiconductor material Indium Antimonide (InSb) is of band gap 0.17 eV and Aluminum Nitride (AlN) is of band gap 6.28 eV. Semiconductor materials are mixed to form semiconductor alloys. Alloy semiconductors are classified as ternary alloys and quaternary alloys. Semiconductors materials are mixed to form semiconductor alloys. Alloying is an effective way to modify the band structure. Ternary alloys are generally formed with three materials where the materials are not compound themselves, but alloys of two binary compounds with one common element.

Ternary alloys have two elements from one column, one from another column of the periodic table. However, ternary compounds are of limited interest, but it is possible to access to continuous ranges of band-gaps in the ternary alloys. Semiconductor alloy formation is chosen in specific way to achieve specific optical or electronic properties.

Quaternary alloys are mixture of four elements, and they are formed by the mixes of four binaries. Among the four binary elements, two elements are from one column, two from the other column or it is also possible that three elements from the one column and one from the other column. Pseudo-binary or ternary alloy is the mixture of two compound semiconductors, where only compound element is changed and other two elements are kept unchanged. The distance between minima of conduction band and maxima of valence band is denoted as band gap. Each semiconductor has direct or indirect band gap as its inherent property. Band gap is the property of lattice constant, which is changed by varying lattice constant. The lattice constant can be changed by various ways, e.g., by mixing another semiconductor, applying pressure by heating and cooling (strain), and reducing crystal size to the quantum dot level. Germanium is an indirect band gap semiconductor material, which is converted to a direct band gap semiconductor by combination of tensile strain and n type doping. For Si, because of large gap between direct and indirect band gap, there is still no method to achieve it, however, band adjustment is possible by the application of strain. It is possible to obtain quantum dot of Si and Ge in certain size and direct band gap may be achievable [2]. As for example, MoS_2 is a direct band gap material in its atomic layer, however the bulk material is indirect type.

For the direct band gap semiconductor materials, momentum of the electrons in the conduction band is equal to the momentum in the valence band. Hence, no addition energy is required to recombine into excitons. These are useful for oscillator, amplifiers, and photonic devices as it increases the efficiency.

Semiconductor is the most widely used material for optoelectronics devices. Primarily, optical properties of optoelectronics devices are determined by the semiconductor band gap and band structure. The monochromatic light with frequency, ν will be emitted from direct band gap semiconductor having band gap is E_g when the electrons will transmit from conduction band to the valence band given the relationship $E_g = h\nu$.

Therefore, $E_g = \frac{hc}{\lambda}$, here, h is the Planck constant, c is the speed of light in vacuum and λ is the photon's wavelength. The value of Planck constant, $h = 6.626 \times 10^{-34}$ joule·s or 4.135×10^{-15} eV.s. The speed of light in vacuum, $c = 10.998 \times 10^8$ m/s. Thus, the value of $(h \times c)$ will be $(6.626 \times 10^{-34}$ joule·s$) \times (10.998 \times 10^8$ m/s$) = 1.9 \times 10^{-25}$ joule·m. This can also be expressed in terms of eV by multiplying $(h \times c) = (4.135 \times 10^{-15}$ eV.s$) \times (10.998 \times 10^8$ m/s$)$ equals to 1.2398×10^{-6} eV.m or 1.2398 eV.µm.

GaAs has the direct band gap of $E_g = 1.42$ eV and the value of $h \times c = 1.24$ eV.µm, hence, emitted light wavelength, $\lambda = \frac{hc}{E_g} = \frac{1.2398 \text{ eV·µm}}{1.42 \text{ eV}} = 0.873$ µm $= 873$ nm in room temperature. Hence, band gap plays an important role in colour of the emitted photon/light from the semiconductor. As the visible spectrum shown in Table 10.1, varies from 400–700 nm, so the direct band gap semiconductor materials required for light-emitting diodes with band gaps between 3.097 eV to 1.77 eV, with $\lambda = \frac{hc}{E_g} = \frac{1.2398 \text{ eV·µm}}{1.43 \text{ eV}}$ or $E_g = \frac{hc}{\lambda} = \frac{1239}{\lambda(\text{nm})}$ (Table 10.2).

Silicon and germanium are indirect by nature and these semiconductors have too small band gaps. III-V elements of the periodic table forms the compound semiconductor with band gaps varies from 0.17 eV (InSb) to 6.28 eV (AlN) as shown in Table 10.3. Except ZnSe and ZnO, Group II–IV compound semiconductors usually are p-type. ZnSe and ZnO are only n-type Group II–IV compound semiconductors. Group II–IV compound materials are difficult to doping due to its non-stoichiometry structure. The non-stoichiometry structure creates chemical imbalance in the compound semiconductors. For example, the ratio of Zn and oxygen (Zn:O) is not (1:1). The amount of oxygen presents is 0.95 and deficiency of oxygen or oxygen vacancy works as n-type. Similarly there is a deficiency of Fe (0.95) in FeO. At the time of doping the deficiency will be compensated and while all the deficiency will be fulfilled no doping is possible. Group IV–IV compound semiconductor, lead sulfide, lead selenide, and lead telluride are narrow band gap materials having high

TABLE 10.1

Visible Light Spectrum versus Colour of the Light Graph

Colour	Red	Orange	Yellow	Green	Blue	Violet
Wavelength, λ (nm)	620–750	590–620	570–590	495–570	450–495	380–450
Frequency (THz)	400–484	484–508	508–526	526–606	606–668	668–789

TABLE 10.2

Periodic Table of Elements Varies from Group II to Group VI

II	III	IV	V	VI
	Boron (B)	Carbon (C)	Nitrogen (N)	Oxygen (O)
	Aluminium (Al)	Silicon (Si)	Phosporus (P)	Sulphur (S)
Zinc (Zn)	Gallium (Ga)	Germanium (Ge)	Arsenic (As)	Selenium (Se)
Cadmium (Cd)	Indium (In)	Tin (Sn)	Antimony (Sb)	Tellurium (Te)
Mercury (Hg)	Thallium (Tl)	Lead (Pb)	Bismuth (Bi)	Polonium (Po)

TABLE 10.3

Classification and Details of Different Types of Semiconductors

Material Name	Classification	Symbol	Bandgap	Direct/Indirect
Boron nitride	III-V	BN	5.5	indirect
Boron nitride, Hexagonal		BN	5.96	quasi-direct
Boron phosphide		BP	2	indirect
Boron arsenide		BAs	1.14	direct
Boron arsenide		$B_{12}As_2$	3.47	indirect
Aluminium nitride		AlN	6.28	indirect
Aluminium phosphide		AlP	10.43	indirect
Aluminium arsenide		AlAs	10.1	indirect
Aluminium antimonide		AlSb	1.58	indirect
Gallium nitride		GaN	3.44	direct
Gallium phosphide		GaP	10.26	indirect
Gallium arsenide		GaAs	1.42	direct
Gallium antimonide		GaSb	0.726	direct
Indium nitride		InN	0.7	direct
Indium phosphide		InP	1.35	direct
Indium arsenide		InAs	0.36	direct
Indium antimonide		InSb	0.17	direct
Zinc oxide	II-VI	ZnO	3.37	direct
Zinc sulfide		ZnS	3.68	direct
Zinc selenide		ZnSe	10.71	direct
Zinc telluride		ZnTe	10.25	direct
Cadmium sulfide		CdS	10.42	direct
Cadmium selenide		CdSe	1.74	direct
Cadmium telluride		CdTe	1.49	direct
Lead sulfide	IV-IV	PbS	0.37	direct
Lead selenide		PbSe	0.27	direct
Lead telluride		PbTe	0.32	direct
Diamond	IV	C	5.45	indirect
Silicon		Si	1.12	indirect
Germinium		Ge	0.66	indirect
Grey tin		Sn	0.08	direct

carrier mobility, high dielectric constant and positive temperature co-efficient of bands gaps. Normally, bands gap increases as temperature decreases. However, for these materials, bands gaps increase with temperature.

10.3 How the Compound Materials Are Chosen?

Crystal structures determine the compatibility and heterostructure feasibility. The lattice spacing varies linearly with alloy composition. The best way is to start is by looking at lattice period of material with energy gap for selection of heterostructure compatibility.

The materials are chosen in such a way that they have minimum difference of lattice constant and the crystal structure determines the heterostructure feasibility. It has been observed that energy gap increases and lattice period decreases as one move up periodic table. The lattice spacing varies linearly with alloy composition and the narrower gap implying higher electron mobility.

Heterostructure are mainly three types depending on the alignment of smaller and larger bandgap materials. Band alignments determine characteristics as shown in Figure 10.1. In type I structure, the band gap of the smaller band gap materials aligns inside the larger band gap is called straddling type heterostructure. Most of the devices are designed based on this type of heterostructure. In the type III structure, smaller band gap materials align completely outside of the larger band gap materials is called broken gap type. Device fabrication is not possible from the broken gap type materials. In type II structure, a part of the small band gap materials is outside and certain percentage is inside the larger band gap materials and is called staggered type heterostructure. The conduction band and the valance of the smaller band gap semiconductor are denoted as E_{C1} and E_{V1}. Similarly, the conduction band and the valance of the larger band gap semiconductor are denoted as E_{C2} and E_{V10}.

Binary compound semiconductor formation could be done by various ways and the choices are with column III-V, column II-VI, column IV-VI, and column I-VII. If a material named A is chosen from column III and another material is chosen from column V, then the compound material will be formed like $A_{III}B_V$. The compound materials formed with column I and column VII of the periodic table are mainly insulators. Compound or alloy material can also have formed using column IV and column IV materials of the periodic table or elemental semiconductors (diamond, Si, Ge), like silicon carbide (SiC) and silicon germinium, (SiGe). SiGe is an indirect band gap compound semiconductor and carrier mobility is lower than GaN and GaAs. Therefore, it is not suitable for optoelectronic and high frequency applications. Band gap of SiGe is much lower; hence, it is unable to offer higher breakdown voltage, which is required for high voltage applications of the power devices. Among the high band gap materials for power industry, diamond (C) has the band gap of 5.5 eV and SiC has two band gaps in the crystalline form, cubic crystalline (10.3 eV) and hexagonal crystalline (3.0 eV). SiC can emit bright blue light, critical electric field is eight times higher than Si, thermal conductivity is three times higher than Si, good match of mechanical and thermal properties was a good choice as semiconductor material for harsh environment. SiC is the only compound semiconductor that forms a native oxide insulator (SiO$_2$) by a chemical reaction with oxygen. Although it must be noted that during oxidation of SiC, excess carbon at the interface of SiO$_2$ and SiC create high interfacial traps. These traps causes reduced conductance and hence, the electron channel mobility is

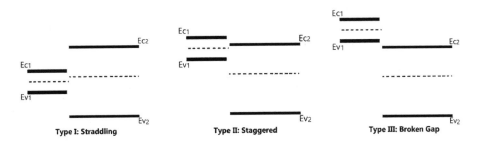

FIGURE 10.1
Different types of heterostructure types and relationship between narrow-wide band gaps.

also reduced for SiC MOSFETs. SiC does not form any 2-dimension electron-gas structure and much slower than Si and have lower ON-currents. Besides that, silicon carbide is not available as natural mineral and extensive furnace techniques are needed to produce the compound from Si, which makes the device very expensive.

To grow heterostructure, different band gap materials are required and ternaries do not fulfill the purpose. AlGaAs is an important exception as it is intrinsically lattice matched to GaAs. Besides, GaAs substrate is always available at discrete band gap. Therefore, it was the first material to use in heterostructure. Due to band gap engineering techniques and epitaxial growth developments, superior materials are available. The direct band gap compound semiconductor, gallium arsenide has a band gap of 1.42 eV but it is also important for light emitting diodes and key substrate material.

A graded band gap is obtainable by varying Al content in AlGaN/GaN. Increase in Al content increase the band gap of AlGaN. A ternary compound is formed like $A_xB_{1-x}C$, it implies that an alloy semiconductor is formed using 3 elements—A, B and C. The x atoms of element B is replaced by x atoms of A and the x is denoted by mole fraction. For alloy semiconductor, $Al_xGa_{(1-x)}N$, as mole fraction, x increases the band gap of the material also increases. A and B atoms are mixed in group of III sub-lattice and all group V lattice sites are occupied by element C. Lattice parameter and band gap energy of alloy composition is calculated by Vegard's Law by a given formula $E_g(A_xB_{1-x}C) = x E_g(AC) + (1-x) E_g(BC)$. A curvature correction is required due to by bowing parameter in band gap, b, arises due to disorder at the time of alloying. The formula is modified by $E_g(A_xB_{1-x}C) = x E_g(AC) + (1-x) E_g(BC) - b x(1-x)$. The bowing parameter, b is generally considered as 1.

10.4 Why GaN in Semiconductor Technology?

Generally the white light is generated by mixture of the three basic colors—red, green and blue. The RED and GREEN light is generated by GaP and BLUE is generated only by GaN. Therefore, GaN is very important for white light and wider spectral range in light emitting diodes (LEDs). GaN led to discovery of p-type GaN, p–n junction high performance blue/UV-LEDs and long-life violet-laser diodes. High brightness GaN LEDs completed the range of primary colours and used for daylight full colour LED displays. Due to high band gap of GaN (3.49 eV), transistors fabricated using GaN compared to Silicon (1.1 eV)- and GaAs (1.42 eV)-based transistors work in higher temperature up to 400°C whereas Silicon transistor works at temperature up to 150°C. Due to its higher band gap, the GaN transistors have higher breakdown voltage which is essential for higher output power generation in case of radio frequency (RF) applications. RF devices are essential for wireless communication, military applications like radars and missile seekers. GaN transistors make ideal power amplifiers at microwave frequencies and also showagreement asTHz devices. HEMTs have the advantages of high mobility and low noise for high-frequency (HF) and high-power (HP) applications compared with conventional Silicon MOSFETs. GaN based RF transistors are also used as microwave source for microwave oven replacing magnetrons.

In an energy and data-driven world, there is a huge requirement for small-size, compact, faster devices in the semiconductor power industry. GaN-based devices require one quarter of size, one quarter of weight, one quarter of power and less expensive in silicon-based solution compared to other devices. GaN shows high sheet charge, hence large current density. GaN-based devices are ten times faster than Silicon, therefore, it will be a suitable choice for 5G wireless technology. GaN gains additional boost from high electron mobility when it forms a 2-dimension electron gas (2DEG) at GaN/AlGaN interface. It doubles the electron mobility from 900 $cm^2V^-s^1$ to 2000 $cm^2V^-s^1$. GaN-based transistors have lower parasitic capacitor, therefore, transistors have higher cut off frequency. Silicon technology has been the primary option for the semiconductor industry since 1960s, with MOSFETs being the key electronic devices. However, some specific applications like LEDs, laser diodes and RF devices require the use of compound semiconductors due to its high electron mobility and band gap. Therefore, GaN is the most important semiconductor materials after Silicon. For GaN-based devices, very thin layer (1 nm) of single crystal formation is made possible by epitaxial growth, which is not possible for other bulk material. GaN is a very hard, mechanically stable wide band gap semiconductor materials with high capacity of thermal conductivity and high breakdown voltage as a power device. Despite lattice mismatch, GaN has been deposited as a thin film on sapphire, silicon carbide or zinc oxide using metal organic vapour phase epitaxy (MOVPE). Sapphire is very hard and even can damage the floor if it is dropped on the floor. Therefore, GaN-based devices fabricated on sapphire substrate is very reliable. Due to maximum lattice mismatch with zinc oxide, it is hardly used as a substrate. It is also possible to grow GaN on silicon nowadays. GaN on SiC can withstand up to 1000°C, which is great for temperature management for high-power applications, hence, it is more reliable.

A variety of power amplifier technologies with different III-V materials of suitable band gap such as GaAs (or GaAs/InGaP) heterojunction bipolar transistors (HBT), SiC MESFETs, and GaN High Electron Mobility transistors (HEMTs) are available to technology developers. GaN material has an electron affinity for AlN and InN, therefore, ternary epitaxial alloys are formed by band gap engineering. InGaN on GaN is harder to grow than AlGaN on GaN as the lattice parameter shows the misfit strain.

The combined figure-of-merit (CFOM) evident from Table 10.4 hows that GaN has superior attributes for high-power, high-frequency, high-temperature applications. With a large band gap of 3.49 eV, a breakdown field of 3.3×10^6 V/cm, electron mobility of 1350 $cm^2V^{-1}s^{-1}$ and its ability of forming a 2-dimensional electron gas (2DEG) at a heterojunction

TABLE 10.4

Physical Properties of Semiconductors

	Si (SiGe/Si)	GaAs (AlGaAs/ InGaAs)	InP (InAlAs/ InGaAs)	4H SiC	GaN (AlGaN/ GaN)	Diamond
Bandgap (eV)	1.1	1.42	1.35	3.26	3.49	5.45
Electron-mobility (cm²/Vs)	1500 (2800)	8500 (8000)	5400 (10000)	700	900 (>2000)	4000
Saturated (peak) Electron velocity (x10⁷cm/s)	1.0 (1.0)	1.0 (10.1)	1.0 (10.3)	10.0 (10.0)	1.5 (10.7)	10.8
2DEG sheet electron density (cm⁻²)	2×10^{12}	$<3 \times 10^{12}$	$<4 \times 10^{12}$	N.A.	$1–2 \times 10^{13}$	NA
Critical breakdown field (MV/cm)	0.3	0.4	0.5	10.0	3.3	10
Thermal conductivity (W/cm-K)	1.5	0.5	0.7	4.5	>10.0	10.2

interface due to its spontaneous polarization properties, with a typical carrier density of $>1 \times 10^{13}$ cm^{-2} and high electron saturation velocities up to 10.5×10^7 cm/s has made GaN one of the best choices for usage in the high-power and high-frequency regime operation.

HEMT, also known as heterostructure field-effective transistor (HFET) or modulation-doped field-effective transistor (MODFET) is a field-effect transistor incorporating a junction between two materials with different band gaps (i.e., a heterojunction) as the channel instead of a doped region (as is generally the case for MOSFET). HEMT devices are able to operate at higher frequencies than ordinary transistors, up to millimeter wave frequencies, and are used in high-frequency products such as cell phones, satellite television receivers, voltage converters, and radar equipment. To utilize all the advantages of GaN-based HEMTs in our daily life, a strong focus is needed not only on fabrication but also on the device/circuit modeling and design.

Enhancement mode GaN FETs (e-GaN FETs) is now replacing the power MOSFET applications where switching speed and power conversion efficiency is critical. The e-GaN FETs are built by growing a thin layer of GaN on a standard Si wafer which allows maintaining low cost as Silicon power MOSFET but with superior electrical performance.

GaN n-type doping is performed by silicon or oxygen and magnesium (Mg) is used for p-type doping. However, the Si and Mg doping makes GaN compound brittle, and it tends to have high dislocation density as dopants intrude a tensile stress. While doping with suitable transition metal such as manganese, GaN works as a promising magnetic semiconductor for spintronics device materials.

Figure 10.2 shows a heterojunction formed by the junction of narrow band gap (GaN) and wide band gap (Al$_x$Ga$_{(1-x)}$N) semiconductors. Electrons in general have potential and kinetic energy. The potential energy is because of the electrostatic field in the nuclei. Free electron level or vacuum level is the position with zero kinetic energy. Nuclei have positive charge by which it attracts the negative charged electrons to the nuclei. The positive energy means free particle with certain kinetic energy, whereas negative energy presents particles at certain energy level of the atom. Certain energy is required to remove the electron from the nuclei. It is a minimum distance where there is no attractive force between the electrons and the nuclei of the material once it is pulled out of the material. The distance is denoted as vacuum level. The amount of energy required to pull the electron from Fermi level to vacuum level is called work function (ϕ), and energy to pull the electron from conduction band to vacuum level is denoted as electron affinity (χ).

The intrinsic type GaN and an n-type Al$_x$Ga$_{(1-x)}$N semiconductor material has been chosen in this context to understand the heterostructure system. Once the junction has been formed, the materials will be aligned according to their Fermi level and both the conduction band and valence bands will be aligned accordingly. At room temperature, few electrons will be there in the conduction band of both the materials. The electrons in the conduction band of the Al$_x$Ga$_{(1-x)}$N are in the higher energy level and would like to go to the conduction band of the GaN to achieve the lowest possible energy once the junction is formed. Therefore, electron in the vicinity of the junction will move more towards the conduction band of the GaN. For Al$_x$Ga$_{(1-x)}$N, as it is losing electrons to its material, it will be less n-type or will become p-type. Hence, the distance between its Fermi level to the conduction band will increases and similarly for GaN materials will become more n-type as the electron concentration increases here. The electron concentration linearly decreases with the distance from the junction and similar phenomena will occur for Al$_x$Ga$_{(1-x)}$N, region near the junction will be more p-type and with the distance it will decrease. As a result of this band-bending, a huge number of electrons will be available on the GaN-side and a triangular potential well will be formed where the electron movements is possible in 2 dimensions

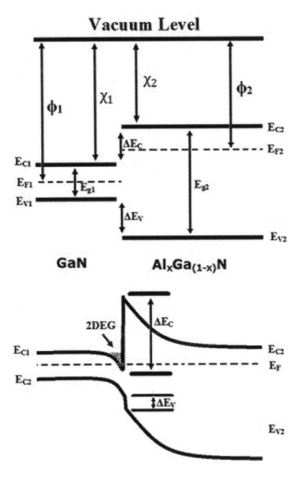

FIGURE 10.2
Band Diagram of (a) narrow bandgap (GaN) and wide bandgap (AlGaN) semiconductors (b) Formation of 2DEG at the interface of wide and narrow bandgap semiconductors. ΔEc and ΔEv represent the conduction and valence band offset, while χ denotes the electron affinity.

only. This is called as 2-dimension electron gas (2DEG) and the carrier mobility is much higher in this portion of the material as it is undoped. The property of the materials will be different than its bulk counterpart as the carrier mobility in one direction is restricted. Similarly, by the restrictions of carrier mobility in 2 directions, 1 dimension electron gas (1-DEG) and restrictions of carrier mobility in all 3 directions, 0 dimension electron gas (0-DEG) can be formed. In this way, only by restricting carrier mobility optical and electrical properties of the devices are altered.

For HEMT devices, when a voltage is applied between the source and drain terminals, the electrons will from the source to drain, resulting in terminal currents. The device will remain in the ON-state in the zero gate bias voltage and to switch off the device, it is required to deplete the 2DEG formed in junction of $Al_xGa_{(1-x)}N$ and GaN by application of negative gate voltage in the gate terminal. The negative voltage depletes or removes the channel and switch off the device. Thus, based on gate voltage, one can turn the device on and off.

10.5 Challenges in GaN HEMT Modeling

With aggressive scaling, the integration of millions of transistors and other passive devices on a single chip has become more challenging for process engineers. Thus, in the past few years, the importance of device modeling in electronics industry has grown tremendously. In today's market, to meet the huge demand for electronics products, not only progress in fabrication process is important, but compact and accurate device models are also critical for prediction of advanceddevice characteristics. For accurate prediction of device electrical behaviors, device-model development requires correct understanding on the layout, fabrication, and physical operations.

In contrast to Silicon transistors, the operation of III-V-based HEMTs is determined by the conductivity of the "2-Dimensional Electron Gas" (2DEG) located near the heterointerface. Though the channel in a Silicon transistor is similar in many respects to that in a heterostructure, the values of certain parameters force us to consider them otherwise. In GaN HEMTs, the small electron effective mass in GaN gives rise not only to strong degeneracy effects under a moderate level of the sheet concentration of the channel electrons but also to a strong effect of the transversal quantization of channel electrons. In a Silicon transistor, both of these factors are less pronounced. So, it is rational to have a different approach for HEMT modeling compared to Silicon transistor by incorporating a coupled Poisson-Schrödinger (PS) solution.

The first approach towards HEMT device modeling is to have a self-consistent solution of PS equation by using the finite-difference method (FDM). Real space is divided into discrete mesh points and the wave function is solved within those discrete spacing. The one-dimensional, one-electron Schrodinger equation is given by

$$-\frac{\hbar^2}{2}\frac{d}{dx}\left(\frac{1}{m^*(x)}\frac{d}{dx}\right)\psi(x)+V(x)\psi(x)=E\psi(x) \tag{10.1}$$

where $\psi(x)$ is the electron wave function, E is the total energy, V is the potential energy, $\hbar$ is reduced Planck's constant, and m^* is the effective mass. Applying the Poisson's equation at the heterointerface, one can write

$$\frac{d}{dx}\left(\varepsilon_{\text{uid}}(x)\frac{d}{dx}\right)\phi(x)=-\frac{q\left[N_D(x)-n(x)\right]}{\varepsilon} \tag{10.2}$$

where ε_{uid} is the dielectric constant (uid=**unintentionally** doped, in this case refer to the GaN buffer), ϕ is the electrostatic potential, $N_D(x)$ is the ionized donor concentration, and $n(x)$ is the electron density distribution. In a quantum well of arbitrary potential energy profile, the potential energy V is related to the electrostatic potential ϕ as follows:

$$V(x)=-q\phi(x)+\Delta E_c(x) \tag{10.3}$$

where ΔE_c is the conduction band offset at the heterointerface. The wave function $\psi(y)$ in (10.1) and electron density $n(x)$ in (10.2) are related by

$$n(x)=\sum_{k=1}^{n}\psi_k^*(x)\psi_k(x)n_k \tag{10.4}$$

where n is the number of bound states, and n_k is the electron occupation for each state. Based on the Fermi-Dirac statistics, the electron concentration for each state can be expressed by:

$$n_s = \frac{m^*}{\pi\hbar^2}\int_{Ek}^{\infty}\frac{dE}{1+e^{(E-E_f)/kT}} = \frac{m^*kT}{\pi\hbar^2}\ln\left[1+e^{\frac{E_f-E_k}{kT}}\right]$$ (10.5)

where E_k is the Eigen energy. Iterative solution of (10.1–10.5) will give the 2DEG concentration profile as well as the potential in a heterostructure. This approach is implemented in mostly available Poisson-Schrödinger solver as in most commercially available TCAD simulators. However, this procedure is time consuming and hence not applicable for circuit designers.

A simpler way to solve this problem is by the assumption of a triangular quantum well at the AlGaN/GaN heterointerface. The potential well in which the 2DEG is trapped is triangular in shape close to the heterojunction, although it flattens out in the buffer. Assuming that the well is perfectly triangular, one can use the relation

$$V(x) = qFx$$ (10.6)

where q is the electronic charge and F is the electric field. After solving (10.1) using (10.6), one can express the quantized energy levels as

$$E_n = c_n\left[\frac{(qF\hbar)^2}{2m}\right]^{1/3} = c_n\left[\frac{\hbar^2}{2m}\left(\frac{q^2 n_s}{\varepsilon_{\text{uid}}}\right)^2\right]^{1/3}, \quad n = 0,1,2,3,\dots$$

$$= \gamma_n n_s^{2/3}, \quad n = 0,1,2,3,\dots$$ (10.7)

where c_n can be obtained from the so called Airy function and is equal to 10.338 for the lowest energy level. From WKB theory, c_n can be approximately expressed as

$$c_n \approx \left[\frac{3}{2}\pi\left(n-\frac{1}{4}\right)\right]^{2/3}$$ (10.8)

Yoshida [3] made a comparison between the self-consistent 2-D model, 3-D Fermi model and the quantum mechanical model with triangular well approximation. The classical approach using Fermi statistics was found reasonably accurate although computationally difficult due to the Fermi integral. The triangular well model may need some tuning to exactly match the self-consistent solution by replacing F in (10.7) with jF, where j is the tuning parameter ranging between 0 and 1. It was shown that the classical Boltzmann statistics always over-estimates the capacitance in moderate- as well as strong-inversion region. The Fermi statistics shows a closer match with the coupled P-S solution in the strong inversion but still overestimates the capacitance value in the moderate-inversion.

For compact modeling of HEMT devices, instead of relying on the coupled P-S solution, most literatures follow the work of Delagebeaudeuf and Linh [4] whereby only 2 sub-bands are considered, and the 2DEG density n_s is given by

$$n_s = Dv_{\text{th}}\ln\left[\left(1+e^{(E_f-E_0)/v_{\text{th}}}\right)\left(1+e^{(E_f-E_1)/v_{\text{th}}}\right)\right]$$ (10.9)

$$E_0 = \gamma_0 n_s^{2/3} \tag{10.10}$$

$$E_1 = \gamma_1 n_s^{2/3} \tag{10.11}$$

where D is the density of states of the 2DEG associated with a single quantized energy level, with E_0 and E_1 being the two lowest sub-bands in the triangular well with respect to the conduction-band edge at the heterointerface, γ_0 and γ_1 are the constants derived from (10.7) related to lowest and second sub-band and $v_{th} = kT/q$ is the thermal voltage. The energy-band diagram of a generic AlGaN/GaN metal-insulator-semiconductor (MIS)-HEMT is shown in Figure 10.3 [5], in which physical parameters are shown by the respective labels. Φ_{MB} and Φ_{SB} represent the metal-to-insulator and insulator-to-GaN cap barrier height, respectively. T_{cap}, T_{ox}, d, and e represent the thickness of insulator, GaN cap, AlGaN barrier and AlGaN undoped spacer layer, respectively. The generic MIS-HEMT includes a cap layer with the same material as the unintentionally doped (UID) body, which automatically becomes a HEMT if T_{cap}, T_{ox}, Φ_{MB}, and Φ_{SB} are set to zero. By applying Gauss Law at the 2DEG interface, n_s in (10.9) can be related to the gate (V_g) and channel (V_c) voltages as

$$n_s = \frac{C_d}{q}\left(V_g - V_{off} - \phi_s\right) = \frac{C_d}{q}\left(V_g - V_{off} - V_c - E_f\right) \tag{10.12}$$

where the Fermi potential, E_f in eV, is referenced to the bottom of the triangular well (conduction-band edge at $x = 0$), and the surface potential, ϕ_s in V, is referenced to the Fermi level in the body where the channel voltage $V_c = V_s$ or V_d at the source or drain ends ($y = 0$ or gate length L_g), respectively, given by

$$\phi_s\left(V_c\right) = V_c + E_f\left(V_c\right) \tag{10.13}$$

where C_d is an effective capacitance (per unit area) between the gate and 2DEG, and V_{off} is the cutoff voltage below which it is in "subthreshold" conduction. By assuming the AlGaN barrier layer being fully depleted and ignoring the electron charge in the cap layer, C_d and V_{off} can be expressed as

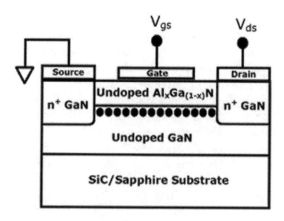

FIGURE 10.3
The cross-section of a typical AlGaN/GaN HEMTstructures.

$$C_d = \frac{\varepsilon_b}{d\left(1 + \dfrac{\varepsilon_b T_{ox}}{\varepsilon_{ox}d} + \dfrac{\varepsilon_b T_{cap}}{\varepsilon_{cap}d}\right)} \tag{10.14}$$

$$V_{off} = \Phi_{MB} - \Phi_{SB} - \left(\frac{T_{ox}}{\varepsilon_{ox}} + \frac{T_{cap}}{\varepsilon_{cap}}\right)qN_b\left(d - e\right)$$
$$- \frac{qN_b}{2\varepsilon_b}\left(d - e\right)^2 - \frac{q\sigma d}{\varepsilon_b} \tag{10.15}$$

where N_b is the doping concentration of the AlGaN barrier layer, ε_b and ε_{ox} are the permittivity of the barrier layer and insulator, respectively, and σ is the polarization-induced charge density at the heterointerface.

10.6 Available Compact Models of AlGaN/GaN HEMT

An alternative approach of device modeling is the physics-based compact model that analyzes the cause of the behavior in the devices based on semiconductor physics such as material characteristics and carrier transport. The equations in physics-based compact models are physical, often with a much smaller number of parameters than empirical models. These physical parameters can not only be extracted from one's measurement, but also be taken from other independent sources, which saves time and effort in model construction.

Currently, there are two industry standard GaN HEMT compact models: the ASM-HEMT model [6–8], and the MIT-Virtual Source model [9,10].The ASM-HEMT (**A**dvanced **S**PICE **M**odel for GaN **HEMT**s) model is a surface-potential model whereby the variation of the Fermi potential $E_f(V_g)$ in the triangular quantum well is divided into three regions of operation such as strong-inversion, moderate-inversion and subthreshold region as shown in Figure 10.4. Regional solutions of these three regions are mathematically derived and finally unified drain-current and charge-voltage equation are proposed. In this model, a new algorithm was proposed to obtain the open-form Poisson and Schrodinger's equations that can be implemented in standard circuit simulator. Apart from current model and charge model, the ASM-HEMT model captures most of the device characteristics required by circuit simulation, such as gate leakage, self-heating, trapping and noise.

The Virtual Source model, on the other hand, was first proposed at MIT for highly scaled silicon-based FETs with quasi-ballistic mode of transport and later extended to drift-diffusive transport for GaN HEMTs. The Virtual Source model calculates the density of carriers, which flow in the channel to form the transistor current model and integrates the carrier distribution along the channel to form the transistor charge model. The model is physical based on the drift-diffusive transport theory and employs only a small number of fitting parameters to form the model (Figure 10.5).

Apart from the above-mentioned compact models, other research groups have also demonstrated physics-based models [11–14]. One such model was developed at NTU-Singapore (Xsim) [11,15–17], where the unified reginal modeling approach was being used to develop

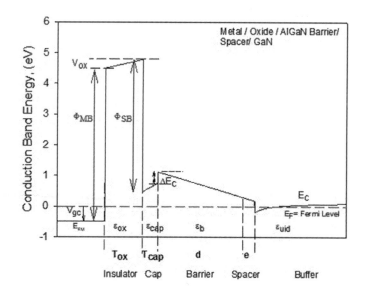

FIGURE 10.4
Band diagram of (a) MIS-HEMT structure obtained from coupled Poisson-Schrodinger solution. Fermi potential (in volts) is referenced to the conduction-band edge at the heterointerface ($x = 0$).

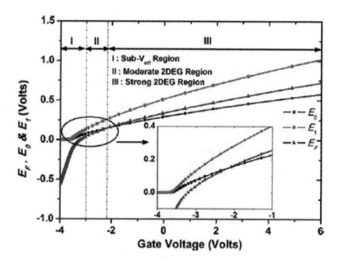

FIGURE 10.5
Variation of the Fermi potential w.r.t the gate voltage in a generic GaN HEMT device. (From Khandelwal, S. et al., *IEEE Trans. Electron Devices*, 58, 3622–3625, 2011.)

the surface potential-based model. Based on this model, apart from the Fermi potential variation, the 2DEG capacitance C_{gg} can be expressed as

$$C_{gg} = q \frac{dn_s}{dV_g} = C_d \left(1 - \frac{dE_f}{dV_g} \right) \tag{10.16}$$

Furthermore, n_s related to E_f and E_0 is given by (10.9), from which dn_s/dV_g can be written as

$$\frac{dn_s}{dV_g} = \frac{D}{qe^{n_s/Dv_{th}}}\left(e^{n_s/Dv_{th}} - 1\right)\left(\frac{dE_f}{dV_g} - \frac{dE_0}{dV_g}\right) \tag{10.17}$$

From (10.10), dE_0/dV_g can be written as

$$\frac{dE_0}{dV_g} = \frac{2}{3}\gamma_0 n_s^{-1/3}\frac{dn_s}{dV_g} \tag{10.18}$$

Combining (10.15–10.17), the gate capacitance C_{gg} can be expressed as,

$$C_{gg} = \frac{C_{K1}}{1 + C_{K1}C_{K2}} \tag{10.19}$$

$$C_{K1} = \frac{D\left(e^{n_s/Dv_{th}} - 1\right)}{e^{n_s/Dv_{th}}} \tag{10.20}$$

$$C_{K2} = \frac{1}{C_g} + \frac{2\gamma_0}{3}n_s^{-1/3} \tag{10.21}$$

Most of the existing compact models for HEMT devices follow the n_s expression with two sub-bands in the triangular well shown in Figure 10.6 given by (10.12). Through Gauss' law, n_s in (10.12) is related to the gate (V_g) and channel (V_c) voltages as

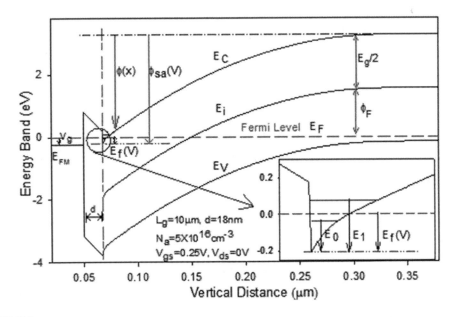

FIGURE 10.6
Energy-band diagram of AlGaN/GaN heterostructure for $N_a = 5 \times 10^{16}$ cm^{-3}, showing the relation between actual surface potential and the Fermi potential (E_f). (From Yoshida, J., *IEEE Trans. Electron Devices*, ED-33, 154–156, 1986.)

$$n_s = \frac{C_d}{q}\left(V_g - V_{off} - V_c - E_f\right) = \frac{C_d}{q}\left(V_g - V_{off} - \phi_s\right) \qquad (10.22)$$

The surface potential ϕ_s in V, is referenced to the Fermi level in the body with channel voltage $V_c = V_s$ or V_d at the source or drain ends ($y = 0$ or L_g), respectively, given by

$$\phi_{s1}\left(V_c\right) = V_c + E_f\left(V_c\right) \qquad (10.23)$$

Although these models are scalable with most of the primary device parameters, they fail to capture the effect of GaN buffer doping (N_a) with corrects built-in physics. There are quite a few reports about non-desirability of "undoped or unintentionally doped" buffer in GaN HEMTs [18,19]. Inclusion of deep-level dopants in the GaN buffer is necessary to control bulk leakage/short-channel effects (SCEs) [20,21]. Thus, for short-channel HEMT devices, capturing model dependency on N_a is very much necessary. A comparison of E_f w.r.t V_g computed from (10.12–10.14) and that from coupled P-S numerical simulation in TCAD of a GaN HEMT with $N_a = 5 \times 10^{16}$ cm^{-3} is shown in Figure 10.7. The E_f solution considering Boltzmann approximation as used for metal-oxide-semiconductor (MOS) system is also shown along with the first-order derivative of E_f w.r.t V_g. There are two main differences between (10.12) and coupled P-S solution: firstly, a considerable V_{off}-shift is observed when doping is considered compared to (10.12), which does not have N_a dependent term and secondly, the first order derivatives w.r.t V_g in the subthreshold region are different. Although one can fit (10.12) with coupled P-S solution by tuning V_{off}, the discrepancy in the first-order derivative in the subthreshold region cannot be fixed. As (10.12) does not consider any bulk charge term based on N_a and is fully based on inversion charge, the first-order derivative in the subthreshold region is always ~1. Whereas, coupled P-S solution considering both bulk charge and inversion charge results in surface potential slope less than unity in the below-V_{off} region. Also, the referencing of surface potential to the Fermi level in the body as shown in (10.22) is also different from the conventional MOS system

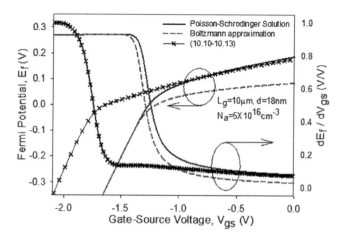

FIGURE 10.7
Simulated Fermi potential E_f versus gate-to-source voltage V_{gs} from TCAD using Boltzmann statistics and coupled P-S solution compared with the E_f computed from (10.10–10.13). Corresponding derivatives of the E_f solutions w.r.t. V_{gs} is shown on the right-hand axiswhere the slope is less than unity in the subthreshold region for P-S solution in contrast to E_f from (10.10–10.13). (From Yoshida, J., *IEEE Trans. Electron Devices*, ED-33, 154–156, 1986.)

where surface potential is referenced to the intrinsic Fermi level E_i or the conduction band E_c. The surface potential referencing in (10.22) is physically inaccurate as it will never capture the effect of GaN buffer doping on the surface potential as shown in Figure 10.6 where the band energy diagram across the AlGaN/GaN heterostructure interface is shown for $N_a = 5 \times 10^{16}$ cm^{-3}. This discrepancy is accurately addressed by the NTU-Xsim HEMT model by referencing the actual surface potential to the conduction band E_c in the body, so that it takes into consideration the term ϕ_F which is the difference between the Fermi level E_F and the intrinsic potential E_i.

One important point to address is the source and drain access regions in HEMT devices. Conventionally, the source and drain resistance are modeled using a sub-circuit approach in Verilog-A. The schematic diagram is shown in Figure 10.8, where the intermediate voltages are termed as V_{si} and V_{di} at the source and drain side respectively. Due to the high field effects, the series resistance R_s and R_d is not a linear function but depends on the velocity saturation effects as

$$R_s = \frac{L_{gs}}{qW\mu_{rs}\sigma}\left(1 + \xi_{rs}\left(\frac{\mu_{rs}\left(V_{si} - V_s\right)}{v_{sats}L_{gs}}\right)^2\right)^{\frac{1}{2}} \tag{2.24}$$

$$R_d = \frac{L_{gd}}{qW\mu_{rd}\sigma}\left(1 + \xi_{rd}\left(\frac{\mu_{rd}\left(V_d - V_{di}\right)}{v_{satd}L_{gd}}\right)^2\right)^{\frac{1}{2}} \tag{2.25}$$

where L_{gs} (L_{gd}) is the gate to source (gate to drain) access region length, μ_{rs} (μ_{rd}) is the source (drain) access region mobility, v_{sats} (v_{satd}) is the saturation velocity in the source (drain) access region, ξ_{rs} and ξ_{rd} are tuning parameters.

Although only the buffer doping and access region resistance modeling were discussed in the above section, several important aspects must be taken into consideration for HEMT model development such as short-channel effects/DIBL, self-heating effects, velocity saturation and mobility degradation effects. One of the most critical issues for AlGaN/GaN HEMTs is the so-called "current collapse" [22,23]: the presence of traps—located either

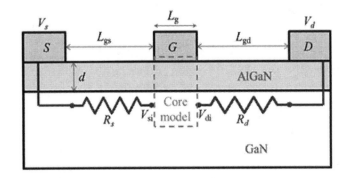

FIGURE 10.8
Schematic of sub-circuit approach for R_s/R_d model. (From Yoshida, J., *IEEE Trans. Electron Devices*, ED-33, 154–156, 1986.)

under the gate or in the gate–drain access region—which results in arecoverable decrease in drain current when exposed to high negative gate voltages (gate-lag), or to high drain voltage levels (drain-lag) in OFF-state. A detailed model development for current collapse effect has been reported by Syamal et al. [24] for interested readers.

10.7 Simulation of DC Characteristics

To explore the device characteristics of new material system, commercial TCAD simulators are useful to great extent. In this section, the different aspects of GaN HEMT TCAD simulation are discussed. Although, this chapter focuses on SILVACO-based-TCAD simulations [25], the outline is similarly applicable for other commercially available simulators.

A basic GaN HEMT simulation deck begins with MESH statements and corresponding meshing density. Extensive care should be taken in the gate-drain edge since due to the peak electric field simulation studies. After mentioning the four terminals (drain/gate/source/substrate), comes the MODEL statements. Two key parameters to be mentioned here are the MOBILITY and POLARIZATION statements. For GaN material system, based on the works of Albrecht et al. [26], the ALBRCT model is available for electron/holes. For high field mobility simulations, a dedicated model known as GANSAT is available which is based on a fit to Monte Carlo data for nitride-based materials. Polarization in wurtzite materials is characterized by two components, spontaneous polarization, P_{sp} and piezoelectric polarization, P_{pi}. The polarization enters into the simulation as a positive and negative fixed charges appearing at the top (most negative Y coordinate) and bottom (most positive Y coordinate) of the layer in question. By default, the positive charge is added at the bottom and the negative charge is added at the top. You can modify the sign and magnitude of this charge by specifying POLAR.SCALE in the REGION statement. The default value for POLAR.SCALE is 1.0.

Although the GaN based devices have the advantage of high electron density and output current, the high current flow generates a lot of heat which is known as self-heating. Self-heating is a serious concern in GaN devices. Due to self-heating, channel temperatures can reach several hundred degrees above the ambient base temperature. The temperature increases can significantly change the temperature dependent material properties like band-gap and mobility which lead to degradation of device performance. The reduction in mobility leads to a reduction in current due to increased operating voltage. This decreases the maximum power density and also increases the gate leakage. A simple transverse and output characteristics of GaN-HEMTS from TCAD simulation are shown in Figures 10.9 and 10.10. As seen in Figure 10.10b, with self-heating effects, the drain current tends to decrease with increasing drain-to-source voltage in the high V_{ds} region.

Once the basic simulation deck is created, the various GaN-based effects such as "current-collapse" can be studied by specifying the insulator interface traps using the "INTTRAP" statement. The key parameters that need to be mentioned are the trap activation energy (from conduction-band edge), the density of the interface traps (/cm^2) and the trap capture cross-section. Another way to study the effect of traps is to by focusing on bulk traps, which can be incorporated in both AlGaN barrier of GaN buffer by using the "DOPING TRAP" statements.

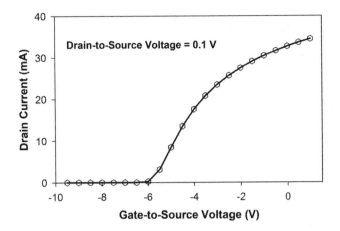

FIGURE 10.9
Transverse characteristics simulated using SILVACO-TCAD for a generic AlGaN/GaN HEMT.

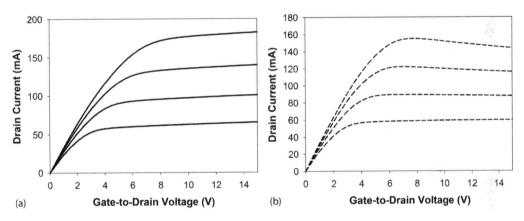

FIGURE 10.10
Output characteristics simulated using SILVACO-TCAD for a generic AlGaN/GaN HEMT without (a) and with (b) self-heating effects.

10.8 Analysis of RF Performance

GaN HEMTs offers high-breakdown voltage, so that with large drain voltages, high output impedance per watt of RF power can be achieved, resulting in lower loss matching circuits. The high-saturated drift velocity leads to high-saturation current densities and watts per unit gate periphery. This leads to lower capacitances per watt of output power-making GaN HEMTs suitable for high speed switch-mode amplifiers [27].

In general, RF transistor are characterized by two important parameters; the cut-off frequency (f_T) and the maximum oscillation frequency (f_{max}). The cut-off frequency is defined for which the current gain equal to 1(0 dB), whereas the maximum oscillation frequency corresponds to the maximum frequency of use of the transistor, wherein the power gain is 1 (0 dB). Mathematically, f_T and f_{max} can be derived as shown below from the small-signal equivalent circuit shown in Figure 10.11a.

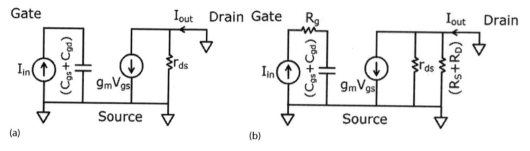

FIGURE 10.11
Small-signal equivalent circuit without (a) and with (b) source/drain resistance R_s/R_d.

Applying KVL, the gate-to-source voltage V_{gs} can be expressed as

$$V_{gs} = \frac{I_{in}}{j\omega(C_{gs} + C_{gd})} \tag{10.26}$$

$$I_{out} = g_m V_{gs} \tag{10.27}$$

$$V_{gs} = \frac{I_{out}}{g_m} \tag{10.28}$$

where C_{gs} and C_{gd} represent the gate-to-source and gate-to-drain capacitances and g_m is the small-signal transconductance of the transistor.

By comparing Equations (10.25) and (10.27)

$$\frac{I_{out}}{g_m} = \frac{I_{in}}{j\omega(C_{gs} + C_{gd})} \tag{10.29}$$

$$\frac{I_{out}}{I_{in}} = \frac{g_m}{j\omega(C_{gs} + C_{gd})} \tag{10.30}$$

As the cut-off frequency is defined for which the current gain equal to 1 or $\frac{I_{out}}{I_{in}} = 1$,

$$1 = \left| \frac{g_m}{(C_{gs} + C_{gd})} \right| \tag{10.31}$$

$$\omega_T = \frac{g_m}{(C_{gs} + C_{gd})} \tag{10.32}$$

$$f_T = \frac{1}{2\pi} \left(\frac{g_m}{(C_{gs} + C_{gd})} \right) \tag{10.33}$$

As shown in Figure 10.8, HEMTs are associated with source and drain access region resistances, and thus the equivalent circuit modified is shown in Figure 10.11b. Mathematically, f_t and f_{max} can then be defined as given in the Equations 10.33 and 10.34.

$$f_T = \frac{g_m}{2\pi \times \left[(C_{gs} + C_{gd}) + g_m C_{gd} (R_S + R_D) \right]} \tag{10.34}$$

$$f_{max} = \frac{1}{2} \frac{f_T}{\sqrt{2\pi \times f_T \times C_{gd}(R_g + R_s) + \dfrac{R_g + R_s}{r_{ds}}}} \tag{10.35}$$

The effect of the source/drain resistance is critical in analyzing the RF performance of GaN HEMTs and requires further studies of subsequent s-parameters and high frequency noise behaviours.

10.9 Conclusion

A detailed overview of GaN as an alternative material to Silicon was presented in this chapter. In last 10 years, the microelectronics research community is exploring GaN-based system extensively due to five main advantages such as (a) high breakdown voltages, due to high bandgap, (b) high current density, due to higher 2DEG, (c) high temperature operation, due to good thermal conductivity, (d) lower noise, due to less carrier scattering and (e) high cut-off frequencies, due to high saturated electron velocity and low-field mobility. There are inherent challenges associated with this technology such as the effect of traps and defects which appears in the form of "current-collapse" or "DC-RF dispersion," and this requires further studies and optimization of the substrate-growth as well as device-design.

The International Technology Roadmap for Semiconductors [28] explains that the current trend has two different paths: The first one focuses on the conventional "miniaturization" and its associated benefits in terms of performances and is labeled as "More Moore." The second trend is characterized by functional diversification of semiconductor-based devices which alongside also contribute to the miniaturization of electronic systems, although they do not necessarily scale at the same rate as the one used in digital functionality. Accordingly, this trend is designated as "More-than-Moore." One important subject of interest of "More-than-Moore" is RF communication and security, where GaN-based systems comes into the picture. The current state-of-the-art research on GaN HEMTs show promising capabilities for hybrid designs in future generation III-V/Si co-integrated ULSI systems.

References

1. J. T. Clemens, "Silicon microelectronics technology," *Bell Labs Tech. J.*, vol. 2, no. 4, pp. 76–102, 2002.
2. S. Tolbert, A. Herhold, C. Johnson, and A. Alivisatos, "Comparison of quantum confinement effects on the electronic absorption spectra of direct and indirect gap semiconductor nanocrystals," *Phys. Rev. Lett.*, vol. 73, no. 24, pp. 3266–3269, 1994.

3. J. Yoshida, "Classical versus quantum mechanical calculation of the electron distribution at the N-Algaas/Gaas heterointerface," *IEEE Trans. Electron Devices*, vol. ED-33, no. 1, pp. 154–156, 1986.

4. D. Delagebeaudeuf and N. T. Linh, "Metal-(n) AlGaAs-GaAs two-dimensional electron gas FET," *IEEE Trans. Electron Devices*, vol. 29, no. 6, pp. 955–960, 1982.

5. B. Syamal, "Compact modeling for GaN HEMT devices," Nanyang Technological University, Singapore, 2017.

6. S. Khandelwal, N. Goyal, and T. A. Fjeldly, "A physics-based analytical model for 2DEG charge density in AlGaN/GaN HEMT devices," *IEEE Trans. Electron Devices*, vol. 58, no. 10, pp. 3622–3625, 2011.

7. S. Khandelwal and T. A. Fjeldly, "A physics based compact model of I–V and C–V characteristics in AlGaN/GaN HEMT devices," *Solid. State. Electron.*, vol. 76, pp. 60–66, 2012.

8. S. Ghosh, A. Dasgupta, S. Khandelwal, S. Agnihotri, and Y. S. Chauhan, "Surface-potential-based compact modeling of gate current in AlGaN/GaN HEMTs," *IEEE Trans. Electron Devices*, vol. 62, no. 2, pp. 443–448, 2015.

9. U. Radhakrishna, L. Wei, D. S. Lee, T. Palacios, and D. Antoniadis, "Physics-based GaN HEMT transport and charge model: Experimental verification and performance projection," in *Technical Digest – International Electron Devices Meeting, IEDM*, pp. 13–16, 2012.

10. U. Radhakrishna, D. Piedra, Y. Zhang, T. Palacios, and D. Antoniadis, "High voltage GaN HEMT compact model: Experimental verification, field plate optimization and charge trapping," *Tech. Dig. – Int. Electron Devices Meet. IEDM*, pp. 814–817, 2013.

11. B. Syamal, S. Ben Chiah, and X. Zhou, "GaN HEMT compact model for circuit simulation," *Electron Devices and Solid-State Circuits (EDSSC), 2015 IEEE International Conference*, no. 3, pp. 535–538, 2015.

12. X. Cheng, M. Li, and Y. Wang, "An analytical model for current-voltage characteristics of AlGaN/GaN HEMTs in presence of self-heating effect," *Solid State Electron.*, vol. 54, no. 1, pp. 42–47, 2010.

13. N. Dasgupta and A. Dasgupta, "A new spice mosfet level 3-like model of hemt's for circuit simulation," *IEEE Trans. Electron Devices*, vol. 45, no. 7, pp. 1494–1500, 1998.

14. J. Zhang, B. Syamal, X. Zhou, S. Arulkumaran, and G. I. Ng, "A compact model for generic Mis-hemts based on the unified 2deg density expression," *IEEE Trans. Electron Devices*, vol. 61, no. 2, pp. 314–323, 2014.

15. X. Zhou, S. B. Chiah, B. Syamal, A. Ajaykumar, X. Liu, and H. Zhou, "Compact modeling of III-V/Si FETs," in *Proceedings – 2014 IEEE 12th International Conference on Solid-State and Integrated Circuit Technology, ICSICT 2014*, 2014.

16. X. Zhou, G. Zhu, G. H. See, K. Chandrasekaran, S. Ben Chiah, and K. Y. Lim, "Unification of MOS compact models with the unified regional modeling approach," *J. Comput. Electron.*, vol. 10, no. 1–2, pp. 121–135, 2011.

17. X. Zhou, J. Zhang, B. Syamal, S. Ben Chiah, H. Zhou, and L. Yuan, "Unified regional modeling of GaN HEMTs with the 2DEG and DD formalism," in *ICSICT 2012 – 2012 IEEE 11th International Conference on Solid-State and Integrated Circuit Technology, Proceedings*, 2012.

18. V. Desmaris et al., "Comparison of the DC and microwave performance of AlGaN/GaN HEMTs grown on SiC by MOCVD with Fe-doped or unintentionally doped GaN buffer layers," *IEEE Trans. Electron Devices*, vol. 53, no. 9, pp. 2413–2417, 2006.

19. Y. C. Choi, L. F. Eastman, and M. Pophristic, "Effects of an Fe-doped GaN buffer in AlGaN/GaN power HEMTs on Si substrate," in *ESSDERC 2006 – Proceedings of the 36th European Solid-State Device Research Conference*, 2006, pp. 282–285.

20. M. J. Uren et al., "Intentionally carbon-doped algan/gan HEMTs: Necessity for vertical leakage paths," *IEEE Electron Device Lett.*, vol. 35, no. 3, pp. 327–329, 2014.

21. H. Tang, J. B. Webb, J. A. Bardwell, S. Raymond, J. Salzman, and C. Uzan-Saguy, "Properties of carbon-doped GaN," *Appl. Phys. Lett.*, vol. 78, no. 6, pp. 757–759, 2001.

22. A. Sleiman, A. Di Carlo, G. Verzellesi, G. Meneghesso, and E. Zanoni, "Current collapse associated with surface states in GaN-based HEMT's. Theoretical/experimental investigations," *Simul. Semicond. Process. Devices*, pp. 81–84, 2004.

23. A. Chini et al., "Experimental and numerical correlation between current-collapse and Fe-doping profiles in GaN HEMTs," in *IEEE International Reliability Physics Symposium Proceedings*, pp. 2–5, 2012.

24. B. Syamal, X. Zhou, S. B. Chiah, A. M. Jesudas, S. Arulkumaran, and G. I. Ng, "A comprehensive compact model for GaN HEMTs, including quasi-steady-state and transient trap-charge effects," *IEEE Trans. Electron Devices*, vol. 63, no. 4, pp. 1478–1485, 2016.

25. Silvaco, Inc., *Atlas User's Manual*, Silvaco, Inc., pp. 567–1000, 2016.

26. J. D. Albrecht, R. P. Wang, P. P. Ruden, M. Farahmand, and K. F. Brennan, "Electron transport characteristics of GaN for high temperature device modeling," *J. Appl. Phys.*, vol. 83, no. 9, p. 4777, 1998.

27. R. S. Pengelly, S. M. Wood, J. W. Milligan, S. T. Sheppard, and W. L. Pribble, "A review of GaN on SiC high electron-mobility power transistors and MMICs," *IEEE Transactions on Microwave Theory and Techniques*, vol. 60, no. 6 PART 2. pp. 1764–1783, 2012.

28. W. Arden, M. Brillouët, P. Cogez, M. Graef, B. Huizing, and R. Mahnkopf, "More-than-Moore," ITRS White Paper, p. 31, 2010.

11

Breakdown Voltage Improvement Techniques in AlGaN/GaN HEMTs

Vimala Palanichamy

CONTENTS

11.1 Effect of Passivation

In high electron mobility transistors (HEMTs), which are also called modulation-doped field-effective transistor (MODFETs), the modulation doping is achieved by doping a wide band gap semiconductor layer grown adjacent to a narrow band gap semiconductor to form a junction between two materials with different band gaps as the channel instead of doped region. The two dissimilar compound semiconductor materials, which are grown one above the other, form a heterostructure, thus it is also called a heterostructure FETs. AlGaN/GaN high-electron-mobility transistors (HEMTs) have received much attention for their ability to operate at high-power levels [1]. They are very useful components for the development of base stations in the telecommunications networks and for civil, military and space radar applications. There are several economic and technological stakes, which require the development of suitable techniques for failure analysis on GaN-based HEMTs. The failure analysis techniques implemented for GaAs components are not directly transposable to the GaN components [2–6] because of the different band gaps of the two semiconductor materials.

These materials require the development of new methodology for physical analysis, such as specific polishing techniques and FIB cross-section preparation, new chemical decoration and delayering techniques.

When the component presents a passivation-related problem, traps located at the semiconductor surface contribute to the creation of a virtual gate, which modifies the distribution of the electric field in the structure. This causes a reduction in the off-state breakdown voltage and maximum output power (P_{max}).

11.1.1 Passivation

Passivation refers to a material to become "passive," i.e., the material is less affected by the environmental conditions. Passivation involves creation of an outer layer of shield material that is applied as a micro coating, created by chemical reaction with the base material, or allowed to build from spontaneous oxidation in the air.

As a technique, passivation is the use of a light coat of a protective material, such as metal oxide, to create a shell against corrosion. Passivation protects the semiconductor surfaces from electrical and chemical contaminants [7,8]. Passivation can occur only in certain conditions and is used in microelectronics to enhance silicon. The technique of passivation strengthens and preserves the appearance of metallics.

Semiconductor devices such as gallium arsenide field effect transistors (GaAs FET) [9], HEMTs are commonly used for its high-voltage and high-speed applications such as military and commercial microwave applications. It is necessary that these devices last a considerable amount of the time and are immune to the deteriorating effects of elements within the environment in which they are used. Without such immunity, oxygen, hydrogen, water, etc. will operate to modify the electrical characteristics of these devices overtime. For example, semiconductor devices exposed to hydrogen generally demonstrate a shift or a gradual drift in the pinch-off voltage and also in the optimum gate voltage, which reduces the maximum transconductance and the maximum output power of the device.

To cater to these issues, it is common practice to encapsulate (passivates) semiconductor devices (HEMTs), [10–14] by an inert material to isolate the device from its immediate environment and to protect the device from oxygen, water, etc. within the environment. In general, all Group III-V FETs employ some form of encapsulation process to protect these

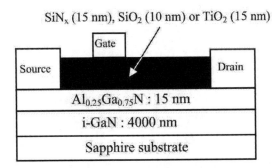

FIGURE 11.1
Schematic cross section of AlGaN/GaN HEMT Structure.

devices from airborne or space-borne contaminants, particulates, and humidity. Figure 11.1 shows the passivation of SiN/SiO2/TiO2 on AlGaN/GaN HEMT device.

This encapsulation process also passivates these semiconductor devices by terminating dangling bonds created during manufacture of the semiconductor devices and by adjusting the surface potential to either reduce or increase the surface leakage current associated with these devices. Silicon nitride (SiN) is commonly used to produce a passivation layer on semiconductor devices, such as GaAs metal semiconductor FETs (MESFETs), to prevent oxidation of the surface of these devices [15–19]. Silicon nitride is one of the most widely used materials for passivation layers on GaAs semiconductor devices due to the fact that this material is extremely chemically stable and has excellent barrier properties.

In this passivating process silicon nitride, silicon monoxide, silicon dioxide, silicon oxynitride, or polymide can be used to produce a passivation layer on the surface of a GaAs semiconductor device to isolate exposed surfaces of the GaAs device from an external environment.

11.1.2 Methods of Passivation

Lots of methodologies of deposition on passivation layer is carried out in HEMT devices like plasma-enhanced chemical vapor deposition (PECVD) [20–24], metal-organic chemical vapor deposition (MOCVD), E-beam evaporation method, atomic layer deposition, low-pressure chemical vapor deposition (LPCVD), etc.

1. **Plasma Enhanced Chemical Vapor Deposition (PECVD):**

    The most common technique used to deposit silicon nitride on GaAs semiconductor devices is the plasma-enhanced chemical vapor deposition (PECVD) technique shown in Figure 11.2. PECVD technique provides plasma of reactant gases including silicon and nitrogen to a chamber in which a semiconductor is disposed. The gases are then reacted within the chamber to deposit a layer or film of silicon nitride on the surface of the semiconductor. While the gases may be reacted within the chamber at a temperature in the range of approximately 150°C–350°C, most PECVD techniques react to the gases at temperature ranges between 200°C and 350°C to deposit silicon nitride on Group III-V semiconductors.

    PECVD process that deposits a silicon nitride passivation layer on HEMT devices is to prevent oxidation of the exposed semiconductor surfaces of the devices. This process uses ammonia to silane gas ratio in the range of 1.62–2.05 to

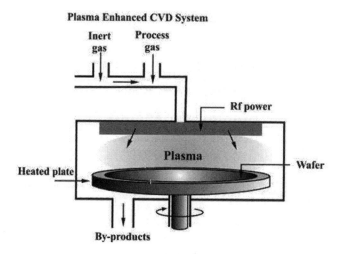

FIGURE 11.2
Process of Plasma Enhanced Chemical Vapor Deposition Technique.

produce a silicon nitride film or passivation layer that is under tensile stress and has a nominal index of refraction of 2.0. However, it is generally recognized that silicon nitride passivation layers have undesirable characteristics and, in particular, that silicon nitride passivation layers tend to change the threshold and/or reverse breakdown voltages of the semiconductor devices. Silicon nitride layers tend to produce an undesirable reduction in the reverse breakdown voltage. This reduction in the reverse breakdown voltage reduces the effectiveness of such devices.

2. Metal-Organic Chemical Vapor Deposition (MOCVD):

This is a technique for depositing thin layers of atoms onto a semiconductor wafer. Using MOCVD we can build up many layers, each of a precisely controlled thickness, to create a material which has specific optical and electrical properties. Using this technique, it's possible to build a range of semiconductor photo detectors and lasers. The principle of MOCVD is quite simple. Atoms to be passivated are combined with complex organic gas molecules and passed over a hot semiconductor wafer. The heat breaks up the molecules and deposits the desired atoms on the surface, layer by layer. By varying the composition of the gas, you can change the properties of the crystal at an almost atomic scale. It can grow high quality semiconductor layers (as thin as a millionth of a millimetre) and the crystal structure of these layers is perfectly aligned with that of the substrate.

3. E-beam evaporation method:

E-beam or electron beam evaporation is a form of physical vapor deposition shown in Figure 11.3. The target material to be used as a coating is bombarded with an electron beam from a charged tungsten filament to evaporate and convert it to a gaseous state for deposition on the material to be coated. Taking place in a high vacuum chamber, these atoms or molecules in a vapor phase then precipitate and form a thin film coating on the substrate.

E-beam evaporation, which is a thermal evaporation process and sputtering are the two most common types of physical vapor deposition, or PVD. Of these

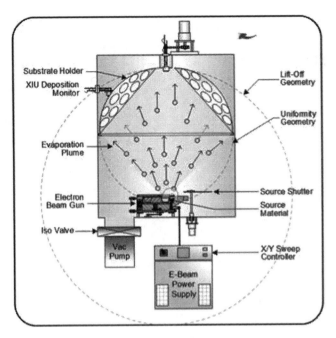

FIGURE 11.3
Process of E-beam evaporation method.

two processes, The E-beam deposition technique has several clear advantages for many types of applications.

It permits the direct transfer of energy with the electron beam to the target material to be evaporated making it ideal for metals with high melting points. Electron beam evaporation can yield significantly higher deposition rates—from 0.1 nm per minute to 100 nm per minute resulting in higher density film coatings with increased adhesion to the substrate.

4. Atomic layer deposition (ALD):

Atomic layer deposition (ALD) is a vapor phase technique used to deposit thin films onto a substrate shown in Figure 11.4. The process of ALD involves the surface of a substrate being exposed to alternating precursors, which do not overlap but instead are introduced sequentially. In each alternate pulse, the precursor molecule reacts with the surface in a self-limiting way. This ensures that the reaction stops once all of the reactive sites on the substrate have been used. A complete ALD cycle is determined by the nature of the precursor-surface interaction. The ALD cycle can be performed multiple times to increase the layers of the thin film, depending on the requirement. The process of ALD is often performed at lower temperatures, which is beneficial when working with substrates that are fragile, and some thermally unstable precursors can still be employed with ALD as long as their decomposition rate is slow. A wide range of materials can be deposited using ALD, including oxides, metals, sulfides, and fluorides, and there is a wide range of properties that these coatings can exhibit, depending on the application.

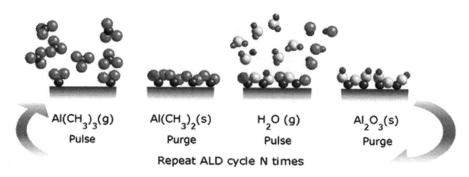

FIGURE 11.4
Process of Atomic Layer Deposition Method.

The ALD process is widely used as it provides ultra-thin nano layers in an extremely precise manner on a variety of substrates, including micron to sub-micron size particles. The nano layers achieved with ALD are by nature conformal and pinhole free.

5. **Low pressure chemical vapor deposition (LPCVD):**

LPCVD is a chemical vapor deposition technology that uses heat to initiate a reaction of a precursor gas on the solid substrate. This reaction at the surface is what forms the solid phase material. Low pressure (LP) is used to decrease any unwanted gas phase reactions, and also increases the uniformity across the substrate. The LPCVD process can be done in a cold- or hot-walled quartz tube reactor. The tube is evacuated to low pressures, which can range from 10 mTorr to 1 Torr. Once the tube is under vacuum, the tube is then heated up to deposition temperature, which corresponds to the temperature at which the precursor gas decomposes. Temperatures can range from 425°C to 900°C depending on the process and the reactive gases being used. Gas is injected into the tube, where it diffuses and reacts with the surface of the substrate creating the solid phase material. Any excess gas is then pumped out of the tube and goes through an abatement system. LPCVD films are typically more uniform, lower in defects, and exhibit better step coverage that films produced by PECVD and PVD techniques.

11.1.3 SiO$_2$ and Si$_3$N$_4$ Passivation on HEMT Structure

The performance of the HEMT device is improved after passivation. Bernat et al, [6,25] Investigated on the performance of AlGaN/GaN/Si HEMTs before and after passivation.

SiO$_2$ and Si$_3$N$_4$ passivation is applied for undoped and doped HEMT structures, and analysis of DC as well as small-signal and large-signal RF performance of devices is carried. Different impact of passivation on device performance is found.

AlGaN/GaN heterostructures were grown on 2-in (1 1 1) Si substrates and an undoped as well as modulation doped AlGaN layers were grown. In the former case a 25 nm thick undoped AlGaN barrier layer was grown on top of GaN, shown schematically in Figure 11.5a. In the later case the structure consisted of a 4 nm undoped AlGaN spacer, followed by a 10 nm thick Si-doped AlGaN carrier supply layer, a 6 nm undoped AlGaN barrier layer and a 4 nm undoped GaN cap layer, as shown in Figure 11.5b. The GaN cap

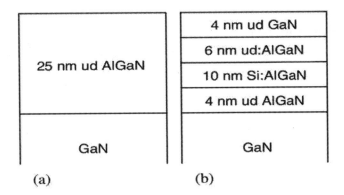

FIGURE 11.5
Schematic diagram of AlGaN/GaN HEMT structure with (a) undoped and (b) doped AlGaN barrier layer.

was applied in order to reduce the gate leakage current. The device processing consisted of conventional HEMT fabrication steps. Both types of samples were then subjected to a variety of measurements before passivation was applied.

Room temperature Hall effect, DC characteristics, as well as small-signal and large-signal RF characteristics of devices were measured. After that two types of passivation layers were applied. SiO_2 and Si_3N_4, both with a thickness of 100 nm, were prepared by plasma enhanced chemical vapor deposition (PECVD) at 150°C and 300°C, respectively. The layer thickness and refraction coefficient were controlled by ellipsometry. The same measurements as described above were performed on SiO_2 and Si_3N_4 passivated devices.

11.1.4 Outcomes of Passivation

11.1.4.1 Hall Effect Measurements

For AlGaN/GaN heterostructures with a 25 nm thick undoped barrier layer expect a sheet carrier concentration of about 1×10^{13} cm^2 without doping, i.e. only due to the polarization charge.

An increase of sheet carrier concentration ns and nearly the same or slightly lower carrier mobility μn were observed on all samples investigated after SiO_2 or Si_3N_4 passivation was applied. However, room temperature Hall effect measurements on undoped and un passivated are listed in Table 11.1. It shows that the passivation is more effective for undoped samples than doped samples and passivation using Si_3N_4 shows higher increase of the carrier concentration than using SiO_2.

11.1.4.2 DC Characteristics

Figures 11.6a–c and 11.7a–c illustrate typical output and transfer characteristics of undoped and doped AlGaN/GaN/Si HEMTs with 0.3 μm gate length before and after passivation. The I–V characteristics with the gate biased from +1 V down in steps of –1 V are shown.

The drain saturation current of undoped devices without passivation is Ids = 0.45 A/mm and increased to 0.54 and 0.68 A/mm after passivation with SiO_2 and Si_3N_4, respectively, as shown in Figure 10.5a and b. Similarly, the peak extrinsic transconductance of undoped devices increased from 170 mS/mm for un passivated samples to 196 and 215 mS/mm after passivation with SiO_2 and Si_3N_4, respectively, as follows from Figure 10.7a and b. It should

TABLE 11.1

Room Temperature Hall Data Obtained on Undoped (ud) and Doped (d) AlGaN/GaN Hetero Structures Without (w/o) and with (w) SiO_2 and Si_3N_4 passivation

Sample	ns (cm^{-2})	μn $(cm^2/V\ s)$
ud, w/o	6.44×10^{12}	1330
ud, w-SiO_2	6.87×10^{12}	1330
ud, w/o	7.12×10^{12}	1150
ud, w-Si_3N_4	9.13×10^{12}	1080
d, w/o	9.20×10^{12}	1270
d, w-Si_3N_4	9.86×10^{12}	1220

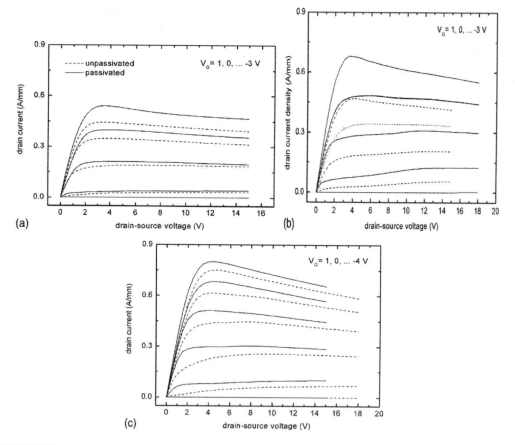

FIGURE 11.6

Output Characteristics of AlGaN/GaN/Si HEMTs before and after passivation: (a) undoped, passivated with SiO_2 (b) undoped, passivated with Si_3N_4, and (c) doped, passivated with Si_3N_4.

be mentioned that the gate voltage of the peak gm has slightly changed, to higher value after SiO_2 passivation and in opposite direction after Si_3N_4 passivation.

The results show that the passivation with Si_3N_4 has higher impact on the DC performance than the passivation with SiO_2 and quantitative improvements are in agreement with changes of the sheet carrier concentration resulting from Hall effect measurements (Table.11.1).

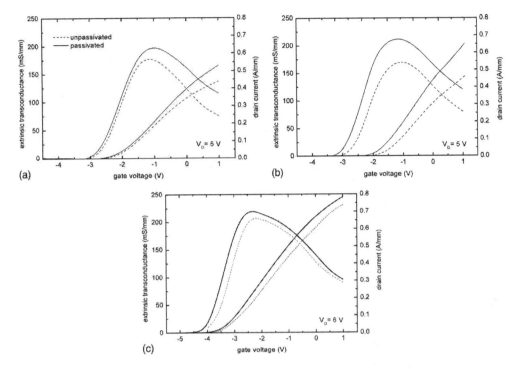

FIGURE 11.7
Transfer Characteristics of AlGaN/GaN/Si HEMTs before and after passivation: (a) undoped, passivated with SiO_2 (b) undoped, passivated with Si_3N_4, and (c) doped, passivated with Si_3N_4.

11.1.4.3 Small-Signal RF Performance

Small-signal characterization of the devices was carried out by on-wafer S-parameter measurements using an HP 8510C network analyzer. An extrinsic current-gain cut off frequency f_T and a maximum frequency of oscillation f_{max} were evaluated for the same devices before and after passivation. Figure 11.8a shows the current-gain cut off frequency as a function of gate bias for undoped samples investigated. A remarkable decrease of f_T values is found after SiO_2 passivation, the peak f_T value decreased from 18.6 to 9 GHz. On the other hand, passivation with Si_3N_4 significantly improved current-gain cut off frequency in whole range of applied gate biases.

The peak f_T value increased from 18.4 to 28.8 GHz after Si_3N_4 passivation of undoped sample. Figure 11.8b shows the f_T versus VG dependence for doped sample before and after Si_3N_4 passivation. Also, in this case an increase of the current-gain and cut off frequency are observed. The peak f_T value increased from 26.8 to 31 GHz. The maximum frequency of oscillation f_{max} follows the changes of f_T after passivation, i.e., the $f_{max} - f_T$ ratio is not affected by passivation. A decrease as well as an increase of f_T and f_{max} after passivation was encountered. To overcome this different passivating material can be used with different dielectric constants.

11.1.4.4 Microwave Power Measurements

The output power Pout, gain G and power-added efficiency (PAE) as a function of the input power Pin were measured and the output density was evaluated. Figure 11.9 shows

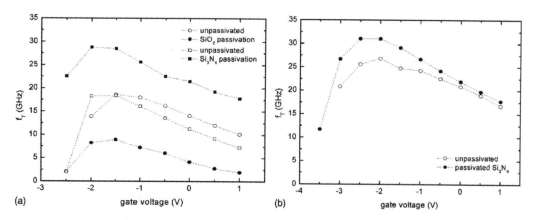

FIGURE 11.8

Current-gain cut off frequency as a function of gate bias for 0.3 μm gate length AlGaN/GaN/Si HEMTs before and after passivation: (a) undoped device, passivated with SiO_2 and Si_3N_4, and (b) doped device, passivated with Si_3N_4.

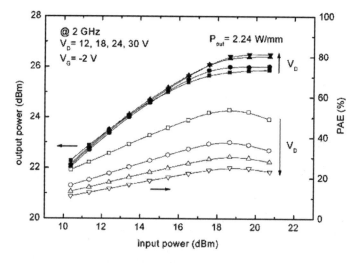

FIGURE 11.9

Output power and power-added efficiency at 2 GHz as a function of drain bias for 0.5 μm gate length un passivated AlGaN/GaN/Si HEMT.

Pout and PAE as a function of Pin for un passivated HEMT with 0.5 and 200 μm gate length and width, respectively. The output power increases and power-added efficiency decreases with increased drain bias. The peak output power density of 2.24 W/mm is comparable to 1.8 and 3.3 W/mm obtained on AlGaN/GaN/Si HEMTs. Figure 11.10a and b shows typical results of the peak output power density for various drain biases evaluated on the same devices with 0.3 μm gate length before and after passivation. A comparison of SiO_2 and Si_3N_4 passivation is illustrated on undoped devices in Figure 11.10a. The peak output power density before passivation is nearly the same for samples investigated. However, only a half and more than doubled power densities follow from SiO_2 and Si_3N_4 passivation, respectively.

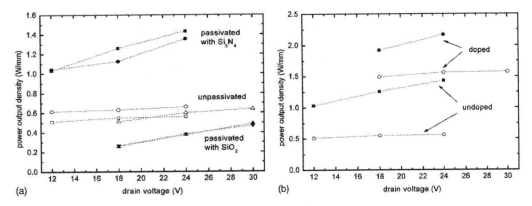

FIGURE 11.10
Peak 2 GHz power output density dependence on drain bias for 0.3 μm gate length AlGaN/GaN/Si HEMTs before and after passivation: (a) undoped devices, passivated with SiO_2 and Si_3N_4, and (b) undoped and doped devices passivated with Si_3N_4.

This is similar result as obtained for cut off frequencies evaluated from small-signal RF measurements in Figure 11.8a. An influence of Si_3N_4 passivation on undoped (squares) and doped (dots) devices is shown in Figure 11.10b. For both types of devices an increase of the output power after passivation is found. However, the doped devices show higher power before as well as after passivation. This difference is connected with quality of samples used as well as principal difference in power capability of undoped versus doped devices.

11.2 Field Plate Technique

The high electron mobility transistor (HEMT) is used to provide very high levels of performance at microwave frequencies for high speed applications. The HEMT offers a combination of low noise figure combined with the ability to operate at the very high microwave frequencies. Accordingly, the device is used in areas of RF design where high performance is required at very high RF frequencies. This is related to the nature of the 2DEG and the fact that there are less electron collisions. Due to this their noise performance they are widely used in low noise small-signal amplifiers, power amplifiers, oscillators and mixers operating at frequencies up to 60 GHz [26]. HEMT devices are used in a wide range of RF design applications including cellular telecommunications, Direct broadcast receivers—DBS, radar, radio astronomy, and any RF design application that requires a combination of low noise and very high frequency performance [27].

Different types of breakdown enhancement techniques are carried out Schottky source/drain technique is employed to improve the breakdown voltage and frequency, Si_3N_4 passivation technique is applied to reduce the effect of polarization variation and to smoothen the field distribution at drain end of the gate region to improvise breakdown voltage. Further improvement of breakdown voltage is achieved by using Field plate technique.

Figure 11.11 shows a 3 dimensional view of Al_2O_3/AlInN/AlN/GaN FP-MOS–HEMT, SiN passivation and Field plate stacking is employed to improvise the breakdown voltage in the HEMT structure.

In general, the field plates introduce an increase in parasitic capacitances, thus resulting reduction in frequency and power gain. In order to ease the undesired effect, an

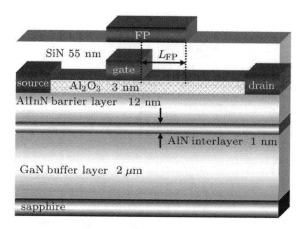

FIGURE 11.11
Cross section of the 3nm-Al_2O_3/AlInN/AlN/GaN FP-MOS–HEMT.

extension in gate metal towards the drain helps to smoothen the gate peak electric field alongside the channel. Along with improvising breakdown voltage field plate structure reduces current collapse effect, because it surpasses the high field trapping effect. Field plate techniques are best suited for reshaping the electric field distribution in the channel by a capacitive action and to reduce its peak value on the drain side of the gate edge, further increase in the breakdown voltage, and will reduce high field trapping effect and is optimized with a proper thickness of passivation layer and Field plate length. As the thickness reduces the capacitance effect in the channel and field plate region increases, this being a tradeoff, FP enhances the power capability of the device. For larger thickness there is less effect of the FPs.

FP towards drain is of great necessity, although this will deteriorate the breakdown characteristics to some extent Lower unity current gain frequency due to increase in the C_{gd} (gate connected FP) and C_{ds} (source connected FP). Different field plate structures are gate field plate, floating field plate, multiple field plate, source field plate technique and source-connected air-bridge field plate (AFP).

11.2.1 $Al_{0.2}Ga_{0.8}N$/GaN Field Plated HEMT Structure

Figure 11.12 shows the field-plated HEMT cross-sectional schematic. In order to increase the thermal conductivity and reduce lattice mismatching in the device a Silicon Carbide material is used as the substrate [28]. The six-layer layered structure starting with an undoped GaN Buffer layer grown over the substrate is used to decrease the effect of lattice mismatching and associated trap effect. Buffer layer is followed by a GaN channel, undoped AlGaN spacer, n-doped AlGaN Barrier and GaN cap layer [29]. The mole fraction of the ternary AlGaN material is fixed to 0.26 to obtain a higher polarization effect to improve the 2DEG sheet carrier density. The energy band discontinuity along with the high piezoelectric and spontaneous polarization at the GaN channel/AlGaN spacer interface creates a Two-Dimensional Electron Gas Channel (2DEG) inside the GaN channel. The function of the cap layer on top of the barrier layer is to decrease the overall capacitance as well as the metal semiconductor contact resistance.

Moreover, the cap layer gives better gate control over the channel and increases the breakdown voltage by suppressing the gate field. In addition to these a passivation layer

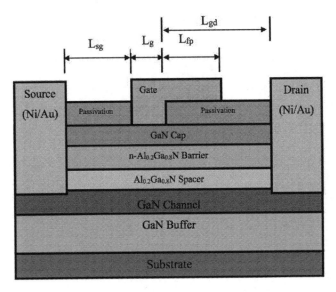

FIGURE 11.12
Structure of field plated AlGaN/GaN HEMT Device.

(Si3N4) is incorporated at the top of GaN Cap layer to prevent the device from external polarization and also to reduce the peak electric field at the drain end of the gate region. Device parameters are given in Table 11.2.

Finally, to improve the breakdown voltage characteristics of the device the drain side edge of the gate is extended towards the drain by a length of LFP, which is called as the field plate (FP). The role of FP is to reduce the field distribution in the drain end of the gate region.

A stacked layer of Ni/Au Schottky contact is used as the source and drain contact. The lower On-resistance (R_{on}) and high field suppressed source carrier injection associated with the Schottky Source Drain (SSD) HEMT further improves the breakdown voltage characteristics of the FP-HEMT device [30].

TABLE 11.2

Device Parameters of Field Plated HEMT Structure

Layer	Thickness	Al Concentration
Field plate (L_{FP})	0.2 μm	—
Si$_3$N$_4$ passivation	50 nm	—
GaN cap	3 nm	—
AlGaN barrier	16 nm	0.26
AlGaN spacer	2 nm	0.26
GaN channel	5 nm	—
GaN buffer	2 μm	—
Gate to drain distance (L_{gd})	5 μm	—
Gate length (L_g)	0.8 μm	—

11.2.2 Effect of the Field Plate Technique in HEMT

The following section describes the various analysis made on the field plated HEMT device. The section describes the DC, RF and breakdown voltage characteristics of the device in detail.

11.2.2.1 DC Characteristics Analysis

The improvement in breakdown voltage with various field plate dimensions is analyzed and an optimum dimension for high power and high frequency is achieved for the GaN/AlGaN HEMT. Figure 11.13 shows the output current Id characteristic of the device. The device is simulated with drain voltage sweep from 0 to 50 V with different gate voltage. The simulated result for the Field Plated (FP) HEMT shows a peak drain current of 0.91 A/mm against 1 A/mm non-field plated (NFP) HEMT at gate voltage of 2 V. The drain current of NFP-HEMT device simulation is validated using experimental data at a gate voltage of 1 V [31]. The result also shows current improvement over NFP-HEMT device. The effect of field plate is clearly shown in the DC transfer characteristic of the HEMT. Figure 11.14 shows a comparative analysis to obtain drain current variation between FP and NFP-HEMT device. The device without field plate clearly shows a higher drain current of 0.86 A/μm at a gate voltage of 2 V compared with 0.8 A/μm of the field plated HEMT at 50 V drain bias.

The transconductance analysis of the FP and NFP device is also done and the interpretation is given. Figure 11.15 shows the transconductance variation of the device FP and NFP-HEMT device. The variation in transconductance between FP and NFP device is sparingly small, 224 and 225 mS/mm, respectively. The threshold voltage of the devices are −2.1 V and −2.6, respectively. The reduction in transconductance and drain current is the result of additional depletion layer extension due to the gate field plate.

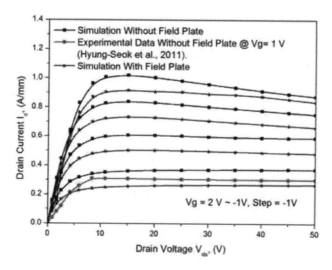

FIGURE 11.13
I_d-V_{ds} characteristics for $L_{gd} = 5$ μm and field plate length 3 μm.

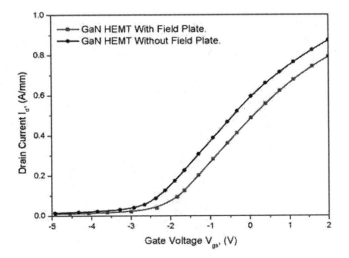

FIGURE 11.14
Transfer characteristics of the GaN/AlGaN HEMT with and without field plate at an $L_{gd} = 5$ µm and $L_{FP} = 3$ µm.

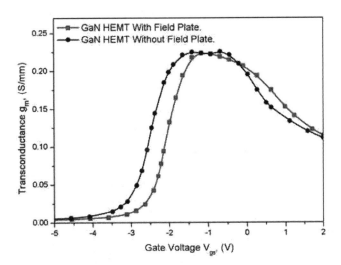

FIGURE 11.15
Transconductance characteristics of GaN/AlGaN HEMT with $L_{gd} = 5$ µm and $L_{FP} = 3$ µm.

11.2.2.2 Electric Field and Breakdown Voltage

An AlGaN/GaN high-electron-mobility transistor (HEMT) with a novel source-connected air-bridge field plate (AFP) is shown in Figure 11.16a–c. The device features a metal field plate (FP) that jumps from the source over the gate region and lands between the gate and drain. The fabrication process is based on a commercially available radio-frequency GaN-on-SiC technology. With reference to the below mentioned HEMT structure.

Figure 11.17a explains the surface electric-field distribution along the AlGaN/GaN interface for the AFP structure exhibits lower electric-field strength near both the drain-side gate edge and the FP edge. This clearly shows that the devices with an AFP still have room to accommodate a higher drain voltage and lower gate-edge electric-field-induced leakage.

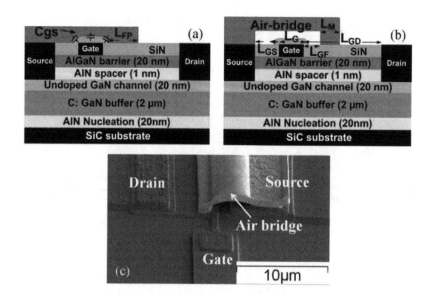

FIGURE 11.16

(a) Schematic cross section of AlGaN/GaN HEMT with a conventional F_P, where L_{FP} is the FP length. (b) AlGaN/GaN HEMT with novel AFP, where L_{GS}, L_G, L_{GF}, L_M, and L_{GD} are the gate–source distance, gate length, gate-to-air-bridge field-plate distance, air-bridge footprint width, and gate–drain distance (drift region), respectively. (c) SEM view of fabricated novel AFP HEMT.

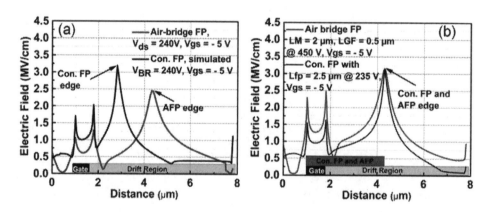

FIGURE 11.17

(a), (b). Simulated Electric Field Strength for different device parameters.

The electric-field distributions for the HEMTs near breakdown with the same effective FP length are as shown in Figure 11.17b. The low dielectric constant of air can better modulate the surface electric-field distribution, allowing the device with an AFP to support a higher breakdown voltage even with the same device dimensions.

The measured breakdown voltages (VBR) as a function of LM with different LGF are as shown in Figure 11.18a. Three samples from different wafers were measured. It can be

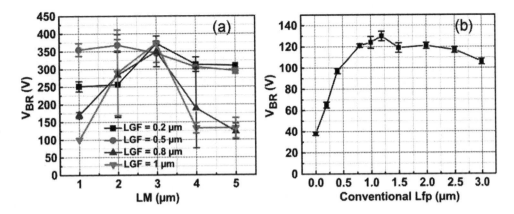

FIGURE 11.18

(a) Measured breakdown voltage (V_{BR}) as a function of L_M with different L_{GF} for AFP devices and (b) breakdown voltage as a function of conventional L_{FP} for conventional devices.

observed that breakdown voltage (2 mA/mm) has the best device dimension tolerance as a function of L_M with $L_{GF} = 0.5$ μm. The breakdown voltage degrades slightly from 375 V for $L_M = 2$ μm and $L_{GF} = 0.5$ μm to 325 V for $L_M = 5$ μm and $L_{GF} = 0.5$ μm. The breakdown voltage of the device with a conventional FP is also shown in Figure 11.18b. The longest LFP fabricated was 3 μm. The breakdown voltage increases from 37 V without an FP to 125 V with a conventional LFP length of 1.2 μm. No further increase in breakdown voltage was observed after this point.

When compared to a similar-sized HEMT device with a conventional FP structure, the AFP not only minimizes the parasitic gate-to-source capacitance but also exhibits a higher OFF-state breakdown voltage and one order of magnitude lower drain leakage current. In a device with a gate-to-drain distance of 6 μm and a gate length of 0.8 μm, a three-times-higher forward-blocking voltage of 375 V was obtained at $V_{GS} = -5$ V. In contrast, a similar-sized HEMT with a conventional FP can only achieve a breakdown voltage no higher than 125 V using this process, regardless of device dimensions. The specific on-resistance for the device with the proposed AFP is 0.58 mΩ cm² at $V_{GS} = 0$ V, which compares favorably with 0.79 mΩ cm² for the best device with a conventional FP [32].

11.2.2.3 CV Measurements

Figure 11.19 shows the C–V measurements at 1 MHz of the gate capacitors of the HEMT and the FP-MOS–HEMT with a diameter of 130 μm for the device structure shown in Figure 11.11. A sharp transition from the depletion region to the accumulation region in the profile of the FP-MOS–HEMT, which is similar to that of the HEMT, can be observed, which shows a high-quality interface between the Al_2O_3 and the AlInN. In the accumulation region the total capacitance obtained for FP-MOS-HEMT & HEMT are 78.6 and 95.1 pF respectively. Which clearly shows the increased effect of FP in C-V characteristics [33].

11.2.2.4 Small-Signal RF Performance

S-parameters of the 100 m FP and nFP-HEMT devices were measured from 1 to 40 GHz using a Cascade Micro tech on-wafer probing system with an HP8510XF vector network

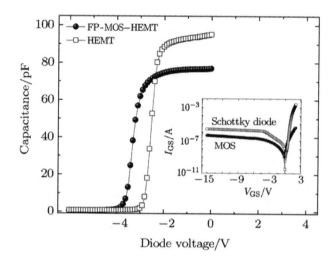

FIGURE 11.19
C–V curves at 1 MHz on the gate capacitors of the HEMT and the FP-MOS–HEMT with a diameter of 130 μm. The inset shows the I–V characteristics of the gate Schottky diode and the gate Al$_2$O$_3$/HEMT MOS structures.

analyzer. Additionally, a standard load-reflection-reflection-match (LRRM) calibration method was used to calibrate the measurement system, and the calibrated reference planes were at the tips of the corresponding probes. The parasitic effects (mainly capacitive) from the probing pads have been carefully removed from the measured S-parameters using the same method as Yamashita et al. and the equivalent circuit model is in Figure 11.20a and b shows the extracted RF Gm and drain-to-gate capacitance (C_{gd}), gate-to-source capacitance (C_{gs}) and cutoff frequency (f_T) as functions of the drain bias for the FP and nFP devices. In order to have a fair comparison, the gate biases for generating the three figures were chosen at the peak DC transconductance to accommodate for the slight shift in the threshold voltage for device with field plate.

As expected, the extracted C_{gd} of FP device exhibited a major increase compared with that of nFP one. Moreover, the gate-to-source capacitance showed a minor increase with increasing drain bias. Figure 11.16 shows the dependence of f_T on the drain bias for FP and nFP devices. Clearly, the FP devices showed a lower f_T due to the major increase in C_{gd} compared to the nFP devices. However, it is also relevant that the dependence of f_T on drain bias is very weak for FP devices.

11.2.2.5 RF Power Performance

The devices were tuned for the maximum output power. Figure 11.21 shows the output power, gain, and PAE as a function of input power for both devices at 2 GHz when biased at 30 V with a 15 mA/mm current density. The optimum load reflection coefficients were 0.886 9.9o and 0.885 7.5o for FP and nFP devices, respectively.

A higher output power of 25.36 dBm and PAE of 43% were achieved for the FP device. Meanwhile, P1dB for the FP and nFP devices when biased at a 30 V drain bias were 18.52 and 16.26 dBm, respectively. The softer gain compression of the FP device is believed to be the reason for the linearity improvement at high output power levels even beyond P1dB. Such superior performance is mainly attributed to the reduction of trapping in surface states due to the field plate structure [34].

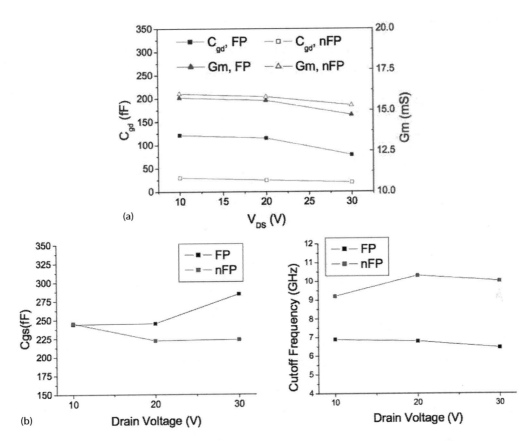

FIGURE 11.20
Dependence of C_{gd}, C_{gs} and ft on drain bias for FP and nFP devices.

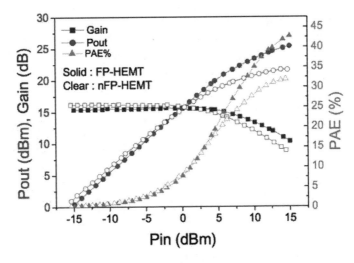

FIGURE 11.21
The measured RF power performance at 2GHz when both devices were biased at 30V.

11.3 Asymmetric ALGAN/GAN HEMTs for High-Power Applications

Asymmetric AlGaN/GaN HEMT are extremely important in micro and terahertz wave communication. AlGaN/GaN-based HEMTs are currently the mainstream in power devices, attracting increasing attempts to improve its high-frequency performance to meet the requirement of the μ-wave and even THz wave applications. Increase in the operating frequency is not much easier task by shrinking the gate length likewise on InP-based HEMT enhances RF and DC characteristics in these devices, which is explained later in this topic. The bottle-neck is the negative fixed charge accumulated at the gate-foot edge as well as the gate-head edge near the drain side on the device surface and/or in the AlGaN and GaN layers, resulting in uneven distribution of electric field in the AlGaN/GaN heterojunction. These peak electric fields should be responsible for the current collapse when the gate bias (V_g) is beyond the breakdown voltage. As the gate length (L_g) is shrunk down, the negative fixed charge existing at the gate edges increases. Consequently, the device is more easily broken down at even lower gate bias. Although the use of the field plates directly grown on the top of a SiN passivation layer, can alleviate such electric field crowding, the cost is the loss of high frequency performance due to the large plate area. Due to frequent current collapse in AlGaN/GaN-based high-electron mobility transistors causing the peak electric field on the device surface around the gate edges and are affected more on scaling the gate length down to sub 100 nm for the high frequency operation. To improve the device performance various asymmetric devices are proposed.

An air spaced field plates was proposed to reduce the gate capacitance to the drain (C_{gd}) in hopes to upturn the operation frequency. Unfortunately, when the gate length is further shrunk well below 100 nm, the current collapse becomes serious once again and the restriction on high operation frequency of AlGaN/GaN HEMTs with T shaped gates becoming inevitable [35].

To incur these, researchers from state key lab of ASIC and System, Fudan University has proposed a new configuration of an asymmetric gate model is proposed, which is a T shape gate with a longer upward arm overhanging the drain side which is shown in the Figure 11.22. To carry out the technical study of such an asymmetric gate, 3D greyscale electron beam lithography (EBL) assisted by Monte Carlo simulation was successfully applied for the asymmetric gates with the gate-length of 80 nm and the arm length of 1.5 μm. It has been recorded that asymmetric T shape gate can suppress the electric field on the gate edge efficiently as well as reduce the parasitic capacitance between the gate and the drain, leading to higher operation frequency. It is proved that the proposed asymmetric gate clearly holds at least two important advantages over the traditional field plates.

1. The long arm overhangs the gate-to-drain region, efficiently alleviating the peak electric-fields on the gate edge and suppressing the current collapse. As the result, the breakdown voltage decreases most slowly in the three gate structures when the gate length is shrunk down below 100 nm.

2. The increased air space between the gate and the substrate significantly reduces the parasitic capacitance, giving rise to an increase in both the current gain cut-off frequency (f_T) and the maximum oscillation frequency (f_{max}).

It is therefore believed that such a specific gate configuration is hopefully applicable for high-frequency operation by using shorter gate-length in GaN-based HEMTs.

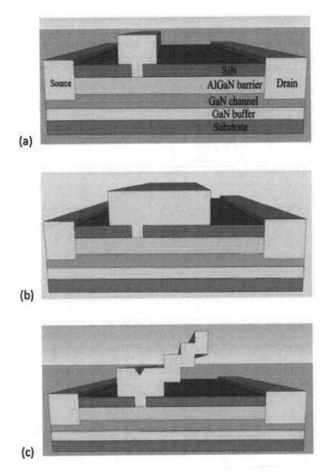

FIGURE 11.22
Schematic of the HEMT devices (a) The traditional *T* shape gate; (b) The *T* gate with broad field plates and (c) the proposed asymmetric *T* shape gates.

Numerical simulation was carried out using the software of Silvaco to calculate the resultant fields in 2DEG and to prove the advantage of asymmetric gate in elimination of the current collapse in GaN HEMTs. For comparison, three different configurations of the gates were included in the simulation shown in Figure 11.22. They are: (a) a T shape gate with the head-width of 300 nm, (b) a field plate on the SiN layer with the head-width of 1.5 μm, (c) an asymmetric *T* shape gate with the whole head including the long arm of 1.5 μm in length.

The electric field strength at 20 nm below the top surface in the barrier was quantitatively calculated, as shown in Figure 11.23. Strong electric field is observed in the vicinity of the gate foot as well as the head-edge for all the three gates. At the foot edge, the *T* shape gate (blue line) suffers the strongest field (4.12 MV/cm), the field plate (black line) feels a reduced field of 2.27 MV/cm, and the asymmetric T-shape gate (red line) experiences a field between these two as 2.6 MV/cm, suggesting that the upward long arm partially alleviates the electric field at this position. At the head-edge, however, both the *T* shape gate and the field plate have to stand the same strong field around 2.8–3.2 MV/cm while the asymmetric *T* shape gate just 1.8–2.5 MV/cm. Also, the electric field under the asymmetric *T* shape gate changes relatively slowly, comparing with that of the field plate, which varies rapidly.

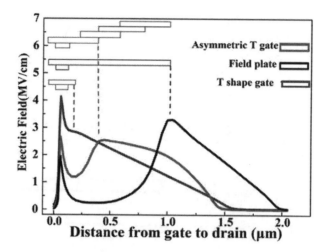

FIGURE 11.23
The simulated electric field strength in the channel layer at the hetero-interface. The x-coordinate is the distance from the center of foot to drain electrode.

Figure 11.24 represents the SEM micrographs for the generated 3D templates in PMMA/ PMMA-MAA bilayer with three kinds of asymmetric designs. The most important variables to decide the 3D shapes are the step width and the step number, respectively, which are controlled by the exposure dose under the certain design. Figure 11.24a is for the bilayer step with 1 μm head-width, Figure 11.24b the trilayer steps with 1.3 μm head-width, while

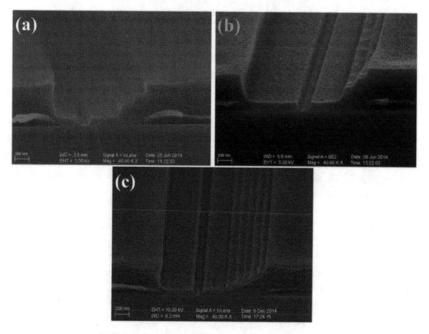

FIGURE 11.24
The micrographs by a scanning electron microscope (SEM) in cross-sectional views for the asymmetric *T* shape profiles in PMMA/copolymer bi layer with three different charge modulations. (a) Asymmetric profile with 1 μm head-widths. (b) Asymmetric profile with 1.3 μm head-widths. (c) Asymmetric profile with 1.5 μm head-widths.

Figure 11.24c the multistep with 1.5 μm head-width. Metallization of Cr/Au (10/300 nm) was undertaken by thermal evaporation in a Kurt Leskert evaporator. A lift-off process in warm acetone cleared away the unwanted metals.

With the increased distance from the gate-head to the drain and designed upward arm, the asymmetric T gate should also own the benefits in the reduction of parasitic capacitance between the gate and the drain (C_{gd}), resulting in the improvement of high frequency. The advantages of high-breakdown voltage and high-frequency characteristic owing to the low-peak electric field and low parasitic capacitance between the gate and the drain is critical for the asymmetric T-shaped gate.

11.3.1 Asymmetric InAlAs HEMT Structure

Figure 11.25 shows the vertical schematic cross sectional view of the novel asymmetric GaAs MHEMT [36]. The epitaxial layers from bottom to top consist of a semi-insulating GaAs substrate over which a 300 nm thick $In_xAl_{1-x}As$ graded buffer layer is employed. Above the graded buffer layer, a 300 nm thick $In_{0.52}Al_{0.48}As$ buffer layer is placed, over which a silicon δ-doping sheet with a doping density of $2 \times 10^{12}/cm^2$ is inserted. The 10 nm thick intrinsic $In_{0.75}Ga_{0.25}As$ (3 nm)/InAs (5 nm)/$In_{0.75}Ga_{0.25}As$ (2 nm) channel layers are inserted between lower and upper $In_{0.52}Al_{0.48}As$ spacer layers. The thicknesses of the lower and upper $In_{0.52}Al_{0.48}As$ spacer layers are 3 and 2 nm, respectively. A silicon δ-doping plane with a doping concentration of $5 \times 10^{12}/cm^2$ is placed above the upper spacer layer, over which a 2 nm thick $In_{0.52}Al_{0.48}As$

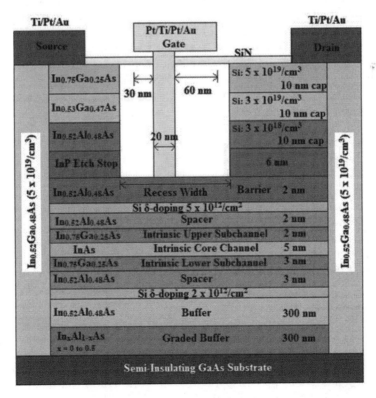

FIGURE 11.25
Schematic diagram of $L_g = 20$ nm, novel asymmetric GaAs MHEMT.

Schottky barrier layer is placed. The multi layer cap region consist of 6 nm thick InP etch-stop layer, on top of which a 10 nm thick *n*-type $In_{0.52}Al_{0.48}As$ layer with a silicon doping density of $3 \times 10^{18}/cm^3$. Above the $In_{0.52}Al_{0.48}As$ cap layer, a 10 nm thick n+-type $In_{0.53}Ga_{0.47}As$ layer with a silicon doping density of $3 \times 10^{19}/cm^3$. On top of the $In_{0.53}Ga_{0.47}As$ cap layer, a 10nm thick n+-type $In_{0.75}Ga_{0.25}As$ layer with a silicon doping density of $5 \times 10^{19}/cm^3$. The novel MHEMT also consist of a cavity structure in the gate recess region and over the cap a 20 nm thick SiN layer is used for transistor passivation. The source and drain regions are formed using $In_{0.52}Ga_{0.48}As$ material with a silicon doping density of $5 \times 10^{19}/cm^3$. The S/D contacts are formed using Ti/Pt/Au (25/25/200 nm) metal layers and the Γ-gate is formed by Pt/Ti/Pt/Au (7/50/50/150 nm) metal layers. The gate-source spacing and the gate-drain spacing employed in the novel asymmetric MHEMT structure are 100 and 400 nm, respectively. The gate foot is just 4 nm away from the channel layer which helps to reduce the short-channel effects. The silicon impurities in the δ-doping planes are assumed to be in the tight Gaussian distribution with a diffusion length of 2 nm. The intrinsic Schottky barrier layer and InP etch stop layer effectively avoids the kink effects which leads to the enhancement of DC-RF and breakdown performance parameters. The upper and lower silicon δ-doping planes provide excellent amount of charge density in the quantum well which improves the DC and RF parameters such as gm_max and IDS_max. In order to investigate the influence of gate recess width on the DC-RF and break-down performance parameters of the novel MHEMT, the recess width is varied between 150 and 350 nm.

The proposed MHEMT consists of stack of thin semiconductor layers of extremely small thicknesses which should be carefully controlled while physically fabricating the device. The highly advanced thin film growth techniques such as MBE and MOCVD can be used for manufacturing this complicated epitaxial MHEMT heterostructure. Many researchers have successfully fabricated such kind of complicated HEMT structures. The deviation in grading and thicknesses of epitaxial layers during fabrication will have an impact on device performance. The increase in the amount of indium in the channel may leads to the enhancement of RF and DC performances. However, the reduction in the amount of indium in the channel has an adverse effect on the RF and DC performance of the proposed MHEMT. Similarly, the increase in recess width in the gate region helps to improve the breakdown performance; however, this limits the RF and DC performances. On the other hand, the decrease of recess width in the gate region during etching process may lead to the enhancement of DC and RF performance parameters at the cost of poor breakdown performance. Therefore, in order to achieve a good combination of RF and DC and break-down characteristics, it is necessary to carefully control the grading of semiconductors and their thicknesses during physical manufacturing.

Figure 11.26 shows the variation of g_m and I_{DS} with respect to V_{GS} for 50 nm gate length MHEMT with a gate width equal to 2×10 µm. The 50 nm gate length novel GaAs MHEMT achieved a gm_max and IDS_max of 3200 mS/mm and 1050 mA/mm respectively at a $V_{DS} = 0.7$ V. For $L_g = 50$ nm, the novel MHEMT shows 7% improvement in gm_max and 24% improvement in IDS_max. This excellent g_m and I_{DS} values mainly arises from the fact that reduction in S/D resistances due to the use of n+ doped InGaAs S/D regions. Other factors contribute to this outstanding DC parameters are the use of silicon δ-doping sheets on either side of the channel which improves the electron density in the channel, reduced parasitic due to the reduced source-drain spacing and higher electron mobility in the channel due to the strain induced in the channel because of the lattice mismatch between n+ doped $In_{0.52}Ga_{0.48}As$ S/D regions and the composite channel layers. The electron mobility and 2DEG density in the quantum well obtained for 50 nm gate length novel GaAs MHEMT are 12,900 cm^2/Vs and $5.6 \times 10^{12}/cm^2$ respectively.

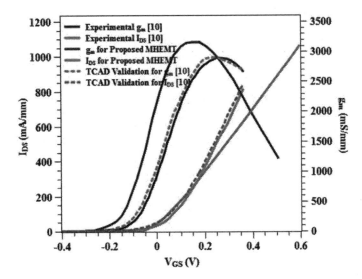

FIGURE 11.26
The dependence of I_{DS} and gm on V_{GS} for $L_g = 50$ nm and $W = 2 \times 10$ μm novel asymmetric GaAs MHEMT.

11.4 Effect of Parasites in HEMTs

GaN on-Si devices show better static DC performance, their RF performance is lower than that of the GaN-on-sapphire devices. This is because that the GaN on Si devices suffer from parasitic loading effects induced by the conductive substrate. The influence of substrate conductivity on small-signal performance has been reported by Chumbes et al. For power performance, they attributed the lower than expected output power to the conductive substrate without a detailed analysis. To gain an insight into the influence of the parasitic loading effects on device power performance, we constructed a large-signal model for the GaN on Si HEMTs. The conductive substrate not only limits the output power but also severely degrades the operating power gain and PAE [37].

Nowadays more recent research has begun to focus on reducing the parasitic device elements such as access resistance and gate fringing capacitance, which become crucial for short gate length device performance maximization. Adopting a self-aligned T-gate architecture is one method used to reduce parasitic device access resistance, but at the cost of increasing parasitic gate fringing capacitances. As the device gate length is then reduced, the magnitude of the static fringing capacitances will have a greater impact on performance. To better understand the influence of these issues on the dc and RF performance of short-gate length InP pHEMTs.

Beyond the material and lithographic issues that require attention for high-speed device realization, extrinsic device elements such as access resistance and fringing capacitance are found to impact on device performance, restricting the potential performance of the short gate length system. Also, for shorter gate length devices (100 nm and below), effort must be made to appropriately scale the device geometry to ensure efficient operation and reduce short-channel effects. More recently, this has been combined with intricate fabrication techniques in an effort to suppress the deteriorative effects of parasitic elements on the well-scaled short gate length system.

11.4.1 Parameter Extraction Method

11.4.1.1 *Extraction of Parasitic Capacitances and Inductances*

A new method to extract parasitic capacitances and inductances for high electron mobility transistors (HEMTs) is proposed in this section. Compared with the conventional extraction method, the depletion layer is modeled as a physically significant capacitance model, and the extrinsic values obtained are much closer to the actual results. In order to simulate the high-frequency behavior with higher precision, series parasitic inductances are introduced into the cold pinch-off model which is used to extract capacitances at low frequency and the reactive elements can be determined simultaneously over the measured frequency range. The values obtained by this method can be used to establish a 16-elements small-signal equivalent circuit model under different bias conditions. The results show good agreements between the simulated and measured scattering parameters up to 30 GHz.

The open test structure method and the pinch-off cold-FET method have been proposed to extract parasitic capacitances. The first method requires a specific test structure and the accuracy of this technique is high. A direct extraction method for extrinsic capacitances of PHEMTs is proposed, which was based on a scalable small-signal model under pinch-off condition, the voltage of the gate was set under the threshold voltage and the drain-to-source bias was set to zero.

In order to reduce the errors particularly at the high frequency when extracting the parasitic capacitances, series inductors are added into the reduced pinch-off cold-FET equivalent circuit. The capacitors and inductors can be determined simultaneously over the measured frequency range. Then, there are no limitations and assumptions at all on the depletion layer capacitances under pinch-off bias condition. The depletion layer extension is described into three capacitors. Eventually, the high precision of the values obtained by the proposed method have been validated by the open-short test structure method. Figure 11.27 shows the small-signal electrical equivalent circuit topology for this device. The equivalent circuit can be categorized into extrinsic part and intrinsic part. L_g, L_s, and L_d represent the gate, source and drain electrode inductances; R_s and R_d are the source and drain resistances, and R_g is the gate distributed resistance; C_{pg} and C_{pd} are the gate and drain pad extrinsic capacitances; C_{gs}, C_{gd}, and C_{ds} represent the gate-source, gate-drain, and

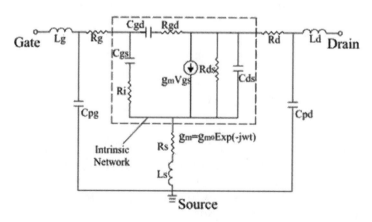

FIGURE 11.27
Small-signal equivalent circuit for the HEMTs.

drain-source capacitances, respectively; R_i, R_{gd}, and R_{ds} are the charge and output resistances; t and g_m are the intrinsic delay and intrinsic transconductance, correspondingly. C_{gs}, C_{gd}, C_{ds}, R_i, R_{gd}, R_{ds}, t, and g_m are intrinsic elements which are emphasized by the dashed frame in the above Figure. The rest are extrinsic elements which are assumed to be as independent [38,39].

11.4.1.2 *Extraction of Parasitic Resistance*

The parasitic source and drain resistances severely affect the performance of high electron mobility transistor (HEMT), because the transconductance, noise figure, power consumption and frequency performance could be degraded by the parasitic resistances. So, the accurate measurement of the parasitic source and drain resistances becomes important for device modeling, characterization, and process monitoring. Many techniques and methods for measuring parasitic resistances of HEMTs were based on the end-resistance technique. But this method needs a complicated procedure requiring accurate measurement of various D.C. device parameters, such as threshold voltage, built-in voltage and gate diode ideality factor. And a variation in the above method is employed, the ratio of floating-drain voltage to gate current could be used as a measure of the end-resistance under the conditions of forward-biased gate and floating drain, then the difference between parasitic source and drain resistances, R_S and R_D, could be obtained from the end-resistances, while an additional measurement of channel resistance and an accurate determination of the threshold voltage were required to obtain the sum of R_S and R_D, and hence R_S and R_D.

In this approach the channel voltage drop is much less than nkT/q, where n is the ideality factor of Schottky gate. But it is violated in this method; therefore an extended end-resistance measurement technique is proposed where a drain current is applied instead of the floating drain configuration with drain current, I_D, much larger than gate current, I_G, different from the above technique with I_D much smaller than I_G. In this modified approach, the small-signal end-resistance, defined as a derivative of the V_D with respect to the I_G, could be used to obtain the R_S, without any other measurement. In this method an accurate modeling of the forward-biased Schottky-gate current is required whenever a probing gate or drain is used, which leads to the following problems.

- Many assumptions and approximation techniques should be adopted to get a concise solution for practical application due to the complication of solving the distributed gate current along the channel length, which resulted in the theoretical errors and limited the application range, such as $I_D \ll I_G$ or $I_D \gg I_G$, $V_D \ll nkT/q$ or $V_D > nkT/q$.
- Due to the exponential dependence of gate current on gate voltage, derivatives or differential calculation should be used to obtain the results, which certainly accompanied significant effect of measurement noise.
- Large enough forward-biased gate current required in some cases possibly degrades the performance of devices, where the small gate current was applied using lock-in amplifier techniques to avoid the possible damage of large gate current on gate region.

Different from the end-resistance technique, the floating-gate transmission line (FGTL) technique was proposed. In order to avoid modeling the exponential dependence of gate current on gate voltage. However, this technique was reported to be unlikely to be correct for HEMTs. We think that the assumption in FGTL technique that no current flowing through gate barrier with floating gate configuration and the floating gate sampling the voltage in the middle of the channel are not correct due to a potential drop existing along the channel length, to obtain the parasitic resistances several devices with different gate lengths are required by FGTL technique [40,41], so some errors should be caused by the non-uniformity between different devices. A measurement technique based on a single device is preferable in many practical applications.

11.4.2 Methods to Reduce Parasitic in HEMT

The high electron mobility transistors (HEMTs) have potentials for applications in high-speed digital communication and microwave fields due to their superior transport properties. InP-based InAlAs/InGaAs HEMTs have demonstrated high operating frequency, low microwave and millimeter wave noise, as well as high-gain performance, all of which are very attractive for millimeter and sub-millimeter wave applications, due to the high sheet carrier density, high peak drift velocity, and high low-field electron mobility. In recent years, many efforts have been made to improve the high-frequency characteristics of InAlAs/InGaAs HEMTs. A record maximum oscillation frequency (f_{max}) exceeding 1 THz and a record current-gain cutoff frequency (f_T) of 644 GHz have been reported in the sub 50 nm InP HEMTs and the 30 nm InAs HEMTs, respectively. These remarkable results stem from the combination of harmonious size scaling and an increase of the indium content in the channel, improving the carrier transport properties. However, the reduction of the gate length increases technology difficulties and leads to a lower yield of the HEMTs. Furthermore, the HEMTs with an indium rich channel usually suffer from a serious kink effect, low breakdown voltage, and high output conductance, leading to a much higher electric field under the gate area. Therefore, it is desirable to improve HEMTs' frequency characteristics without sacrificing their breakdown voltages and DC characteristics at a certain gate length, which is of great importance for their applications in low noise amplifiers (LNA), power amplifiers (PA), and high-speed circuits. Other efforts have been made to improve the performance of HEMTs, such as optimizing the side-recess spacing and reducing the parasitic resistances of the source and the drain.

11.4.2.1 Reduction of Parasitic Resistance in Asymmetrical Recessed Gate

The characteristics of 100 nm gate-length $In_{0.52}Al_{0.48}As/In_{0.53}Ga_{0.47}As$ InP-based HEMTs are studied in this section. An asymmetrical recessed gate and minimized parasitic resistances improved the extrinsic maximum transconductance $g_{m.max}$ and the frequency response of the HEMTs. An InP etching-stopper layer was designed to avoid the kink effect. A lattice-matched $In_{0.53}Ga_{0.47}As$ channel was adopted to improve the f_T and f_{max}. A non-alloyed Ohmic contact technique was used to produce a very low contact resistance and provide sharply defined Ohmic edges, and the gate recess process was optimized to trade off DC characteristics and RF performance. We improved the off-state breakdown voltage (BVDG) of the device to 5.92 V by increasing the distance between the gate and the drain. High f_T/f_{max} values of 249/415 GHz and an extrinsic maximum transconductance $g_{m.max}$ of 1051 mS/mm were achieved at room temperature. These outstanding results are very promising for improving the performance of LNAs, PAs, and high-speed circuits [42].

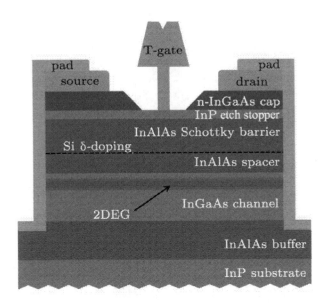

FIGURE 11.28
Schematic cross-sectional view of the InP-based HEMT.

A cross-sectional view of the InP HEMT is shown in Figure 11.28. The epitaxial layer structures were grown by molecular beam epitaxy (MBE) on a 3-inch semi-insulating (100) InP substrate. The epitaxial structure of the HEMTs employed in our study was designed and optimized to balance the tradeoff between high-frequency characteristics and high breakdown voltage. From bottom to top, the epitaxial layers consisted of a 500 nm InAlAs buffer, a 15 nm $In_{0.53}Ga_{0.47}As$ channel layer, a 3 nm InAlAs spacer layer, an 8 nm InAlAs Schottky barrier layer, a 4 nm InP etching stopper layer, and a 20 nm Si-doped composite InGaAs cap layer. A Si-doped plane was inserted between the Schottky barrier layer and the spacer layer to supply the electrons for current conduction. The thick barrier layer was designed to decrease the gate leakage current, and finally enhance the channel current density and the off-state drain–gate breakdown voltage (BVDG). A 10 nm Si-doped $In_{0.6}Ga_{0.4}As$ layer and a 10 nm Si-doped $In_{0.53}Ga_{0.47}As$ transition layer formed the composite cap layer to facilitate the formation of non-alloyed Ohmic contacts. All InAlAs layers were lattice matched with the InP substrate. Hall measurements made at room temperature showed a two-dimensional (2DEG) sheet density of 3.266×10^{12} cm^{-2} and a carrier mobility (μ_n) of 8000 cm^2/(Vs).

11.4.2.1.1 DC Characteristics by HP4142 SPA

DC properties were characterized by using a HP4142 semiconductor parameter analyzer at room temperature. Figure 11.29a shows the current–voltage (I–V) characteristics of the HEMT with $L_g = 97$ nm and gate width $W_g = 2 \times 50$ µm at room temperature. The gate–source voltage (V_{GS}) is increased from −0.6 to 0.4 V with step of +0.1 V, and the drain– source voltage (V_{DS}) changes from 0 to 1.7 V. The HEMT presents good pinch-off characteristics, which is attributed to the enhancement of the gate modulation by using the recessed gate technique. Moreover, the kink effect of the device is negligible due to the introduction of the InP etching-stopper layer as the surface passivation, which makes the pinning Fermi level close to the conduction band minimum; thus, the initial density in the channel is high enough to decrease the impact of impact ionization and the traps in the buffer or barrier.

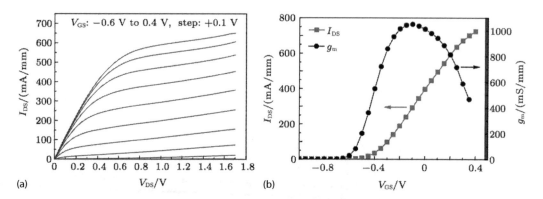

FIGURE 11.29
(a) I-V characteristics (b) Dependency of transconductance and drain current on the gate bias of $V_{DS} = 1.8$ V.

The extrinsic transconductance and the drain current of the HMET at $V_{DS} = 1.8$ V are shown in Figure 11.29b. The pinch-off voltage is about −0.6 V. A maximum extrinsic transconductance $g_{m.max}$ of 1051 mS/mm is achieved at $V_{GS} = -0.1$ V. It has a measured full channel current of 724 mA/mm at V_{GS} of 0.4 V and saturation drain to source current I_{DSS} of 396 mA/mm at V_{GS} of 0 V.

11.4.2.2 Adopting a Self-Aligned T-Gate Architecture

Adopting a self-aligned T-gate architecture is one method used to reduce parasitic device access resistance, but at the cost of increasing parasitic gate fringing capacitances. As the device gate length is then reduced, the magnitude of the static fringing capacitances will have a greater impact on performance. To better understand the influence of these issues on the dc and RF performance of short gate length InP pHEMTs, Figures of merit for these devices include transconductance greater than 1.9 S/mm, drive current in the range 1.4 A/mm, and f_T up to 490 GHz. Simulation of the parasitic capacitances associated with the self-aligned gate structure then leads a discussion concerning the realistic benefits of incorporating the self-aligned gate process into a sub-50 nm HEMT system.

The self-aligned T-gate process, successfully adopted into a short gate length HEMT process, allows for a reduction in the device access resistance by reducing the physical separation between the metalized ohmic contacts and the intrinsic gate region. Conversely, by bringing the ohmic contacts in closer proximity to the gate, the magnitude of the gate fringing capacitance will be larger with the self-aligned architecture than with standard device geometry. A tradeoff therefore exists with the self-aligned gate process dependent on the increased performance achieved through access resistance reduction, compared with the deterioration in performance that results from increased fringing capacitance for a particular gate-length node. The material layers used for this structure are presented in Figure 11.30. These layers resemble a typical InP pHEMT layer structure, with semi-insulating InP substrate, $In_{0.52}Al_{0.48}As$ buffer layer, pseudo morphic 15 nm $In_{0.7}Ga_{0.3}As$ channel, 15 nm $In_{0.52}Al_{0.48}As$ barrier, and a 20 nm bulk-doped $In_{0.53}Ga_{0.47}As$ cap layer. Instead of a single delta doping layer positioned within the barrier layer, however, an additional layer is added closer to the barrier/cap layer interface. As has been described elsewhere, this double doping technique allows the use of a no alloyed ohmic process by tailoring the conduction band profile vertically through the structure and hence favoring better vertical conduction through the layers.

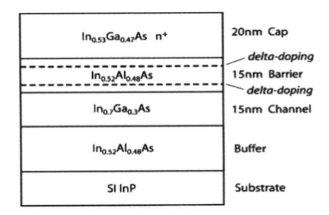

FIGURE 11.30
DDD InP pHEMTs material layer structure.

11.4.2.2.1 DC Characteristics by Agilent 4155 SPA

DC characterization of the completed 50 nm self-aligned and standard devices was performed using an Agilent 4155 semiconductor parameter analyzer (SPA). The output response of both device types is given in Figure 11.31a, with the extrinsic transconductance response for each shown in Figure 11.31b. Extremely high drive current is observed for both types of device, with the self-aligned exhibiting an increase of ~15% over the standard with an I_{DS} close to 1.4 A/mm at a gate bias of +0.2 V. Both devices, however, demonstrate a similar threshold voltage of −0.68 V V_{GS}. In addition, a little kink is observed in the output characteristics for either device, which we attribute to a high carrier concentration within the vicinity of the gate as a result of using the DDD material. The self-aligned device is also found to outperform the standard when comparing the transconductance curves for each. A peak gm of 1.6 S/mm is measured for the standard device, compared to an extremely high figure of more than 1.9 S/mm for the self-aligned device, corresponding to an increase of ~19% in gm by moving to a self-aligned gate structure [43].

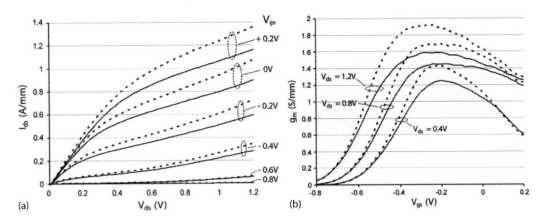

FIGURE 11.31
(a) Self-aligned (broken line) and standard (continuous line) device output characteristics, i.e., I_{ds} versus V_{ds} for fixed V_{gs}. (b). Device extrinsic transconductance characteristics, i.e., g_m versus V_{gs} for fixed V_{ds}.

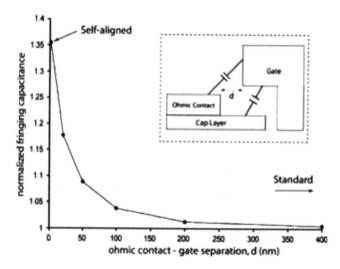

FIGURE 11.32

Simulated parasitic gate capacitance between gate and ohmic contact/cap layer for varied gate ohmic contact separation d. Capacitance values are normalized to that simulated for d = ∞, i.e., between cap layer and gate.

To better estimate the increase in gate fringing capacitance as a function of ohmic contact and T-gate separation, three-dimensional capacitance simulation of the structure treating the gate as one electrode and the cap/ohmic as the other was performed. The geometry of the simulated structure was chosen to best emulate the source side of our 50 nm T-gate devices, i.e., a vertical ohmic contact to T-gate head separation of 15 nm with a 45 nm thick ohmic contact and a 20 nm deep, 30 nm long recess etch. The aim of this simulation was not to calculate discreet values for the fringing capacitances for a given device geometry, but rather to better understand the trend in the increase in such capacitance as a function of ohmic contact gate separation. The results of this simulation are shown in Figure 11.32. The simulated capacitance is shown normalized to that for a structure with an infinite gate to ohmic contact separation. As the ohmic contact is brought closer to the gate, the additional fringing capacitance remains insignificant until at a separation of 100 nm, an increase of approximately 4% is observed. As the separation is reduced below 100 nm, more electric field lines begin to terminate on the ohmic contact, leading to higher capacitance until at a separation of 0 nm, i.e., that for the self-aligned gate structure, an increase of ~36% is observed. Although this simulation considers only the effective capacitance through the air between gate and ohmic contact, we find considerable agreement with the increase in fringing capacitance extracted from our equivalent standard and self-aligned circuit data, i.e., an increase of ~42% to 44%.

11.5 Scalability

III and V group compound semiconductors have recently emerged as promising materials for future silicon industry. Because of the excellent electron transport properties. For future III-V based field-effect transistors (FETs) to enter the CMOS roadmap, they will have dimensions compatible with the 22 nm CMOS node or beyond. Therefore, scalability is of critical importance in the semiconductor devices, although the performance of the device is enhanced by integrating high-k material in the gate-stacked architecture. There are several challenges to overcome these issues. In the meantime scalability of these FETs can be studied in high electron mobility transistors (HEMTs), though being an promising device structure for the millimeter wave applications, but they can also constitute an excellent model system to study issues of critical importance of future III-V MOSFETs, such as intrinsic carrier transport, quantum capacitance, band-to-band tunneling (BTBT), impact ionization, and the impact of parasitic resistances onto device performance. Device Scaling rules were obtained by systematically varying the dimensions of the HEMT power cells. The total gate width of HEMTs and the architecture of the power cells are of particular interest for power applications [44–46].

Recent research on InAlAs/InGaAs HEMTs has demonstrated that these devices exhibit excellent logic characteristics down to about 60 nm. This has been accomplished by scaling down the InAlAs barrier (t_{ins}) less than 5 nm. Further lateral scaling resulted in scaling of the channel thickness (t_{ch}) also. However, as tch scales down, carrier transport in the channel deteriorates, mainly as a consequence of the increased carrier scattering mechanisms. InAs is a very attractive material with an electron mobility as high as 20,000 cm^2/Vs at room temperature. The introduction of InAs is expected to mitigate the deleterious effects of channel thickness scaling. Many researches are carried out in scaling the HEMT device structures. It has been clearly proved that scalability of the HEMT devices yields a better performance rate.

GaN HEMTs has been very attractive for power switching devices because of both the high critical electric field of GaN, and the very high electron mobility of the AlGaN/GaN heterojunction. A hybrid MOS-HEMT incorporates HEMT with a GaN MOS-gated channel, which gives the advantage low gate leakage and normally-off operation mode. GaN MOS-HEMT has been demonstrated by our group with best specific on-resistance and breakdown voltage. Due to the lower electron mobility of GaN MOS channel than that in the channel in HEMT, the downscaling of MOS channel is essential to the improvement of specific on resistance, while other undesirable short-channel effects must be prevented [47,48].

In 2010, Dae-Hyun Kim and Jesus A. del Alamo experimentally demonstrated scaling behavior of Sub-100 nm InAs HEMTs on InP substrate. These devices are specially designed for scalability and combine a thin InAlAs barrier with a thin channel. According to the proposed device, it is an InAlAs/InGaAs HEMT structure with a composite channel that includes a strained InAs sub channel. The device structural parameters are shown in Table 11.3.

Device layers consists of, from bottom to top, a 500 nm buffer, a 10 nm composite channel, a 3 nm spacer, a Si δ-doping with doping concentration of 5×10^{12}/cm^2, a 8 nm barrier, a 6 nm InP etch stopper and a set of n+ doped sub cap layers.

Fabrication is initiated with mesa isolation using H_3PO_4-based wet etchant and is followed by Ni/Ge/Au (10/45/150 nm) source and drain ohmic contacts with a 2 μm spacing with alloying at 320°C for 30 s in N_2 ambient. E-beam lithography is used to define T-shaped gates with various dimensions. Using this method, the channel length is varied by adjusting e beam dose and scalability of the device is studied.

Gate recessing is performed in three different stages. First, an isotropic etching of the InGaAs/InAlAs multilayer cap was performed in the mixture of citric acid and H_2O_2 (20:1). This was followed by anisotropic etching of InP layer using low-damaged Ar-based plasma. Finally, time-controlled wet etching of the InAlAs barrier using diluted citric acid

TABLE 11.3

Device Parameters for InAlAs/InGaAs HEMT
Structure with a Composite Channel

n+ Cap	$In_{0.65}Ga_{0.35}As$	10 nm
	$In_{0.53}Ga_{0.47}As$	25 nm
	$In_{0.52}Ga_{0.48}As$	15 nm
Stopper	InP	6 nm
Barrier	$In_{0.52}Ga_{0.48}As$	8 nm
δ-doping	Si	—
Spacer	$In_{0.52}Ga_{0.48}As$	3 nm
Channel	$In_{0.53}Ga_{0.47}As$	2 nm
	InAs	5 nm
	$In_{0.53}Ga_{0.47}As$	3 nm
Buffer	$In_{0.52}Al_{0.48}As$	500 nm

solution was performed to further thin down the insulator thickness (t_{ins}). We calibrated the depth of the remaining InAlAs barrier through scanning transmission electron microscope (STEM) inspection and obtained at t_{ins} of about 4 nm [49].

Figure 11.33a shows the STEM inspection of the device, which corresponds to a $L_g = 40$ nm device with insulator thickness of 4 nm. In addition, the device insulator thickness is made thicker by omitting the third recess step. An insulator thickness of 10 nm was measured by STEM inspection, as shown in Figure 11.33b, which corresponds to a $L_g = 45$ nm device. This is slightly different from the MBE-grown nominal barrier thickness of 11 nm, probably as a result of the finite etching selectivity of InP against InAlAs during Ar-based plasma and the device fabrication was completed by the evaporation and lift-off of Ti/ Pt/Au (20/20/300 nm) Schottky gate metal stack. Devices with various gate lengths were fabricated, from 340 nm down to 40 nm. Figure 11.33c shows a STEM image of the cross section of a $L_g = 45$ nm T-gate device. The evaporated gate metal partially overlaps with the InP etch stopper at both edges of the InAlAs recessed region. As a result, the metallurgical

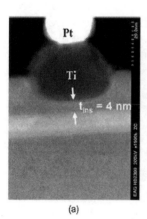
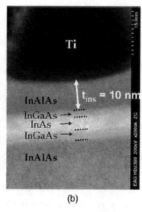
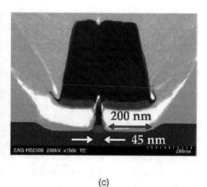

FIGURE 11.33
STEM images of the cross section of the fabricated device. (a) Close up of gate region for an $L_g = 40$ nm and $t_{ins} = 4$ nm device. (b) Close up of gate region for $t_{ins} = 10$ nm device with $L_g = 45$ nm. (c) Gate view of an $L_g = 45$ nm gate length device showing $L_{side} = 200$ nm.

L_g seems to be larger than 45 nm. Figure 11.33c also shows that the side-recess-spacing length (L_{side}) in these devices was set to be around 200 nm to ensure optimum logic operation. This was done by adjusting the recess etching time of the InGaAs/InAlAs capping layers. The source and drain spacing (L_{SD}) was 2 μm in these devices. Future III-V HFETs will require a scaling of L_{SD} with L_g since a reduction in the device pitch is the main driving force for logic technology.

Figure 11.34 shows typical output characteristics of InAs HEMTs with (a) $t_{ins} = 10$ nm and (b) $t_{ins} = 4$ nm for various gate lengths. It is found that devices with smaller insulator thickness exhibit better current driving capability and an enhanced positive V_T. These devices also show better scalability. Looking at output conductance (g_o) for both types of 40 nm InAs HEMTs, it is seen that the devices with $t_{ins} = 4$ nm show slightly better output conductance than those with $t_{ins} = 10$ nm, which is a result of the reduction in t_{ins}. Improved short-channel effects (SCEs) that arise from a thinner insulator are evident in the sub threshold characteristics.

Figure 11.35 shows sub threshold characteristics of both types of 40 nm devices at $V_{DS} = 50$ mV and 0.5 V, together with gate leakage current (I_G). It is clearly seen that a 4-nm barrier device exhibits superior sub threshold swing (S) and drain-induced barrier lowering (DIBL, change in V_T with V_{DS}) and V_T shifts positive as the barrier is thinned down. A tradeoff with scaling the insulator thickness that is evident in Figure 11.35 is an increase of gate leakage current in the forward and reverse gate bias condition. A consequence of this is that the minimum sub threshold current is about an order of magnitude higher for the thin barrier device. It is interesting to note that in both types of devices, the minimum sub threshold current is dominated by Schottky gate leakage, not by band-to-band tunneling (BTBT). This will be a serious concern when a high mobility channel material is utilized with a significantly smaller band gap than Si and might ultimately limit the scalability of the device. In this device design, BTBT is not a significant concern in spite of the narrow band gap of the channel. This might be due to electron quantization in the thin InAs sub channel layer, which is expected to enlarge the effective band gap (E_g). However, future device designs with much reduced gate leakage current might show more pronounced BTBT currents. The improvement in transistor performance with t_{ins} scaling can be seen more clearly in the transconductance (g_m) characteristics.

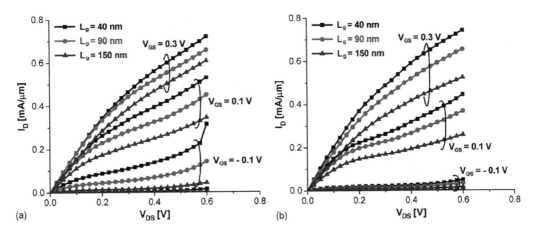

(a) (b)

FIGURE 11.34
Output characteristics of InAs HEMTs with (a) $t_{ins} = 10$ nm and (b) $t_{ins} = 4$ nm for various values of L_g.

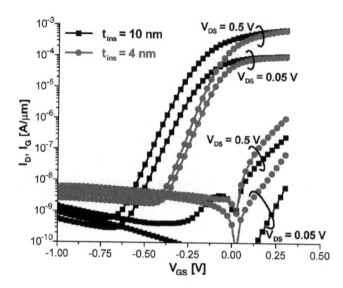

FIGURE 11.35
Sub threshold and gate current characteristics of 40-nm InAs HEMTs with t_{ins} = 10 nm (black) and t_{ins} = 4 nm (red), at V_{DS} = 0.05 V and 0.5 V.

Figure 11.36 plots g_m of both types of InAs HEMTs for various values of L_g at V_{DS} = 0.5 V. For thicker barrier devices with t_{ins} = 10 nm (in Figure 11.36a), as L_g scales down, g_m initially increases, and then, it actually gets worse beyond a L_g of about 60 nm. We can also see a significant negative shift in V_T as L_g scales down. These are clear indications of severe SCEs. On the contrary, t_{ins} = 4 nm devices (in Figure 11.36b) exhibit much less VT shift as L_g scales down. More importantly, the peak g_m continues to scale gracefully down to L_g = 40 nm.

Figure 11.37 shows measured R_S^* as a function of L_g for InAs HEMTs with t_{ins} = 4 nm. Since $R * S$ is the sum of the actual source resistance (RS) plus half of the channel resistance (Rch), RS can be extracted by linear extrapolation of the measured $R * S$ to L_g = 0. It is seen that in InAs HEMT, RS does not degrade. The reason for excellent value of RS in InAs

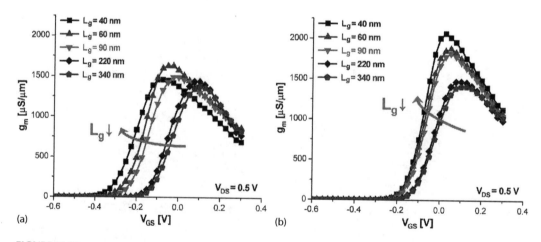

FIGURE 11.36
Transconductance characteristics of 40-nm InAs HEMTs with (a) t_{ins} = 10 nm and (b) t_{ins} = 4 nm, for various values of L_g, at V_{DS} = 0.5 V.

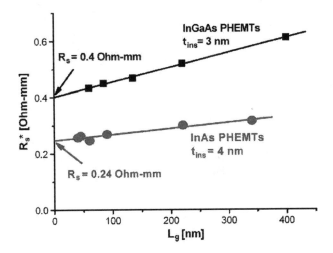

FIGURE 11.37
Measured RS* as extracted by the gate current injection technique as a function of L_g for InAs HEMTs with $t_{ins} = 4$ nm and $In_{0.7}Ga_{0.3}As$ HEMTs with $t_{ins} = 3$ nm. The extracted values of Rs for both devices are shown.

HEMTs is likely to arise from a more anisotropic etching profile of the InAlAs barrier after the three-step recess process. With the help of RS the intrinsic transconductance (g_{mi}) is evaluated to understand the intrinsic performance potential of the technology.

Figure 11.38 shows g_{mi} as a function of L_g for InAs HEMTs with both $t_{ins} = 4$ nm and 10 nm at $V_{DS} = 0.5$ V, together with $In_{0.7}Ga_{0.3}As$ HEMTs with $t_{ins} = 3$ nm. For long values of L_g, g_{mi} degrades slightly when thinning down t_{ins}. This might be due to an increased probability of carrier scattering with thinned InAlAs barrier surface, which deteriorates carrier transport properties. However, as L_g scales down, the superior scalability of the intrinsic transconductance becomes evident for devices with $t_{ins} = 4$ nm. In comparison

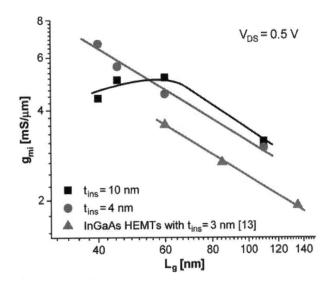

FIGURE 11.38
Extracted intrinsic transconductance (g_m) as a function of L_g for InAs HEMTs with $t_{ins} = 4$ nm and 10 nm, and $In_{0.7}Ga_{0.3}As$ HEMTs with $t_{ins} = 3$ nm, at $V_{DS} = 0.5$ V.

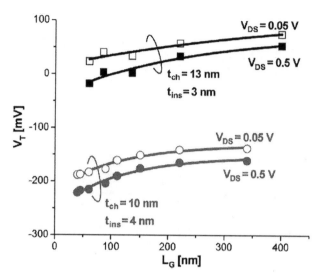

FIGURE 11.39
V_T roll-off behavior of InAs HEMTs with $t_{ins} = 4$ nm and $In_{0.7}Ga_{0.3}As$ HEMTs with $t_{ins} = 3$ nm.

with the reference $In_{0.7}Ga_{0.3}As$ devices with $t_{ins} = 3$ nm, InAs HEMTs exhibit far higher values of g_{mi}, arising from better carrier transport properties due to the InAs sub channel.

Figure 11.39 shows VT roll-off behavior of InAs HEMTs with $t_{ins} = 4$ nm and $In_{0.7}Ga_{0.3}As$ HEMTs with $t_{ins} = 3$ nm, at $V_{DS} = 0.05$ and 0.5 V, using the V_T definition of $I_D = 1$ μA/μm. Both sets of devices exhibit excellent V_T roll-off behavior down to L_g of sub 100 nm regime due to their thin insulator.

Figure 11.40 summarizes (a) DIBL and (b) sub threshold swing, as a function of L_g for $t_{ins} = 4$ nm InAs devices, together with those of $t_{ins} = 3$ nm $In_{0.7}Ga_{0.3}As$ devices at $V_{DS} = 0.5$ V. This figure shows how thinning down tch improves SCEs. In particular, the $t_{ch} = 10$ nm devices with $t_{ins} = 4$ nm show excellent V_T roll-off < 60 mV, DIBL < 80 mV/V, and S < 70 mV/ dec, down to the $L_g = 40$ nm regime.

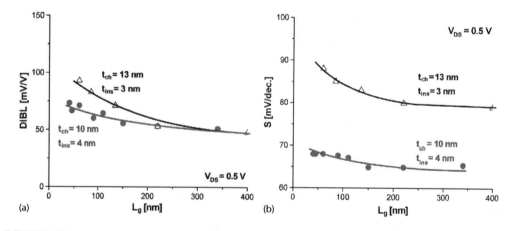

FIGURE 11.40
(a) DIBL and (b) sub threshold swing as a function of L_g, for InAs HEMTs with $t_{ins} = 4$ nm and $t_{ch} = 4$ nm, and $In_{0.7}Ga_{0.3}As$ HEMTs with $t_{ins} = 3$ nm and $t_{ch} = 13$ nm.

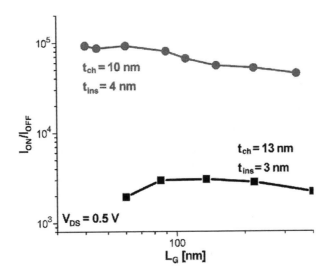

FIGURE 11.41
I_{ON}/I_{OFF} as a function of L_g, for InAs HEMTs with $t_{ins} = 4$ nm and $t_{ch} = 4$ nm, and $In_{0.7}Ga_{0.3}As$ HEMTs with $t_{ins} = 3$ nm and $t_{ch} = 13$ nm.

Figure 11.41 shows I_{ON}/I_{OFF} ratios for the $t_{ch} = 10$ nm devices with $t_{ins} = 4$ nm, together with the $t_{ch} = 13$ nm $In_{0.7}Ga_{0.3}As$ devices with $t_{ins} = 3$ nm, all at $V_{DS} = 0.5$ V. As mentioned earlier, V_T was defined as $I_D = 1$ μA/μm, and then, I_{ON} and I_{OFF} were defined at 2/3VCC swing above V_T and 1/3VCC swing below V_T, respectively. Not only do thin-channel InAs devices exhibit much higher values of I_{ON}/I_{OFF} ratio, but their I_{ON}/I_{OFF} ratios are also maintained as L_g scales down to 40 nm. This stems from the combination of the use of high mobility InAs channel, thinning down the channel and the barrier, and process optimization that prevents R_s degradation. Indeed, 40 nm InAs HEMTs with $t_{ins} = 4$ nm show excellent I_{ON}/I_{OFF} of 9×10^4. Microwave performance of the fabricated InAs HEMTs was characterized from 0.5 to 40 GHz. We used an on-wafer open/short deem bedding method to subtract both pad capacitance and inductance components.

Figure 11.42 shows the H21, Maximum-Stable-Gain/Maximum-Available-Gain (MSG/MAG), and U of 40 nm InAs HEMTs with $t_{ins} = 4$ nm, at $V_{GS} = 0.2$ V and $V_{DS} = 0.5$ V. It is seen that the 40 nm InAs devices exhibit excellent f_T of 491GHz and f_{max} of 402 GHz, even at $V_{DS} = 0.5$ V. This high f_T arises partly from the excellent transport properties associated with the incorporation of the InAs sub channel. We have benchmarked our InAs HEMTs against state-of the-art Si CMOS.

Speed performance can be evaluated through an estimation of the logic gate delay (CV/I). Figure 11.43 shows the CV/I of the $t_{ch} = 10$ nm InAs HEMTs having $t_{ins} = 4$ nm at $V_{DD} = 0.5$ V as well as the $t_{ch} = 13$ nm $In_{0.7}Ga_{0.3}As$ devices having $t_{ins} = 3$ nm, as a function of L_g. Here, we also include those of Si CMOS, which are typically obtained at $V_{DD} = 1.1$ to 1.3 V. the proposed InAs HEMTs device model exhibit significantly better CV/I than Si-CMOS in spite of the lower voltage of operation. Besides, CV/I scales gracefully down to the $L_g = 40$ nm regime. The advantage in CV/I that we observed with respect to Si CMOS is not as large as one would expect from the ratio of motilities in these two materials. Indeed, what is relevant in CV/I is the carrier velocity and the intrinsic capacitance. Measurements have shown that the electron velocity in InAs HEMTs is about a factor of two times higher than in Si, even at the lower voltages.

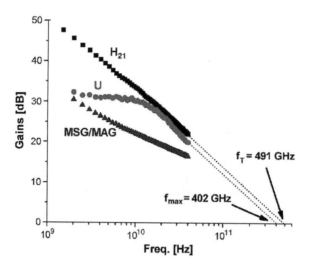

FIGURE 11.42
H_{21}, U, and MSG/MAG of 40-nm InAs HEMTs at $V_{GS} = 0$ V and $V_{DS} = 0.5$ V.

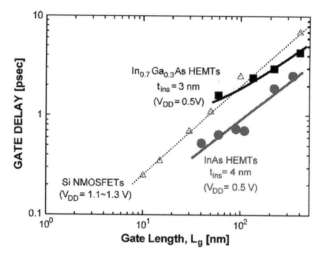

FIGURE 11.43
Gate delay (CV/I) of InAs HEMTs with $t_{ins} = 4$ nm and $t_{ch} = 4$ nm, and In$_{0.7}$Ga$_{0.3}$As HEMTs with $t_{ins} = 3$ nm and $t_{ch} = 13$ nm at $V_{DD} = 0.5$ V, together with that of advanced Si MOSFETs at $V_{DD} = 1.1$ V to 1.3 V.

References of Section 11.1

1. K. F. Brennan. *Introduction to Semiconductor Devices for Computing and Telecommunication Applications*, Cambridge University Press, Chapter 11.
2. D. Ducatteau et al. "Influence of passivation on high-power AlGaN/GaN HEMT devices at 10GHz," *12th GAAS Symposium*, Amsterdam, the Netherlands (2004).
3. J. C. Gallagher et al. "Effect of surface passivation and substrate on proton irradiated AlGaN/GaN HEMT transport properties," *ECS Journal of Solid State Science and Technology* 6 (11) S3060–S3062 (2017).

4. A. Gupta et al. "Effect of surface passivation on the electrical characteristics of nan scale AlGaN/GaN HEMT," *InIOP Conference Series: Materials Science and Engineering* 225, 012095 (2017).

5. S. Yagi et al. "Effects of surface passivation films on AlGaN/GaN HEMT with MIS gate structure," *Physica Status Solidi c* 5 (6), 2004–2006 (2008).

6. H.-K. Lin et al. "An alternative passivation approach for AlGaN/GaN HEMTs," *Solid-State Electronics* 54, 552–556 (2010).

7. M. Asif et al. "Improved DC and RF performance of InAlAs/InGaAs InP based HEMTs using ultra-thin 15 nm ALD-Al2O3 surface Passivation," *Solid State Electronics* 142, 36–40 (2018).

8. C. Wang et al. "Comparison of SiO2-based double passivation scheme by e-beam evaporation and PECVD for surface passivation and gate oxide in AlGaN/GaN HEMTs," *Microelectronic Engineering* 109, 24–27 (2013).

9. United States Patent Schmitz et al. Patent No: US6, 316,820 B1, November 13, 2001.

10. H. Hanawa et al. "Effects of buffer leakage current on breakdown characteristics in AlGaN/GaN HEMTs with a high-k passivation layer," *Microelectronic Engineering* 147, 96–99 (2015).

11. W. A. Sasangka et al. "Improved reliability of AlGaN/GaN-on-Si high electron mobility transistors (HEMTs) with high density silicon nitride passivation," *Microelectronics Reliability* 76–77, 287–291 (2017).

12. T. Hashizume et al. "State of the art on gate insulation and surface passivation for GaN-based power HEMTs," *Materials Science in Semiconductor Processing* 78, 85–95 (2018).

13. Z. Bai et al. "Study on the electrical degradation of AlGaN/GaN MIS-HEMTs induced by residual stress of SiNx Passivation," *Solid-State Electronics* 133, 31–37 (2017).

14. B. K. Jebalin et al. "The influence of high-k passivation layer on breakdown voltage of Schottky AlGaN/GaN HEMTs," *Microelectronics Journal* 46, 1387–1391 (2015).

15. C. Ostermaier et al. "Reliability investigation of the degradation of the surface passivation of InAlN/GaN HEMTs using a dual gate structure," *Microelectronics Reliability* 52, 1812–1815 (2012).

16. W. Lu et al. "A comparative study of surface passivation on AlGaN/GaN HEMTs," *Solid-State Electronics* 46, 1441–1444 (2002).

17. B. Sun et al. "ALD Al2O3 passivation of Lg = 100 nm metamorphic InAlAs/InGaAs HEMTs with Si-doped Schottky layers on GaAs substrates," *Solid State Electronics* 138, 40–44 (2017).

18. V. Desmaris et al. "Influence of oxynitride (SiOxNy) passivation on the microwave performance of AlGaN/GaN HEMTs," *Solid-State Electronics* 52, 632–636 (2008).

19. L. Sileo et al. "Gallium arsenide passivation method for the employment of high electron mobility transistors in liquid environment," *Microelectronic Engineering* 97, 333–336 (2012).

20. https://www.researchgate.net/publication/316064826_Cellulose_composite_for_electronic_devices/figures?lo=1.

21. http://www.anu.edu.au/CSEM/machines/MOCVD.htm.

22. http://www.semicore.com/news/89-what-is-e-beam-evaporation.

23. http://forgenano.com/uncategorized/atomic-layer-deposition/.

24. http://lnf-wiki.eecs.umich.edu/wiki/Low_pressure_chemical_vapor_deposition.

25. J. Bernat et al. "Effect of surface passivation on performance of AlGaN/GaN/Si HEMTs," *Solid-State Electronics* 47, 2097–2103 (2003).

References of Section 11.2

26. S. Gao et al. "MMIC class-F power amplifiers using field-plated GaN HEMTs," *IEE Proceeding-Microwaves Antennas and Propagation* 153 (3), 259–262 (2006).

27. A. Fletcher et al. "Design and modeling of HEMT using field plate technique," *Innovations in Electrical, Electronics, Instrumentation and Media Technology (ICEEIMT), 2017 International Conference IEEE* pp. 157–159 (2017).

28. Analysis of breakdown voltage improvement using gate field plate engineering technique in AlGaN/GaN HEMT. Chapter 5.

29. Z. Kai et al. "Field plate structural optimization for enhancing the power gain of GaN-based HEMTs," *Chinese Physics B* 22 (9), 097303 (2013).
30. C.-Y. Chianga et al. "Effect of field plate on the RF performance of AlGaN/GaN HEMT devices," *Physics Procedia* 25, 86–91 (2012).
31. W. M. Waller et al. "Impact of silicon nitride stoichiometry on the effectiveness of AlGaN/GaN HEMT field plates," *IEEE Transactions on Electron Devices* 64 (3) (2017).
32. G. Xie et al. "Breakdown-voltage-enhancement technique for RF-based AlGaN/GaN HEMTs with a source-connected air-bridge field plate," *IEEE Electron Device Letters* 33 (5), 670–672 (2012).
33. C. Wang et al. "Development and characteristic analysis of a field-plated Al2O3/AlInN/GaN MOS—HEMT," *Chinese Physics B* (2011).
34. S. A. Ahsan et al. "Capacitance modeling of a GaN HEMT with gate and source field plates," *IEEE Compound Semiconductor Week and International Conference on Indium Phosphide and Related Materials* (2015).

References of Section 11.3

35. J. Shao et al. "Nanofabrication of 80 nm asymmetric T shape gates for GaN HEMTs," *Microelectronic Engineering* 189, 6–10 (2018).
36. J. Ajayan et al. "Investigation of DC-RF and breakdown behaviour in $Lg = 20$ nm novel asymmetric GaAs MHEMTs for future submillimetre wave applications," *AEU-International Journal of Electronics and Communications* 84, 387–393 (2008).

References of Section 11.4

37. N. Malbert et al. "Characterization and modeling of parasitic effects and failure mechanisms in AlGaN/GaN HEMTs," *Microelectronics Reliability* 49, 1216–1221 (2009).
38. H. Zhang et al., "Extraction method for parasitic capacitances and inductances of HEMT models," *Solid-State Electronics* 129, 108–113 (2017).
39. G. Meneghesso et al. "Parasitic effects and long term stability of InP-based HEMTs," *Microelectronics Reliability* 40, 1715–1720 (2000).
40. S. Oh et al. "Effect of parasitic resistance and capacitance on performance of InGaAs HEMT digital logic circuits," *IEEE Transactions on Electron Devices* 56 (5), 1161–1164 (2009).
41. Z. Yang et al. "A novel method for measuring parasitic resistance in high electron mobility transistors," *Solid-State Electronics* 100, 27–32 (2014).
42. W. Li-Dan et al. "100-nm T-gate InAlAs/InGaAs InP-based HEMTs with $f_T = 249$ GHz and $f_{max} = 415$ GHz," *Chinese Physics B* 23 (3) 038501 (2014).
43. D. A. Moran et al. "50-nm self-aligned and 'Standard' T-gate InP pHEMT comparison: The influence of parasitics on performance at the 50-nm node," *IEEE Transactions on Electron Devices* 53 (12) 2920–2925 (2006).

References of Section 11.5

44. M. E. Hoque et al. "Scalable HEMT model for small signal operations," *International Conference on Electro magnetics in Advanced Applications (ICEAA)* (2010).
45. R. Lossy et al. "Uniformity and scalability of AlGaN/GaN HEMTs using stepper lithography," *Physica Status Solidi (a)* 188 (1), 263–266 (2001).
46. A. Caddemi et al. "Impact of the self-generated heat on the scalability of HEMTs," *Microelectronic Engineering* 82, 143–147 (2005).
47. B. M. Paine. "Scaling DC life tests on GaN HEMT to RF conditions," *Microelectronics Reliability* 55, 2499–2504 (2015).
48. Z. Li. "Channel scaling of hybrid GaN MOS-HEMTs," *Solid-State Electronics* 56, 111–115 (2011).
49. D.-H. Kim and J. A. del Alamo, "Scalability of sub-100 nm InAs HEMTs on InP substrate for future logic applications," *IEEE Transactions on Electron Devices* 57 (7) (2010).

12

InP/InAlAs/InGaAs HEMTs for High Speed and Low Power Applications

Nilesh Kumar Jaiswal and V. N. Ramakrishnan

CONTENTS

12.1 Introduction

As CMOS device scaling is approaching its physical and peak performance, future downscaling of CMOS in accordance with Moore's law and to meet the demands of the ITRS (International Technology Roadmap for Semiconductors) roadmap will involve new materials for the gate dielectrics and the high mobility channels as well as novel structures. High Electron Mobility Transistor (HEMT) is laterally developed III-V compound semiconductor which uses heterojunction for its operation and performance. The increasing interest of high-speed communication and high-frequency systems leads to low cost,

TABLE 12.1

Transport Property of Main FET Channel Materials

Property	Silicon	GaAs	$In_{0.53}Ga_{0.47}As$	InAs	InSb
Energy gap (eV)	1.12	1.43	0.75	0.356	0.175
Electron effective mass (m_o)	0.19	0.072	0.041	0.027	0.013
Pure material electron mobility ($cm^2V^{-1}s^{-1}$)	1500	8500	14000	30000	780000
Electron mobility at 1×10^{12} cm^{-2} ($cm^2V^{-1}s^{-1}$)	600	4600	7800	20000	30000
Electron saturation velocity (cm^*s^{-1})	1.0×10^7	1.2×10^7	8×10^6	3.0×10^7	5.0×10^7
Electron mean free path (nm)	28	80	106	194	226
Intrinsic carrier concentration (cm^3)	1.6×10^{10}	1.1×10^7	5.0×10^{11}	1.3×10^{15}	1.9×10^{16}

and power consumption, high-level integration devices. On the other hand, with superior material properties and band-engineering, compound semiconductors have enabled new device concepts like Resonant Tunnelling Diode (RTD), HEMT, and Heterojunction Bipolar Transistor (HBT). These devices have demonstrated excellent characteristics for demanding applications that range from high speed, low-noise microwave and millimeter-wave circuits to fast digital and optical electronics. For the above applications, InP/InAlAs/InGaAs HEMT devices perform better than Si-based devices. GaAs was presented by Takashi Mimura shows more advantages than Si-based transistors [1]. To separate the ionized donors from channel to enhance the electron mobility 2DEG gas is formed in the channel layer which is grown epitaxially.

In the past three decades, the HEMT devices have undergone explosive research and development. A variety of semiconductor material systems have been introduced. The transport properties like electron mobility and peak velocity are the key material features for high speed performance. Table 12.1 summarizes the related transport parameters for silicon and main compound semiconductor channel materials [2].

The group III-V compound semiconductor FET and bipolar transistors technology requirements roadmap truncates at the following expected ends of scaling: GaAs PHEMT in 2015, GaAs power Metamorphic HEMT (MHEMT) in 2020, and InP power HEMT in 2016. However, the low noise GaAs MHEMT and InP HEMT, InP HBT, and GaN HEMT will continue with physical scaling [3] as shown in Figure 12.1a–d.

12.2 InP/InAlAs/InGaAs HEMT Structure

In recent years, there has been increasing interest in InP-based heterostructure MOSFETs with high mobility InGaAs channel. Applying oxides with high dielectric constant (high-k) instead of SiO_2 as the gate oxide on III-V materials can reduce the gate leakage current at the same Equivalent Oxide Thickness (EOT), thus reduces the power consumption. There has been considerable interest in developing InP-type transistors such as HBTs, MOSFETs, and HEMTs using various technologies such as gate, source/drain, and material engineering ever since some advantages of the intrinsic properties such as high electron mobility, high thermal conductivity and low voltage operation in order to gain advantages over GaAs, silicon and SiGe based semiconductor technologies. III-V compound semiconductors such as GaAs, InP, InAs, etc. can be studied as an alternative channel material in the future nanoscale applications.

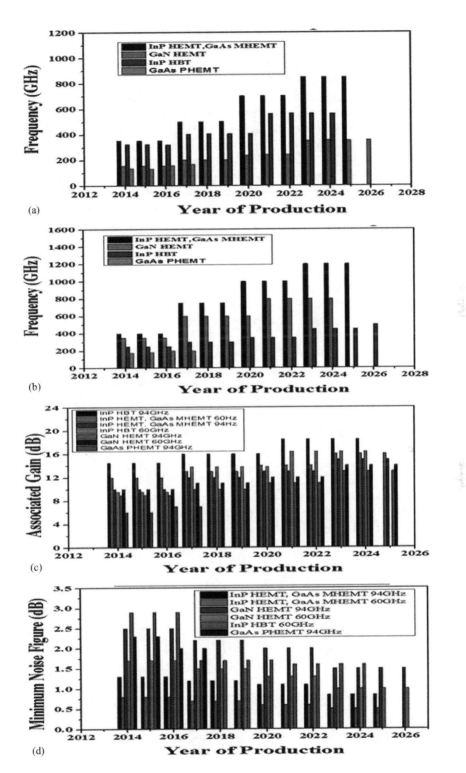

FIGURE 12.1
(a) III-V fT roadmap, (b) III-V fMAX roadmap, (c) III-V associated gain roadmap, and (d) III-V noise figure roadmap.

The epi-structure of InP/InAlAs/InGaAs HEMT consists of InP substrate, InAlAs-buffer, InGaAs-channel, InAlAs- spacer, InAlAs-Schottky barrier layer, and a heavily doped InGaAs cap layer as shown in Figure 12.2. In MOSFETs, a layer of silicon through process steps forms the source, drain and channel, with the source and drain regions are heavily doped which offer free electrons (or holes) to the lightly doped channel. In these, when electrons move from source to drain the mobility of electrons decreases through collisions with the impurities (dopants) in the channel. But HEMTs overcome this mobility degradation due to collision with impurities through the use of high mobility electrons generated using the heterojunction of a heavily doped wide bandgap n-type donor (supply layer (example InAlAs)) and an undoped narrow bandgap channel layer with no dopant impurities (example InGaAs). The mobile electrons generated in the supply layer (example InAlAs) drop completely into an undoped channel layer (example InGaAs), which results in the depletion of InAlAs donor layer, because the heterojunction formed by semiconductors with different band-gaps creates a quantum well in the conduction band on the InGaAs material where the mobile electrons are able to move freely and quickly by avoiding collision with the dopants since the InGaAs channel layer is un-doped, and also these conducting electrons cannot escape from the quantum well. This results in the formation of a very thin layer of high mobility electrons with very high density, in other words "high electron mobility." The conductivity of the channel layer in the HEMTs can be altered by applying a potential at its gate terminal like other conventional FET devices.

The buffer layer isolates the defects of substrate from the channel. The spacer layer isolates the mobile electrons from the positively charged silicon donor atoms, which helps to minimize the impurity scattering which is the primary cause for mobility degradation in conventional MOSFETs. The separation of electrons from the donor atoms by spacer layer leads to the enhancement of mobility of electrons in the quantum well. The access resistance can be minimized by using a heavily doped n-type cap layer.

The InGaAs/InAlAs materials offer advantages over conventional Pseudomorphic HEMT (PHEMT) [4–5], which lead to improved device performance. The discontinuity present at the InGaAs/InAlAs heterojunction interface leads to higher conduction band. The value has been determined to be 0.5 eV compared to 0.25 eV for GaAs/AlGaAs heterojunction and approximately 0.4 eV for InGaAs/AlGaAs pseudomorphic structure. From Table 12.2, The cost and fabrication yield of HEMT has better advantages compared to InP HEMT.

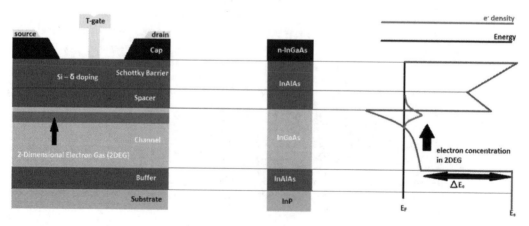

FIGURE 12.2
Layer structure of a typical InP/InAlAs/InGaAs HEMT.

TABLE 12.2

Comparison of Lattice-Matched InP HEMT and Metamorphic GaAs HEMT

Property	InP HEMT	MHEMT
Substrate Availability, Cost	4-inch now, higher cost	6-inch now
MBE Growth Time	+ ½ hours	−½ hours
Process Difficulty Yield	− Higher breakage, more difficult/ slower backside process	+ Lower breakage, standard GaAs backside process
Performance, Impedence Characteristics	No difference	No difference
Achievable Channel In Content	53%–80%	30%–80%
Thermal Resistance	+ InP has 50% higher thermal conductivity than GaAs	Comparable to GaAs PHEMT, effect of buffer unclear
Reliability	Proven for low noise, unproven for power	Excellent initial data for low noise, power unknown

Source: Schleeh, J. et al., Cryogenic ultra-low noise amplification: InP PHEMT vs. GaAs MHEMT, *International Conference on Indium Phosphide and Related Materials (IPRM)*, 2013.

The HEMTs have shown good performance in terms of operation at high frequencies, high speed and low noise and have become one of the important semiconductor transistor. The superior microwave performance capability of HEMTs made it useful for advanced military system applications and stimulated the development of new monolithic integrated circuits. InP-based HEMTs have become one of the promising candidates for microwave and millimeter-wave frequency applications. Since 1992, several works have shown that InP-based HEMTs can exhibit a high cutoff frequency f_T in the sub- millimeter-wave frequency range (300 GHz–3 THZ) when the gate length L_g is reduced [7]. To further increase the frequency and speed performance of the device, the dimensions were reduced to nanoscale. But, reduction in gate length (L_g) accompanied short-channel effects which deteriorate the microwave performance of HEMT. Thus, a high aspect ratio must be maintained to avoid such affects. However, this is also limited by a leakage current at the gate and as a result, double-gate HEMT (DG-HEMT) [8] emerged as a solution and provided a better charge control and improved frequency performance as shown in Figure 12.3.

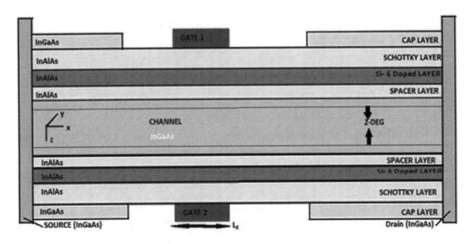

FIGURE 12.3

Schematic device structure of InAlAs/InGaAs DGHEMT.

DG-HEMT with nanoscale dimensions is a quantum heterostructure that requires a detailed quantum mechanical treatment. The channel here comprises symmetric double triangular quantum wells, which are separated by a barrier formed due to two similar interfaces on either side of the channel. These quantum wells provide confinement to the carriers and necessitate the prevailing quantum effects to be considered. The 1D Schrodinger equation accounting for the potential profile of a channel comprises of a nanoscale symmetric Double Triangular Quantum Well (DTQW) separated by a barrier for double heterostructure double gate InAlAs/InGaAs HEMT at equilibrium and at different electric fields is analytically solved in the published literature [9]. However, channel thickness being an essential parameter needs to be further studied. Hence, the work published in this paper focuses on the effect of variation in the channel thickness, different barrier widths and well widths on eigen energies for the DTQW system formed in the channel can be studied [10]. The ground and first excited energy states for various barrier and well widths are evaluated from a transcendental equation and reveal its importance.

12.3 Physics of InP/InAlAs/InGaAs HEMT

12.3.1 Operation of HEMTs

Under equilibrium conditions, in a modulation-doped structure electrons are transferred from their parent donors in the higher band-gap into the lower bandgap material, leading to considerable band bending and formation of a 2DEG. Because of the spatial separation of the conducting electrons from donor impurities, this heterojunction system exhibits high electron mobility, especially at low temperatures. The electric potential (band diagram) and charge distribution follows the Poisson's equation while the electron wave function satisfies the Schrodinger equation. Theoretically the potential and charge profile can be calculated by self-consistently solving the two equations. Based on the assumption that the charge depleted from the donor layer is accumulated in the 2DEG, the Fermi level is constant across hetero-interface, and the interfacial potential at the channel can be approximated by a triangular well, the electron charge (n_s) stored at the interface in a modulation doped structure is [11]:

$$n_s = \frac{\varepsilon}{qd}[V_g - (\phi_b - V_{p^2} + E_{fi/q} - E_C/q)] \tag{12.1}$$

where ε is the dielectric constant of the barrier layer, q is electron static charge, d is sum of the thickness of undoped barrier d_i and doped barrier d_d, V_g is the applied gate to source voltage, ϕ_b is the Schottky barrier height of the gate metal deposited on the barrier layer, $V_{p^2} = \frac{qN_d d_d^2}{2\varepsilon}$ while N_d is the donor concentration in the barrier layer, E_{fi} is the Fermi level with respect to the conduction band edge in the channel layer, and ΔE_c is the conduction band offset between the barrier and channel.

The interface change in Equation 12.1 can be expressed in a more general form, which can be used for planar doped structure as [12]:

$$n_s = \frac{\varepsilon}{q} \frac{V_g - V_{th}}{d_B + d_S + \Delta d} \tag{12.2}$$

where d_B and d_S are the thickness of the barrier and spacer respectively, Δd is the average distance between 2DEG and interface. The threshold voltage is given as:

$$V_{th} = \phi_b - \frac{\Delta E_c}{q} - V_{p^2} + \frac{E_{fi}}{q} \tag{12.3}$$

for modulation doped structure

$$V_{th} = \phi_b - \frac{\Delta E_c}{q} - V_{p^2} + \frac{q}{\varepsilon} N_\delta d_B \tag{12.4}$$

for δ doped structure.

12.3.2 Current Voltage Model

When the drain bias is applied to the field effect transistor, the lateral electric field induces a channel voltage that varies along the channel. Based on assumption that the electron density at the channel voltage and electron velocity-field model, analytical expression of drain current and voltage relation can be deducted. The simplest form of velocity and field model is known as two-piece model where the electron velocity is linear with field and then saturate at high field. The result shows an abrupt change between constant mobility regime to constant velocity regime. A more accurate picture with smoother transition uses Si-like field characteristics expressed as [11]:

$$v = \frac{\mu F_{(z)}}{1 + \mu F_{(z)} / v_{sat}} \tag{12.5}$$

where $F_{(z)}$ represents electric field in the channel, μ is the mobility and v_{sat} the saturation velocity.

Using Equation 12.5, we can obtain the expression for normalized drain current as [11]:

$$I_d(V_d) = \frac{\varepsilon \mu v_{sat}}{d + \Delta d} \left| \frac{V_{geff} V_d - V_d^2 / 2}{v_{sat} L_g + \mu V_d} \right| \tag{12.6}$$

where V_d and V_{geff} are drain bias and effective gate bias, respectively.

12.3.3 Small-Signal Analysis

The small-signal response of high-speed transistors is normally analyzed by two port scattering parameters. The measurement is done by on-wafer probing with a network analyzer in certain frequency range. A general small-signal equivalent circuit of HEMT device [13] is shown in Figure 12.4. The intrinsic circuit elements in the dashed line are in π-shaped configuration. One can deduce Y parameters from the scattering parameters to extract the numeric values for the circuit elements.

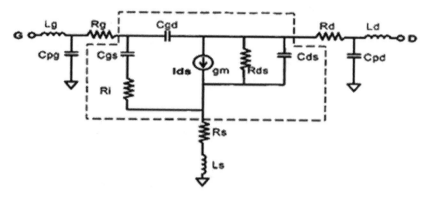

FIGURE 12.4
A classic 8-element small-signal equivalent circuit model of a typical HEMT.

12.4 Buried Platinum Technology

To achieve enhancement mode operation in InP/InAlAs/InGaAs a gate metal with high work function and Schottky barrier is required. The commonly used metals are titanium (work function = 4.1 eV), nickel (work function = 5.15 eV), palladium (work function = 5.1 eV) and (work function = 5.65 eV). Among these, platinum metal has the highest Schottky barrier height (0.83 eV) on InAlAs material [14]. By using techniques like in situ Ar^+ ion cleaning of InAlAs layer, the Schottky barrier height can be increased up to 1.09 eV after gate annealing. Also platinum sinking is used for minimizing gate to channel separation which is essential for suppressing short-channel effects. This is because platinum metal has the ability to react with InAlAs material to form $PtAs_2$ alloy. The reduction of gate to channel separation will provide better control over 2DEG by the gate, which results in enhanced electron velocity under the gate. Improvement in electron velocity under the gate can improve the cut off frequency of the device [15–17].

To fabricate enhancement-mode HEMT (E-HEMT) in InAlAs/InGaAs material system, a gate metal with a high Schottky barrier is needed. The most commonly used gate metal, Ti/Pt/Au, has a Schottky barrier of 0.65 eV on InAlAs, which is only 0.15 eV larger than the conduction band discontinuity (ΔE_c) at the InAlAs/InGaAs hetero-junction, and not enough to deplete the channel at zero gate bias. Therefore, Pt has emerged as the optimum choice for the gate metal used in the fabrication of E-HEMT's in InP-based materials [18].

12.5 Double Delta Doping Technology

The HFETs based on InP/InAlAs/InGaAs material systems are well known to have an excellent microwave and millimetre wave performance [19–20]. The standard HFET with a single-top delta doping layer generally gives rise to a high electron mobility, high electron saturation velocity, and a low minimum noise figure. A double delta-doped or uniform-doped channel design is preferred to achieve larger carrier concentration and therefore higher drain current capability and hence larger output power [21]. On the other hand, double delta-doped or uniform-doped channel transistors usually suffer from lower electron mobility [22]. The thickness of the delta doping layer can calculate which is a reasonable estimation of the delta doping layer in real devices since the ideal delta doping pulse with high doping concentration slightly widens to a few nm layers during the MBE growth [19].

If delta position close to the bottom heterojunction, the conduction band changes its shape and becomes deeper along the bottom heterointerface. The amount of electrons occupying the first energy subband is larger than that of the second subband even at lower gate biases. The position of the electrons occupying the first energy level is unaffected with the increasing gate bias. In all three cases, the electrons in the first subband may be scattered by a strong Coulomb potential because they share the same space with the parent ionized donors, and this degrades the transport properties of the channel. Thus one should not expect high mobility of electrons occupying the first energy level. Those electrons occupying the second subband are moved toward the middle of the channel with the increasing gate bias. This reduces the effect of the Coulomb interaction with the parent ionized donors, which may help to improve the mobility of electrons residing in the second subband. With constant delta doping concentration and varying only the delta doping position inside the channel, the threshold voltage of the devices is not affected. The total channel charge density decreases for a fixed delta doping position and reducing only the channel thickness. The maximum transconductance of the device is hardly influenced by varying the delta doping position. The total gate capacitance of the device is strongly affected by delta layer positioning inside the InGaAs channel. The total gate capacitance of the device is strongly affected by delta layer positioning inside the InGaAs channel [23].

The main advantage of using silicon δ-doping in between barrier and spacer layer is that it requires only a few atomic layer thicknesses and this heavily doped layer provides electrons to the channel as shown in Figure 12.5. There are two different types of silicon δ-doping strategy is available. The first is the single silicon δ-doping layer which can be either located above or below the channel [24]. In the second method, two layers of silicon δ-doping can be included either on the same side of the channel or on either side of the channel [25], and this second strategy is also called double δ-doping strategy. Increasing the silicon δ-doping concentration increases the two-dimensional electron charge densities in the channel, which results in enhanced transconductance, drive current and cutoff frequencies. The silicon δ-doping also provides good device linearity compared with

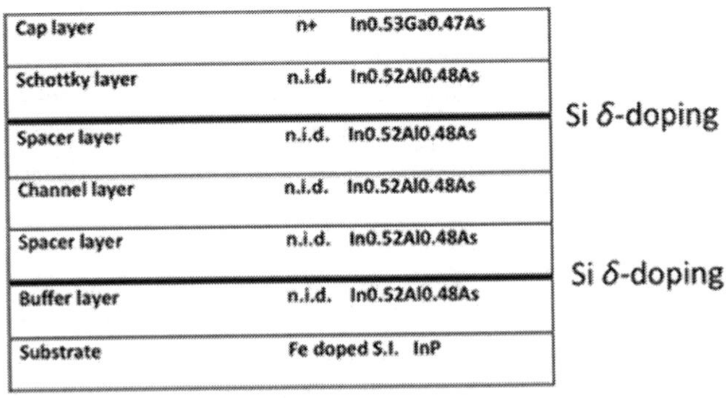

Cap layer	n+	In0.53Ga0.47As	
Schottky layer	n.i.d.	In0.52Al0.48As	Si δ-doping
Spacer layer	n.i.d.	In0.52Al0.48As	
Channel layer	n.i.d.	In0.52Al0.48As	
Spacer layer	n.i.d.	In0.52Al0.48As	Si δ-doping
Buffer layer	n.i.d.	In0.52Al0.48As	
Substrate	Fe doped S.I. InP		

FIGURE 12.5
Double-delta doping InP-HEMT structure for power applications.

conventional uniform doping techniques. Silicon δ-doping technique also helps to minimize the access resistance. The various research groups found that double δ-doping strategy can greatly minimize the kink effects.

12.6 DC, RF and Breakdown Voltage Characteristics

Successful device development relies on the crucial processes of measurement and metrology that provide benchmarks for the performance of both epitaxial material and the devices fabricated on it. Accurate measurement provides an understanding of the underlying phenomena that dictate device performance, allowing device fabrication to be tailored to the improvement of these metrics. Understanding of the materials structure on which the device is fabricated is therefore invaluable, since it defines ultimate performance to such a large degree. Techniques that characterize the material and contacts made to it are hence of great importance to the device engineer, in addition to device DC and RF measurements [26–27].

In this section, some basic techniques are presented to characterize semiconductor material. In particular, the van der Pauw method for extracting carrier mobility and the Transmission Line Method for determining contact and sheet resistances are described. There is then a discussion of the process of characterizing devices, both at DC and RF [28–31].

12.6.1 Device Characterization

Both DC and RF measurements make use of three-signal probes in a ground-signal-ground configuration, at the heart of the measurement system. The probes are mounted on precision manipulator arms that allow three-dimensional positioning of the probes to attain reliable contacts. A general system setup requires a network analyser, able to supply and analyse a range of signals of variable frequency, matched in specification to the probes and cabling. It is these components which define the measurement range of the system. A semiconductor parameter analyser (SPA) is also required to provide the DC measurement and bias capability. The systems are configured such that the RF components are disabled during DC measurement, with the DC bias injected directly to the probes. During RF measurements, the DC bias is used to provide the desired operating conditions for the device, whose response to an input spectrum generated by the network analyser is then measured. Both systems are connected using a General Purpose Interface Bus (GPIB). Parameters can be passed, and one system may control the measurement timing, while computer control can additionally be exerted over GPIB.

The measurement setup used in this work varied, since the systems were upgraded during the course of this project. Initial device measurements were taken using Picoprobe probes, mounted on a Karl Suss probe station using a Wiltron Vector Network Analyser (VNA) with an Agilent 4155B SPA. The system was capable of RF measurement upto 60 GHz.

This setup was later replaced by a semi-automatic probe station able to make measurements under automated control at a given sample location, then move to another site and repeat measurements: a far more flexible and efficient setup. The new system also uses Picoprobe probes, connected via Agilent frequency extender arms to an Agilent E8361A PNA network analyser and N2560 test set, capable of measurements up to 110 GHz [32].

The manipulators are mounted on a Cascade Microtech Summit 12000 semi-automatic probe station, which is able to move between sample sites, but is incapable of altering probe separation. The system is connected to a control computer running Cascade Microtech Nucleus software. This computer also runs the Cascade MicrotechWinCal software which allows the configuration and management of RF measurements. DC measurement capabilities are provided by an Agilent B1500A SPA, which runs Agilent EasyEXPERT measurement software. All systems are connected by GPIB, and by the end of this project the system could be configured for fully automated multiple measurements at either DC or RF for a variety of bias conditions, allowing large volumes of data to be efficiently extracted.

12.6.1.1 DC Measurements

In general, HEMT-output characteristics depend on the application of a variable drain bias, modulated by the gate voltage.

The B1500A provides a method of measuring the drain current whilst sweeping the drain voltage and keeping the source earthed. The gate voltage can then be stepped to a new value and the drain voltage sweep repeated. By repeating this process, the complete device characteristics can be extracted. Figure 12.6a shows general HEMT I–V characteristics as measured by the system. Gate current resulting from leakage through the imperfect Schottky contact is also usually measured. The system is also able to extract additional measurements from the measured data. These data can then be exported to various file formats for processing [27].

The transconductance can be extracted using similar processes. The drain current is measured while the gate voltage is swept, then the drain voltage is incremented. The SPA is then used to calculate the derivative of the drain current with respect to the gate voltage: the transconductance. The profiles for a typical device are shown in Figure 12.6a and b.

The I–V profile, peak drain current, gate leakage current and transconductance hence allow a great deal to be concluded about the quality of the devices, material and their processing. From these measurements, the DC metrics of the devices can be easily compared and extracted. All these measurements can then be repeated across many sample locations using the autoprober.

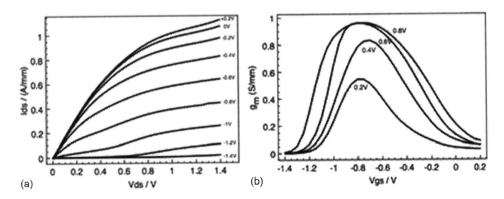

FIGURE 12.6
(a) I_{ds}/V_{ds}, V_{gs} characteristics, (b) Transconductance/V_{gs} characteristics.

12.6.1.2 RF Characteristics

RF characteristics rely on the treatment of a circuit as a multi-port device, whereby each port has a corresponding input and output signal. Single device measurements such as those taken in this work treat the HEMT as a two-port network [27]. At RF frequencies, signals behave like waves on a transmission line according to Maxwell's equations, where fractions of each applied signal are transmitted or reflected, depending on the load and matching conditions as well as the line itself. Applying an input signal at one port results in a corresponding output on both ports as a consequence of the transmission and reflection of the original signal. These transmitted and reflected signals can be characterised in magnitude and phase over a given frequency spectrum, for a given range of bias conditions. The general setup is shown in Figure 12.7.

12.6.2 Breakdown

The breakdown characteristics of InP/InAlAs/InGaAs HEMTs can be classified as off-state breakdown and on-state breakdown depending on the bias conditions of gate and drain. The off-state breakdown is the examination of maximum voltages the device can sustain in channel-off condition ($V_{GS} < V_T$). The off state breakdown mechanism is believed to be a combined effect of gate-to-channel thermionic-field emission and impact ionization in the channel [33]. Electrons from the inverse biased gate were first injected to the InAlAs barrier by thermionic-field emission. Then due to the large electric field and conduction band offset between barrier and channel, the electrons gained energy when entering the channel and relaxed by impact ionization before being collected by source and drain contacts. Somerville et al. [34] demonstrated that the most significant variables in determining the off-state breakdown voltage of power HEMTs are the 2DEG carrier concentration (N_s), the gate Schottky barrier height (ϕ_b), and temperature. Bahl et al. [33] also found that such breakdown voltages have a negative temperature coefficient and decrease with the increase of In mole fraction of the InGaAs channel. So the off-state breakdown is believed to be two-step process dominated by thermionic-field-emission at low electric field and impact ionization at high electric field. This is agreed by Dickmann et al. [35] in the investigation of gate leakage current at different temperatures. However, a more recent analysis showed that the impact ionization coefficient in InGaAs may have an abnormal positive temperature coefficient [36].

On the other hand, the on-state breakdown is characterized in open-channel condition when gate is biased beyond threshold. Impact ionization was found to be the dominant mechanism for the breakdown by the measurement of temperature dependence of on-state breakdown voltage [37]. Some phenomenological models [38–39] for specific experiments

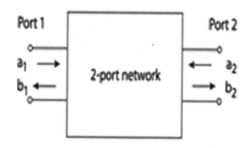

FIGURE 12.7
General overview of a two-port network and its input/output signals.

have been proposed to correlate breakdown voltage with impact ionization coefficient, but it is still doubtful whether they can be applied to general InAlAs/InGaAs HEMTs.

In the on-state operation of InAlAs/InGaAs HEMTs, gate leakage current increases drastically under high drain to source bias and forms a bell shape in the curve. Impact ionization is found responsible for this phenomena. When drain bias is high, hot carriers in the channel are able to gain enough energy to generate electron-hole pairs by impact ionization. The generated electrons will be collected by the drain contact and cause an increase in the drain current. At the mean time, accumulated holes in the channel will drift parallel to the channel to recombine with electrons on the source contact [40]. When the electric field is sufficiently high, some holes will be able to tunnel through the heterojunction barrier and reach the gate and cause severe increase in the gate current. Hence the total gate leakage current consists of the electrons tunnelling current I_{Ge} through the reverse biased Schottky diode and the hole tunnelling current I_{Gh} caused by impact ionization [41]:

$$I_G = I_{Ge} + I_{Gh} = I_{Ge} + |I_D \cdot \alpha(E) \cdot L_{eff}(E) \cdot T_h(E)| \tag{12.7}$$

Here $\alpha(E)$ is the electric-field-dependent impact ionization coefficient, $L_{eff}(E)$ is the effective length of high field region between gate and drain and $T_h(E)$ is the tunnelling probability of holes from channel to the gate contact. The value of $T_h(E)$ can be approximated through optical measurement. Both of these components decrease substantially with temperature.

12.7 Scalability

Usually the gate metal consists of several layers of different metals like titanium (Ti), platinum (Pt), gold (Au), palladium (Pd), molybdenum (Mo), etc. The most commonly used gate metal stacks are Ti/Pt/Au or Ti/Pd/Au or Ti/Mo/Au or Pt/Ti /Pt/Au, etc, [15–16]. The T-shaped gate provides excellent mechanical strength in comparison with conventional gate structures and it also exhibits extremely low gate resistance due to the large cross sectional area of the T-gate head. The T-gate architecture also improves the scalability of the HEMT devices. The T-gate structure is widely used in HEMT design in order to improve the frequency performance of the device. The T-gate structure has a very short foot width and larger head width. The ultra-short foot width enhances the RF performance of the device and larger head area of the T-gate provides high gain at very high frequencies; i.e., the gate length plays a vital role in defining HEMT performance.

12.7.1 Gate Capacitances and Resistance

One obvious effect of the reduction of gate length is the linearly correspondent reduction in gate capacitances. Considering the gate and channel as a parallel-plate capacitor,

$$C_{parallel} = \frac{\epsilon\, WL}{d} \tag{12.8}$$

where d defines the gate-channel separation, ϵ is the dielectric constant of the material between the gate and channel (though in reality this will be an InGaAs/InAlAs heterostructure), W is width, and L length of the gate stripe. Consequently, the intrinsic gate capacitance will reduce with the gate length, L. The gate capacitance, clearly, will be far more complex than this simple scenario, due to the effects of fringing electric fields, dopant planes, non-uniform electron densities in the channel, and the presence of free carriers or trapped charges throughout the layer structure. The general trend for intrinsic capacitance reduction, however, remains valid.

Less desirably, gate resistance increases linearly with reducing gate length, as Equation 12.7 shows.

$$R_g = \frac{\rho W}{A} \tag{12.9}$$

where ρ is the resistivity of the metal, W is gate width, and A is cross-sectional gate area.

As a consequence, it is crucial to effective device scaling to adopt an advanced gate scaling strategy, which does not result in correspondingly increasing gate resistance.

12.7.2 Non-equilibrium Electron Transport

The long-channel electron transport, where carrier velocity is governed by low-field mobility and velocity saturation effects. As the gate length is reduced, however, non-linear effects become important due to the short timescales associated with the transit of the intrinsic region. The magnitude of the drain current is therefore directly dependent on carrier velocity, whilst there is a direct correlation of high frequency performance to electron velocity, as previously discussed. The term "non-equilibrium" refers to the relative energy imbalance between free electrons and the material in which they are moving. In equilibrium, electrons in a semiconductor have energy equal to that of the surrounding crystal lattice. As electrons acquire more energy as a result of high electric fields, they have greater energy than the thermal energy of the lattice, so are often known as "hot" electrons. Electron heating can result in various interesting transport phenomena apart from the low-field, mobility-limited velocity regime.

12.7.3 Effective Channel Length

While the intrinsic channel region is dominant in defining device performance, it is crucial to note that the actual channel length is in practice somewhat different from the defined gate length. The real channel length will be defined by the region under the influence of the electric fields induced by the gate. Since the gate and channel approximate a parallel-plate capacitor, fringing electric fields will extend beyond the physical gate length. The result is the extension of the region of the channel under the influence of the gate: the effective gate length of the device [42]. This is often known as the depletion length since it implies the region of the channel that can be pinched off and depleted of carriers. This extension effect occurs for all gate lengths, but since the extension can be tens of nanometres, dependent on geometry, it becomes increasingly apparent in short gate length devices where the extension may form an appreciable percentage of the effective gate length.

Additionally, the extent to which fringing fields define the depletion length is affected by the aspect ratio of the channel. If the gate-channel separation is large, the fringing fields will extend further, increasing effective gate length. The modulation efficiency will also drop as a consequence of the increased separation. In the extreme case of extension, the depletion region becomes semi-circular and entirely dominated by the fringing field elements. As a consequence, regardless of the physical gate length defined lithographically, the effective gate length will always remain longer by a finite extension. The consequence of this, considering previous derivations, will be to suppress drive current, transconductance and high frequency performance over that expected for a lithographic gate length. The physical situation is exacerbated when surface states are considered. The role of surface states in defining the pinned Fermi level in unpassivated semiconductors were discussed. The effect of surface states, however, is more complex than defining the surface potential, as suggested by their probable role in defining the kink effect. Surface states represent traps and surface charges which can significantly affect electrostatics and the ability to control conduction in the vicinity of the pinned surface. Additionally, these states may have varying corresponding activation energies and response times, with the consequence that their behaviour may vary with bias conditions and operating frequency of a device incorporating unpassivated surfaces.

A normal HEMT layout features a gate formed in a recessed region where the highly doped cap layer has been etched away. As a consequence, the device barrier layer is exposed on both sides of the gate and surface states are present adjacent to the gate on both sides. These surface states lead to trapped charge in the regions adjacent to the gate, which can be modulated by the gate's fringing fields and causes their extension. The variable charge present at the surface then influences the channel electron density, increasing the effective gate length [43].

12.7.4 Vertical Scaling

It is critical to note that the HEMT must be scaled vertically as the gate is scaled laterally, since the aspect ratio of the gate length to gate-channel separation is of importance in maintaining the channel electrostatics [44]. A given gate-channel separation may therefore be known to be optimal for a device of a given gate length, yielding both favourable electron dynamics in the channel and the capacity to control the electron population. If the barrier layer is too thin, the pinned surface may have an overly significant impact on the channel population, whilst a thick barrier reduces the effect of the Schottky barrier height variation by the application of a gate voltage. If the gate length is reduced without corresponding epitaxial scaling, the effective electric field induced by the gate voltage acts on a reduced volume of the channel. As a consequence, the channel population is modulated less efficiently and transconductance decreases, to the detriment of potential high-frequency performance.

Proportional vertical scaling of the device architecture to include the thickness of the barrier, spacer and channel is therefore important to maintain adequate channel population control and compensate the effects of the reduced gate length. As is clear, in the fully scaled case, the gate has full control over the drain current for all gate lengths over the bias range, whilst transconductance increases with decreasing gate length. In the laterally scaled case, transconductance drops off with decreasing gate length and drain current is not easily modulated over the well behaved bias range.

12.7.5 Limits to Scaling

As the transistor is vertically scaled, the eventual limit to scaling becomes tunnelling through the barrier, which affects both gate leakage and the ability to correctly pinch off the channel. Direct field emission through the barrier is expected to be problematic as the gate-channel distance is scaled to dimensions as small as a few nanometres [45]. Similarly, tunnelling may prove to limit the lateral scaling of FETs as well, as the depletion region may become sufficiently small as to be directly tunnelled by channel electrons by thermionic processes under large electric fields [46–47]. This is expected to begin to be problematic at gate lengths shorter than 10 nm, though this must be interplayed with the effective gate length argument of Ferry et al. cited previously [48]. At the ultimate scaling limit, traditional FET geometries will cease to be sufficient as a result of the loss of electrostatic control over the channel [46]. It is therefore expected that fin-gate or double-gate geometries [49–51] may take precedence in the medium term, whilst an ultimate solution may be a wrap-gate nanowire geometry [52–54].

12.8 Kink Effects

One method of improving the performance of HEMTs on InP or GaAs substrates is to increase the indium content in the channel. However, increasing indium concentration in the channel also creates some difficulties like kink effects on the output characteristics of the HEMTs due to impact ionization occur in the channel. The variation in indium concentration affects the key parameters of the HEMTs like transconductance, drain current, cut off and maximum oscillation frequencies. The impact of indium concentration on the performance of InP HEMT was demonstrated by [55]. Therefore, it is necessary to optimize the indium content in the channel which ideally varies from 53% to 100% depending up on the applications.

Many of the InAlAs/InGaAs HEMTs exhibit an abnormal drain current increase at certain drain bias in output characteristics. It is known as kink effect. The additional current component does not allow saturation of drain current so that the devices show relatively high output conductance. Kink also causes reduced gain and excess noise and is an unstable factor in circuit design.

The origin of kink effect was first linked with traps in the InAlAs buffer layer [56–57]. The traps capture energetic electrons and release them when drain bias increases. The initial efforts to eliminate kink focused on improve the quality of InAlAs buffer layer. For example, InGaAs and InP was incorporated into the InAlAs buffer to lower the trap density [58]. Furthermore low temperature (150°C) growth of InAlAs buffer layer was used to control the electron trapping behavior in the buffer [59–60]. Some other authors attributed kink phenomena to impact ionization process in the InGaAs channel [61,63]. When the drain bias increases to certain value impact ionization is initiated at the drain-side gate edge and extends all the way to the drain contact. More recently Somerville et al. [62] presented a dynamic model for kink effect and attribute the current increase to the threshold voltage shift caused by impact ionization. Under high electric field, the impact ionization will generate electron-hole pairs in the channel layer. Part of the holes can tunnel through the barrier layer and contribute to the gate leakage current. The remaining holes will flow to the extrinsic source of the transistor and reach equilibrium through recombination. The accumulated holes raise the potential of the channel on the source side and result in a shift of the intrinsic device's threshold voltage.

Some methods tried to change the impact ionization process in InAlAs/InGaAs HEMTs to reduce the kink phenomena. InGaAs/InP composite channel structure was a way to increase the band gap of channel and to reduce impact ionization [67]. A direct ohmic structure on InGaAs channel was successful in changing current paths in the device and relax the electric field at the recess edge of the drain side [65–66]. An InP etching stop layer inserted between contact cap and barrier layer has been proven to reduce the kink effect by changing high electric field distribution [68]. The InP etch stop layer also helped to prevent oxidation to InAlAs barrier and to improve device long-term reliability.

The kink effect in HEMTs is an undesirable non-linearity which introduces a spurious increase in drain current with its onset at given gate and drain bias conditions. The increase is undesirable due to its unpredictability and additionally induces reduced gain and increased noise at high frequencies [69]. The physical basis for the kink has been the subject of some debate. The current understanding [70] involves the phenomena of trapping, including that due to surface states, and impact ionisation effects similar to those at work during breakdown; in effect, the kink is related to the variable charging of surface states and other interface and buffer traps. The kink is therefore temporally dependent. Neglecting impact ionisation, channel charge is imaged with an opposing polarity by charged surface states, interface charges and dopant ions. With the onset of impact ionisation, itself a field-dependent effect, holes are generated by the interactions of high-energy electrons at the drain end of the gate. These holes are attracted to the source, accumulating around the source in the channel, resulting in a steady-state channel hole distribution which modifies the potential profile in the vicinity of surface states, increasing electron density and reducing source resistance [71]. To satisfy the surface Fermi pinning (Section 12.7.3), the hole quasi-Fermi level bends, resulting in hole current flow to the surface and into the buffer. The charge at these locations therefore changes, and must be balanced by a corresponding change in the channel electron population, raising the channel potential and resulting in a kink voltage. The kink effect is therefore most pronounced for wide recesses, where there is a greater total number of surface states to pin the Fermi level; as indeed experimentally evidenced [72]. Many devices now use indium phosphide as a cap to the barrier, which is etched as part of the gate process, in order to reduce the effect of Fermi level pinning of the barrier [73]. It has, however, also been noted that control of field distribution and doping levels also significantly affect the influence of these mechanisms on the presence of the kink.

12.9 Short-Channel Effects

Performance enhancements of HEMTs are mainly achieved by scaling down the device gate length using high resolution electron beam lithography (EBL). Improvements of such characteristics as gate capacitance, transconductance and transit frequency are expected and they should be connected with high nonstationariety in short devices: most electrons are traveling through the gate-controlled active zone without suffering any scattering, or suffering very few. Under these conditions of ballistic transport, one can reach very high drift velocities, much higher than stationary ones. In order to study quantitatively the short-channel effects on drift velocity and transit frequency of a pseudomorphic HEMT, two devices with different gate lengths have been studied using our self consistent many particle ensemble Monte Carlo (MC) simulator. A three valley nonparabolic band structure is used for each consistent material. The electron trajectories are traced in two dimensions in real space subject to all the principal scattering agent. Polar optical phonon, deformation potential and impurity scattering mechanisms are present, as well as the action of the self-consistent two-dimensional electric field profile. The MC procedure is coupled with a two-dimensional Poisson solver. Transport of holes and impact ionization are not taken into account in this study. All simulations assume room temperature operations.

Theoretically, it has been shown that the short-channel effect has two main origins; one is the two-dimensional nature of potential distribution in the channel, and the other is the space-charge-limited current that flows through the substrate under the channel.

T.W. Kim et al. have experimentally demonstrated the impact of channel thickness on the DC and RF performance of InGaAs HEMTs [74] and found that reducing the channel thickness leads to the reduction of transconductance (g_m) and on current but greatly improve the short-channel effect.

12.10 Applications of InP/InAlAs/InGaAs HEMTs

In the last five decades, the HEMT devices have been implemented in a variety of material systems such as AlGaN/GaN, AlGaAs/GaAs, InP/InAlAs/-InGaAs and GaAs/InAlAs/InGaAs etc. AlGaN/GaN HEMTs are widely used for high-power, high-breakdown voltage applications whereas InP/InAlAs/InGaAs HEMTs are considered to be attractive candidates for low-power, high-speed, low-noise and low-voltage operations.

The future is bright for InGaAs/InP devices, HBTs, DHBTs and HEMTs in high speed electronics, radar, communications, physics and other disciplines. InP and InP on GaAs based HBTs, DHBTs, and HEMTs/mHEMTs will continue their march towards high speed, higher integration, higher power efficiency, lower noise, and lower cost. Indium antimonide-based QWFETs will soon emerge into the real world and enable a new generation of ultra-low power and high-speed systems. InP-based HEMTs will continue to morph themselves into other kinds of FETs (e.g., MOS-HEMTs) that will exploit the unique properties of 2DEG in InP/GaAs materials system.

One of the typical applications of InP-based HEMTs is in high-speed integrated circuits (ICs) of optical communication systems. Recently, Murata et al. reported the development of a 100 Gbit/s logic IC using InP-based HEMTs [75]. The designed test chip contains a selector circuit and a D-type flip-flop with the operating speed of D-F/Fs constructed with 562-GHz-HEMTs, which corresponds to the operating speed of 135 Gbit/s. Thus, the ultrahigh of the HEMTs carries promise for the development of ultrahigh-speed ICs for future optical communication systems as well as sub-millimeter-wave applications.

On-wafer noise parameter measurements are reported on InP HEMTs at cryogenic temperatures and the temperature-dependence of the equivalent noise temperatures of drain and source are determined. The influence of some physical mechanisms such as impact ionization on small-signal parameters and Noise Figure are given in detail. A cryogenic LNA with modelling approach was presented and temperature-dependent noise parameters of InP/InAlAs/InGaAs HEMTs with a noise temperature of 10 K at 32 GHz were reported [76]. Due to its superior noise performances at elevated frequency, InP/InAlAs/InGaAs pseudomorphic HEMT is a good choice for high bit rate optical or millimeter-wave circuit applications. The application of mixed InP-InAlAs or InGaP-InAlAs drastically reduces gate leakage current due to impact ionization in AlInAs/InGaAs HEMTs. Also, it is reported that HEMT devices are very sensitive test structures to investigate the defects induced by a growth interruption [77]. The frequency dispersion effect is investigated with environment temperature dependence of the intrinsic delay time. In InAlAs/InGaAs HEMTs. The intrinsic delay time characterizes the minimum time in which carrier through the channel layer under the gate. The intrinsic delay time is found to be 25.6 ps at room temperature, whereas at 77 K, it is 5.15 ps, which is nearly six times faster. The device performance is maximised with an idealized channel layer at cryogenic temperatures [78].

The InP-based HEMT has several performance characteristics, which allows it to be a promising useful power device. The thermal conductivity is higher. InP HEMTs have high current handling capability due to high 2DEG density and velocity. The high gain of the device at mm-wave frequencies is an advantage because under large-signal conditions, gain is reduced significantly from its small-signal value. InAlAs/InGaAs/InP HEMTs have demonstrated the favourable physical properties of the $In_x Ga_{1-x} As$ channel, namely high electron mobility and velocity and large conduction band discontinuity, into performance unmatched by any other transistor. The device is well suited to high-frequency analog applications. Noise performance is superior to any other technology. The development of InP MMICs based on this new device has begun.

InP/GaAs HEMTs have been chosen as one of the most promising candidates for future high-speed logic applications where the HEMTs functions as a switch that toggles between "ON" and "OFF" states. A high on current (I_{ON}) is desirable for fast switching and a low off current (I_{OFF}) is required for minimizing standby power consumption. Ideally, shorter gate length devices are preferable for ultrahigh radio frequency (RF) applications, and also improving cut off and maximum oscillation frequencies for future THz applications. Therefore, in order to improve the DC and RF performance of HEMTs in terms of its drain current, transconductance, cut off frequency and maximum oscillation frequency, the gate length must be reduced down to its technological limits. However, reducing the gate length increases the short-channel effects which in turn affects the frequency performance of the HEMTs. To avoid these problems, the epitaxial layer structure and architecture of the device have to be properly designed and optimized to achieve the desired level of performance. This optimization of device geometry and layer structure can be done with the help of computer-aided simulations with no waste of money. To optimize the device dimensions simulations can be performed using TCAD tool.

References

1. T. Mimura et al., The early history of the High Electron Mobility Transistor (HEMT), *IEEE Transactions on Microwave Theory and Techniques*, vol. 50, no. 3, 2002.
2. T. Ashley et al., Indium antimonide based technology for RF applications, In *Compound Semiconductor Integrated Circuit Symposium*, 2006.
3. The International Technology Roadmap for Semiconductors 2.0, 2015 Edition.
4. D. Hoare et al., Monte Carlo simulation of PHEMTs operating up to terahertz frequencies, *International Journal of Electronics*, vol. 83, no. 4, pp. 429–440, 1997.
5. T. Watanabe et al., Terahertz imaging with InP high-electron-mobility transistors, in *Proceedings of SPIE: The International Society for Optical Engineering*, p. 80230P, 2011.
6. J. Schleeh et al., Cryogenic ultra-low noise amplification: InP PHEMT vs. GaAs MHEMT, *International Conference on Indium Phosphide and Related Materials (IPRM)*, 2013.
7. T. Zimmer, D. O. Bodi , J. M. Dumas , N. Labat, A. Touboul and Y. Danto, Kink effect in HEMT structures: A trap-related semi-quantitative model and an empirical approach for spice simulation, *Solid-State Electronics*, vol. 35, 1543, 1992.
8. N. Verma et al., Quantum simulation of a double gate double heterostructure in AlAs/InGaAs HEMT to analyze temperature effects," *UKSim-AMSS International Conference on Modelling and Simulation (UKSim), 17th*, 2015.

9. J. Jogi et al., Quantum modeling of electron confinement in double triangular quantum well formed in nanoscale symmetric double-gate InAlAs/InGaAs/InP HEMT, *International Semiconductor Device Research Symposium (ISDRS)*, 2011.

10. N. Verma et al., Quantum modeling of nanoscale symmetric double gate InAlAs/InGaAs/InP HEMT, *Journal of Semiconductor Technology and Science*, vol. 13, no. 4, 2013.

11. S. N. Mohammad et al., MODFETs: Operation, status and applications, in *Compound Semiconductor Electronics: The Age of Maturity*. Shur, ed. World Scientific, 1995.

12. S. M. Sze. *High Speed Semiconductor Devices*. John Wiley & Sons, 1990.

13. M. Berroth and M. Bosch, High-frequency equivalent circuit of GaAs FETs, *IEEE Transactions on Microwave Theory Technology*, vol. 39, no. 2, 224–229, 1991.

14. N. Harada et al., Pt-based gate enhancement-mode InAlAs/InGaAs HEMTs for large-scale Integration, *3rd International Conference Indium Phosphide and Related Materials*, 1991.

15. K. J. Chen, T. Enoki, High performance InP based enhancement mode HEMTs using non-alloyed ohmic contacts and Pt-based buried gate technologies, *IEEE Transactions on Electron Devices*, vol. 43, no. 2, 252, 1996.

16. D. H. Kim, f_T1/4 688 GHz and f_{max}1/4 800 GHz in L_g 1/4 40 nm In0.7Ga0.3As MHEMTs with $g_{m,max}$ > 2.7 mS/mm, in *Proceedings of IEDM Technical Digest*, 319, 2011.

17. D. H. Kim, J. A. del Alamo, 30-nm E-mode InAs PHEMTs for THz and future logic applications, in *Proceedings of IEDM Technical Digest*, p. 719, 2008.

18. I. Adesida et al., Enhancement-mode InP-based HEMT devices and applications, *International Conference on Indium Phosphide and Related Materials*, 1998.

19. R. Lee Ross et al., *Pseudomorphic HEMT Technology and Applications*, Springer, 1996.

20. C. Kadow et al., n/sup +/-InAs-InAlAs recess gate technology for InAs-channel millimeter-wave HFETs, *IEEE Transactions on Electron Devices*, vol. 52, 2005.

21. L. Y. Chen et al., On an InGaP/InGaAs double channel pseudomorphic high electron mobility transistor with graded triple delta-doped sheets, *IEEE Transactions on Electron Devices*, vol. 55, no. 11, 2008.

22. Y. C. Lin et al., Device linearity comparison of uniformly doped and δ-doped In0.52Al0.48As/In0.6Ga0.4As metamorphic HEMTs, *IEEE Electron Device Letters*, vol. 27, no. 7, 2006.

23. Y. C. Lin et al., A delta-doped InGaP/InGaAs pHEMT with different doping profiles for device-linearity improvement, *IEEE Transactions on Electron Devices*, vol. 54, no. 7, 1617–1625, 2007.

24. S. M. Sze et al., *Physics of Semiconductor Devices*, 3rd, Wiley, New York, 2007.

25. K. Kalna et al., Role of multiple delta doping in PHEMTs scaled to sub-100 nm dimensions, *Solid-State Electronics*, vol. 48, 1223–1232, 2004.

26. A. E. Parker et al., Measurement and Characterization of HEMT Dynamics, *IEEE Transactions on Microwave Theory and Techniques*, vol. 49, no. 11, 2001.

27. T. Saranovac et al., Pt gate sink-in process details impact on InP HEMT DC and RF performance, *IEEE Transactions on Semiconductor Manufacturing*, vol. 30, no. 4, 2017.

28. A. Goetzberger, R. Scarlett and W. Shockley, Research and investigation of inverse epitaxial UHF power transistors, oai.dtic.mil, 1964.

29. M. Lijadi, F. Pardo, N. Bardou and J. Pelouard, Floating contact transmission line modelling: An improved method for ohmic contact resistance, *Solid State Electronics*, vol. 49, 1655–1661, 2005.

30. H. Murrmann and D. Widmann, Current crowding on metal contacts to planar devices, *IEEE Transactions on Electron Devices*, vol. 16, 1022–1024, 1969.

31. G. Reeves and H. Harrison, Obtaining the specific contact resistance from transmission line model measurements, *Electron Device Letters*, vol. 3, 111, 1982.

32. www.keysight.com, N5290A/900-hz-to-110-ghz pna-mm-wave-system.

33. S. Bahl, J. Alamo, J. Dickmann and S. Schildberg. Off-state breakdown in InAlAs/InGaAs MODFETs. *IEEE Transactions on Electron Devices*, vol. 42, no. 1, 15–22, 1995.

34. M. Somerville and J. Alamo, A model for tunneling-limited breakdown in high power HEMTs. In *IEDM Technical Digest*, 1996.

35. J. Dickmann et al., Breakdown mechanisms in pseudomorphic In0.52Al0.48As/InxGa1−xAs high electron mobility transistors on InP. i: off-state, *The Japanese Journal of Applied Physics*, vol. 34, no. 1, 66–71, 1995.

36. G. Meneghesso et al., Effects of channel quantization and temperature on off-state and on-state breakdown in composite channel and conventional InP-based HEMTs: On state, In *IEDM Technical Digest*, 1996.

37. J. Dickmann et al., Breakdown mechanisms in pseudomorphic In0.52Al0.48As/InxGa1−xAs high electron mobility transistors on InP. ii, *Journal of Applied Physics*, vol. 34, no. 4A, 1805–1808, 1995.

38. U. Auer et al., A consistent physical model for the gate-leakage and breakdown in InAlAs/InGaAs HFETs, In *11th International Conference on Indium Phosphide and Related Materials*, 1999.

39. M. Somerville et al., On-state breakdown in power HEMTs: Measurements and modeling, *IEEE Transactions on Electron Devices*, vol. 46, no. 6, 1087–1093, 1999.

40. C. Heedt et al., Drastic reduction of gate leakage in InAlAs/InGaAs HEMT's using a pseudomorphic InAlAs hole barrier layer, *IEEE Transactions on Electron Devices*, vol. 41, no. 10, 1685–1690, 1994.

41. R. Reuter et al., Investigation and modeling of impact ionization with regard to the RF and noise behavior of HFET, *IEEE Transactions on Microwave Theory and Techniques*, vol. 45, no. 6, 977–983, 1997.

42. J. Hauser et al., Characteristics of junction field effect devices with small channel length-to-width ratios, *Solid-State Electronics*, vol. 10, 577, 1967.

43. T. Suemitsu et al., Improved recessed gate structure for sub-0.1-μm-gate InP-based High Electron Mobility Transistors, *Japanese Journal of Applied Physics*, vol. 37, 1365–1372, 1998.

44. K. Kalna et al., Scaling of pseudomorphic high electron mobility transistors to decanano dimensions, *Solid-State Electronics*, vol. 46, 631, 2002.

45. D. K. Ferry et al., Full-band CMC simulations of terahertz HEMTs, *Journal of Physics: Conference Series*, vol. 109, 012001–1–6, 2008.

46. M. Lundstrom, Device physics at the scaling limit: What matters? in *IEEE International Electron Devices Meeting*, 33.1.1–33.1.4, 2003.

47. H. Wong, Beyond the conventional transistor, *IBM Journal of Research and Development*, vol. 46, 133–168, 2002.

48. R. Akis et al., The upper limit of the cutoff frequency in ultrashort gate-length InGaAs/InAlAs HEMTs: A new definition of effective gate length, *Electron Device Letters*, vol. 29, 306–308, 2008.

49. A. Rahman et al., Novel channel materials for ballistic nanoscale MOSFETs-bandstructure effects, in *IEEE International Electron Devices Meeting*, 4, 2005.

50. N. Wichmann et al., InAlAs/InGaAs double-gate HEMTs with high extrinsic transconductance, in *International Conference on Indium Phosphide and Related Materials*, 295, 2004.

51. N. Wichmann et al., Fabrication and characterization of 100-nm In0.53Ga0.47As–In0.52Al0.48As double-gate HEMTs with two separate gate controls, *Electron Device Letters*, vol. 26, 601, 2005.

52. T. Bryllert et al., Vertical high-mobility wrap-gated InAs nanowire transistor, *Electron Device Letters*, vol. 27, 323–325, 2006.

53. A. Forchel et al., Nanowire-based one-dimensional electronics, *Materials Today*, vol. 9, 28–35, 2006.

54. V. Schmidt et al., Realization of a silicon nanowire vertical surround-gate field-effect transistor, *Small*, vol. 2, 85–88, 2006.

55. I. Thayne, Advanced III-V HEMTs, *III-Vs Review*, vol. 16, no. 8, 48–51, 2003.

56. A. S. Brown, U. K. Mishra, L. E. Larson and S. E. Rosenbaum. *Institute of Physics Conference Series*, vol. 96, 445–448, 1988.

57. A. Georgakilas et al., Suppression of the kink effect in InGaAs/InAlAs HEMTs grown by MBE by optimizing the InAlAs buffer layer, *Fourth International Conference on Indium Phosphide and Related Materials*, 97–100, 1992.

58. J. B. Kuang et al., "Low frequency and microwave characterization of submicron-gate In0.52Al0.48As/In0.53Ga0.47As/In0.52Al0.48As heterojunction metal semiconductor field-effect transistors grown by molecular-beam epitaxy, *Journal of Applied Physics*, vol. 66, no. 12, 6168–6174, 1989.

59. A. S. Brown, U. K. Mishra, J. A. Henige and M.J. Delaney, *IEEE GaAs IC Symposium*, pp. 143–146, 1989.

60. A. S. Brown, C. S. Chow, M. J. Delaney, C. E. Hooper, J. F. Jensen, U. K. Mishra, L. D. Nguyen and M. S. Thompson. *IEEE GaAs IC Symposium*, pp. 143–146, 1989.

61. M. Somerville, A. Ernst and J. Alamo. A new dynamic model for the kink effect in InAlAs/InGaAs HEMTs. In *IEDM Technical Digest*, 1998.

62. G. G. Zhou et al., High output conductance of InAlAs/InGaAs/InP MODFET due to weak impact ionization in the InGaAs channel, *IEDM Technical Digest*, vol. 24, no. 3, 247–250, 1991.

63. G.-G. Zhou et al., I-V kink in InAlAs/InGaAs MODFETs due to weak impact ionization process in the InGaAs channel, *Conference Proceedings of Sixth International Conference on Indium Phosphide and Related Materials*, pp. 435–438, 1994.

64. T. Enoki et al., InGaAs/InP double channel HEMT on InP, *Proceedings of 4th IPRM Conference*, pp. 14–17, 1992.

65. K. Sawada et al., Elimination of kink phenomena and drain current hysteresis in InP-based HEMTs with a direct ohmic structure, *IEEE Transactions on Electron Devices*, vol. 50, 310–314, 2003.

66. G. Meneghesso et al., Improvement of DC, low frequency and reliability properties of InAlAs/InGaAsInP based HEMTs by means of an InP etch stop layer, In *IEDM Technical Digest*, 1998.

67. M. Somerville et al., Direct correlation between impact ionization and the kink effect inInAlAs/InGaAs HEMTs, *Electron Device Letters*, vol. 17, 473–475, 1996.

68. M. Somerville et al., A physical model for the kink effect in InAlAs/InGaAs HEMTs, *IEEE Transactions on Electron Devices*, vol. 47, 922, 2000.

69. T. Suemitsu et al., An analysis of the kink phenomena in InAlAs/InGaAs HEMT's using two-dimensional device simulation, *IEEE Transactions on Electron Devices*, vol. 45, 2390–2399, 1998.

70. G. Zhou et al., High output conductance of InAlAs/InGaAs/InP MODFET due to weak impact ionization in the InGaAs Channel, in *IEEE International Electron Devices Meeting*, pp. 247–250, 1991.

71. W. Kruppa and J. Boos, Low-frequency transconductance dispersion in InAlAs/InGaAs/InP HEMT's with single- and double-recessed gate structures, *IEEE Transactions on Electron Devices*, vol. 44, 687–692, 1997.

72. D.-H. Kim and J. del Alamo, 30-nm InAs Pseudomorphic HEMTs on an InP Substrate with a current-gain cutoff frequency of 628 GHz, *Electron Device Letters*, vol. 29, 830–833, 2008.

73. T. Suemitsu et al., 30-nm two-step recess gate InP-Based InAlAs/InGaAs HEMTs, *IEEE Transactions on Electron Devices*, vol. 49, 1694–1700, 2002.

74. T. W. Kim et al., 60 nm self-aligned-gate InGaAs HEMTs with record high-frequency characteristics, *IEEE International Electron Devices meeting*, 2010.

75. K. Murata et al., 100-Gbit/s logic IC using 0.1-/spl mu/m-gate-length InAlAs/InGaAs/InP HEMTs, *International Electron Devices Meeting*, 2002.

76. T. Gaier et al., Noise performance of a cryogenically cooled 94 GHz InP MMIC amplifier and radiometer, *SPIE*, vol. 2842, 1996.

77. T. Enoki et al., Ultrahigh-speed integrated circuits using InP-Based HEMTs, *The Japan Society of Applied Physics*, vol. 37, 1998.

78. S. Nakano et al., Analysis of intrinsic delay time in InAlAs/InGaAs high-electron-mobility transistors at cryogenic temperature, TENCON 2017, IEEE Region 10 Conference, 2017.

13

A Study of the Elemental and Surface Characterization of AlGaN/GaN HEMT by Magnetron Sputtering System

Roman Garcia-Perez, Karen Lozano, Jorge Castillo, and Hasina F. Huq

CONTENTS

13.1 Introduction

This article focuses on the fabrication of AlGaN/GaN heteroepitaxies for High Electron Mobility Transistors (HEMTs). Samples were developed using magnetron sputtering with multiple combinations of parameters, power (50 and 70 W) and pressure (3, 9, and 15 mT). Silicon (Si) and sapphire (Al_2O_3) wafers were used as substrates. A study of the elemental and surface characterization was conducted to measure the atomic percentages of the compounds and to observe the topography of the device. The elemental composition and the crystal lattices of the alloys were characterized by X-ray Photoelectron Spectroscopy (XPS) and X-ray Diffraction (XRD). While the topography was of the surfaces was captured using Scanning Electron Microscopy (SEM) and Atomic Force Microscope (AFM).

13.2 Background

Nitride composite nanomaterials are currently at the forefront of semiconductor research due to their wide variety of applications. The Nitrides AlGaN and GaN are used as thin films in high power devices to improve physical performance at smaller scale [1–4]. Both compositions are considered semiconductors because their band gaps range from 3.4 to 6.1 eV. These systems promote the formation of a Two-Dimensional Electron Gas (2DEG) that increases the electron mobility of the device when these alloys are grown on standard crystalline wafers [5–7]. Another advantage of these systems is that they possess low levels of toxicity; therefore, they have a positive contribution to the environment. These nitrides possess strong chemical atomic bonding orbiting at the 1s, 2p, and 3d levels for N, Al, and Ga, respectively [8–10]. All of the above-mentioned characteristics make these nitrides systems attractive to be implemented as thin-films in devices for medical monitoring, environmental awareness, telecommunication, RF systems, and optoelectronics even at quantum scale to reach wider regions of the electromagnetic spectrum. Moreover, the AlGaN/GaN based devices serves are used as biosensors, Light Emitting Diodes (LEDs), and High Electron Mobility Transistors (HEMTs) [11–20]. In these devices, triangular quantum wells are formed between the two nitrides usually with wurtzite crystal structure. This well increases electron concentration and forms Two-Dimensional Electron Gas (2DEG), thus increasing mobility for the potential applications of high-speed devices [21–25]. Less leakage current is obtained in the layer interface of the device due to the 2DEG sheet, which is slightly embedded in the third dimension. There are plenty of methods to fabricate these two alloys, most of them focus on high vacuum at high temperatures, for example, the Chemical Vapor Deposition (CVD) and the Molecular Beam Epitaxy (MBE), but the harsh environment of the processes induces impurities on the adhesion with the substrates. This affects the 2DEG performance and results in low-quality depositions breaking uniformity in large areas of the wafer [26,27]. To overcome these issues, the specimens are usually subjected to post processing to reduce imperfections in the surface during the thin-film coating. The efforts to accomplish AlGaN/GaN on different substrates by deposition methods are still active even though crystalline GaN-bulk wafers are now on the market. The fact is that until large wafers of GaN become commercially available, nitride coating methods are still of lower cost. For that reason, Si wafers are immobile in the business of semiconductors because of the expensive price of bulk crystal alloys. However, the great lattice mismatch that it has with the nitrides makes it difficult to control the nucleation in the heterojunction layers. To explore the power capabilities of the nitrides on Si it is necessary to limit the leakage current through the device and that means an increased adhesion of the thin-films to the substrates [28–33]. Moreover, the interest of new technologies that uses 0- to 3-dimension structures calls for less aggressive deposition methods. Stretchable electronics is one example of these emerging areas of interest in the research of semiconductors with higher bandgap. The HEMTs will go beyond the possibilities of what the MOSFETs offer in physical properties and prices. There had been advances in lowering the heat used by the common methods to protect the morphology of the surfaces against the formation of trapping holes. Lower temperature results in important benefits, it prevents the structure from melting or breaking. To demonstrate the excellent theoretical power capability of the nitride alloys, the fabrication of the devices must maintain low impurities in the composition and less surface imperfections to expand the lifetime and to decrease the leakage current, main factors that are limiting the applications of the HEMTs [34–40]. In this paper, we report the preparation of the epitaxy AlGaN/GaN

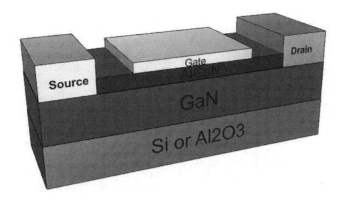

FIGURE 13.1
AlGaN/GaN HEMT device.

on Si and Al$_2$O$_3$ wafers at room temperature. A basic diagram of the device is shown in Figure 13.1. The deposition of the epitaxies on the sapphire substrate is expected to have better adhesion compared to silicon because the lattice mismatch with GaN is lower, but still is considered large at 13.8% of difference [41].

13.3 Materials and Methods (Preparation of GaN and AlGaN)

The layers HEMTs are deposited by an RF magnetron sputtering system [42]. This Physical Vapor Deposition (PVD) configuration operates with direct current (DC) and radio frequency (RF) at low pressures <15 mT and erodes atoms from a planar target with 99.9% of purity hold by a gun. This process uses plasma and it traps free electrons with a generated magnetic field close to the source, it hits the target after collision with ionized particles produced by noble gasses, then atoms fall freely down to the substrate. The plasma aids in the development of a highly uniform deposition by creating a path from the source to the substrate. The magnets behind the cathode and under the base, in conjunction with the rotational speed of the specimens' holder, promote focusing the path of the atoms over the wafers [43]. However, plasma also promotes the adhesion of atmospheric particles to the main chamber, which base pressure is n × 10^{-7} Torr at normal state. To remove impurities, a pre-deposition process is conducted where the turbo pump removes the foreign particles present within the substrate and target [44]. A series of steps are followed, as shown in Table 13.1, to clean the gun and target (bias striking), to clean the substrate (bias cleaning), to prepare the target for the deposition (target striking), and to set up the deposition conditions (sputtering deposition). These four steps prepare the main chamber to conduct the experiments.

The RF mode is set up at the standard 13.56 MHz and the formation of the thin films starts. This process requires at least three hours of deposition, though the growth is non-uniform due to the lattice mismatch of the alloys with the wafers. For that reason, the deposition of GaN and AlGaN is extended to five hours each layer. Through time, the thin-films starts to growth enough to enhance the nucleation of the epitaxies due to the

TABLE 13.1

Prior Deposition Processes: Bias Striking and Cleaning, Target Striking, Deposition: Coating Layer

Id Layer Input Values	Bias Strike		Bias Cleaning		Strike		Sputtering	
Pressure	30 mT		3 mT		30 mT		3mT >	
Gasses Argon / Nitrogen	18 sccm	0 sccm	18 sccm	0 sccm	3 sccm	9 sccm	3 sccm	9 sccm
Power Set Point	30W		50W		30W		50W	
Time	60s		180s		120s		18,000s	

TABLE 13.2

Different Pressures Tested with a Constant Power, Argon 3 sccm, Nitrogen 9 sccm, Rotation at 70%, Delay Time 60 s, Coating Time 5-hours, and RF Gun ON

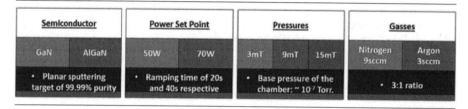

Semiconductor		Power Set Point		Pressures			Gasses	
GaN	AlGaN	50W	70W	3mT	9mT	15mT	Nitrogen 9sccm	Argon 3sccm
• Planar sputtering target of 99.99% purity		• Ramping time of 20s and 40s respective		• Base pressure of the chamber: ~ 10^{-7} Torr.			• 3:1 ratio	

transition of the lattice configuration to wurtzite. Table 13.2 shows the parameters used during a series of experiments. The morphology and elemental composition were studied. The lowest possible pressure to generate plasma in this system is 3 mT when working with any of the mentioned materials. A 3:1 ratio of nitrogen and argon gas is used to enhance the nitride properties of the device and to desorb contaminants during the deposition. Moreover, a power set point of 50 and 70 W are given to analyze the impact of the magnetic field and the target for erosion.

13.3.1 Elemental Characterization

13.3.1.1 X-Ray Photoelectron Spectroscopy (XPS)

The atomic composition of the elements in these alloys determines the physical characteristics of the material. Moreover, it provides information to enhance the adhesion of the epitaxies in future depositions, the existence of impurities, if at all, can be also identified. The XPS measurements were conducted in a Thermo Scientific™ K-Alpha™ system with a monochromatic aluminum Kα radiation source in ultra-high vacuum conditions. The system is capable to reach low pressures of n × 10^{-10} mBar and it can analyze up to nine specimens and several points in one single run [45]. To characterize the thin-films from the surface down to the end of the epitaxies a depth profile was conducted with etching levels >100. At every level, the system analyzes the binding energies of the composition. The ternary alloy $Al_xGa_{1-x}N$ can be molded to be applied at different wavelengths ranging from the infrared to the ultra violet [46].

13.3.1.2 X-Ray Diffraction (XRD)

Samples were analyzed using a Brucker D-8 XRD with Cu K Alpha X-ray source. The system is configured for powder experiments but is possible to get the analysis of the thin-films by rotating the specimens and trying different positions until the elemental picks are identified. Several samples at different pressure and power were tested to locate the angles of the crystals that conform the compounds. The picks were found in similar angular positions, but the counts are different because of the variation in composition. The rate deposition has effects on the crystal sizes of the thin-films. Hence, the angles are slightly moved due to differences in the diffraction. This is strongly correlated with the topography of the surface, rather than the lattice configuration or the atomic elemental percentage of the compounds [47,48].

13.3.2 Surface Characterization

The topography of AlGaN and GaN plays a dominant role, the roughness of the surfaces affects the electron mobility in the HEMT [49]. The surface images of the films are captured by two microscopes that scan large areas and measure structural aspects of the topography.

13.3.2.1 Scanning Electron Microscopy (SEM)

Several areas of the wafers at large scale were examined by a ZEISS Sigma VP microscope. Eighteen samples were analyzed with a nominal voltage of 10 kV, a magnification of ~140.00 K ×, and silicon tape was used to hold the samples to the sample holder. The topography of the samples was analyzed as is, meaning in the absence of extra sputtering treatments. The high voltage allowed to observe low-conductive substrates as sapphire, but it also increased the electric charge on the surfaces burning and affecting the resolution of the image [50].

13.3.2.2 Atomic Force Microscope (AFM)

The topography of the wafers was inspected using a Veeco Dimension 3100 AFM. The scan sizes were set up from 10 nm² to 115 μm² and applied on four different areas of the epitaxies. The analysis was conducted using a silicon probe model Tap 150 Al-G from TED PELLA. The tip radius is <10 nm and the chip size have an area of 3.4 × 1.6 × 0.3 mm. The tip sensed the surface of the specimen by slightly touching it back and forward in a mode called tapping as in Figure 13.2, where the red dashes mark the analyzed area between the epitaxies. The series of experiments provided accurate approximation of thicknesses at the interfaces. During the deposition, a Kapton tape was used as a mask to cover some areas of the wafer and leave them uncoated. Moreover, several points of the surface were scanned to capture 3D images of the topography [51].

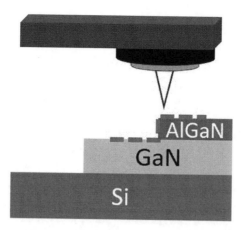

FIGURE 13.2
AFM step-height measurements.

13.4 Results

The deposition of AlGaN and GaN epitaxies are achieved at room temperature on top of silicon and sapphire wafers. After five hours of deposition, uniform nitride alloy thin-films were formed. At first, the GaN layers inherit the topography of the wafers, which was not favorable for the formation of wurtzite crystals. However, the quality of the heterojunction improves over time and deposition rate. The deposition rate is a function of several parameters such as working time, rotation, process gas, pressure, and temperature. At low deposition rates, the surface of the thin films changes the lattice and becomes wurtzite as shown in Figure 13.3. Samples were analyzed after three hours of deposition showing a

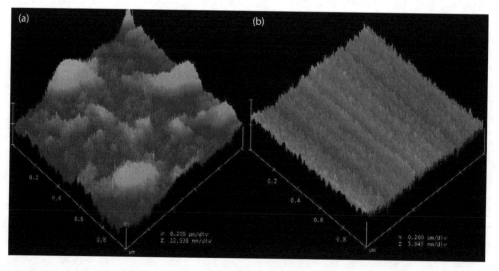

FIGURE 13.3
1 μm^2 AFM surface images: (a) GaN/Si at 9 mT and 70 W; (b) GaN/Al$_2$O$_3$ at 3 mT and 70 W.

heterogeneous structure. It is found that, as the time increases the nucleation of the GaN layer on the substrates becomes clearer. Also, the chemical characterization of the thin-film was more accurate compared to the results when less than three hours of deposition was used. For that reason, when using room temperature deposition of GaN the recommended minimum time for optimum results is three hours. Otherwise, the deposited surfaces do not show evidence of topography nor chemical changes. The growth of the alloys improved once there was a wurtzite lattice established. The deposition time was increased to five hours and surface and elemental composition were characterized. Figure 13.3 shows the grain formation and overall topography. For sapphire in Figure 13.3b, the layer keeps a similar lattice and uniformity as the original surface, moreover, there is no sign of grains formation in silicon substrates.

The deposition rate of AlGaN increases if deposited on GaN since both have a wurtzite crystal structure after sputtering. It is expected, that the growth of this epitaxy could be improved because of the similar properties of nitrides. In the case of the silicon wafer as in Figure 13.4a, certain trajectories are observed in a light gray color that can lead to the formation of grains if the deposition time increases. The epitaxy grows on sapphire shown in Figure 13.4b, reveals a rough surface with spikes structures having step-heights of <4 nm. The surface of AlGaN inherits the deposited GaN structures which defines the formation of crystals.

When the epitaxies are constructed with both layers at room temperature, the surface of the substrate plays an important role for the addition of the wurtzite grid. For deposition to be successful, a base preferably of the same crystal network is needed. These results demonstrate the challenges associated with deposition of these alloys at room temperature.

However, it was reported by Huq et al. that formation of wurtzite grains with diameter greater than 100 nm were observed if GaN is deposited at high temperatures [52]. GaN based layer was grown at 700°C for five hours, this allowed formation of AlGaN crystals on the surface deposited at room temperature and 70 W. In Figure 13.5, the top layer inherits a granular formation in silicon while higher coalescence of crystals is observed in sapphire in Figure 13.6.

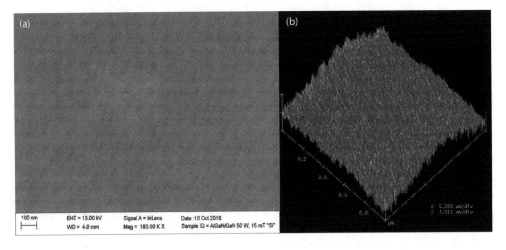

FIGURE 13.4
(a) AlGaN/GaN/Si surface by SEM at 15 mT and 50 W; (b) AlGaN/GaN/Al$_2$O$_3$ surface by AFM at 15 mT and 50 W.

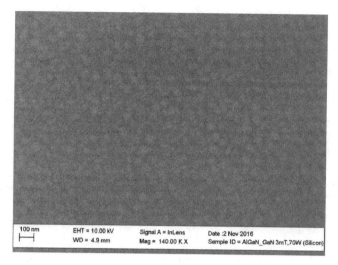

FIGURE 13.5
SEM image of AlGaN on GaN grains at 3 mT and 70 W.

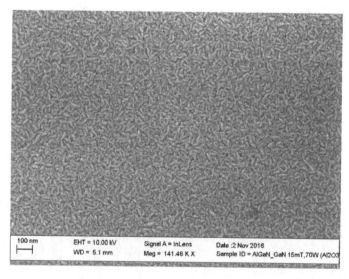

FIGURE 13.6
SEM image of coalesced AlGaN crystals on GaN at 15 mT and 70 W.

The XPS analysis indicates that Ga3d, Al2p, and N1s are part of the AlGaN composition and close to the same binding energies. Ga is near to 21 eV, Al to 75eV, and N to 398 eV in all the cases. However, the counts/s change for Ga and N at the interface level when the layer moves to GaN leaving Al behind. The depth profile was programmed so that the X-ray could drill the surface several etching levels down to the original substrate. Figure 13.7a depicts the elemental characterization of AlGaN on Si at room temperature, single elements are shown in Figure 13.7b–d. In Table 13.3, the atomic percentage of the composition is shown. It is possible to compare the concentration of the elements on silicon

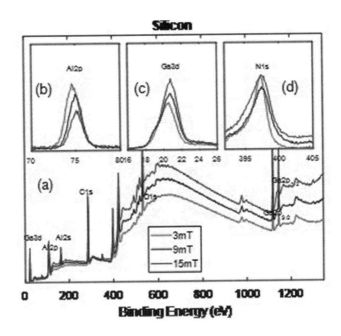

FIGURE 13.7
XPS analysis: (a) Surface survey, (b) Al2p, (c) Ga3d, and (d) N1s.

TABLE 13.3

Atomic Percentage of AlGaN at Room Temperature and 50 W on Silicon/Sapphire

	Ga 3d %		Al 2p %		N 1s %	
3mT	27.87	32.56	18.75	19.06	53.37	48.38
9mT	32.52	32.03	13.71	13.17	53.77	54.80
15mT	27.10	28.20	14.49	14.72	58.40	57.09
	Si	Al$_2$O$_3$	Si	Al$_2$O$_3$	Si	Al$_2$O$_3$

and sapphire for all selected pressures. Overall, difference can be considered negligible. The compositions are as follows: Ga 27.10%~32.56%, Al 13.17%~19.06%, and N 48.38%~58.4%. These results show that the ternary alloy Al$_x$Ga$_{1-x}$N by magnetron is tunable by changing the power. It is deducted that half percent of the composition is nitrogen while gallium is the second predominant element and aluminum is present in less than 20% within the alloy at room temperature with nominal power of 50 W.

Impurities like carbon and oxygen are found at the surfaces as in Figure 13.7a. The atomic percentages of these impurities varied depending on the etching level analyzed. These unwanted particles are detected to come from the atmosphere and the targets surfaces due to air exposure before mounting them into the guns. Furthermore, the oxygen levels are found to be greater in the GaN layers compared to the AlGaN layers as shown in Figure 13.8.

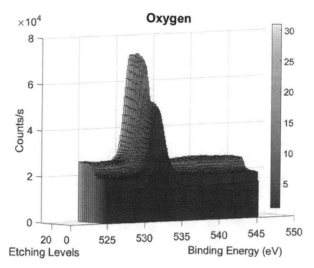

FIGURE 13.8

Oxygen levels on deposited thin-film at 50 W, 15 mT, and room-temperature on silicon.

The atmospheric particles are originated at the etching level 0 of the depth-profile which is the surface of the epitaxies [53]. Then, the etching levels signals from 2 to 15 are linked to the AlGaN layer. Finally, the shift signal which in this case rises even higher, is correlated to the next layer of GaN. The color bar indicates the etching levels of each signal of the depth-profile. Table 13.3 presents the peaks related to the elements of the ternary alloy. It is possible to deduct some configurations of $Al_xGa_{1-x}N$. However, those configurations are limited by the magnetron deposition from one single target.

A more in-depth surface inspection is conducted by the AFM and shown in Figure 13.9 for scan sizes of (a) for 1 μm², (b) 200 nm², and (c) 10 nm². The granular structures shown with a dimension of 1 μm² were analyzed using a step-height method, where the average of several peak heights varied from 0.7 to 1.7 nm for silicon and from 0.5 to 2.2 nm for sapphire, depending on the selected scanned area. The X and Z axis deviation is configurated similar to the work presented by Chang et al. [54]. Moreover, smaller structures are

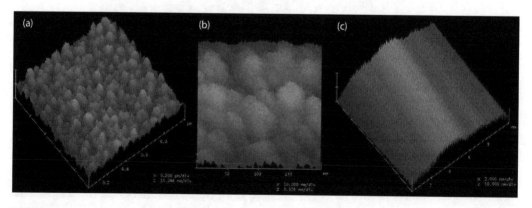

FIGURE 13.9

(a) 1 μm² AlGaN/GaN/Si at 3 mT and 70 W; (b) 200 nm² AlGaN/GaN/Al₂O₃ at 15 mT and 70 W; (c) 10 nm² AlGaN/GaN/Si at 15 mT and 70 W.

detected on top of the grains using a scanned dimension of 200 nm^2, with step-heights ranges of 0.06–0.09 nm for silicon and an average of 0.1 nm for sapphire. The same effect of hills coalescing to a flat surface with other structures was observed by Pakuła et al. [55].

Finally, pressure versus thickness (nm) graphs with 50 and 70 W are plotted to study their correlation after 5-hours of deposition. The results are shown in Figure 13.10a for GaN on Si and Al$_2$O$_3$, 13.10b for AlGaN on GaN, and 13.10c for AlGaN on 700°C GaN. The rate of deposition on Al$_2$O$_3$ is greater when compared to Si. Here the deposition rate at 50 and 70 W on silicon is <300 nm, which is lower when compared to sapphire at <400 nm. The power has an impact on the thickness of the thin films, but nevertheless, in these heterojunctions the most important factor for the optimum rates is the lattice mismatch. In Figure 13.10b and c, the formation of the wurtzite structures on the wafers from the GaN layer enhances the deposition rate of similar nitrides. Compared to Figure 13.10a the second layer of the epitaxy possesses greater thicknesses. In this case, the AlGaN adhesion layer is highly influenced by the power and the lattice as well. However, the pressure is equally correlated in all the depositions as shown in Figure 13.10b and c, the greater thicknesses are found at 3 mT while the lowest at 15 mT. The rate of the epitaxial growth depends on the above-mentioned factors. The surface topography and temperature are strongly correlated to the purity of the composition. As an example, in Figure 13.10c the thicknesses of HEMTs epitaxies are much greater than the one observed in Figure 13.10b at <600 nm.

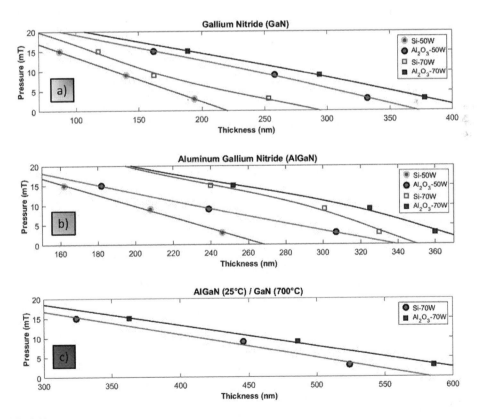

FIGURE 13.10

Thicknesses: (a) GaN at room temperature; (b) AlGaN on GaN at room temperature; and (c) AlGaN at room temperature on GaN sputtered at 700°C.

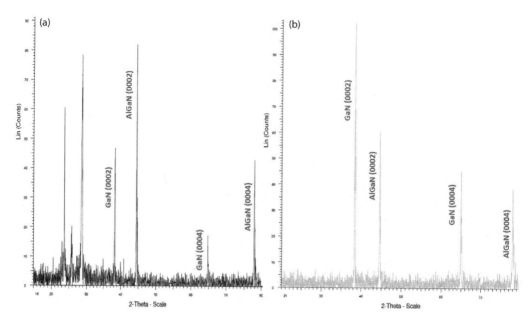

FIGURE 13.11
XRD analysis (a) AlGaN at room temperature on GaN/Si; (b) AlGaN at room temperature on GaN/Al$_2$O$_3$.

The XRD results show that the AlGaN compound crystals are present on the surface of the heteroepitaxies. In Figure 13.11a and b, the results of the silicon and sapphire substrates are shown. In both samples, the direction of the crystals was found at the lattices 0002 and 0004. Where GaN is phased with AlN to form the AlGaN peaks, which are located closed to the 2-theta angles 45 and 78. Similar results were discussed by Chandolu et al. and Wang et al. [56,57]. In Figure 13.11a, the counts of the GaN peaks are lower compared to the AlGaN and it is the opposite in Figure 13.11b. The non-similar crystal configuration as in Figure 13.11a, did not allow the first layer to grow greater than shown in Figure 13.11b, where the similitude of the structures cubic and wurtzite made the rate deposition to increase. Hence, the GaN is prominent in the sapphire samples when they are compared with the silicon samples using the same parameters. Also, the counts of the peaks vary depending on the thicknesses of the samples. The wurtzite crystals are identified in the surfaces corroborating the results shown in the AFM and SEM figures. In the XRD experiments, the sizes of the wurtzite structures are the predominant factor for better graphing quality [58].

Lastly, it was observed that the constant non-stop use of high power at 70 W was not appropriate for the deposition of AlGaN. The deposition rate increased when those parameters at room temperature were used, but the surface of the target changed with use. Figure 13.12 shows an aggravated surface due to several depositions with duration of five hours each, non-stopping one after another. Moreover, the small carbonized boulders on top of the target's surface fell down eventually, wasting the duration of the experiment and contaminating the deposition chamber, as shown in Figure 13.13. Another problem was that some areas of the substrate were covered during the deposition, creating a mask that affected the uniform adhesion of the erosion. These types of problems are related to surface geometry changes with use. The halo shown in Figure 13.12 can develops a "race-track" depression on the surface of the target which eventually interferes with the flux

FIGURE 13.12
Aggravated surface of the AlGaN target due to constant use.

FIGURE 13.13
Deposition chamber contaminated by boulder particles from target.

pattern and the deposition rate [59]. However, it is safe to use high power for only a few hours of deposition, then the surface of the target should be cleaned in order to avoid the formation of carbonized particles on the target.

13.5 Conclusions

Several depositions of the AlGaN and GaN thin films are achieved at room temperature by RF magnetron sputtering. The growth of GaN as the base layer for the HEMT enhances the deposition rate of AlGaN. It is also found that GaN layer is helpful for growing wurtzite crystals. The atomic percentage concentration of the compounds and the thicknesses are correlated to each other. It is clear that the GaN deposition induces a higher concentration of wurtzite in the surface that aids the deposition rate of the upper layer AlGaN. In addition, high temperature depositions of GaN grew greater thicknesses compared to the room temperature depositions, even the ones at 3 mT. For that reason, the tendency of the AlGaN topography prior thin-films deposition can be predicted when using magnetron sputtering, because now there is enough information to achieve controlled thicknesses and uniformity on the deposited surfaces. Moreover, the temperature has proven to impact the chemical composition of the base layer, in this case by reducing the oxygen when using 700°C compared to 25°C [52]. The formation of a thicker layer of GaN with wurtzite crystal will enhance AlGaN layers thickness. The depositions of the nitrides are uniform at large areas using the magnetron sputtering system due to the rotation. However, some of the results reported in this study show that the deposition at 70 W aggravated the surface of the AlGaN target with carbonized particles. So, further investigation is needed to protect the target from poisoning and breaking. In conclusion, the team reports an optimized sputtering method to fabricate AlGaN/GaN HEMT with wurtzite crystal structures.

Acknowledgments

"This project is supported in part by the National Science Foundation (NSF) under Grant No. 1229523, and by the US Army Research Office W911NF-14-1-0100." The authors acknowledge support received from National Science Foundation under PREM grant DMR 1523577. The authors acknowledge the partial participation of this project as an oral presentation at the 2nd International Conference on Advanced Functional Materials (ICAFM) 2017. The authors acknowledge the technical support of Hilario Cortes and Edgar Muñoz for their skills using the SEM and XPS instruments.

References

1. Lidow, A. *GaN Transistors for Efficient Power Conversion.* Chichester, UK: John Wiley & Sons, 2015.
2. Medjdoub, F., and K. Iniewski. *Gallium Nitride (GaN): Physics, Devices, and Technology.* Boca Raton, FL: CRC Press, Taylor & Francis Group, 2016.
3. Pelá, R. R., C. Caetano, M. Marques, L. G. Ferreira, J. Furthmüller, and L. K. Teles. "Accurate band gaps of AlGaN, InGaN, and AlInN alloys calculations based on LDA-1/2 approach." *Applied Physics Letters* 98(15) (2011): 151907.
4. Pan, L., X. Dong, J. Ni, Z. Li, Q. Yang, D. Peng, and C. Li. "Growth of compressively-strained GaN films on Si(111) substrates with thick AlGaN transition and AlGaN superlattice buffer layers." *Physica Status Solidi (c)* 13(5–6) (2016): 181–185.
5. Mishra, U.K., P. Parikh, and Y.-F. Wu. "AlGaN/GaN HEMTs-an overview of device operation and applications." *Proceedings of the IEEE* 90(6) (2002): 1022–1031.
6. Nanjo, T., M. Takeuchi, A. Imai, Y. Suzuki, M. Suita, K. Shiozawa, Y. Abe, E. Yagyu, K. Yoshiara, and Y. Aoyagi. "AlGaN channel HEMT with extremely high breakdown voltage." *MRS Proceedings* 1324 (2011): n.p.

7. Nepal, N., J. Li, M. L. Nakarmi, J. Y. Lin, and H. X. Jiang. "Temperature and compositional dependence of the energy band gap of AlGaN alloys." *Applied Physics Letters* 87(24) (2005): 242104.

8. Xu, X., V. Jindal, F. Shahedipour-Sandvik, M. Bergkvist, and N. C. Cady. "Direct immobilization and hybridization of DNA on group III nitride semiconductors." *Applied Surface Science* 255(11) (2009): 5905–5909.

9. Higashiwaki, M., S. Chowdhury, B. L. Swenson, and U. K. Mishra. "Effects of oxidation on surface chemical states and barrier height of AlGaN/GaN heterostructures." *Applied Physics Letters* 97(22) (2010): 222104.

10. Dumont, J., E. Monroy, E. Muñoz, R. Caudano, and R. Sporken. "Investigation of metal–GaN and metal–AlGaN contacts by XPS depth profiles and by electrical measurements." *Journal of Crystal Growth* 230(3–4) (2001): 558–563.

11. Florian, C., R. Cignani, A. Santarelli, and F. Filicori. "Design of 40-W AlGaN/GaN MMIC high power amplifiers for C-band SAR applications." *IEEE Transactions on Microwave Theory and Techniques* 61(12) (2013): 4492–4504.

12. Li, B., X. Tang, J. Wang, and K. J. Chen. "Optoelectronic devices on AlGaN/GaN HEMT platform." *Physica Status Solidi (a)* 213(5) (2016): 1213–1221.

13. Monroy, E., F. Guillot, S. Leconte, L. Nevou, L. Doyennette, M. Tchernycheva, F. H. Julien, E. Baumann, F. R. Giorgetta, and D. Hofstetter. "Latest developments in GaN-based quantum devices for infrared optoelectronics." *Journal of Materials Science: Materials in Electronics* 19(8–9) (2007): 821–827.

14. Liu, Z. J., T. Huang, J. Ma, C. Liu, and K. M. Lau. "Monolithic integration of AlGaN/GaN HEMT on LED by MOCVD." *IEEE Electron Device Letters* 35(3) (2014): 330–332.

15. Tanuma, N., M. Tacano, S. Yagi, and J. Sikula. "Low-frequency-noise monitoring of current collapse in AlGaN/GaN high-electron-mobility transistors on sapphire substrate." *Journal of Statistical Mechanics: Theory and Experiment* 2009(1) (2009): n.p.

16. Shmidt, N., A. Usikov, E. Shabunina, A. Chernyakov, A. Sakharov, S. Kurin, A. Antipov, I. Barash, A. Roenkov, H. Helava, and Y. Makarov. "Degradation of external quantum efficiency of AlGaN UV LEDs grown by hydride vapor phase epitaxy." *Physica Status Solidi (c)* 12(4–5) (2015): 349–352.

17. Muth, J. F., J. D. Brown, M. A. L. Johnson, Z. Yu, R. M. Kolbas, J. W. Cook, and J. F. Schetzina. "Absorption coefficient and refractive index of GaN, AlN and AlGaN alloys." *MRS Proceedings* 537 (1998): n.p.

18. Wagner, J., H. Obloh, M. Kunzer, M. Maier, K. Köhler, and B. Johs. "Dielectric function spectra of GaN, AlGaN, and GaN/AlGaN heterostructures." *Journal of Applied Physics* 89(5) (2001): 2779–2785.

19. Ariyawansa, G., M. B. M. Rinzan, M. Alevli, M. Strassburg, N. Dietz, A. G. U. Perera, S. G. Matsik et al. "GaN/AlGaN ultraviolet/infrared dual-band detector." *Applied Physics Letters* 89(9) (2006): 091113.

20. Kumar, C. S. S. R. *Semiconductor Nanomaterials.* Weinheim, Germany: Wiley-VCH, 2010; Chapter 3.3.1.

21. Pulfrey, D. L. *Understanding Modern Transistors and Diodes.* Cambridge, UK: Cambridge University Press, 2010; Chapter 11.3.1.

22. Karumuri, N., S. Turuvekere, N. DasGupta, and A. DasGupta. "A continuous analytical model for 2-DEG charge density in AlGaN/GaN HEMTs valid for all bias voltages." *IEEE Transactions on Electron Devices* 61(7) (2014): 2343–2349.

23. Aminbeidokhti, A., S. Dimitrijev, A. K. Hanumanthappa, H. A. Moghadam, D. Haasmann, J. Han, Y. Shen, and X. Xu. "Gate-voltage independence of electron mobility in power AlGaN/GaN HEMTs." *IEEE Transactions on Electron Devices* 63(3) (2016): 1013–1019.

24. Yigletu, F. M., S. Khandelwal, T. A. Fjeldly, and B. Iniguez. "Compact charge-based physical models for current and capacitances in AlGaN/GaN HEMTs." *IEEE Transactions on Electron Devices* 60(11) (2013): 3746–3752.

25. Paszkiewicz, B. K., T. Szymanski, and M. Tlaczala. "Theoretical stress calculations in polar, semipolar and nonpolar AlGaN/GaN heterostructures of different compositions." *Crystal Research and Technology* 51(5) (2016): 349–353.

26. Kaushik, J. K., V. R. Balakrishnan, D. Mongia, U. Kumar, S. Dayal, B. S. Panwar, and R. Muralidharan. "Investigation of surface related leakage current in AlGaN/GaN high electron mobility transistors." *Thin Solid Films* 612 (2016): 147–152.

27. Bradley, S. T., A. P. Young, L. J. Brillson, M. J. Murphy, W. J. Schaff, and L. E. Eastman. "Influence of AlGaN deep level defects on AlGaN/GaN 2-DEG carrier confinement." *IEEE Transactions on Electron Devices* 48(3) (2001): 412–415.

28. Chen, Jr T., C.-W. Hsu, U. Forsberg, and E. Janzén. "Metalorganic chemical vapor deposition growth of high-mobility AlGaN/AlN/GaN heterostructures on GaN templates and native GaN substrates." *Journal of Applied Physics* 117(8) (2015): 085301.

29. Yoon, Y. J., J. H. Seo, M. S. Cho, H.-S. Kang, C.-H. Won, I. M. Kang, and J.-H. Lee. "TMAH-based wet surface pre-treatment for reduction of leakage current in AlGaN/GaN MIS-HEMTs." *Solid-State Electronics* 124 (2016): 54–57.

30. Chen, J., H. Wakabayashi, K. Tsutsui, H. Iwai, and K. Kakushima. "Poly-Si gate electrodes for AlGaN/GaN HEMT with high reliability and low gate leakage current." *Microelectronics Reliability* 63 (2016): 52–55.

31. Swain, R., K. Jena, and T. Ranjan Lenka. "Modeling of forward gate leakage current in MOSHEMT using trap-assisted tunneling and Poole–Frenkel emission." *IEEE Transactions on Electron Devices* 63(6) (2016): 2346–2352.

32. Mosbahi, H., M. Gassoumi, H. Mejri, M. A. Zaidi, C. Gaquiere, and H. Maaref. "Electrical characterization of AlGaN/GaN HEMTs on Si substrate." *Journal of Electron Devices* 15 (2012): 1225–1231.

33. Moereke, J., E. Morvan, W. Vandendaele, F. Allain, A. Torres, M. Charles, and M. Plissonnier. "Leakage current paths in isolated AlGaN/GaN heterostructures." *IEEE Transactions on Semiconductor Manufacturing* 29(4) (2016): 363–369.

34. Tiwari, A., and A. N. Nordin. *Advanced Biomaterials and Biodevices*. Hoboken, NJ: Wiley, 2014; Chapter 4.3.

35. Trung, T. Q., and N.-E. Lee. "Materials and devices for transparent stretchable electronics." *Journal of Materials Chemistry C* (2017): n.p.

36. Rogers, J. A., R. Ghaffari, and D.-H. Kim. *Stretchable Bioelectronics for Medical Devices and Systems*. Basel, Switzerland: Springer, 2016.

37. Wang, H., H. Xu, C. Wu, A. M. Soomro, H. Guo, T. Wei, S. Li, J. Kang, and D. Cai. "Family of Cu@metal nanowires network for transparent electrodes on n-AlGaN." *Physica Status Solidi (a)* 213(5) (2016): 1209–1212.

38. van den Brand, J., M. De Kok, M. Koetse, M. Cauwe, R. Verplancke, F. Bossuyt, M. Jablonski, and J. Vanfleteren. "Flexible and stretchable electronics for wearable health devices." *Solid-State Electronics* 113 (2015): 116–120.

39. Chou, Y.C., D. Leung, I. Smorchkova, M. Wojtowicz, R. Grundbacher, L. Callejo, Q. Kan et al. "Degradation of AlGaN/GaN HEMTs under elevated temperature lifetesting." *Microelectronics Reliability* 44(7) (2004): 1033–1038.

40. Dutta, G., N. Dasgupta, and A. Dasgupta. "Low-temperature ICP-CVD SiNx as gate dielectric for GaN-based MIS-HEMTs." *IEEE Transactions on Electron Devices* 63(12) (2016): 4693–4701.

41. Redwing, J. M., M. A. Tischler, J. S. Flynn, S. Elhamri, M. Ahoujja, R. S. Newrock, and W. C. Mitchel. "Two-dimensional electron gas properties of AlGaN/GaN heterostructures grown on 6H–SiC and sapphire substrates." *Applied Physics Letters* 69(7) (1996): 963–965.

42. "ATC Orion Series Sputtering Systems." *AJA International*, March 2, 2017. ATC-Orion 5 UHV with Load-Lock.

43. Mattox, D. M. *Handbook of Physical Vapor Deposition (PVD) Processing*. Oxford, UK: William Andrew, 2010; Chapter 7.3.

44. Mattox, D. M. *Handbook of Physical Vapor Deposition (PVD) Processing*. Oxford, UK: William Andrew, 2010; Chapter 5.7.

45. "K-Alpha™ X-ray photoelectron spectrometer (XPS) system." *Thermo Fisher Scientific*, March 7, 2017.
46. Nunez-Gonzalez, R., A. Reyes-Serrato, A. Posada-Amarillas, and D. H. Galvan. "First-principles calculation of the band gap of $Al_xGa_{1-x}N$ and $In_xGa_{1-x}N$." *Revista Mexicana de Fisica S* 54(2) (2008): 111–118.
47. Cullity, B. D. *Elements of X-ray Diffraction*. Utgivningsort okänd: Scholar's Choice, 2015.
48. "Bruker: Descripción general de D8 ADVANCE." *Bruker.com*, February 17, 2017. Accessed May 2, 2017.
49. Li, Q., J. Zhang, L. Meng, J. Chong, and X. Hou. "Mobility limitations due to dislocations and interface roughness in AlGaN/AlN/GaN heterostructure. *Journal of Nanomaterials* 2015 (2015): 1–6.
50. "ZEISS." *Scanning Electron Microscope-SEM*, Accessed March 7, 2017. https://www.zeiss.com/microscopy/int/products/scanning-electron-microscopes.html.
51. Misumi, I., S. Gonda, T. Kurosawa, Y. Azuma, T. Fujimoto, I. Kojima, T. Sakurai, T. Ohmi, and K. Takamasu. "Reliability of parameters of associated base straight line in step height samples: Uncertainty evaluation in step height measurements using nanometrological AFM." *Precision Engineering* 30(1) (2006): 13–22.
52. Huq, H. F., R. Y. Garza, and R. Garcia-Perez. "Characteristics of GaN thin films using magnetron sputtering system." *Journal of Modern Physics* 7(15) (2016): 2028–2037.
53. He, C., Q. Wu, X. Wang, Y. Zhang, L. Yang, N. Liu, Y. Zhao, Y. Lu, and Z. Hu. "Growth and characterization of ternary AlGaN alloy nanocones across the entire composition range." *ACS Nano* 5(2) (2011): 1291–1296.
54. Huang, W.-C., C.-M. Chu, Y.-Y. Wong, K.-W. Chen, Y.-K. Lin, C.-H. Wu, W.-I Lee, and E.-Y. Chang. "Investigations of GaN growth on the sapphire substrate by MOCVD method with different AlN buffer deposition temperatures." *Materials Science in Semiconductor Processing* 45 (2016): 1–8.
55. Pakuła, K., R. Bożek, K. Surowiecka, R. Stępniewski, A. Wysmolek, and J. M. Baranowski. "Low density GaN quantum dots on AlGaN." *Physica Status Solidi (b)* 243(7) (2006): 1486–1489.
56. Chandolu, A., S. Nikishin, M. Holtz, and H. Temkin. "X-ray diffraction study of AlN/AlGaN short period superlattices." *Journal of Applied Physics* 102(11) (2007): 114909.
57. Wang, W., H. Wang, W. Yang, Y. Zhu, and G. Li. "A new approach to epitaxially grow high-quality GaN films on Si substrates: the combination of MBE and PLD." *Scientific Reports* 6(1) (2016): n.p.
58. Iliopoulos, E., K.F. Ludwig, and T.D. Moustakas. "Complex ordering in ternary wurtzite nitride alloys." *Journal of Physics and Chemistry of Solids* 64(9–10) (2003): 1525–1532.
59. Mattox, D. M. *Handbook of Physical Vapor Deposition (PVD) Processing*. Oxford, UK: William Andrew, 2010; Chapter 7.7.

14

Metamorphic HEMTs for Sub Millimeter Wave Applications

J. Ajayan and D. Nirmal

CONTENTS

14.1 Introduction to MHEMT Technology

The signals whose frequency ranges between 30 and 300 GHz are called millimeter waves and the signals whose frequency ranges between 300 GHz and 3 THz are known as sub-millimeter waves. Millimeter and sub-millimeter frequencies are gaining attention due to their wide variety of applications including safety and security (detection of concealed weapon or plastic explosives), avionics (landing aid, control of runway, foreign object debris, helicopter wire detection) and communication technologies. The millimeter wave frequencies such as 94, 140, 220 GHz and sub-millimeter wave frequencies such as 340, 410, 480, 660, and 850 GHz are extremely important in communication technologies because the local maxima exist in these frequencies when signals are transmitted through atmosphere [1–80]. Increasing the operating frequency of microwave monolithic integrated circuits (MMICs) and terahertz monolithic integrated circuits (TMICs) have the following benefits:

1. Significant reduction in the size of the components especially antennas in high speed wireless communication systems

2. Ease of integration and deployment

3. Large bandwidth that makes them attractive for high-speed wireless and optical communication systems

4. Waves can penetrate through rain, fog, dust, snow, clothes, packing materials, making them suitable for atmospheric sensing and radiometry applications

GaAs metamorphic HEMTs (MHEMTs) are considered as the cost-effective alternative to InP HEMTs for low noise, low power, low operating voltage, high gain and high frequency applications. There is no doubt that lattice-matched InP HEMTs are superior in performance in terms of higher drain current (I_D), transconductance (g_m), cut-off frequency (f_T) and maximum oscillation frequency (f_{max}) compared with GaAs MHEMTs due to the outstanding thermal conductivity of InP substrate. However, GaAs MHEMTs are preferred by the semiconductor industry over InP HEMTs for the mass production of S-MMICs because InP substrates are fragile/brittle, smaller in size and it is very difficult to etch. For the same diameter, InP wafer cost is four times higher than GaAs wafers. InP HEMTs are difficult to manufacture in larger volume due to smaller wafer sizes. On the other hand, GaAs wafers are robust and are also cost effective due to the availability of GaAs wafers with large diameter [81–100].

14.2 GaAs MHEMT for Sub-millimeter Wave Applications

GaAs MHEMTs are considered as an alternative solid state transistor technology for InP HEMTs and they are suitable for sub-millimeter wave monolithic microwave integrated circuits (S-MMIC) like medium-power amplifiers, low-noise amplifiers (LNA), cryogenic low-noise amplifiers for radiometric receivers, cascode amplifiers, voltage-controlled oscillators (VCOs), space and military communication systems, high-speed point-to-point radio links, satellite cross-links, high-resolution imaging radar systems for safety, security and health, cryogenic radiometric receivers for radio astronomy, sensors for collision avoidance radar, destructive materials sensing and testing, and remote sensing, up and down conversion mixers, spectroscopy, radio telescope and medical imaging etc. Schottky diode detectors with MHEMT technology can be used for space applications. MHEMTs can also be used for developing frequency dividers which play a key role in the frequency synthesizers and phase locked loop (PLL) applications. Low-noise amplifiers, medium-power amplifiers and power amplifiers are key components in microwave transmitters and receivers and MHEMT technology is highly preferred to manufacture these devices mainly due to their high-gain, low-noise figure and linearity characteristics. The output power of these amplifiers operated at millimeter wave and sub-millimeter wave frequencies depends upon the breakdown voltage of the HEMTs. The higher breakdown voltage characteristics of asymmetric MHEMTs make them suitable for implementing high power amplifiers. The millimeter and sub-millimeter wave automotive radars are considered as crucial components in next generation adaptive cruise control systems and the key components in these automotive radars are power amplifiers, driver amplifiers, low-noise amplifiers and frequency multipliers and all these components have to be integrated into the single chip, and MHEMT technology is highly attractive for the fabrication of automotive radar systems due to its low cost and high robustness. The word metamorphic means that the lattice constant of the GaAs semiconductor substrate is varied by introducing a buffer to a value required by the semiconductor layer on top of the substrate. MHEMTs can also be used to manufacture sensors for imaging the environment and high-resolution radars for military applications. Light weight, compactness and wide bandwidth are the three most important requirements for S-MMICs applications [101–203].

14.3 Asymmetric MHEMT for High Power Applications

Symmetric MHEMTs provide high speed, very low noise and high gain but suffer from poor breakdown voltage characteristics. Increasing the concentration of indium (In) in the InGaAs channel material, reduction of gate length to the maximum possible extend and reducing the contact resistances are the straight forward methods to enhance the speed performance of MHEMTs. However, increasing the indium concentration in the InGaAs channel and decreasing the gate length drastically degrade the ON and OFF state breakdown voltages of MHEMTs. There are many techniques available now to improve the breakdown voltages of the HEMTs. One such method to improve the breakdown voltage characteristics of MHEMTs is to employ a wide band gap channel like InAsP or InP, but this limits the DC parameters of MHEMTs like drain current and transconductance. The reduction of transconductance leads to the degradation of transistor gain. Hence improvement of breakdown voltages in short gate length MHEMTs without the degradation of transistor gain is a big challenge. Dong Xu et al. [85–90] experimentally demonstrated that an asymmetric MHEMT with InAs inserted composite channel and double silicon delta (δ) doping technique can significantly improve the breakdown voltages without reducing transconductance, drain current and transistor gain. Conventional symmetric MHEMTs use T-shaped gates, but Γ-gate is highly preferred in asymmetric MHEMTs designed for achieving higher breakdown voltages because it can effectively reduce the device parasitics. Adding a passivation layer can further improve the breakdown voltages by reducing the trapping effects. The schematic of an asymmetric MHEMT is shown in Figure 14.1.

The gate recess width plays a critical role on the ON and OFF state breakdown voltage characteristics of MHEMTs. A wide gate recess is highly desirable for achieving higher breakdown voltages.

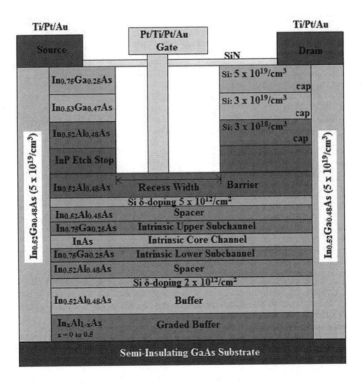

FIGURE 14.1
Schematic diagram of an asymmetric MHEMT. (Based on Ajayan, J. et al., High speed low power Full Adder circuit design using current comparison based domino, in *Devices, Circuits and Systems (ICDCS), 2014 2nd International Conference on*, IEEE, 2014, pp. 1–5; Ajayan, J. et al., *AEU-International Journal of Electronics and Communications*, 84, 387–393, 2018.)

The gate recess width influences the ON and OFF state breakdown voltages which is evident from Figure 14.2a and b. The OFF-state breakdown voltage (BV_{OFF}) is defined as the gate-drain voltage at which a gate leakage current of 1 mA/mm is obtained with floating source. Figure 14.2a and b clearly indicates that reducing the gate recess width results in the degradation of BV_{OFF} and BV_{ON} respectively. This may be due to the self heating effects in the device. When the gate recess width is reduced, the device parasitic decreases and that result in the enhancement of drain current. The increase in drain current increases the channel temperature which further generates electron-hole pairs. The generation of electron-hole pair further improves the drain current that again increases the heat in the active channel layer. This self-heating process reduces the breakdown voltage of the HEMTs. ON state breakdown voltage (BV_{ON}) is defined as the drain to source voltage at which the gate current becomes 1 mA/mm at a particular V_{GS} for peak transconductance. Figure 14.3 shows the impact of gate recess scaling on the source resistance of asymmetric MHEMTs. The downscaling of gate recess width helps to minimize the source and drain parasitic resistances which in turn minimize the ON-resistance of the device. Figure 14.2d plots the variation of drain current as a function of drain voltage for two different gate lengths and the plot reveals that

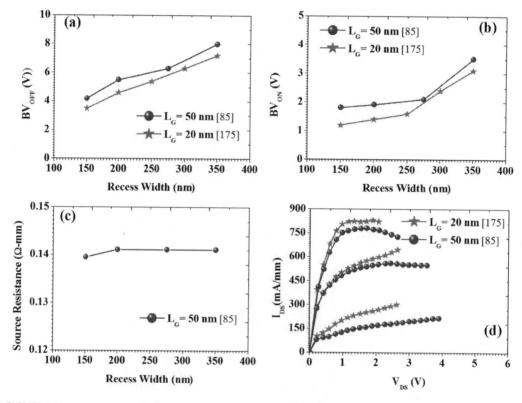

FIGURE 14.2
The influence of recess width on the (a) OFF-state breakdown voltage (BV_{OFF}), (b) ON-state breakdown voltage (BV_{ON}). (c) Source resistance (R_S). (d) Drain bias dependence of drain current [85,175].

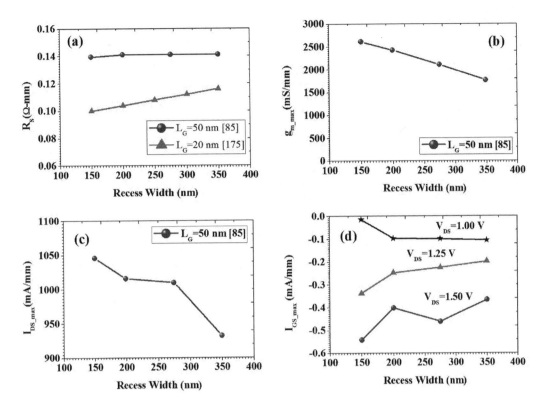

FIGURE 14.3
The influence of gate recess width on the (a) source resistance for $L_G = 50$ nm and $L_G = 20$ nm (b) peak transconductance (c) peak drain current and (d) maximum gate leakage current for the asymmetric MHEMTs [85,175].

reducing the gate length enhances the drain current, that may be due to the reduction of ON resistance. From Figure 14.2a and b it is evident that lower gate length devices exhibit poor breakdown voltages.

The influence of gate length and gate recess width downscaling on the source resistance of asymmetric MHEMT is shown in Figure 14.3a. The reduction of gate length and gate recess width helps to minimize the device parasitic (Figure 14.3a), which results in the increase of transconductance and drain current (Figure 14.3b and c). The influence of drain bias and gate recess width scaling on the gate leakage current characteristics of asymmetric MHEMT is shown in Figure 14.3d. At higher drain voltages, gate leakage current decreases. By employing an asymmetric gate recess, the gate-drain bridge capacitance can be minimized which enables the reduction of short-channel effects. Reduction of gate to channel distance and source/drain resistances can significantly improve the transconductance and drain current and this can be made possible by adopting a buried platinum metal gate technology. The reduction of source/drain spacing is found to be very effective in improving transconductance and drain current due to the reduction of source and drain contact resistances. Self-aligned ohmic contact fabrication process can be used to reduce the source/drain contact resistances.

14.4 DC and RF Characteristics

GaAs/InAlAs/InGaAs MHEMTs have better noise figure and high output power compared with InP/InAlAs/InGaAs lattice-matched HEMTs. InP substarte is smaller in size (2-inch or 3-inch) and they are easily breakable. However, GaAs wafers are available in large size up to 6-inch and they are less brittle compared to InP wafers. When InAlAs/InGaAs material system is transferred to GaAs substrate lattice mismatch will occur. Therefore, suitable buffer layers must be inserted between the channel and GaAs substrate layers. A thick InAlAs layer with graded indium concentration can effectively minimize the lattice mismatch between GaAs substrate and active channel layer. When comes to the DC performance, a transconductance (g_m) of over 3000 mS/mm and drain current (I_{DS}) of over 800 mA/mm were reported for GaAs MHEMTs using high-mobility InGaAs or InAs inserted composite channel. The cut off frequency (f_T) and maximum oscillation frequency (f_{max}) are the two most important parameters used to characterize the RF performance of HEMTs. An f_T of over 600 GHz and an f_{max} of over 1000 GHz were reported for MHEMTs with extremely low-noise figure.

The epitaxial layer details of an advanced symmetrical MHEMT are shown in Figure 14.4. Usually, the gate metal is made of either Ti/Pt/Au or Pt/Ti/Pt/Au metal layers. The source/drain contacts can be formed by using Au/Ge/Ni/Au or Ge/Au/Ni/Au or Ni/Ge/Au or Ge/Ni/Au or Ti/Pt/Au metal layers depending up on applications. Burying

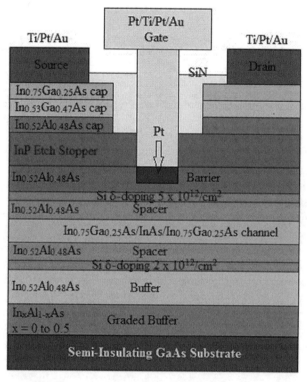

FIGURE 14.4
The vertical layer structure of an advanced symmetrical MHEMT.

platinum metal into the InAlAs barrier layer is found to be very effective in suppressing short-channel effects and it has the added benefits of improved DC and RF performance. Platinum sinking process also helps to achieve E-Mode operation of the device by reducing gate-channel distance. Employing a passivation layer protects the device from external damages and surface contamination, which also minimizes the surface trapping effects on the DC and RF performances of the devices. The InAs inserted composite channel provides high 2DEG density in the channel with outstanding electron mobility that is highly preferable for achieving good DC and RF characteristics.

Figure 14.5a indicates that electron mobility in the quantum well of MHEMTs is inversely proportional to InAlAs barrier layer thickness. That is, electron mobility in the quantum well of MHEMTs increases with decrease in barrier layer thickness. Reducing the barrier layer thickness helps to obtain a very short gate-channel distance, which is essential for improving transconductance (g_m) of MHEMTs (Equation 14.1).

$$g_{m\_max} = \frac{W_g \cdot v_{sat} \cdot \varepsilon}{d_{GC}} \tag{14.1}$$

where:

g_{m_max} is the peak transconductance
W_g is the width of the gate
v_{sat} is the saturation electron drift velocity
ε is the permittivity of the semiconductor
d_{GC} is the effective gate-channel distance

The other factors which affect the electron mobility in the channel of MHEMTs are 2DEG density and indium concentration in the channel. A channel with higher indium concentration shows higher electron mobility. However, increase in 2DEG density in the channel ($n_s(x)$) limits the electron mobility due to various scattering effects. Figure 14.5b shows the trend in electron mobility as a function of sheet charge density and indium concentration

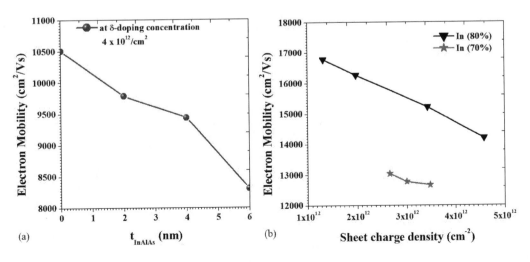

FIGURE 14.5
(a) Dependence of barrier thickness on the electron mobility characteristics of MHEMTs. (b) Dependence of indium concentration and sheet charge density on the electron mobility characteristics of MHEMTs [59].

in the channel. The gate to source voltage (V_{GS}) dependence of sheet charge density (2DEG) at the interface of a heterojunction can be computed as [204–206]

$$n_S(x) = \frac{\varepsilon_m}{qd}\left[V_{GS} - V_{TH} - E_f\right] \tag{14.2}$$

where,

$$\varepsilon_m = \left(12.9 - 1.64\,m\right)\varepsilon_0 \text{ at } 300°\text{K} \tag{14.3}$$

$$V_{TH} = \Phi_B - \Delta E_C - \frac{q.\delta(n).d_{GC}}{\varepsilon} \tag{14.4}$$

V_{TH} is the threshold voltage
q is the charge of an electron
E_f is the Fermi potential
m is the indium concentration in the channel
ε_0 is the permittivity of free space
Φ_B is the Schottky barrier height
ΔE_C is the conduction band discontinuity at barrier/channel interface
$\delta(n)$ is the delta doping concentration

The major factors that influence the threshold voltage of MHEMTs are Schottky barrier height, conduction band discontinuity between barrier and channel layers, delta doping concentration and gate-channel distance. For obtaining E-Mode operation, the gate metal should be selected in such a way that it should create a large Schottky barrier height with the barrier layer. For achieving large Schottky barrier height, metals with higher work functions are required. Usually platinum is used to achieve E-Mode operation because it can provide a Schottky barrier height of 0.85 eV with InAlAs barrier layer because platinum has higher work function (5.65 eV) compared with other widely used metals like titanium (4.1 eV), palladium (5.1 eV) and nickel (5.15 eV). The other factors influence the threshold voltage of MHEMTs are silicon delta doping concentration and gate-channel distance. Increasing the gate-channel distance shifts the threshold voltage towards negative regime. Also, a higher silicon delta doping concentration leads to the negative shift of threshold voltage. Therefore, to achieve E-Mode operation for MHEMTs it must be essential to reduce the gate-channel distance and silicon delta doping concentration. However, reducing the delta doping concentration has the drawback of reduced transconductance and drain current. Hence, in most of the cases, for the same geometric dimensions D-Mode MHEMTs are superior in DC and RF performance when compared with E-Mode MHEMTs [168–177]. The conduction band discontinuity between barrier and channel layers varies with indium concentration in the channel and temperature which is expressed in Equation 14.5 [204–206].

$$\Delta E_C(m,T) = 0.73\left(E_g^{\text{In}_{0.52}\text{Al}_{0.48}\text{As}}(T) - E_g^{\text{In}_m\text{Ga}_{1-m}\text{As}}(m,T)\right) \tag{14.5}$$

where:

$$E_g^{In_{0.52}Al_{0.48}As}(300\,\text{K}) = 0.52\,E_g^{InAs}(300\,\text{K}) + 0.48\,E_g^{AlAs}(300\,\text{K}) - 0.17472 \tag{14.6}$$

$$E_g^{In_xGa_{1-x}As}(x,300\,\text{K}) = x\,E_g^{InAs}(300\,\text{K}) + (1-x)\,E_g^{GaAs}(300\,\text{K}) - x(1-x)\,0.477 \tag{14.7}$$

$$E_g^{InAs}(T) = 0.417 - 0.276 \times 10^{-3}\left(\frac{T^2}{T+93}\right) \tag{14.8}$$

$$E_g^{AlAs}(T) = 3.099 - 0.885 \times 10^{-3}\left(\frac{T^2}{T+530}\right) \tag{14.9}$$

$$E_g^{GaAs}(T) = 1.519 - 0.5405 \times 10^{-3}\left(\frac{T^2}{T+204}\right) \tag{14.10}$$

The conduction band discontinuity (ΔE_C) between channel and barrier layers significantly influences the carrier transport characteristics in the quantum well. Also, for a fixed delta-doping concentration, the electron mobility in the channel is inversely proportional to the thickness of the barrier layer. The Schottky barrier height also changes with temperature and the temperature dependence of Schottky barrier height with $In_{0.52}Al_{0.48}As$ barrier layer is described in Equation 14.11 [204–206].

$$\Phi_b(T) = \Phi_{b300}\left(E_g^{In_{0.52}Al_{0.48}As}(300°\text{K}) - E_g^{In_{0.52}Al_{0.48}As}(T)\right) \tag{14.11}$$

where Φ_{b300} represents the Schottky barrier height at 300 K. The drain current of a single channel HEMT as a function of $n_s(x)$, electron drift velocity in the channel ($v(x)$) and gate width (W_G) can be expressed as

$$I_D = W_G \cdot q \cdot n_S(x) \cdot v(x) \tag{14.12}$$

$$v(x) = \frac{\mu_0 \cdot E_C(x)}{1 + \dfrac{\mu_0 \cdot E_C(x)}{v_{sat}}} \tag{14.13}$$

where $E_C(x)$ is the channel electric field which varies with respect to channel potential and v_{sat} is the saturation electron drift velocity. Saturation electron drift velocity changes with respect to temperature and the relationship between saturation electron drift velocity and temperature [194] is given in Equation 14.14.

$$v_{sat}\alpha\left(\frac{e^{\frac{E_{op}}{K_BT}} - 1}{e^{\frac{E_{op}}{K_BT}} + 1}\right)^{1/2} \tag{14.14}$$

In Equation 14.14, E_{op} represents optical phonon energy. The saturation electron drift velocity of electrons decreases with respect to increase in temperature. The saturation drift velocity of electrons also depends on the indium concentration in the channel. As the indium content in the channel increases the saturation electron drift velocity of electrons also increases. That is saturation electron drift velocity is directly proportional to the indium concentration in the channel.

The transconductance, drain current, subthreshold current and gate leakage current are the most important DC parameters of a HEMT and for a high-performance HEMT transconductance and drain/current should be maximum and subthreshold current and gate leakage currents should be minimum. Transconductance of MHEMTs varies with respect to V_{GS}, which is evident from Figure 14.6a. Li-Dan et al. [207] developed an analytical expression for computing the transconductance (g_m) of a HEMT, which is given in Equations 14.15 and 14.16).

$$g_{m\_int} = \frac{\mu_n W \varepsilon_n \varepsilon_0}{L_g \cdot d_{GC}} \cdot V_{DS} + \frac{\mu_n W \varepsilon_n \varepsilon_0}{L_g \cdot d_{GC}} \left(V_{GS} - V_T \right) \tag{14.15}$$

$$g_{m\_ext} = \frac{g_{m\_int}}{1 + g_{m\_int} \cdot R_S + g_d \left(R_S + R_D \right)} + \frac{g_{m\_int}}{1 + g_{m\_int} \cdot R_S} \tag{14.16}$$

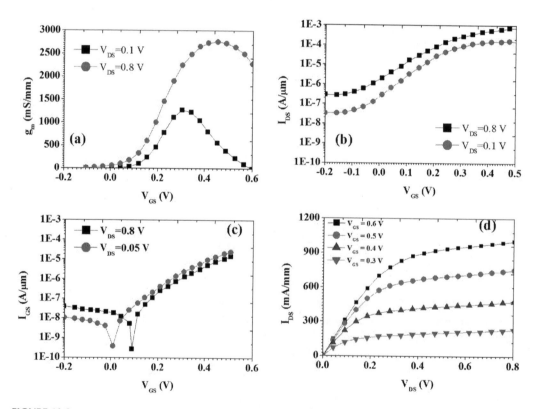

FIGURE 14.6

(a) Transconductance as a function of V_{GS}, (b) Sub-threshold current as a function of V_{GS}, (c) Gate leakage current as a function of V_{GS}, and (d) Drain current as a function of V_{DS} for a $L_G = 40$ nm MHEMT. (From Kim, D.-H. et al, $f_T = 688$ GHz and $f_{max} = 800$ GHz in $L_g = 40$ nm In$_{0.7}$Ga$_{0.3}$As MHEMTs with $g_{m\_max} > 2.7$ mS/μm, in: *Electron devices Meeting (IEDM), 2011 IEEE International, IEEE*, 2011, pp. 13.16. 11–13.16. 14.)

where g_{m_int}, g_{m_ext}, W, ε_n, ε_0, L_g, d_{GC}, R_S, and R_D represents the internal g_m, external g_m, gate width, permittivity of the semiconductor layer between gate and channel, permittivity of air, gate length, gate to channel distance, source resistance and drain resistance, respectively. When V_{GS} changes that affect the sheet charge density in the channel because g_m is directly proportional to sheet charge density in the channel. Figure 14.6b shows the trend in subthreshold current as a function of V_{GS} and V_{DS}. For MHEMTs, higher V_{DS} results in higher subthreshold current. The V_{GS} dependence of gate leakage current for MHEMT is shown in Figure 14.6c. The gate leakage current plays a significant role in determining the breakdown voltage of the MHEMTs. MHEMTs with lower gate leakage current possess higher ON and OFF state breakdown voltages. Under reverse bias conditions, an increase in V_{DS} results in the increase of gate leakage currents. However, under forward bias conditions, an increase in V_{DS} decreases the gate leakage currents. Transistor scaling and indium-rich channel are the two most important requirements for achieving higher DC performance. The gate and drain bias voltage dependence of drain current is plotted in Figure 14.6d.

The saturation drain current for MHEMTs increases with increase in gate to source bias. To achieve higher drain current under ON conditions, the ON resistance of the transistor must be very low. Therefore, to reduce the ON resistance it is essential to minimize the parasitic resistance in the MHEMT such as source resistance, gate resistance, drain resistance and channel resistance. Higher mobility channels such as InGaAs, InAs, and InSb is suitable for designing high performance MHEMTs. To reduce the short-channel effects such as DIBL (drain-induced barrier lowering) and SS (sub-threshold swing) the channel aspect ratio (AR_C) should be greater than unity. The channel aspect ratio can be computed as

$$AR_C = \frac{L_G}{d_{GC} + d_C} \tag{14.17}$$

Equation 14.17 reveals that in order to keep the channel aspect ratio of the HEMTs greater than unity for suppressing short-channel effects, it is important to scale down the gate to channel distance (d_{GC}) and channel thickness (d_C) along with the reduction of gate length (L_G). However, it is challenging to reduce the d_{GC} below 4 nm due to severe gate leakage current, which limits the ON and OFF state breakdown voltages of the MHEMTs. The range of the channel thickness can be in between 10 and 20 nm for better ON state performance of the MHEMTs.

Various gate metal formation techniques that can be employed during MHEMT fabrication play a significant role in determining the ON current because these gate formation techniques decides the gate resistance of the devices. Mainly there are two types of gate metal formation techniques and they are

1. Mesa type gate realization
2. Air type gate metal realization

Like InP HEMTs, GaAs MHEMTs also utilizes an InGaAs/InAlAs quantum well heterostructure. InGaAs/InAlAs heterostructure offers a low Schottky barrier height (ϕ_B) due to their relatively large conduction band discontinuity (ΔE_C). The low ϕ_B of InGaAs/InAlAs heterostructure in MHEMTs have the following limitations.

1. Poor ON and OFF state breakdown voltages
2. Severe kink effects due to the higher impact ionization rate in the narrow band gap channel

One straightforward method to improve the ϕ_B is the proper selection of gate foot metal with higher work function. Therefore, platinum is widely employed at the foot of the gate during gate contact realization over other metals such as gold, silver, nickel, and titanium. Platinum offer a ϕ_B in the range of 0.83–0.85 eV with InAlAs Schottky barrier layer. The use of platinum also helps to achieve E-Mode operation of the MHEMTs. E-Mode transistors have certain advantages over D-Mode transistors in circuit design and they are

1. Single-power supply operation
2. Higher integration and flexibility
3. Low-power dissipation due to relatively low threshold voltages
4. Does not conduct under zero bias conditions

Figure 14.7 shows the influence of Mesa-type and Air-type gate metal realization techniques on the drain current performance of the MHEMTs [84]. From Figure 14.7, it is clear that Air-type gate metal formation helps to achieve higher drain current when compared with Mesa-type gate metal formation. This is mainly due to the Air-type gate metal realization helps to achieve a low-gate resistance compared to its counterpart. Also, Air-type gate formation with platinum sink technology is found to be very effective in suppressing kink effects in the drain current characteristics of the InP/GaAs HEMTs. The kink effects limit the performance parameters of MHEMTs such as transistor gain, increased noise at high frequencies and poor ON and OFF state breakdown voltages.

Air-type gate formation scheme provides better transistor gain compared to mesa type (Figure 14.8a). Annealing process significantly affects the gain characteristics of MHEMTs

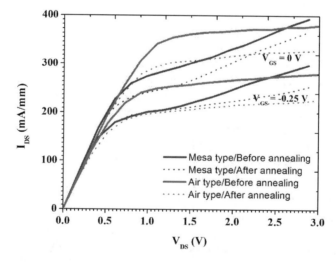

FIGURE 14.7
The influence of various gate metal formation on MHEMT output characteristics. (From Hsu, M. et al., *Semicond. Sci. Technol.*, 22, 35, 2006.)

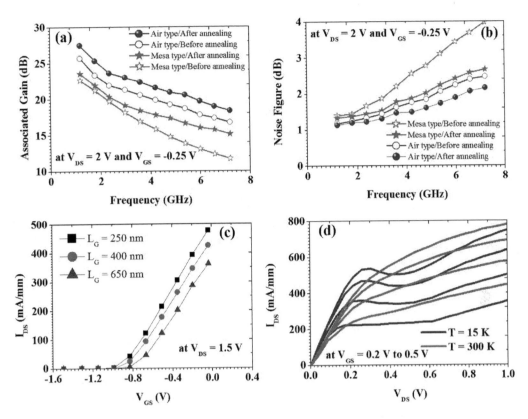

FIGURE 14.8
(a) The impact of gate metal formation techniques and annealing process on the transistor gain [84], (b) The influence of gate metal formation techniques on the noise figure of MHEMTs [84], (c) Influence of transistor scaling on the drain current performance of MHEMTs [152], and (d) Impact of temperature on the drain current characteristics of MHEMTs [25].

and it is observed that MHEMTs after annealing process provides higher associated gain. Figure 14.8b shows the impact of Air-type and Mesa-type gate contact formation techniques on the noise figure characteristics of the MHEMTs. Air type gate contact formation method offers relatively low noise figure than Mesa-type gate contact realization method. The factors that influence the noise figure of MHEMTs are gate parasitic resistance, source parasitic resistance and gate leakage currents. In summary, Air-type gate fabrication process provides higher transistor gain with low noise figure in the case of MHEMTs. The noise figure (NF$_{min}$) of the HEMTs can be computed as [208]

$$NF_{min} = 10\log\left(1 + 2\pi \cdot K_f \cdot f\left(C_{GS} + C_{GD}\right)\sqrt{\left(R_G + R_S\right)/g_m^{int}}\right) \tag{14.18}$$

where K_f, f, C_{GS}, C_{GD}, R_G, R_{sj} and g_m^{int} represents fitting factor, frequency of operation, gate to source parasitic capacitance, gate to drain parasitic capacitance, gate resistance, source resistance and internal transconductance, respectively. By employing a cavity structure at the gate, the noise performance of the HEMTs can be further enhanced. Figure 14.8c shows the trends in drain current with the reduction of gate length. Reducing gate length is essential for enhancing the drain current. Temperature also plays a significant role on the DC performance of MHEMTs, which is plotted in Figure 14.8d. Under low temperature, that is close to 0 K the MHEMTs exhibits severe kink effects. Suemitsu et al. [209,210] has developed an analytical expression for computing kink voltage (V_{kink}) in the quantum well as a function of hole concentration, which is given in Equation 14.19.

$$V_{kink} = \frac{K_B T}{q} \ln\left[\frac{(P_0 + P_1)}{P_0}\right] \qquad (14.19)$$

where K_B, T, P_0, and P_1 represents Boltzmann's constant, temperature, pre-kink hole concentration and excess hole concentration due to kink effects, respectively. The kink current (ΔI_D) as a function of kink voltage and pre-kink transconductance (g_{m0}) is expressed in Equation 14.20 [209,210].

$$\Delta I_D = g_{m0} \cdot V_{kink} \qquad (14.20)$$

The kink current is also a function of drain to source voltage (V_{DS}) and the relationship between kink current and V_{DS} is given in Equation 14.21 [209,210].

$$\Delta I_D = g_{m0} \frac{K_B T}{q} \ln\left[1 + AI_D \exp\left(\frac{-B}{V_{DS} - V_{DS\_sat}}\right)\right] \qquad (14.21)$$

where A and B are temperature dependent constants and I_D represents pre-kink current. The reduction of source resistance is the primary reason for the sudden increase in drain current under kink effects. Impact ionization and traps at the barrier/buffer interface are the primary sources of origin of kink effects in InP/GaAs HEMTs. Impact ionization leads to the generation of holes in the channel that suddenly reducs the ON resistance of the device. This sudden reduction of ON resistance results in the sudden increase in drain current. To eliminate kink effects the following methods can be employed.

1. Use of an InP etch stopper layer under the gate regions
2. Keep the side recess spacing as low as possible
3. Use of a low temperature buffer that can capture the excess holes generated by the impact ionization process takes place in the narrow band gap channel
4. Use of δ-doping strategy can also effectively minimize kink effects

The ON and OFF state breakdown voltages (BV$_{ON}$ and BV$_{OFF}$ respectively) depends on several factors such as indium content in the channel, gate recess spacing, gate length and gate leakage current. Breakdown voltages of the MHEMTs both under ON and OFF conditions decreases with gate length down scaling. The reduction of gate length improves peak transconductance and peak drain current. However, gate length scaling limits the breakdown voltages due to self heating and increased gate leakage currents.

Using a narrow band gap material in the channel leads to impact ionization that results in the sudden increase of drain current. This phenomenon is called kink effect, and this happens mainly due to the impact ionization taken place in the channel. Impact ionization creates holes in the channel which are collected by the gate through the Schottky barrier layer. This undesirable effect leads to the increase of output conductance of the transistors, which in turn severely degrades the breakdown voltages of the devices. The increase of operating temperature also limits the breakdown voltages. The reduction in breakdown voltages in MHEMTs due to the reduction of gate length can be explained as follows. When the transistors are scaled down, the gate to drain separation decreases that increases the electric field strength in the gate-drain regions. This higher electric field enhances the drift velocity of electrons through the channel that results in the increase of drain current. The increase of drain current increases the temperature in the channel. This increase in channel temperature reduces the reliability and life time of the device due to self heating process. The higher electric field strength in the gate/drain area also produces hot electrons in the channel which leads to impact ionization that increases the gate leakage current. Increase in gate leakage current severely degrades the breakdown voltages. The trend in ON and OFF state breakdown voltages when the gate length is scaled down is shown in Figure 14.9a. The impact of gate length down scaling on the peak transconductance, peak drain current and source/contact resistances of the

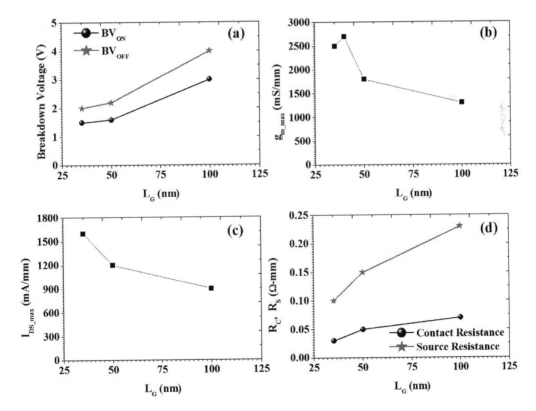

FIGURE 14.9
Impact of gate length scaling on the (a) ON and OFF state breakdown voltages, (b) peak transconductance, (c) peak drain current, and (d) source and contact resistances of the MHEMTs.

MHEMTs are shown in Figure 14.9,b–d respectively. If the gate length is scaled down the parasitic resistances associated with the MHEMTs such as source resistance, drain resistance, channel resistance and contact resistance decrease, which in turn reduce the ON resistance of the device. The overall reduction of ON resistance of the MHEMTs due to the gate length down scaling results in the enhancement of peak transconductance and peak drain current.

Figure 14.10a shows the gate length dependence of C_{GS} and C_{GD} for fixed bias conditions, and the plot shows that these parasitic capacitances decreases with decrease in gate length. This reduction in parasitic capacitances leads to the enhancement of f_T and f_{max} as the gate length of the MHEMTs are scaled down (Figure 14.10b). The relationship between C_{GS}, C_{GD}, and f_T is given in Equation 14.22. f_T is inversely proportional to both C_{GS} and C_{GD} and directly proportional to transconductance of the device. However, Equation 14.22 does not take into account of the influence of parasitic resistances such as gate resistance (R_G), source resistance (R_S), and drain resistance (R_D).

$$f_T = \frac{g_m}{2\pi\left(C_{GS}+C_{GD}\right)}$$

(14.22)

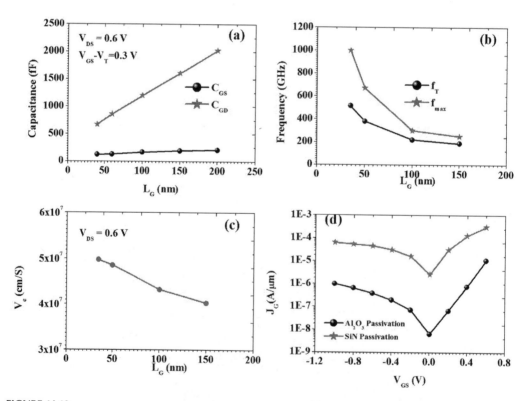

FIGURE 14.10

The impact of gate length down scaling on the (a) Gate to source (C_{GS}) and gate drain (C_{GD}) parasitic capacitances [140], (b) Cut off frequency (f_T) and maximum frequency of oscillation (f_{max}) [140], (c) Electron velocity in the channel [140], and (d) The influence of dielectric passivation materials on the gate leakage currents of the MHEMTs [151].

The relationship between f_T and L_G is expressed in Equation 14.23.

$$f_T = \frac{1}{2\pi\left(\tau_{ex} + \dfrac{L_G}{V_S}\right)} \tag{14.23}$$

$$\tau_{ex} = \tau_{cc} + C_{GD}(R_S + R_D) + \frac{\Delta L}{V_S} \tag{14.24}$$

$$f_T = \frac{g_m}{2\pi} \frac{1}{(C_{GS} + C_{GD})\left(1 + \dfrac{R_S + R_D}{R_{SD}}\right) + g_m C_{GD}(R_S + R_D)} \tag{14.25}$$

where τ_{cc}, V_S, ΔL, R_S, R_D, and R_G represents channel charging time, electron saturation velocity, extention of effective gate length, source resistance and drain resistance, respectively. The source-drain resistance R_{SD} can be computed as [211]

$$R_{SD} = R_{Sheet}(L_{SG} + L_{GD}) + 2R_{contact} + 2R_{side} + R_{sd,gate} \tag{14.26}$$

where R_{Sheet}, $R_{contact}$, L_{SG}, L_{GD}, R_{side}, and $R_{sd,gate}$ represents the sheet resistance, contact resistance, source to gate length, gate to drain length, resistance at the side etched region and intrinsic channel resistance, respectively. The f_{max} of the HEMTs can be computed using the analytical expression developed by Das et al. from the small-signal equivalent circuit, which is given in Equation 7.

$$f_{max} = \frac{f_T}{\sqrt{4g_{ds}\left(R_{in} + \dfrac{R_S + R_g}{1 + g_m R_S}\right) + \dfrac{4C_{gd}}{5C_{gs}}\left(1 + \dfrac{2.5C_{gd}}{C_{gs}}\right)(1 + g_m R_S)^2}} \tag{14.27}$$

where g_{mi} and g_{oi} are intrinsic transconductance and output conductance respectively, C_{gs} and C_{gd} are the parasitic capacitances exist between gate to source and gate to drain terminals respectively and these two capacitances can be split into intrinsic (C_{gsi}, C_{gdi}) and extrinsic (C_{gs_ext}, C_{gd_ext}) components, R_S, R_D, and R_G are the source, drain, and gate resistances respectively. R_{in} is the input resistance of the device. The f_T is inversely proportional to the total delay (τ) encountered by the electron while travelling from source to drain, which can be split into three components: intrinsic delay (τ_i), extrinsic delay (τ_{ext}) and parasitic delay (τ_{par}). The impact of gate length down scaling on the saturation electron drift velocity of electrons in the channel of MHEMTs is shown in Figure 14.10c and from the plot it is evident that a low gate length is required to enhance the electron drift velocity in the channel. Increase in electron velocity can reduce the transit time under the gate which is the ratio of gate length to electron drift velocity of electrons in the channel of MHEMTs.

14.5 Applications of MHEMTs

MHEMT technology is considered as one of the most suitable solid-state transistor technology for microwave (MW), millimeter wave (MMW), sub-millimeter wave (S-MMW) and terahertz (THZ) frequency applications due to their unique properties such as low noise, large bandwidth, high gain, low DC power consumption, high operating frequency and low operating voltage, etc. MHEMT technology is most commonly used in MMW (30–300 GHz) and S-MMW (300–700 GHz) applications in the field of safety, security and communication. MHEMT technology can be used for developing sensors for collision avoidance radars, material detection (plastics and explosives) atmospheric and radio astronomic sensing and environmental sensing and imaging, high resolution radars, etc. An overview of MHEMT applications and their features are given in Table 14.1.

TABLE 14.1

Overview of MHEMT Applications and Their Features

Ref.	L_G (nm)	Application	Features
[178]	35	S-MMIC power amplifier cell	300 GHz bandwidth
[3]	40	D-band low noise amplifier	115–160 GHz band, 20 dB gain, 4 dB NF at 115 GHz
[4]	50	D-band low noise amplifier	97–164 GHz band, 3 dB NF
[6]	50	Broadband balanced medium power amplifier	175–245 GHz band, 13.4 dB gain at 228 GHz
[5]	35	Direct coupled baseband amplifier	53 GHz bandwidth, 21 dB gain, 3.84 dB NF
[7]	70	Ka-band low noise amplifier	20–30 GHz band, 27 dB gain, 1.1 dB NF
[179]	35	Down conversion mixer	530–606 GHz band, Maximum conversion gain of –9.2 dB
[9]	150	Frequency divider	Locking range of 28.8–31.6 GHz
[13]	20	D-band Amplifier	110–170 GHz band, 28 dB gain
[13]	20	H-band amplifier	220–325 GHz band, 24 dB gain
[14]	50	Low noise amplifier	43–90 GHz band, 25 dB gain at 80 GHz, 1.6 dB NF
[21]	50	Low noise amplifier	243 GHz bandwidth, 19.5 dB gain, 6 dB NF
[18]	35	G-band cascode power amplifier	180–200 GHz band, 19 dB gain
[16]	50	G-band low noise amplifier	165 GHz bandwidth, 16 dB gain, 4.4 dB NF
[22]	150	Down conversion mixer	32 dB image rejection ratio, 1.7 dB conversion gain
[19]	100	High resolution RADAR	20 dBm output power, range resolution 3.7 mm
[36]	70	W-band low noise amplifier	70–105 GHz band, 25 dB gain, 2.7 dB NF
[37]	150	Up conversion mixer	–0.2 dB conversion gain
[27]	100	Cryogenic low noise amplifier	4–12 GHz band, 22 dB gain
[25]	100	Cryogenic low noise amplifier	25–34 GHz band, 24.2 dB gain
[28]	100	Mixer	185–202 GHz band, Conversion loss 8 dB
[24]	150	Cryogenic low noise amplifier	30–50 GHz band, 23.1 dB gain
[49]	150	Frequency multiplier	8 dBm output power, 210 mW power dissipation
[52]	150	Automotive RADAR	77 GHz band, 15.5 dBm output power
[48]	70	Mixer	94 GHz, 8.9 dB Conversion loss
[50]	150	4-Bit phase shifter	8–12 GHz, 5.5 dB insertion loss
[47]	150	Transmitter	4.1 ± 1.5 dBm output power, 56.5–64.5 GHz
[47]	150	Receiver	54.5–64.5 GHz, 10.7 dB gain, 13 dB image rejection ratio
[42]	100	Voltage controlled oscillator	Tuning range 3.4 GHz, output power 5.5 dB, 92 GHz band

Abbreviation: NF, Noise Figure.

References

1. Y. Chen, K.-L. Wu, C.-Y. Huang, C.-K. Chang, X. Sun, B. Li, R. Hu, OMMIC 70nm-mHEMT LNA design, in: *Microwave Conference (APMC)*, *2017 IEEE Asia Pacific*, IEEE, 2017, pp. 1192–1195.
2. A. Hülsmann, A. Leuther, I. Kallfass, R. Weber, A. Tessmann, M. Schlechtweg, O. Ambacher, Advanced mHEMT technologies for space applications, in: *The Proceedings of the 20th International Symposium on Space Terahertz Technology*, 2009, pp. 178–182.
3. R. Cleriti, W. Ciccognani, S. Colangeli, A. Serino, E. Limiti, P. Frijlink, M. Renvoisé, R. Doerner, M. Hossain, D-band LNA using a 40-nm GaAs mHEMT technology, in: *Microwave Integrated Circuits Conference (EuMIC)*, *2017 12th European*, IEEE, 2017, pp. 105–108.
4. R. Weber, H. Massler, A. Leuther, D-band low-noise amplifier MMIC with 50% bandwidth and 3.0 dB noise figure in 100 nm and 50 nm mHEMT technology, in: *Microwave Symposium (IMS)*, *2017 IEEE MTT-S International*, IEEE, 2017, pp. 756–759.
5. L. John, T. Merkle, C. Friesicke, A. Tessmann, A. Leuther, R. Lozar, M. Schlechtweg, T. Zwick, Investigation of direct-coupled amplifier topologies for wireless communication systems using normally-on mHEMT technology, in: *Microwave Symposium (IMS)*, *2017 IEEE MTT-S International*, IEEE, 2017, pp. 1129–1132.
6. B. Amado-Rey, Y. Campos-Roca, C. Friesicke, A. Tessmann, H. Massler, S. Wagner, A. Leuther, M. Schlechtweg, O. Ambacher, A broadband 175–245 GHz balanced medium power amplifier using 50-nm mHEMT technology, in: *Microwave Conference (APMC)*, *2016 Asia-Pacific*, IEEE, 2016, pp. 1–4.
7. X. Cheng, L. Zhang, X. Deng, Ka-band low noise amplifier using 70nm mHEMT process for wideband communication, in: *Semiconductor Technology International Conference (CSTIC)*, *2017 China*, IEEE, 2017, pp. 1–2.
8. R. Cleriti, W. Ciccognani, S. Colangeli, E. Limiti, P. Frijlink, M. Renvoise, Characterization and modelling of 40 nm mHEMT process up to 110 GHz, in: *Microwave Integrated Circuits Conference (EuMIC)*, *2016 11th European*, IEEE, 2016, pp. 353–356.
9. W.-L. Chang, C. Meng, K.-C. Tsung, G.-W. Huang, 30-GHz mHEMT divide-by-three injection-locked frequency divider with Marchand balun, in: *Radio and Wireless Symposium (RWS)*, *2015 IEEE*, IEEE, 2015, pp. 68–70.
10. M.L. Bhavsar, R. Sharma, A. Bhattacharya, Monolithic Ka to Ku-band all balanced sub-harmonic resistive MHEMT mixer for satellite transponder, *IEEE Microwave and Wireless Components Letters*, 25 (2015) 316–318.
11. M.L. Bhavsar, A. Bhattacharya, R.K. Arora, Design and comparison of MHEMT and Diode based K-band Sub-harmonically pumped Mixer MMICs, in: *Microwave and RF Conference (IMaRC)*, *2014 IEEE International*, IEEE, 2014, pp. 266–269.
12. K. Yaohui, W. Weibo, G. Jianfeng, C. Chen, 100nm MHEMT transistor technology for W-band amplifier, in: *Antennas and Propagation (APCAP)*, *2014 3rd Asia-Pacific Conference on*, IEEE, 2014, pp. 1339–1341.
13. T. Merkle, A. Leuther, S. Koch, I. Kallfass, A. Tessmann, S. Wagner, H. Massler, M. Schlechtweg, O. Ambacher, Backside process free broadband amplifier MMICs at D-Band and H-Band in 20 nm mHEMT technology, in: *Compound Semiconductor Integrated Circuit Symposium (CSICs)*, *2014 IEEE*, IEEE, 2014, pp. 1–4.
14. P.M. Smith, M. Ashman, D. Xu, X. Yang, C. Creamer, P. Chao, K. Chu, K. Duh, C. Koh, J. Schellenberg, A 50nm MHEMT millimeter-wave MMIC LNA with wideband noise and gain performance, in: *Microwave Symposium (IMS)*, *2014 IEEE MTT-S International*, IEEE, 2014, pp. 1–4.
15. M. Ohlrogge, M. Seelmann-Eggebert, A. Leuther, H. Massler, A. Tessmann, R. Weber, D. Schwantuschke, M. Schlechtweg, O. Ambacher, A scalable compact small-signal mHEMT model accounting for distributed effects in sub-millimeter wave and terahertz applications, in: *Microwave Symposium (IMS)*, *2014 IEEE MTT-S International*, IEEE, 2014, pp. 1–4.

16. M. Karkkainen, M. Kantanen, S. Caujolle-Bert, M. Varonen, R. Weber, A. Leuther, M. Seelmann-Eggebert, A. Alanne, P. Jukkala, T. Narhi, MHEMT $ G $-band low-noise amplifiers, *IEEE Transactions on Terahertz Science and Technology*, 4 (2014) 459–468.

17. F. Thome, H. Massler, S. Wagner, A. Leuther, I. Kallfass, M. Schlechtweg, O. Ambacher, Comparison of two W-band low-noise amplifier MMICs with ultra low power consumption based on 50nm InGaAs mHEMT technology, in: *Microwave Symposium Digest (IMS), 2013 IEEE MTT-S International*, IEEE, 2013, pp. 1–4.

18. Y. Campos-Roca, A. Tessmann, V. Hurm, H. Massler, M. Seelmann-Eggebert, A. Leuther, A G-band cascode MHEMT medium power amplifier, in: *Infrared, Millimeter, and Terahertz Waves (IRMMW-THz), 2013 38th International Conference on*, IEEE, 2013, pp. 1–2.

19. A. Tessmann, A. Leuther, H. Massler, U. Lewark, S. Wagner, R. Weber, M. Kuri, M. Zink, M. Riessle, H.-P. Stulz, A monolithic integrated mHEMT chipset for high-resolution sub-millimeter-wave radar applications, in: *Compound Semiconductor Integrated Circuit Symposium (CSICS), 2013 IEEE*, IEEE, 2013, pp. 1–4.

20. D.-W. Zhang, J. Li, H.-H. Ma, Design of 60GHZ MMIC LNA in mHEMT technology, in: *IET International Radar Conference 2013*, IET, 2013.

21. V. Hurm, R. Weber, A. Tessmann, H. Massler, A. Leuther, M. Kuri, M. Riessle, H. Stulz, M. Zink, M. Schlechtweg, A 243 GHz LNA module based on mHEMT MMICs with integrated wave-guide transitions, *IEEE Microwave and Wireless Components Letters*, 23 (2013) 486–488.

22. Y.-C. Hsiao, C. Meng, J.-Y. Su, S.-W. Yu, G.-W. Huang, 16-GHz mHEMT double-quadrature Gilbert down-conversion mixer with polyphase filters, in: *Wireless Symposium (IWS), 2013 IEEE International*, IEEE, 2013, pp. 1–4.

23. J. Schleeh, H. Rodilla, N. Wadefalk, P.-Å. Nilsson, J. Grahn, Characterization and modeling of cryogenic ultralow-noise InP HEMTs, *IEEE Transactions on Electron Devices*, 60 (2013) 206–212.

24. S.-H. Weng, W.-C. Wang, H.-Y. Chang, C.-C. Chiong, M.-T. Chen, A cryogenic 30–50 GHz balanced low noise amplifier using 0.15-μm MHEMT process for radio astronomy applications, in: *Radio-Frequency Integration Technology (RFIT), 2012 IEEE International Symposium on*, IEEE, 2012, pp. 177–179.

25. B.A. Abelán, M. Seelmann-Eggebert, D. Bruch, A. Leuther, H. Massler, B. Baldischweiler, M. Schlechtweg, J.D. Gallego-Puyol, I. Lopez-Fernandez, C. Diez-Gonzalez, 4–12-and 25–34-GHz cryogenic mHEMT MMIC low-noise amplifiers, *IEEE Transactions on Microwave Theory and Techniques*, 60 (2012) 4080–4088.

26. B. Aja, M. Seelmann-Eggebert, A. Leuther, H. Massler, M. Schlechtweg, J. Gallego, I. López-Fernández, C. Diez, I. Malo, E. Villa, 4–12 GHz and 25–34 GHz cryogenic MHEMT MMIC low noise amplifiers for radio astronomy, in: *Microwave Symposium Digest (MTT), 2012 IEEE MTT-S International*, IEEE, 2012, pp. 1–3.

27. B. Aja, K. Schuster, F. Schafer, J. Gallego, S. Chartier, M. Seelmann-Eggebert, I. Kallfass, A. Leuther, H. Massler, M. Schlechtweg, Cryogenic low-noise mHEMT-based MMIC amplifiers for 4–12 GHz band, *IEEE Microwave and Wireless Components Letters*, 21 (2011) 613–615.

28. Y. Yan, Y.B. Karandikar, S.E. Gunnarsson, B.M. Motlagh, S. Cherednichenko, I. Kallfass, A. Leuther, H. Zirath, Monolithically integrated 200-GHz double-slot antenna and resistive mixers in a GaAs-mHEMT MMIC process, *IEEE Transactions on Microwave Theory and Techniques*, 59 (2011) 2494.

29. J. Lynch, F. Traut, K. Benson, R. Tshudy, An mHEMT Q-band integrated LNA and vector modulator MMIC, in: *Compound Semiconductor Integrated Circuit Symposium (CSICS), 2010 IEEE*, IEEE, 2010, pp. 1–4.

30. C.-I. Kuo, W.C. Lim, H.-T. Hsu, C.-T. Wang, L.-H. Hsu, F. Aizad, G.-W. Hung, Y. Miyamoto, E.Y. Chang, Bonding temperature effect on the performance of flip chip assembled 150nm mHEMT device on organic substrate, in: *Enabling Science and Nanotechnology (ESciNano), 2010 International Conference on*, IEEE, 2010, pp. 1–2.

31. P. Blount, C.J. Trantanell, L. Coryell, R. Lau, Low noise, low power dissipation mHEMT-based amplifiers for phased array application, in: *Phased Array Systems and Technology (ARRAY), 2010 IEEE International Symposium on*, IEEE, 2010, pp. 233–237.

32. A. Tessmann, A. Leuther, H. Massler, V. Hurm, M. Kuri, M. Zink, M. Riessle, R. Lösch, High-gain submillimeter-wave mHEMT amplifier MMICs, in: *Microwave Symposium Digest (MTT), 2010 IEEE MTT-S International*, IEEE, 2010, pp. 53–56.

33. H. Zirath, N. Wadefalk, R. Kuzhuharov, S.E. Gunnarsson, I. Angelov, M. Abbasi, B. Hansson, V. Vassilev, J. Svedin, S. Rudner, Integrated receivers up to 220 GHz utilizing GaAs-mHEMT technology, in: *Radio-Frequency Integration Technology, 2009. RFIT 2009. IEEE International Symposium on*, IEEE, 2009, pp. 225–228.

34. G.-Y. Chen, Y.-S. Wu, H.-Y. Chang, Y.-M. Hsin, C.-C. Chiong, A 60–110 GHz low conversion loss tripler in 0.15-μm MHEMT process, in: *Microwave Conference, 2009. APMC 2009. Asia Pacific*, IEEE, 2009, pp. 377–380.

35. Y.-C. Tsai, J.-L. Kuo, H. Wang, A 40–80-GHz mHEMT single-pole-double-throw switch using traveling-wave concept, in: *Microwave Conference, 2008. APMC 2008. Asia-Pacific*, IEEE, 2008, pp. 1–4.

36. W. Ciccognani, F. Giannini, E. Limiti, P.E. Longhi, Full W-band high-gain LNA in mHEMT MMIC technology, in: *Microwave Integrated Circuit Conference, 2008. EuMIC 2008. European*, IEEE, 2008, pp. 314–317.

37. J.-Y. Su, C. Meng, P.-Y. Wu, Q-band pHEMT and mHEMT subharmonic Gilbert upconversion mixers, *IEEE Microwave and Wireless Components Letters*, 19 (2009) 392–394.

38. E.C. Niehenke, J. Whelehan, D. Xu, D. Meharry, K.G. Duh, P.M. Smith, A Q-band MHEMT 100-mW MMIC power amplifier with 46% power-added efficiency, in: *Microwave Symposium Digest, 2008 IEEE MTT-S International*, IEEE, 2008, pp. 277–280.

39. S.-H. Weng, C.-H. Lin, H.-Y. Chang, C.-C. Chiong, Q-band low noise amplifiers using a 0.15 μm MHEMT process for broadband communication and radio astronomy applications, in: *Microwave Symposium Digest, 2008 IEEE MTT-S International*, IEEE, 2008, pp. 455–458.

40. J. Lee, R.-B. Lai, K.-Y. Lin, C.-C. Chiong, H. Wang, A Q-band low loss reduced-size filter-integrated SPDT switch using 0.15 μm MHEMT technology, in: *Microwave Symposium Digest, 2008 IEEE MTT-S International*, IEEE, 2008, pp. 551–554.

41. Y.H. Baek, S.J. Lee, T.J. Baek, J.H. Oh, S.G. Choi, D.S. Kang, S.D. Kim, J.K. Rhee, Millimeter-wave Broadband Amplifier using MHEMT, in: *Millimeter Waves, 2008. GSMM 2008. Global Symposium on*, IEEE, 2008, pp. 48–51.

42. R. Weber, M. Kuri, M. Lang, A. Tessmann, M. Seelmann-Eggebert, A. Leuther, A PLL-stabilized W-band MHEMT push-push VCO with integrated frequency divider circuit, in: *Microwave Symposium, 2007. IEEE/MTT-S International*, IEEE, 2007, pp. 653–656.

43. K.-W. Lee, K.-L. Lee, X.-Z. Lin, C.-H. Tu, Y.-H. Wang, Subthreshold characteristics and high-frequency performance in InAlAs/InGaAs MHEMT with a liquid phase oxidized InAlAs Gate, in: *Solid-State and Integrated Circuit Technology, 2006. ICSICT'06. 8th International Conference on*, IEEE, 2006, pp. 881–883.

44. C.-S. Lee, W.-C. Hsu, K.-H. Su, J.-C. Huang, C.-H. Liao, High-temperature characteristics of A symmetrically-graded InAlAs/In$_x$Ga$_{1-x}$As/GaAs MHEMT, in: *Solid-State and Integrated Circuit Technology, 2006. ICSICT'06. 8th International Conference on*, IEEE, 2006, pp. 878–880.

45. D.M. Kang, J.Y. Hong, J.Y. Shim, H.S. Yoon, K.H. Lee, A W-band MMIC one-chip set for automotive radar sensor by using a 0.15 μm mHEMT process, in: *European Microwave Integrated Circuits Conference, 2006. The 1st*, IEEE, 2006, pp. 328–331.

46. S.E. Gunnarsson, M. Gavell, D. Kuylenstierna, H. Zirath, 60 GHz MMIC double balanced Gilbert mixer in mHEMT technology with integrated RF, LO and IF baluns, *Electronics Letters*, 42 (2006) 1402–1403.

47. S.E. Gunnarsson, C. Karnfelt, H. Zirath, R. Kozhuharov, D. Kuylenstierna, C. Fager, A. Alping, Single-chip 60 GHz transmitter and receiver MMICs in a GaAs mHEMT technology, in: *IEEE MTT-S International Microwave Symposium Digest*, 2006, pp. 801–804.

48. D. An, S.-C. Kim, J.-D. Park, M.-K. Lee, H.-C. Park, S.-D. Kim, W.-J. Kim, J.-K. Rhee, A novel 94-GHz MHEMT resistive mixer using a micromachined ring coupler, *IEEE Microwave and Wireless Components Letters*, 16 (2006) 467–469.

49. C. Karnfelt, R. Kozhuharov, H. Zirath, I. Angelov, High-purity 60-GHz-band single-chip/spl times/8 multipliers in pHEMT and mHEMT technology, *IEEE Transactions on Microwave Theory and Techniques*, 54 (2006) 2887–2898.

50. A.I. Khalil, M. Mahfoudi, F. Traut, M. Shifrin, J. Chavez, A X-band 4-bit mHEMT phase shifter, in: *Compound Semiconductor Integrated Circuit Symposium, 2005. CSIC'05. IEEE*, IEEE, 2005, p. 4.

51. V. Krozer, T.K. Johansen, T. Djurhuus, J. Vidkjær, Wideband monolithic microwave integrated circuit frequency converters with GaAs mHEMT technology, in: *Microwave Conference, 2005 European*, IEEE, 2005, p. 4, p. 1586.

52. D.M. Kang, J.Y. Hong, J.Y. Shim, J.H. Lee, H.S. Yoon, K.H. Lee, A 77GHz automotive radar MMIC chip set fabricated by a 0.15/spl mu/m MHEMT technology, in: *Microwave Symposium Digest, 2005 IEEE MTT-S International*, IEEE, 2005, p. 4.

53. J.H. Lee, H.S. Yoon, J.Y. Shim, J.Y. Hong, D.M. Kang, H.C. Kim, K.I. Cho, K.H. Lee, B.W. Kim, 0.15/spl mu/m gate length MHEMT technology for 77 GHz automotive radar applications, in: *Indium Phosphide and Related Materials, 2004. 16th IPRM. 2004 International Conference on*, IEEE, 2004, pp. 24–27.

54. G. Moschetti, A. Leuther, H. Maßler, B. Aja, M. Rösch, M. Schlechtweg, O. Ambacher, V. Kangas, M. Geneviève-Perichaud, A 183 GHz metamorphic HEMT low-noise amplifier with 3.5 dB noise figure, *IEEE Microwave and Wireless Components Letters*, 25 (2015) 618–620.

55. J. Jeong, S.W. Kim, W. Choi, K.S. Seo, Y. Kwon, A distributed amplifier with 12.5-dB gain and 82.5-GHz bandwidth using 0.1 μm GaAs metamorphic HEMTs, *Microwave and Optical Technology Letters*, 49 (2007) 2873–2875.

56. Y. Cordier, P. Lorenzini, J.-M. Chauveau, D. Ferre, Y. Androussi, J. DiPersio, D. Vignaud, J.-L. Codron, Influence of MBE growth conditions on the quality of InAlAs/InGaAs metamorphic HEMTs on GaAs, *Journal of Crystal Growth*, 251 (2003) 822–826.

57. S. Bollaert, Y. Cordier, M. Zaknoune, H. Happy, V. Hoel, S. Lepilliet, D. Théron, A. Cappy, The indium content in metamorphic $In_xAl_{1-x}As/In_xGa_{1-x}As$ HEMTs on GaAs substrate: A new structure parameter, *Solid-State Electronics*, 44 (2000) 1021–1027.

58. M. Zaknoune, Y. Cordier, S. Bollaert, D. Ferre, D. Theron, Y. Crosnier, 0.1-μm high performance double heterojunction $In_{0.32}Al_{0.68}As/In_{0.33}Ga_{0.67}As$ metamorphic HEMTs on GaAs, *Solid-State Electronics*, 44 (2000) 1685–1688.

59. H. Geka, S. Yamada, M. Toita, K. Nagase, N. Kuze, Effects of AlGaAsSb electron supply layer for InGaAs/InAlAs metamorphic HEMTs on GaAs substrate, *Journal of Crystal Growth*, 323 (2011) 522–524.

60. H. Happy, S. Bollaert, H. Fouré, A. Cappy, Numerical analysis of device performance of metamorphic In/sub y/Al/sub 1-y/As/In/sub x/Ga/sub 1-x/As (0.3/spl les/x/spl les/0.6) HEMTs on GaAs substrate, *IEEE Transactions on Electron Devices*, 45 (1998) 2089–2095.

61. H.S. Yoon, J.H. Lee, J.Y. Shim, J.Y. Hong, D.M. Kang, K.H. Lee, Extremely low noise characteristics of 0.1/spl mu/m/spl Gamma/-gate power metamorphic HEMT on GaAs substrate, in: *Indium Phosphide and Related Materials, 2005. International Conference on*, IEEE, 2005, pp. 133–136.

62. M. Ferndah, H. Zirath, Residual and oscillator phase noise in GaAs metamorphic HEMTs, in: *Microwave Conference, 2006. APMC 2006. Asia-Pacific*, IEEE, 2006, pp. 472–475.

63. A. Cappy, Y. Cordier, S. Bollaert, M. Zaknoune, Status of metamorphic $In_xAl_{1-x}As/In_xGa_{1-x}As$ HEMTs, in: *GaAs IC Symposium, 1999. 21st Annual*, IEEE, 1999, pp. 217–220.

64. S. Bollaert, Y. Cordier, H. Happy, M. Zaknoune, V. Hoel, S. Lepilliet, A. Cappy, Metamorphic $In_xAl_{1-x}As/In_xGa_{1-x}As$ HEMTs on GaAs substrate: The influence of In composition, in: *Electron Devices Meeting, 1998. IEDM'98. Technical Digest, International*, IEEE, 1998, pp. 235–238.

65. P. Win, Y. Druelle, A. Cappy, Y. Cordier, J. Favre, C. Bouillet, Metamorphic $In_{0.3}Ga_{0.7}As/In_{0.29}Al_{0.71}As$ layer on GaAs: A new structure for high performance high electron mobility transistor realization, *Applied Physics Letters*, 61 (1992) 922–924.

66. C. Whelan, P. Marsh, W. Hoke, R. McTaggart, C. McCarroll, T. Kazior, GaAs metamorphic HEMT (MHEMT): an attractive alternative to InP HEMTs for high performance low noise and power applications, in: *Indium Phosphide and Related Materials, 2000. Conference Proceedings. 2000 International Conference on*, IEEE, 2000, pp. 337–340.

67. W. Deal, Solid-state amplifiers for terahertz electronics, in: *Microwave Symposium Digest (MTT), 2010 IEEE MTT-S International*, IEEE, 2010, pp. 1122–1125.

68. W. Ha, Z. Griffith, D.-H. Kim, P. Chen, M. Urteaga, B. Brar, High performance InP mHEMTs on GaAs substrate with multiple interconnect layers, in: *Indium Phosphide & Related Materials (IPRM), 2010 International Conference on*, IEEE, 2010, pp. 1–4.

69. A. Tessmann, V. Hurm, A. Leuther, H. Massler, R. Weber, M. Kuri, M. Riessle, H.-P. Stulz, M. Zink, M. Schlechtweg, 243 GHz low-noise amplifier MMICs and modules based on metamorphic HEMT technology, *International Journal of Microwave and Wireless Technologies*, 6 (2014) 215–223.

70. L. Samoska, Towards terahertz MMIC amplifiers: Present status and trends, in: *Microwave Symposium Digest, 2006. IEEE MTT-S International*, IEEE, 2006, pp. 333–336.

71. A. Tessmann, A. Leuther, V. Hurm, I. Kallfass, H. Massler, M. Kuri, M. Riessle, M. Zink, R. Loesch, M. Seelmann-Eggebert, Metamorphic HEMT MMICs and modules operating between 300 and 500 GHz, *IEEE Journal of Solid-State Circuits*, 46 (2011) 2193–2202.

72. P. Kangaslahti, B. Lim, T. Gaier, A. Tanner, M. Varonen, L. Samoska, S. Brown, B. Lambrigtsen, S. Reising, J. Tanabe, Low noise amplifier receivers for millimeter wave atmospheric remote sensing, in: *Microwave Symposium Digest (MTT), 2012 IEEE MTT-S International*, IEEE, 2012, pp. 1–3.

73. A. Tessmann, A. Leuther, C. Schworer, H. Massler, W. Reinert, M. Walther, R. Losch, M. Schlechtweg, Millimeter-wave circuits based on advanced metamorphic HEMT technology, in: *Infrared and Millimeter Waves, 2004 and 12th International Conference on Terahertz Electronics, 2004. Conference Digest of the 2004 Joint 29th International Conference on*, IEEE, 2004, pp. 165–166.

74. A. Tessmann, A. Leuther, H. Massler, M. Kuri, M. Riessle, M. Zink, R. Sommer, A. Wahlen, H. Essen, Metamorphic HEMT amplifier circuits for use in a high resolution 210 GHz radar, in: *Compound Semiconductor Integrated Circuit Symposium, 2007. CSIC 2007. IEEE*, IEEE, 2007, pp. 1–4.

75. Y. Cordier, S. Bollaert, M. Zaknoune, J. Dipersio, D. Ferre, InAlAs/InGaAs metamorphic high electron mobility transistors on GaAs substrate: influence of indium content on material properties and device performance, *Japanese Journal of Applied Physics*, 38 (1999) 1164.

76. H. Fourre, F. Diette, A. Cappy, Selective wet etching of lattice-matched InGaAs/InAlAs on InP and metamorphic InGaAs/InAlAs on GaAs using succinic acid/hydrogen peroxide solution, *Journal of Vacuum Science & Technology B: Microelectronics and Nanometer Structures Processing, Measurement, and Phenomena*, 14 (1996) 3400–3402.

77. M. Zaknoune, B. Bonte, C. Gaquiere, Y. Cordier, Y. Druelle, D. Theron, Y. Crosnier, InAlAs/InGaAs metamorphic HEMT with high current density and high breakdown voltage, *IEEE Electron Device Letters*, 19 (1998) 345–347.

78. J.I. Chyi, J.L. Shieh, J.W. Pan, R.M. Lin, Material properties of compositional graded In_xGa_{1-x}As and In_xAl_{1-x}As epilayers grown on GaAs substrates, *Journal of Applied Physics*, 79 (1996) 8367–8370.

79. W. Hoke, P. Lemonias, J. Mosca, P. Lyman, A. Torabi, P. Marsh, R. McTaggart, S. Lardizabal, K. Hetzler, Molecular beam epitaxial growth and device performance of metamorphic high electron mobility transistor structures fabricated on GaAs substrates, *Journal of Vacuum Science & Technology B: Microelectronics and Nanometer Structures Processing, Measurement, and Phenomena*, 17 (1999) 1131–1135.

80. B.-H. Lee, D. An, M.-K. Lee, B.-O. Lim, S.-D. Kim, J.-K. Rhee, Two-stage broadband high-gain W-band amplifier using 0.1-/spl mu/m metamorphic HEMT technology, *IEEE Electron Device Letters*, 25 (2004) 766–768.

81. K. Lee, Y. Kim, K. Lee, Y. Jeong, Process for 20 nm T gate on $Al_{0.25}Ga_{0.75}$As/$In_{0.2}Ga_{0.8}$As/Ga As epilayer using two-step lithography and zigzag foot, *Journal of Vacuum Science & Technology B: Microelectronics and Nanometer Structures Processing, Measurement, and Phenomena*, 24 (2006) 1869–1872.

82. B. Matinpour, N. Lal, J. Laskar, R.E. Leoni, C.S. Whelan, K-band receiver front-ends in a GaAs metamorphic HEMT process, *IEEE Transactions on Microwave Theory and Techniques*, 49 (2001) 2459–2463.

83. H. Ono, S. Taniguchi, T.-k. Suzuki, Indium content dependence of electron velocity and impact ionization in InAlAs/InGaAs metamorphic HEMTs, *Japanese Journal of Applied Physics*, 43 (2004) 2259.

84. M. Hsu, H. Chen, S. Chiu, W. Chen, W.-C. Liu, J. Tasi, W.-S. Lour, Characteristics of mesa- and air-type $In_{0.5}Al_{0.5}As/In_{0.5}Ga_{0.5}As$ metamorphic HEMTs with or without a buried gate, *Semiconductor Science and Technology*, 22 (2006) 35.

85. D. Xu, X. Yang, P. Seekell, L.M.M. Pleasant, L. Mohnkern, K. Chu, R.G. Stedman, A. Vera, R. Isaak, L.L. Schlesinger, 50-nm asymmetrically recessed metamorphic high-electron mobility transistors with reduced source–drain spacing: Performance enhancement and tradeoffs, *IEEE Transactions on Electron Devices*, 59 (2012) 128–138.

86. D. Xu, W.M. Kong, X. Yang, P. Smith, D. Dugas, P. Chao, G. Cueva, L. Mohnkern, P. Seekell, L.M. Pleasant, Asymmetrically recessed 50-nm gate-length metamorphic high electron-mobility transistor with enhanced gain performance, *IEEE Electron Device Letters*, 29 (2008) 4–7.

87. D. Xu, X. Yang, P. Seekell, L.M. Pleasant, R. Isaak, W. Kong, G. Cueva, K. Chu, L. Mohnkern, L. Schlesinger, 50-nm self-aligned high electron-mobility transistors on GaAs substrates with extremely high extrinsic transconductance and high gain, *International Journal of High Speed Electronics and Systems*, 20 (2011) 393–398.

88. D. Xu, W.M. Kong, X. Yang, L. Mohnkern, P. Seekell, L.M. Pleasant, K.G. Duh, P.M. Smith, P.-C. Chao, 50-nm metamorphic high-electron-mobility transistors with high gain and high breakdown voltages, *IEEE Electron Device Letters*, 30 (2009) 793–795.

89. D. Xu, X. Yang, W.M. Kong, P. Seekell, K. Louie, L. Pleasant, L. Mohnkern, D.M. Dugas, K. Chu, H.F. Karimy, Gate-length scaling of ultrashort metamorphic high-electron mobility transistors with asymmetrically recessed gate contacts for millimeter-and submillimeter-wave applications, *IEEE Transactions on Electron Devices*, 58 (2011) 1408–1417.

90. D. Xu, W.M. Kong, X. Yang, P. Seekell, L. Mohnkern, L.M. Pleasant, H. Karimy, K. Duh, P. Smith, P. Chao, Scaling behaviors of 25-NM asymmetrically recessed metamorphic high electron-mobility transistors, in: *Indium Phosphide & Related Materials, 2009. IPRM'09. IEEE International Conference on*, IEEE, 2009, pp. 305–307.

91. M. Kotiranta, S. Türk, F. Schäfer, A. Leuther, J. Goliasch, H. Massler, M. Schlechtweg, Cryogenic 50-nm mHEMT MMIC LNA for 67-116 GHz with 34 K noise temperature, in: *Millimeter Waves (GSMM) & ESA Workshop on Millimetre-Wave Technology and Applications, 2016 Global Symposium on*, IEEE, 2016, pp. 1–3.

92. R. Cleriti, S. Colangeli, W. Ciccognani, M. Palomba, E. Limiti, Cold-source cryogenic characterization and modeling of a mHEMT process, in: *Microwave Integrated Circuits Conference (EuMIC), 2015 10th European*, IEEE, 2015, pp. 41–44.

93. J. Schleeh, H. Rodilla, N. Wadefalk, P.-Å. Nilsson, J. Grahn, Cryogenic ultra-low noise amplification-InP PHEMT vs. GaAs MHEMT, in: *Indium Phosphide and Related Materials (IPRM), 2013 International Conference on*, IEEE, 2013, pp. 1–2.

94. M. Seelmann-Eggebert, F. Schäfer, A. Leuther, H. Massler, A versatile and cryogenic mHEMT-model including noise, in: *Microwave Symposium Digest (MTT), 2010 IEEE MTT-S International*, IEEE, 2010, pp. 501–504.

95. J. Schleeh, H. Rodilla, N. Wadefalk, P.-Å. Nilsson, J. Grahn, Cryogenic noise performance of InGaAs/InAlAs HEMTs grown on InP and GaAs substrate, *Solid-State Electronics*, 91 (2014) 74–77.

96. A. Leuther, A. Tessmann, M. Dammann, H. Massler, M. Schlechtweg, O. Ambacher, 35 nm mHEMT technology for THz and ultra low noise applications, in: *Indium Phosphide and Related Materials (IPRM), 2013 International Conference on*, IEEE, 2013, pp. 1–2.

97. C.-Y. Chiang, H.-T. Hsu, C.-I. Kuo, C.-T. Wang, W.C. Lim, E.Y. Chang, Impact of bonding temperature on the performance of $In_{0.6}Ga_{0.4}As$ Metamorphic High Electron Mobility Transistor (mHEMT) device packaged using Flip-Chip-on-Board (FCOB) technology, in: *Microwave Conference Proceedings (APMC), 2012 Asia-Pacific*, IEEE, 2012, pp. 938–940.

98. D. Smith, G. Dambrine, J.-C. Orlhac, Industrial MHEMT technologies for 80-220 GHz applications, in: *Microwave Integrated Circuit Conference, 2008. EuMIC 2008. European*, IEEE, 2008, pp. 214–217.

99. A. Leuther, A. Tessmann, M. Dammann, C. Schworer, M. Schlechtweg, M. Mikulla, R. Losch, G. Weimann, 50 nm MHEMT technology for G-and H-band MMICs, in: *Indium Phosphide & Related Materials, 2007. IPRM'07. IEEE 19th International Conference on*, IEEE, 2007, pp. 24–27.

100. I. Thayne, K. Elgaid, D. Moran, X. Cao, E. Boyd, H. McLelland, M. Holland, S. Thoms, C. Stanley, 50ánm metamorphic GaAs and InP HEMTs, *Thin Solid Films*, 515 (2007) 4373–4377.

101. M. Chertouk, H. Heiss, D. Xu, S. Kraus, W. Klein, G. Böhm, G. Tränkle, G. Weimann, Metamorphic InAlAs/InGaAs HEMTs on GaAs substrates with composite channels and 350-GHz f_{max} with 160-GHz fT, *Microwave and Optical Technology Letters*, 11 (1996) 145–147.

102. H. Li, C.W. Tang, K.M. Lau, Metamorphic AlInAs/GaInAs HEMTs on GaAs substrates by MOCVD, *IEEE Electron Device Letters*, 29 (2008) 561–564.

103. K. Elgaid, H. McLelland, M. Holland, D. Moran, C. Stanley, I. Thayne, 50-nm T-gate metamorphic GaAs HEMTs with f T of 440 GHz and noise figure of 0.7 dB at 26 GHz, *IEEE Electron Device Letters*, 26 (2005) 784–786.

104. S.-J. Yeon, M. Park, J. Choi, K. Seo, 610 GHz InAlAs/In 0.75 GaAs metamorphic HEMTs with an ultra-short 15-nm-gate, in: *Electron Devices Meeting, 2007. IEDM 2007. IEEE International*, IEEE, 2007, pp. 613–616.

105. M. Chertouk, H. Heiss, D. Xu, S. Kraus, W. Klein, G. Bohm, G. Trankle, G. Weimann, Metamorphic InAlAs/InGaAs HEMTs on GaAs substrates with a novel composite channels design, *IEEE Electron Device Letters*, 17 (1996) 273–275.

106. L. Hai-Ou, H. Wei, T.C. Wah, D. Xiao-Fang, L.K. May, Fabrication of 160-nm T-gate metamorphic AlInAs/GaInAs HEMTs on GaAs substrates by metal organic chemical vapour deposition, *Chinese Physics B*, 20 (2011) 068502.

107. H. Li, Z. Feng, C.W. Tang, K.M. Lau, Fabrication of 150-nm T-gate metamorphic AlInAs/GaInAs HEMTs on GaAs substrates by MOCVD, *IEEE Electron Device Letters*, 32 (2011) 1224–1226.

108. D. Dumka, G. Cueva, W. Hoke, P. Lemonias, I. Adesida, 0.13/spl mu/m gate-length In/sub 0.52/Al/sub 0.48/As-In/sub 0.53/Ga/sub 0.47/As metamorphic HEMTs on GaAs substrate, in: *Device Research Conference, 2000. Conference Digest. 58th DRC*, IEEE, 2000, pp. 83–84.

109. M. Zaknoune, M. Ardouin, Y. Cordier, S. Bollaert, B. Bonte, D. Théron, 60-GHz high power performance $In_{0.35}Al_{0.65}As$-$In_{0.35}Ga_{0.65}As$ metamorphic HEMTs on GaAs, *IEEE Electron Device Letters*, 24 (2003) 724–726.

110. D. Dumka, W. Hoke, P. Lemonias, G. Cueva, I. Adesida, Metamorphic $In_{0.52}Al_{0.48}As/In_{0.53}Ga_{0.47}As$ HEMTs on GaAs substrate with fT over 200 GHz, *Electronics Letters*, 35 (1999) 1854–1856.

111. D. Dumka, W. Hoke, P. Lemonias, R. Schwindt, G. Cueva, I. Adesida, Monolithic integration of InAlAs/InGaAs enhancement and depletion (E/D)-mode metamorphic HEMTs on GaAs substrate, in: *Device Research Conference, 2001*, IEEE, 2001, pp. 49–50.

112. A. Leuther, A. Tessmann, M. Dammann, W. Reinert, M. Schlechtweg, M. Mikulla, M. Walther, G. Weimann, 70 nm low-noise metamorphic HEMT technology on 4 inch GaAs wafers, in: *Indium Phosphide and Related Materials, 2003. International Conference on*, IEEE, 2003, pp. 215–218.

113. K.-S. Lee, Y.-S. Kim, Y.-K. Hong, Y.-H. Jeong, 35-nm Zigzag T-Gate $In_{0.52}Al_{0.48}As/In_{0.53}Ga_{0.47}As$ Metamorphic GaAs HEMTs with an Ultrahigh f_{max} of 520GHz, *IEEE Electron Device Letters*, 28 (2007) 672–675.

114. C.-J. Hwang, L.B. Lok, H.M. Chong, M. Holland, I.G. Thayne, K. Elgaid, An ultra-low-power MMIC amplifier using 50-nm δ-doped $In_{0.52}Al_{0.48}As/In_{0.53}Ga_{0.47}As$ Metamorphic HEMT, *IEEE Electron Device Letters*, 31 (2010) 1230–1232.

115. E.Y. Chang, Y.-C. Lin, G.-J. Chen, H.-M. Lee, G.-W. Huang, D. Biswas, C.-Y. Chang, Composite-channel metamorphic high electron mobility transistor for low-noise and high-linearity applications, *Japanese Journal of Applied Physics*, 43 (2004) L871.

116. Y. Lien, E.Y. Chang, H. Chang, L. Chu, G. Huang, H. Lee, C. Lee, S. Chen, P. Shen, C. Chang, Low-noise metamorphic HEMTs with reflowed 0.1-/spl mu/m T-gate, *IEEE Electron Device Letters*, 25 (2004) 348–350.

117. K.C. Sahoo, C.-I. Kuo, Y. Li, E.Y. Chang, Novel metamorphic HEMTs with highly doped InGaAs source/drain regions for high frequency applications, *IEEE Transactions on Electron Devices*, 57 (2010) 2594–2598.

118. A. Tessmann, A. Leuther, H. Massler, M. Seelmann-Eggebert, A high gain 600 GHz amplifier TMIC using 35 nm metamorphic HEMT technology, in: *Compound Semiconductor Integrated Circuit Symposium (CSICS), 2012 IEEE*, IEEE, 2012, pp. 1–4.

119. X. Wu, H. Liu, H. Li, Q. Li, S. Hu, Z. Xi, J. Zhao, Fabrication of 150-nm $Al_{0.48}In_{0.52}As/Ga_{0.47}In_{0.53}As$ mHEMTs on GaAs substrates, *Science China Physics, Mechanics and Astronomy*, 55 (2012) 2389–2391.

120. J. Xu, H. Zhang, W. Wang, L. Liu, M. Li, X. Fu, J. Niu, T. Ye, 200 nm gate-length GaAs-based MHEMT devices by electron beam lithography, *Chinese Science Bulletin*, 53 (2008) 3585–3589.

121. C.-I. Kuo, H.-T. Hsu, Y.-L. Chen, C.-Y. Wu, E.Y. Chang, Y. Miyamoto, W.-C. Tsern, K.C. Sahoo, RF performance improvement of metamorphic high-electron mobility transistor using $(In_xGa_{1-x}As)_m/(InAs)_n$ superlattice-channel structure for millimeter-wave applications, *IEEE Electron Device Letters*, 31 (2010) 677–679.

122. A. Tessmann, A. Leuther, H. Massler, M. Kuri, R. Loesch, A metamorphic 220-320 ghz hemt amplifier mmic, in: *Compound Semiconductor Integrated Circuits Symposium, 2008. CSIC'08. IEEE*, IEEE, 2008, pp. 1–4.

123. A. Tessmann, I. Kallfass, A. Leuther, H. Massler, M. Schlechtweg, O. Ambacher, Metamorphic MMICs for operation beyond 200 GHz, in: *Microwave Integrated Circuit Conference, 2008. EuMIC 2008. European*, IEEE, 2008, pp. 210–213.

124. A. Leuther, A. Tessmann, I. Kallfass, H. Massler, R. Loesch, M. Schlechtweg, M. Mikulla, O. Ambacher, Metamorphic HEMT technology for submillimeter-wave MMIC applications, in: *Indium Phosphide & Related Materials (IPRM), 2010 International Conference on*, IEEE, 2010, pp. 1–6.

125. A. Tessmann, 220-GHz metamorphic HEMT amplifier MMICs for high-resolution imaging applications, *IEEE Journal of Solid-State Circuits*, 40 (2005) 2070–2076.

126. P. Win, Y. Druelle, P. Legry, S. Lepilliet, A. Cappy, Y. Cordier, J. Favre, Microwave performance of 0.4 μm gate metamorphic $In_{0.29}Al_{0.71}As/In_{0.3}Ga_{0.7}As$ HEMT on GaAs substrate, *Electronics Letters*, 29 (1993) 169–170.

127. D.M. Gill, B. Kane, S.P. Svensson, D.-W. Tu, P. Uppal, N. Byer, High-performance, 0.1 μm InAlAs/InGaAs high electron mobility transistors on GaAs, *IEEE Electron Device Letters*, 17 (1996) 328–330.

128. W. Hoke, T. Kennedy, A. Torabi, C. Whelan, P. Marsh, R. Leoni, C. Xu, K. Hsieh, High indium metamorphic HEMT on a GaAs substrate, *Journal of Crystal Growth*, 251 (2003) 827–831.

129. C.-W. Tang, J. Li, K.M. Lau, K.J. Chen, MOCVD grown metamorphic InAlAs/InGaAs HEMTs on GaAs substrates, *Compound Semiconductor Mantech*, (2006) 24–27.

130. W.-C. Hsu, D.-H. Huang, Y.-S. Lin, Y.-J. Chen, J.-C. Huang, C.-L. Wu, Performance improvement in tensile-strained $In_{(0.5)}Al_{(0.5)}As/In_xGa_{1-x}As/In_{(0.5)}Al_{(0.5)}As$ metamorphic HEMT, *IEEE Transactions on Electron Devices*, 53 (2006) 406–412.

131. D.-W. Tu, S. Wang, J. Liu, K. Hwang, W. Kong, P. Chao, K. Nichols, High-performance double-recessed InAlAs/InGaAs power metamorphic HEMT on GaAs substrate, *IEEE Microwave and Guided Wave Letters*, 9 (1999) 458–460.

132. M. Behet, K. Van der Zanden, G. Borghs, A. Behres, Metamorphic InGaAs/InAlAs quantum well structures grown on GaAs substrates for high electron mobility transistor applications, *Applied Physics Letters*, 73 (1998) 2760–2762.

133. K. Higuchi, M. Kudo, M. Mori, T. Mishima, First high performance InAlAs/InGaAs HEMTs on GaAs exceeding that on InP, in: *Electron Devices Meeting, 1994. IEDM'94. Technical Digest, International*, IEEE, 1994, pp. 891–894.

134. S.-W. Kim, K.-M. Lee, J.-H. Lee, K.-S. Seo, High-performance 0.1 m $In_{0.4}AlAs/In_{0.35}GaAs$ MHEMTs with Ar plasma treatment, *IEEE Electron Device Letters*, 26 (2005) 787–789.

135. B.O. Lim, M.K. Lee, T.J. Baek, M. Han, S.C. Kim, J.-K. Rhee, 50-nm T-gate InAlAs/InGaAs metamorphic HEMTs with low noise and high f_T characteristics, *IEEE Electron Device Letters*, 28 (2007) 546–548.

136. S. Kim, O. Lim, H. Lee, T. Baek, D.-H. Shin, J. Rhee, 50 nm InGaAs/InAlAs/GaAs metamorphic high electron mobility transistors using double exposure at 50 kV electron-beam lithography without dielectric support, *Journal of Vacuum Science & Technology B: Microelectronics and Nanometer Structures Processing, Measurement, and Phenomena*, 22 (2004) 1807–1810.

137. J.Y. Shim, H.S. Yoon, D.M. Kang, J.Y. Hong, K.H. Lee, DC and RF characteristics of 60 nm T-gate MHEMTs with 53% indium channel, in: Indium phosphide & related materials, 2007. IPRM'07. *IEEE 19th International Conference on, IEEE*, 2007, pp. 445–446.

138. M. Kawano, T. Kuzuhara, H. Kawasaki, F. Sasaki, H. Tokuda, InAlAs/InGaAs metamorphic low-noise HEMT, *IEEE Microwave and Guided Wave Letters*, 7 (1997) 6–8.

139. N. Hara, K. Makiyama, T. Takahashi, K. Sawada, T. Arai, T. Ohki, M. Nihei, T. Suzuki, Y. Nakasha, M. Nishi, Highly uniform InAlAs-InGaAs HEMT technology for high-speed optical communication system ICs, *IEEE Transactions on Semiconductor Manufacturing*, 16 (2003) 370–375.

140. D.-H. Kim, B. Brar, J.A. del Alamo, f_T = 688 GHz and f_{max} = 800 GHz in L_g = 40 nm in 0.7 Ga 0.3 As MHEMTs with g_{m_max} > 2.7 mS/μm, in: *Electron devices Meeting (IEDM),2011 IEEE International, IEEE*, 2011, pp. 13.16. 11–13.16. 14.

141. H. Maher, I. El Makoudi, P. Frijlink, D. Smith, M. Rocchi, S. Bollaert, S. Lepilliet, G. Dambrine, A 200-GHz true E-mode low-noise MHEMT, *IEEE Transactions on Electron Devices*, 54 (2007) 1626–1632.

142. H. Maher, P. Baudet, I. El Makoudi, M. Périchaud, J. Bellaiche, M. Renviosé, U. Rouchy, P. Frijlink, A true E-mode MHEMT with high static and dynamic performance, in: *Proc. Tech. Dig. InP and Related Mater. Conf*, 2006, pp. 185–187.

143. K. Eisenbeiser, R. Droopad, J.-H. Huang, Metamorphic InAlAs/InGaAs enhancement mode HEMTs on GaAs substrates, *IEEE Electron Device Letters*, 20 (1999) 507–509.

144. D. Dumka, W. Hoke, P. Lemonias, G. Cueva, I. Adesida, High performance 0.35 m gate-length monolithic enhancement/depletion-mode metamorphic In0: 52 Al0: 48 As/In0: 53 Ga0: 47 As HEMTs on GaAs substrates, *IEEE Electron Device Letters*, 22 (2001).

145. D. Dumka, H. Tserng, M. Kao, E. Beam, P. Saunier, High-performance double-recessed enhancement-mode metamorphic HEMTs on 4-In GaAs substrates, *IEEE Electron Device Letters*, 24 (2003) 135–137.

146. M. Li, C.W. Tang, H. Li, K.M. Lau, Enhancement-mode Lg = 50 nm metamorphic InAlAs/InGaAs HEMTs on GaAs substrates with f_{max} surpassing 408 GHz, *Solid-State Electronics*, 99 (2014) 7–10.

147. W. Ha, K. Shinohara, B. Brar, Enhancement-mode metamorphic HEMT on GaAs substrate with 2 S/mm $ g_ {m} $ and 490 GHz $ f_ {T} $, *IEEE Electron Device Letters*, 29 (2008) 419–421.

148. M. Boudrissa, E. Delos, C. Gaquiere, M. Rousseau, Y. Cordier, D. Theron, J. Jaeger, Enhancement-mode Al/sub 0.66/In/sub 0.34/As/Ga/sub 0.67/In/sub 0.33/As metamorphic HEMT, modeling and measurements, *IEEE Transactions on Electron Devices*, 48 (2001) 1037–1044.

149. H. Li, C.W. Tang, K.J. Chen, K.M. Lau, Enhancement-mode metamorphic InAlAs/InGaAs HEMTs on GaAs substrates with reduced leakage current by CF4 plasma treatment, *Compound Semiconductor Mantech*, (2007).

150. K.-H. Su, W.-C. Hsu, C.-S. Lee, I.-L. Chen, Y.-J. Chen, C.-L. Wu, Comparative studies of δ-doped In0. 45Al0. 55As/ In0. 53Ga0. 47As/ GaAs metamorphic HEMTs with Au, Ti/ Au, Ni/ Au, and Pt/ Au Gates, *Journal of The Electrochemical Society*, 153 (2006) G996–G1000.

151. B. Sun, H. Chang, S. Wang, J. Niu, H. Liu, ALD Al2O3 passivation of Lg = 100 nm metamorphic InAlAs/InGaAs HEMTs with Si-doped Schottky layers on GaAs substrates, *Solid-State Electronics*, 138 (2017) 40–44.

152. C.-K. Lin, J.-C. Wu, W.-K. Wang, Y.-J. Chan, Characteristics of In/sub x/Al/sub 1-x/As/In/sub x/Ga/sub 1-x/As (x = 50%, 60%) metamorphic HEMTs on GaAs substrates, in: Indium Phosphide and Related Materials, 2004. *16th IPRM. 2004 International Conference on, IEEE*, pp. 205–208.

153. C.-C. Huang, T.-Y. Chen, C.-S. Hsu, C.-C. Chen, C.-I. Kao, W.-C. Liu, Comprehensive Temperature-Dependent Studies of Metamorphic High Electron Mobility Transistors With Double and Single $\delta $-Doped Structures, *IEEE Transactions on Electron Devices*, 58 (2011) 4276-4282.

154. E. Douglas, K. Chen, C. Chang, L. Leu, C. Lo, B. Chu, F. Ren, S. Pearton, InAlAs/InGaAs MHEMT degradation during DC and thermal stressing, in: *Reliability Physics Symposium (IRPS), 2010 IEEE International, IEEE*, 2010, pp. 818–821.

155. S. Chen, H. Chou, F. Chou, I. Hsieh, D. Tu, Y. Wang, C. Wu, S. Nelson, Reliability study of 0.15 um MHEMT with Vds≫ 3V bias for amplifier application, in: *ROCS Workshop, 2007. [Reliability of Compound Semiconductors Digest], IEEE*, 2007, pp. 47–63.

156. P. Marsh, C. Whelan, W. Hoke, R. Leoni Iii, T. Kazior, Reliability of metamorphic HEMTs on GaAs substrates, *Microelectronics Reliability*, 42 (2002) 997–1002.

157. M. Dammann, M. Chertouk, W. Jantz, K. Köhler, G. Weimann, Reliability of InAlAs/InGaAs HEMTs grown on GaAs substrate with metamorphic buffer, *Microelectronics Reliability*, 40 (2000) 1709–1713.

158. M. Dammann, A. Leuther, H. Konstanzer, W. Jantz, Effect of gate metal on reliability of metamorphic HEMTs, in: *GaAs Reliability Workshop, 2001. Proceedings, IEEE*, 2001, pp. 87–88.

159. M. Dammann, A. Leuther, R. Quay, M. Meng, H. Konstanzer, W. Jantz, M. Mikulla, reliability of 70 nm metamorphic HEMTs, *Microelectronics Reliability*, 44 (2004) 939–943.

160. N. Hayafuji, Y. Yamamoto, T. Ishida, K. Sato, Degradation mechanism of the AlInAs/GaInAs high electron mobility transistor due to fluorine incorporation, *Applied Physics Letters*, 69 (1996) 4075–4077.

161. N. Hayafuji, Y. Yamamoto, T. Ishida, K. Sato, Reliability improvement of AlInAs/GaInAs high electron mobility transistors by fluorine incorporation control, *Journal of the Electrochemical Society*, 145 (1998) 2951–2954.

162. A. Wakita, H. Rohdin, V. Robbins, N. Moll, C.-Y. Su, A. Nagy, D. Basile, Low-noise bias reliability of AlInAs/GaInAs modulation-doped field effect transistors with linearly graded low-temperature buffer layers grown on GaAs substrates, *Japanese Journal of Applied Physics*, 38 (1999) 1186.

163. M. Hsu, H. Chen, S. Chiou, W. Chen, G. Chen, Y. Chang, W. Lour, Gate-metal formation-related kink effect and gate current on In 0.5 Al 0.5 As/In 0.5 Ga 0.5 as metamorphic high electron mobility transistor performance, *Applied Physics Letters*, 89 (2006) 033509.

164. P.-H. Lai, R.-C. Liu, S.-I. Fu, Y.-Y. Tsai, C.-W. Hung, T.-P. Chen, C.-W. Chen, W.-C. Liu, Comprehensive study of thermal stability performance of metamorphic heterostructure field-effect transistors with Ti/Au and Au metal gates, *Journal of the Electrochemical Society*, 154 (2007) H205–H209.

165. C.-W. Chen, P.-H. Lai, W.-S. Lour, D.-F. Guo, J.-H. Tsai, W.-C. Liu, Temperature dependences of an In0. 46Ga0. 54As/In0. 42Al0. 58As based metamorphic high electron mobility transistor (MHEMT), *Semiconductor Science and Technology*, 21 (2006) 1358.

166. Y. Lin, C. Wu, W. Hsu, T. Wang, C. Lee, Y. Chen, High-temperature thermal stability performance in delta-doped In0. 425Al0. 575As/In0. 65Ga0. 35As metamorphic HEMT, *IEEE Electron Device Letters*, 26 (2005) 5–9.

167. P.-H. Lai, S.-I. Fu, C.-W. Hung, Y.-Y. Tsai, T.-P. Chen, C.-W. Chen, Y.-W. Huang, W.-C. Liu, On the temperature-dependent characteristics of metamorphic heterostructure field-effect transistors with different Schottky gate metals, *Semiconductor Science and Technology*, 22 (2007) 475.

168. J. Ajayan, D. Nirmal, A review of InP/InAlAs/InGaAs based transistors for high frequency applications, *Superlattices and Microstructures*, 86 (2015) 1–19.

169. J. Ajayan, D. Nirmal, 20-nm enhancement-mode metamorphic GaAs HEMT with highly doped InGaAs source/drain regions for high-frequency applications, *International Journal of Electronics*, 104 (2017) 504–512.

170. J. Ajayan, D. Nirmal, 20-nm T-gate composite channel enhancement-mode metamorphic HEMT on GaAs substrates for future THz applications, *Journal of Computational Electronics*, 15 (2016) 1291–1296.

171. J. Ajayan, D. Nirmal, P. Prajoon, J.C. Pravin, Analysis of nanometer-scale InGaAs/InAs/InGaAs composite channel MOSFETs using high-K dielectrics for high speed applications, *AEU-International Journal of Electronics and Communications*, 79 (2017) 151.

172. J. Ajayan, D. Nirmal, 20ánm high performance enhancement mode InP HEMT with heavily doped S/D regions for future THz applications, *Superlattices and Microstructures*, 100 (2016) 526–534.

173. J. Ajayan, D. Nirmal, 22 nm In0. 75Ga0. 25As channel-based HEMTs on InP/GaAs substrates for future THz applications, 半导体学报: 英文版, 38 (2017) 27–32.

174. J. Ajayan, D. Nirmal, S. Sivasankari, D. Sivaranjani, M. Manikandan, High speed low power Full Adder circuit design using current comparison based domino, in: *Devices, Circuits and Systems (ICDCS), 2014 2nd International Conference on, IEEE*, 2014, pp. 1–5.

175. J. Ajayan, T. Ravichandran, P. Mohankumar, P. Prajoon, J.C. Pravin, D. Nirmal, Investigation of DC-RF and breakdown behaviour in $L_g = 20$ nm novel asymmetric GaAs MHEMTs for future submillimetre wave applications, *AEU-International Journal of Electronics and Communications*, 84 (2018) 387–393.

176. J. Ajayan, T. Ravichandran, P. Prajoon, J.C. Pravin, D. Nirmal, Investigation of breakdown performance in $L_g = 20$ nm novel asymmetric InP HEMTs for future high-speed high-power applications, *Journal of Computational Electronics*, 17 (2018) 265–272.

177. J. Ajayan, T. Subash, D. Kurian, 20 nm high performance novel MOSHEMT on InP substrate for future high speed low power applications, *Superlattices and Microstructures*, (2017).

178. A.B. Amado-Rey, Y. Campos-Roca, F. van Raay, C. Friesicke, S. Wagner, H. Massler, A. Leuther, O. Ambacher, Analysis and development of submillimeter-wave stacked-FET power amplifier MMICs in 35-nm mHEMT technology, *IEEE Transactions on Terahertz Science and Technology*, 8 (2018) 357–364.

179. R. Weber, A. Tessmann, H. Massler, A. Leuther, U.J. Lewark, 600 GHz resistive mixer S-MMICs with integrated multiplier-by-six in 35 nm mHEMT technology, in: *Microwave Integrated Circuits Conference (EuMIC), 2016 11th European, IEEE*, 2016, pp. 85–88.

180. C.-C. Chiong, H.-M. Chen, J.-C. Kao, H. Wang, M.-T. Chen, 180–220 GHz MMIC amplifier using 70-nm GaAs MHEMT technology, in: *Radio-Frequency Integration Technology (RFIT), 2016 IEEE International Symposium on, IEEE*, 2016, pp. 1–4.

181. M. Schlechtweg, A. Tessmann, A. Leuther, H. Massler, G. Moschetti, M. Rösch, R. Weber, V. Hurm, M. Kuri, M. Zink, Advanced building blocks for (Sub-) millimeter-wave applications in space, communication, and sensing using III/V mHEMT technology, in: *Millimeter Waves (GSMM) & ESA Workshop on Millimetre-Wave Technology and Applications, 2016 Global Symposium on, IEEE*, 2016, pp. 1–4.

182. T. Messinger, J. Antes, S. Wagner, A. Leuther, I. Kallfass, Wideband 200 GHz injection-locked frequency divide-by-two MMIC in GaAs mHEMT technology, in: *Microwave Symposium (MMS), 2015 IEEE 15th Mediterranean, IEEE*, 2015, pp. 1–4.

183. M. Zaknoune, Y. Cordier, S. Bollaert, Y. Druelle, D. Theron, Y. Crosnier, a High performance metamorphic In0. 32Al0. 68As/In0. 33Ga0. 67 As HEMTs on GaAs substrate with an inverse step InAlAs metamorphic buffer, in: *Device Research Conference Digest*, pp. 34.

184. Y. Cordier, P. Lorenzini, J.-M. Chauveau, D. Ferre, Y. Androussi, J. DiPersio, D. Vignaud, J.-L. Codron, Influence of growth conditions on the structural, optical and electrical quality of MBE grown InAlAs/InGaAs metamorphic HEMTs on GaAs, in: *Molecular Beam Epitaxy, 2002 International Conference on, IEEE*, 2002, pp. 71–72.

185. D. Edgar, N. Cameron, H. McLelland, M. Holland, M. Taylor, I. Thayne, C. Stanley, S. Beaumont, Metamorphic GaAs HEMTs with fT of 200 GHz, (1999).

186. M. Chertouk, H. Heiß, D. Xu, S. Kraus, W. Klein, G. Bohm, G. Trankle, G. Weimann, Metamorphic InAlAs/InGaAs HEMTs on GaAs substrates with composite channels and f/sub max/of 350 GHz, in: Indium phosphide and related materials, 1995. In *Conference Proceedings, Seventh International Conference on, IEEE*, 1995, pp. 737–740.

187. K. Joo, S. Chun, J. Lim, J. Song, J. Chang, Metamorphic growth of InAlAs/InGaAs MQW and InAs HEMT structures on GaAs, *Physica E: Low-Dimensional Systems and Nanostructures*, 40 (2008) 2874–2878.

188. A. Leuther, R. Weber, M. Dammann, M. Schlechtweg, M. Mikulla, M. Walther, G. Weimann, Metamorphic 50 nm InAs-channel HEMT, in: *Indium Phosphide and Related Materials, 2005. International Conference on, IEEE*, 2005, pp. 129–132.

189. M. Dammann, F. Benkhelifa, M. Meng, W. Jantz, Reliability of metamorphic HEMTs for power applications, *Microelectronics Reliability*, 9 (2002) 1569–1573.

190. A. Leuther, A. Tessmann, H. Massler, R. Aidam, M. Schlechtweg, O. Ambacher, 450 GHz amplifier MMIC in 50 nm metamorphic HEMT technology, in: *Indium Phosphide and Related Materials (IPRM), 2012 International Conference on*, IEEE, 2012, pp. 229–232.

191. A. Leuther, A. Tessmann, P. Doria, M. Ohlrogge, M. Seelmann-Eggebert, H. Maßler, M. Schlechtweg, O. Ambacher, 20 nm Metamorphic HEMT technology for terahertz monolithic integrated circuits, in: *European Microwave Integrated Circuit Conference (EuMIC), 2014 9th*, IEEE, 2014, pp. 84–87.

192. J.-C. Lin, P.-Y. Yang, W.-C. Tsai, Simulation and analysis of metamorphic high electron mobility transistors, *Microelectronics Journal*, 38 (2007) 251–254.

193. Y. Zeng, X. Cao, L. Cui, M. Kong, L. Pan, B. Wang, Z. Zhu, High-quality metamorphic HEMT grown on GaAs substrates by MBE, *Journal of Crystal Growth*, 227 (2001) 210–213.

194. H. Ono, S. Taniguchi, T.-K. Suzuki, High-frequency characteristics and saturation electron velocity of InAlAs/InGaAs metamorphic high electron mobility transistors at high temperatures, in: *Indium Phosphide and Related Materials, 2004. 16th IPRM. 2004 International Conference on*, IEEE, 2004, pp. 288–291.

195. B. Amado-Rey, Y. Campos-Roca, C. Friesicke, A. Tessmann, R. Lozar, S. Wagner, A. Leuther, M. Schlechtweg, O. Ambacher, A 280 GHz stacked-FET power amplifier cell using 50 nm metamorphic HEMT technology, in: *Microwave Integrated Circuits Conference (EuMIC), 2016 11th European*, IEEE, 2016, pp. 189–192.

196. F. Thome, O. Ambacher, A 50-nm gate-length metamorphic HEMT distributed power amplifier MMIC based on stacked-HEMT unit cells, in: *Microwave Symposium (IMS), 2017 IEEE MTT-S International*, IEEE, 2017, pp. 1695–1698.

197. F. Thome, A. Leuther, H. Massler, M. Schlechtweg, O. Ambacher, Comparison of a 35-nm and a 50-nm gate-length metamorphic HEMT technology for millimeter-wave low-noise amplifier MMICs, in: *Microwave Symposium (IMS), 2017 IEEE MTT-S International*, IEEE, 2017, pp. 752–755.

198. L.A. Samoska, An overview of solid-state integrated circuit amplifiers in the submillimeter-wave and THz regime, *IEEE Transactions on Terahertz Science and Technology*, 1 (2011) 9–24.

199. R. Appleby, R.N. Anderton, Millimeter-wave and submillimeter-wave imaging for security and surveillance, *Proceedings of the IEEE*, 95 (2007) 1683–1690.

200. S. Koenig, D. Lopez-Diaz, J. Antes, F. Boes, R. Henneberger, A. Leuther, A. Tessmann, R. Schmogrow, D. Hillerkuss, R. Palmer, Wireless sub-THz communication system with high data rate enabled by RF photonics and active MMIC technology, in: *Photonics Conference (IPC), 2014 IEEE*, IEEE, 2014, pp. 414–415.

201. K.B. Cooper, R.J. Dengler, N. Llombart, B. Thomas, G. Chattopadhyay, P.H. Siegel, THz imaging radar for standoff personnel screening, *IEEE Transactions on Terahertz Science and Technology*, 1 (2011) 169–182.

202. A. Leuther, A. Tessmann, I. Kallfass, R. Losch, M. Seelmann-Eggebert, N. Wadefalk, F. Schafer, J.G. Puyol, M. Schlechtweg, M. Mikulla, Metamorphic HEMT technology for low-noise applications, in: *Indium Phosphide & Related Materials, 2009. IPRM'09. IEEE International Conference on*, IEEE, 2009, pp. 188–191.

203. W. Ciccognani, E. Limiti, P.E. Longhi, M. Renvoise, MMIC LNAs for radioastronomy applications using advanced industrial 70 nm metamorphic technology, *IEEE Journal of Solid-State Circuits*, 45 (2010) 2008–2015.

204. I. Vurgaftman, J.R. Meyer, L.R. Ram Mohan, Band parameters for III–V compound semiconductors and their alloys, *Applied Physics Review*, 89(11) (2001) 5815–5864.

205. J.-H. Huang, T.Y. Chang, Flat band current voltage temperature method for band-discontinuity determination and its application to strained $In_x Ga_{1-x} As/In_{0.52} Al_{0.48} As$ hetero-structures, *Journal of Applied Physics*, 76(5) (1994) 2893–2903.

206. Y.P. Varshni, Temperature dependence of the energy gap in semiconductors, *Physica*, 34 (1967) 149–154.

207. W. Li-Dan, D. Peng, S. Yong-Bo, C. Jiao, Z. Bi-Chan, J. Zhi, 100-nm T-gate InAlAs/InGaAs InP-based HEMTs with $f_T = 249$ GHz and $f_{max} = 415$ GHz, *Chinese Physics B*, 23 (2014) 038501.

208. T. Takahashi, M. Sato, Y. Nakasha, T. Hirose, N. Hara, Improvement of RF and noise character-
     istics using a cavity structure in InAlAs/InGaAs HEMTs, *IEEE Transactions on Electron Devices*,
     59 (2012) 2136–2141.
209. T. Suemitsu, T. Enoki, M. Tomizawa, N. Shigekawa, Y. Ishii, Mechanism and structural
     dependence of kink phenomena in InAlAs/InGaAs HEMTs, in: *Indium Phosphide and Related
     Materials, 1997. International Conference on*, IEEE, 1997, pp. 365–368.
210. T. Suemitsu, T. Enoki, N. Sano, M. Tomizawa, Y. Ishii, An analysis of the kink phenomena
     in InAlAs/InGaAs HEMT's using two-dimensional device simulation, *IEEE Transactions on
     Electron Devices*, 45 (1998) 2390–2399.
211. T. Suemitsu, H. Yokoyama, Y. Umeda, T. Enoki, Y. Ishii, High-performance 0.1-μm gate
     enhancement-mode InAlAs/InGaAs HEMT's using two-step recessed gate technology, *IEEE
     Transactions on Electron Devices*, 46 (1999) 1074–1080.

15

Metal Oxide Semiconductor High Electron Mobility Transistors

D. K. Panda, G. Amarnath, and T. R. Lenka

CONTENTS

15.1 Introduction: Background

One of the major factors that limit the performance and reliability of GaN high electron mobility transistors (HEMTs) for high-power radio-frequency (RF) applications is their relatively high gate leakage current that reduces the breakdown voltage and the power-added efficiency while increasing the noise figure (NF).

In order to suppress the gate leakage current, the metal–insulator–semiconductor (MIS) structure can be introduced to the GaNHEMT gate to form the MIS-HEMT. Since a broad category of insulation dielectrics for semiconductor devices includes oxides, the insulated gate is mostly of the metal oxide semiconductor (MOS) structure, thus forming the MOS-HEMT. In this chapter, all the HEMTs with oxide as the gate dielectric are referred to as MOS-HEMTs, with MIS-HEMTs as the general designation of insulated-gate HEMTs. Experiments indicate that GaN MOS-HEMT can improve the gate voltage swing, microwave power performance, and long-term reliability. A brief introduction of GaN MOS-HEMTs with different structures, different gate material system along with the different DC and RF performance of MOS-HEMT are given in this chapter.

15.2 MOS-HEMT Device Structure

The schematic diagram of the GaNMOS-HEMT with various barrier layers of AlInN, AlGaN and AlN is shown in Figure 15.1. The structure consists of Ni metal gate followed by Al_2O_3 as oxide layer of thickness varying from 1 to 5 nm. An unintentionally doped

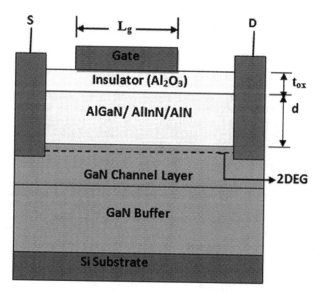

FIGURE 15.1
MOS-HEMT structure.

GaN channel of thickness 500 nm and a GaN semi-insulating buffer layer of thickness 1 μm is considered. In order to study in detail, various materials such as AlInN and AlN are considered as barrier layer along with conventional AlGaN layer as follows.

For an unintentionally doped AlGaN/GaNMOS-HEMT structure the two-dimensional electron gas (2DEG) is formed largely as a result of piezoelectric and spontaneous polarization effects, which arise at the heterointerface of AlGaN and GaN layer. But for scaled devices as the oxide thickness and gate lengths are reduced, barrier thickness scaling is necessary. While attempting to reduce the thickness of the AlGaN barrier to overcome short-channel effects, the sheet carrier density decreases due to proximity of the heterointerface to the negatively charged surface [1]. In order to further increase the 2DEG density as well as the breakdown field in AlGaN structure high Al mole fraction is desirable to increase the strength of polarization. But high Al content will result in poor transport properties in AlGaN/GaN heterostructure [2].

In order to solve this problem, a thin lattice-matched $Al_{0.83}In_{0.17}N$ layer is grown over GaN, which eliminates the strain present in conventional AlGaN-based MOS-HEMT. The bandgap of the AlInN is larger than that of a typical AlGaN and larger spontaneous polarization charge as well. The larger bandgap results in an enhancement of carrier confinement in the well, thus keeping the output resistance of the device high. The larger number of carriers those tend to accumulate in the 2DEG results in a larger sheet carrier density and correspondingly larger current and power densities than conventional AlGaN/GaN system. But in case of AlInN heterostructure, local in-homogeneity in the composition yields regions of increased alloy scattering due to enhanced penetration of the electron wave function into the barrier layer. Therefore, the AlInN barrier layer would be the best if it is uniform [3].

To overcome these problems, AlN/GaN MOS-HEMT can be an alternative candidate for high-power applications. The 2DEG which forms near the AlN/GaN interface can reach up to a concentration of 3×10^{13} cm^{12} for an extremely thin AlN barrier layer even less than 5 nm thickness along with high mobility greater than 1000 $cm^2/V.s$ [4]. However, surface sensitivity and high leakage current in AlN-based device may result poor device performance [5].

15.3 Research Progress of GaNMOS-HEMTs and Different Material System

The introduction of a dielectric between the Schottky-gate and the semiconductor of the GaNHEMT to form an MIS-HEMT can greatly reduce the gate leakage current, thus a higher breakdown voltage and a large range of gate bias within which the device can operate. Both the positive and negative gate voltage swing of the MIS-HEMT increase, owing to the negative shift of the threshold voltage induced by the increased gate-to-channel spacing and gate forward turn-on voltage by the introduction of the gate dielectric. Although there is a drop in the transconductance peak value, the transconductance remains relatively constant within a relatively large range of gate voltage with greatly improved linearity, for which reason MIS-HEMTs enjoy good linearity in microwave power applications. The gate dielectric can also serve as the surface passivation layer to stabilize the surface states of the semiconductor material, thus resulting in effective suppression of current collapse, improvement of the microwave power properties, and long-term reliability of the device.

The first GaN MOS-HEMTs were developed by Khan et al. using SiO_2 deposited by plasma-enhanced chemical vapor deposition [6]. The SiO_2 insulated gate MOS-HEMT and SiN insulated gate MIS-HEMT have a gate current 4–6 orders of magnitude lower than that of the HEMT structure, with stable performance at a forward gate voltage up to 6 V and a temperature up to 300°C [7]. At an even higher temperature of 400°C, the MOS-HEMT devices remain stable under RF (2 GHz) power stress [8]. Researches on field-plated MOS-HEMTs show that the MOS-HEMT devices maintain relatively high-power output and low gate leakage current with long-term RF stress, whereas the HEMTs exhibit visible degradation in power and gate leakage current, which are closely related to the increase of the forward gate current with the stress time. Consequently, MOS-HEMTs have better RF reliability than HEMTs. The field-plated MOS-HEMTs even achieve an output power density of 19 W/mm at 2 GHz and $V_{DS} = 55$ V, with a significantly higher power voltage efficiency (PVE) (the ratio of RF power density to drain voltage) than the HEMT.

Besides good power characteristics, GaN MIS-HEMTs can also exhibit outstanding frequency characteristics. Kuzmik et al. [9] adopted 2 nm Si_3N_4 dielectric layer prepared by catalytic chemical vapor deposition (Cat-CVD) to successfully produce 60 nm gate length MIS-HEMTs with a cut-off frequency f_T of 163 GHz [10]. The MIS-HEMT with a gate length of 30 nm exhibits an f_T of 181 GHz and an f_{max} of 186 GHz [11].

The gate dielectric with a higher dielectric constant can increase the gate capacitance, thereby leading to a better control of the channel charge and less deleterious effect of the gate dielectric introduction on transconductance, as well as a notable advantage in device scaling. Besides SiO_2 (dielectric constant, $\varepsilon = 3.9$) and SiN ($\varepsilon = 7.0$), Al_2O_3 ($\varepsilon = 9.0$) is also widely employed as GaN MIS-HEMT gate dielectrics. Al_2O_3 dielectric layers can be prepared by the oxidization of aluminum deposited on nitride material surface [12] or the ALD [13]. Apart from current collapse suppression and improved gate voltage swing, breakdown voltage, and current output, GaN MOS-HEMTs with Al_2O_3 as gate dielectric are found to have a higher channel electron mobility and even a greater electron saturation velocity than HEMTs, which are attributed to the passivation effect of the Al_2O_3 dielectric layer, offering better transport properties, hence resulting in the occurrence of an even higher MOS-HEMT transconductance than that of the HEMT [14]. Nicollian and Brews [15] investigated the temperature dependence of the forward gate leakage current in MOS-HEMT with Al_2O_3 gate dielectric layer and found that the Fowler-Nordheim tunnelling is the dominant mechanism in high electric field and low temperature (<0°C) operations, whereas the trap-assisted tunnelling plays a major role under moderate electric field and high temperature (>0°C) conditions [16]. In 2011, we reported recessed-gate MOS-HEMTs using Al_2O_3 gate dielectric with an output power density of 13 W/mm and a record-high power-added efficiency (PAE) of 73% under a drain voltage $V_{DS} = 45$ V [17].

Some high-K dielectrics with a dielectric constant greater than 20 have also found successful applications in GaN MIS-HEMTs. We put forward MOS-HEMTs prepared by the deposition of stack gate HfO_2 (3 nm)/Al_2O_3 (2 nm) dielectric using atomic layer deposition (ALD). This dielectric incorporates the high-K properties of HfO_2 and the favourable interface properties of Al_2O_3, and besides the advantages of conventional MOS-HEMTs, the prepared device with 1 μm gate length also exhibits excellent frequency characteristics, with an f_T and f_{max} of 12 and 34 GHz, respectively. The direct employment of the HfO_2 dielectric layer also offers good passivation effect [18]. A breakdown voltage of 400 V was achieved in the MIS-HEMTs prepared by Fujitsu Limited using 10 nm sputtered Ta_2O_5 gate dielectric, and the 1 mm gate width device achieved a power density of 9.4 W/mm, an efficiency of 62.5%, and a linear gain of 23.5 dB at 2 GHz. The output power degradation was less than 0.5 dB after 150 h stress test under P3dB (the output power corresponding to the gain 3 dB lower than the maximum) operation [19]. The feasibility of using ZrO_2 and Pr_2O_3 as gate dielectrics for GaN MOS-HEMTs has also been verified [20].

The gate dielectrics Sc_2O_3 [21], NiO [22], low temperature GaN dielectrics [23], and AlN [24] have also been reported for GaN MOS-HEMTs. In recent years, there have been increasing studies on the application of the MIS structure in N-polar GaNHEMTs. N-polar GaN heterostructures possess lower ohmic contact resistance and better quantum confinement but suffer from substantial point defects in GaN and thus a large gate leakage current; therefore, the introduction of the MIS structure can greatly improve the device properties. Currently, the typical gate dielectric for N-polar MIS-HEMTs is SiN, and a PAE of 71% and an output power of 6.4 W/mm [25] have been reported. MIS-HEMTs have become a very important category of nitride semiconductor electronic devices.

15.4 Selection of High-K Gate Dielectrics

A high-K dielectric is defined as the dielectric whose permittivity K is greater than that of the SiO_2 ($K = 3.9$). As the earliest gate dielectric chosen for GaN MOS-HEMTs, SiO_2 enjoys mature preparation technologies and a good insulating ability but suffers a low dielectric constant, which is a major drawback. The improvement of transconductance by using high-K dielectrics is in favour of better GaN MIS-HEMT frequency characteristics and a greater frequency range in microwave power applications.

The basic properties of common gate dielectric materials are listed in Table 15.1. Among the widely employed gate dielectrics in GaN MIS-HEMTs are Si_3N_4 and Al_2O_3. With a comparatively low permittivity of 7 among the high-K dielectrics, Si_3N_4 is extensively employed as passivation layer for nitride devices with rather mature preparation process. Al_2O_3 dielectrics of the same thickness are superior in both gate capacitance and insulating ability to Si_3N_4, with the only drawback of a poorer passivation effect.

TABLE 15.1

Basic Properties of Common Gate Dielectric Materials

Gate Dielectric	Dielectric Constant	Bandgap(ev)	Crystal Structure
SiO_2	3.9	8.9	Amorphous
Si_3N_4	7	5.1	Amorphous
Al_2O_3	9	8.7	Amorphous
SiON	3.9–7	—	Amorphous
HfO_2	25	5.9	Tetragonal
TiO_2	80	3.5	Tetragonal
ZrO_2	25	5.8	Tetragonal
Y_2O_3	15	5.9	Cubic
La_2O_3	30	4.3	Hexagonal
Ta_2O_5	26	4.3	Orthogonal

A good growth method for Al_2O_3 dielectric is the ALD, which is a very important technique for high-K film preparation. There are also a few reports on the GaN MIS-HEMT prepared using the high-K dielectrics, such as HfO_2. Owing to its very high permittivity of 25, a 20-nm-thick HfO_2 layer is equivalent to a 7.2-nm-thick Al_2O_3 dielectric under identical gate capacitance. A thicker dielectric offers better insulation and less difficulty in thickness control. However, the crystallization of a part of the HfO_2 dielectric may occur when it is exposed to high temperature, owing to its relatively low crystallization temperature (375°C), thus inducing device degradation. Presently, in the Hf-based dielectric researches, the addition of other elements such as aluminum or silicon into HfO_2 is adopted to increase its stability, but the complicated process and increased defects in the dielectric pose many problems that are yet to be solved.

According to Table 15.1, although with large dielectric constants, high-K gate dielectrics generally have relatively small bandgaps, leading to a large leakage current due to small conduction band discontinuity, when used as gate dielectrics in direct contact to the AlGaN barrier layer. This, coupled with the interface problem between the high-K dielectric and AlGaN barrier layer, as well as the surface passivation characteristics, restricts their applications. Therefore, the structure design of the high-K gate dielectrics becomes the key to applications of the dielectrics in AlGaN/GaN MOS-HEMTs.

15.5 DC and RF Characteristics of GaN MOS-HEMT

In this section the different DC and RF characteristics GaN MOS-HEMT are analysed and performance of the device is compared with AlGaN/GaN HEMT having same gate length.

Figure 15.2 shows the output current-voltage (I-V) characteristics of the AlGaN/GaN HEMT and MOS-HEMT. The Schottky-gate device has a maximum drain current of 423 mA/mm^2 at $V_{GS} = 0$, while the MOS-HEMT devices have 539 mA/mm^2 drain currents, respectively. Besides, the HEMTs and MOS-HEMTs were completely pinched-off at a gate voltage of –5 and –6.7 V, respectively.

Figure 15.3 shows the I_{DS} versus V_{GS} curves of HEMT and MOS-HEMT devices. From a comparison of these device performances, it can be seen that the HEMTs have lower I_{DS}

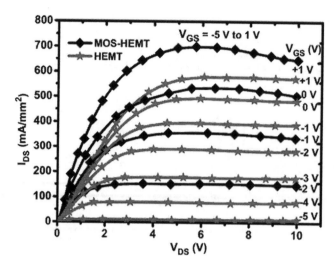

FIGURE 15.2
DC I_D versus V_{DS} characteristics at $V_{GS} = 1$ to -6 V of the AlGaN/GaN HEMT and Al_2O_3 MOS-HEMT.

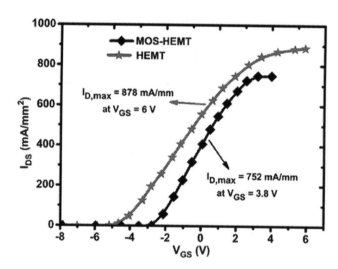

FIGURE 15.3
I_{DS} versus V_{GS} curve for the AlGaN/GaN HEMT and Al_2O_3 MOS-HEMT at the V_{DS} bias is 7 V.

of 752 mA/mm² at $V_{GS} = 3.8$ V, but for MOS-HEMTs, its reaches 878 mA/mm² at 6 V gate bias. In this sense, the good quality of both Al_2O_3 and Al_2O_3/HEMT interface has rendered a higher applicable gate bias, which result in a higher driving current capacity of MOS-HEMTs as compared to HEMTs. Moreover, the drain current at the same gate bias is also higher for MOS-HEMT. This difference arises, thereby making the MOS-HEMT channel depletion for the same gate voltage smaller than that for the HEMT.

In Figure 15.4 a slight transconductance decrease in MOS-HEMTs compared to HEMTs from 173 to 135 mS/mm² was observed, which is consistent with a further separation between the control gate and the 2DEG channel with the presence of an additional Al_2O_3 layer in MOS-HEMTs. However, due to the high dielectric constant of Al_2O_3, the degradation in $g_{m, max}$ of MOS-HEMT is only 21.7% relative to that of HEMT.

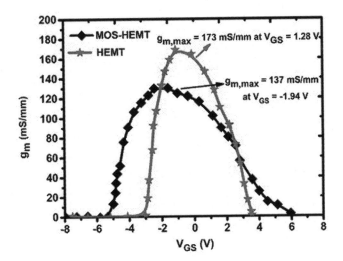

FIGURE 15.4
Transconductance G_m versus gate-source bias V_{GS} at the same drain bias $V_{DS} = 7$ V in the saturation region.

Figure 15.5 shows the gate leakage performance of the both HEMTs and MOS-HEMTs with the same device dimensions, from which the leakage current of MOS-HEMTs is found to be significantly lower than that of the Schottky-gate HEMTs. The gate leakage current density of MOS-HEMTs is almost 3 orders of magnitude lower than that of the HEMTs. Such a low gate leakage current should be attributed to the large band offsets in the Al_2O_3/HEMT and a good quality of both the reactive-sputtered Al_2O_3 dielectric and the Al_2O_3/HEMT interface. This tends to an increase of the two terminal reverse breakdown voltage (about 25%) and of the forward breakdown voltage (about 30%). This confirms that the Al_2O_3 dielectric thin film acts as an efficient gate insulator.

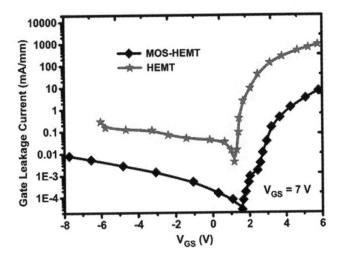

FIGURE 15.5
Gate leakage currents for the AlGaN/GaN HEMT and Al_2O_3 MOS-HEMT with the same device dimensions.

In order to investigate the breakdown behaviour of Al_2O_3- gate dielectric device, the off-state three-terminal drain-source breakdown characteristics of the HEMT and Al_2O_3MOS-HEMT were simulated, the results are as shown in Figure 15.6; the devices were simulated at gate voltage V_{GS} of −8 V. The breakdown voltage BVDS is defined as the drain voltage at a gate current of 1 mA/mm², which is consistent with the rapidly increased currents caused by avalanche breakdown. The Al_2O_3 MOS-HEMT shows a higher breakdown voltage as compared to the conventional HEMT. The high breakdown voltage is related to the utilization of the Al_2O_3 gate dielectric to reduce the leakage current.

Figure 15.7 shows I_{DS} versus V_{GS} transfer curves for GaN MOS-HEMTs and HEMTs with the different drain voltages from 4 to 7 V. With increasing the drain voltage, both of the HEMT and MOS-HEMT devices have higher maximum drain current, except the HEMT at V_{DS} is 7 V. In addition, at forward gate bias beyond +2 V, high drain current drops for the Schottky-gate HEMT was observed as compared to the MOS-HEMT. This is because the high Schottky-gate leakage current of HEMT, with results in the degradation of the I_D. On the other hand, the slope of I_D curve of MOS-HEMT is lower than the conventional HEMT; however, when increasing the gate bias to the positive voltage, the drain current increases

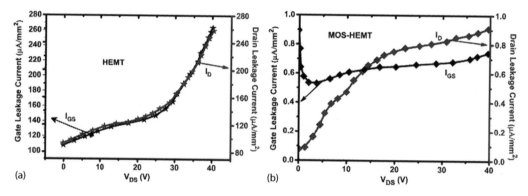

(a) (b)

FIGURE 15.6
Off-state drain-source breakdown characteristics of regular-HEMTs (a) and MOS-HEMT (b).

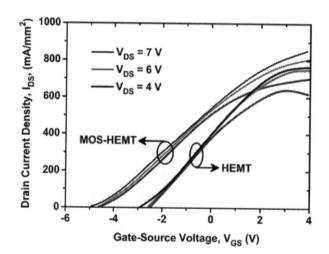

FIGURE 15.7
I_{DS} versus V_{GS} curve for the AlGaN/GaN HEMT and MOS-HEMT at the V_{DS} bias is from 4 to 7 V.

at a stable rate in a large gate bias region. This is because Al_2O_3-gate dielectric with larger bandgap that can afford much higher forward gate bias.

The characteristics of the G_m dependence on the gate-bias of Al_2O_3-gate and Schottky-gate AlGaN/GaN HEMTs with the different drain voltages from 4 to 7 V are shown in Figure 15.8. With the increase of the drain voltage, both of the HEMT and MOS-HEMT devices show almost have the similar maximum transconductance. The maximum drain current depends G_m versus V_{GS} curve, the MOS-HEMT device has a lower maximum G_m value, but a flatter G_m distribution as compared to that of regular HEMT. It represents that the drain current increased in a stable rate in a wider range of gate bias region, this may be due to that the gate leakage current was suppressed in the MOS-HEMT.

At high frequencies, the HEMTs are characterized by two important parameters: the cut-off frequency f_T for which the modulus of the current gain is equal to 1 (0 dB), and the maximum oscillation frequency f_{max} for which the unilateral power gain is equal to 1 (0 dB) [7]. Results are given in Figure 15.9. f_T (HEMT) 8.38 GHz f_{max} (HEMT) 13.52 GHz f_T(MOS-HEMT) 13.59 GHz f_{max} (MOS-HEMT) 28.07 GHz.

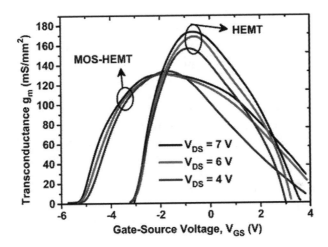

FIGURE 15.8
G_m versus V_{GS} curve for the AlGaN/GaNHEMT and MOS-HEMT at the V_{DS} bias is from 4 to 7 V.

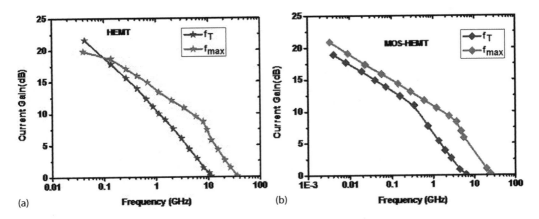

FIGURE 15.9
Current gain as a function of frequency for AlGaN/GaN HEMT (a) and MOS-HEMT (b).

15.6 Advantages of MOS-HEMT

As the state-of-the-art in compound semiconductor device technology advances, microwave devices used in high power electromagnetic and military radar systems are being altered with semiconductor power amplifiers. High-power density, cut-off frequency, maximum-oscillation frequency and low-noise figure are essential factors for high power/high frequency microwave devices enabling compact die sizes and more simply realized input and output matching networks. Wide bandgap semiconductor microwave devices like GaN-based MOS-HEMTs have potential to operate at power densities more times larger than GaAsFET, Si-LDMOS, and silicon carbide (SiC) devices. MOS-HEMTs have other advantages like high breakdown voltages (>200 V), high saturated electron velocity, good thermal conductivity, low parasitic capacitances and low turn-on resistances, high cut-off frequencies. Due to the high breakdown voltages (>200 V), MOS-HEMTs can operate in situations that are not readily viable with other device technologies, i.e., low output capacitance per unit power (resulting from high-power density), high drain operating voltage, good thermal dissipation, and high peak efficiency.

Also, these GaN-based MOS-HEMTs have been considered to be the prime choice in order to realize superior performance in high power amplifiers (HPAs) designs. Recently, these HPAs have become essential for wider bandwidth operation and high linearity operations at lower power consumption. While LDMOS or HEMT have initially been broadly used as high-power amplifier devices, MOS-HEMT provides the following advantages. First, higher power-added efficiency (PAE), which saves electrical power consumption and reduce the cost and size of HPAs due to the small amount of heat dissipation. Next, high operating voltage, MOS-HEMT operates at higher supply voltage similar to power feeder voltage range which is generally used for radar and satellite communication equipment. Generally, the design of amplifier becomes more critical as the low device impedances. GaN-based MOS-HEMTs show higher impedance than different devices. Thus, the HPA designers can utilize the advantages of MOS-HEMTs to increase the performance of HPAs with wide coverage of frequency bands and higher PAE with respect to the required HPA performance.

References

1. Higashiwaki, M., T. Matsui and T. Mimura. 2006. AlGaN/GaN MIS-HEMTs with f_T of 13 GHz using Cat-CVD SiN gate-insulating and passivation layers. *IEEE Electron Device Letters* 27(1):16.
2. Nakamura, F., S. Hashimoto, M. Hara, et al. 1998. AlN and AlGaN growth using low-pressure metal-organic chemical vapor deposition. *Journal of Crystal Growth* 195(4):280.
3. Medjdoub, F., M. Alomari, J. F. Carlin, et al. 2008. Barrier layer scaling of InAlN/GaN HEMTs. *IEEE Electron Device Letters* 95(5):422.
4. Dabiran, A. M., A. M. Wowchak, A. Osinsky, et al. 2008. Very high channel conductivity in low-defect AlN/GaN high electron mobility transistor structures. *Applied Physics Letters* 93(8):082111.
5. Taking, S., A. Banerjee, H. Zhou, X. Li, A. Khokhar, R. Oxland, I. McGregor, S. Bentley, F. Rahman, I. Thayne, et al. 2010. Surface passivation of AlN/GaN MOS-HEMTs using ultra-thin Al_2O_3 formed by thermal oxidation of evaporated aluminium. *Electronics Letters* 46(4):301–302.

6. Khan, M. A., X. Hu, G. Sumin, A. Lunev, J. Yang, R. Gaska, and M. S. Shur. 2000. AlGaN/GaN metal oxide semiconductor heterostructure field effect transistor. *IEEE Electron Device Letters* 21(2):63–65.

7. Khan, M. A., G. Simin, J. Yang, J. Zhang, A. Koudymov, M. S. Shur, R. Gaska, X. Hu, and A. Tarakji. 2003. Insulating gate III-N heterostructure field-effect transistors for high power microwave and switching applications. *IEEE Transactions on Microwave Theory and Techniques* 51(2):624–633.

8. Kordos, P., D. Gregusova, R. Stoklas, K. Cico, and J. Novak. 2007. Improved transport properties of Al_2O_3/AlGaN/GaN metal-oxide-semiconductor heterostructure field-effect transistor. *Applied Physics Letters* 90(12):123513-1.

9. Kuzmik, J., G. Pozzovivo, S. Abermann, J.-F. Carlin, M. Gonschorek, E. Feltin, N. Grandjean, E. Bertagnolli, G. Strasser, and D. Pogany. 2008. Technology and performance of InAlN/AlN/GaN HEMTs with gate insulation and current collapse suppression using ZrO_2 or HfO_2. *IEEE Transactions on Electron Devices* 55(3):937–941.

10. Liu, Z. H., G. I. Ng, S. Arulkumaran, Y. K. T. Maung, and H. Zhou. 2011. Temperature dependent forward gate current transport in atomic-layer-deposited Al_2O_3/AlGaN/GaN metal-insulator-semiconductor high electron mobility transistor. *Applied Physics Letters* 98(16):163501 (3 p.).

11. Marso, M., G. Heidelberger, K. M. Indlekofer, J. Bernat, A. Fox, P. Kordo, and H. Luth. 2006. Origin of improved RF performance of AlGaN/GaN MOSHFETs compared to HFETs. *IEEE Transactions on Electron Devices* 53(7):1517–1523.

12. Matulionis, A. 2004. Comparative analysis of hot-phonon effects in nitride and arsenide channels for HEMTs. Device Research Conference—Conference Digest, 62nd DRC, June 21–23, 2004, Notre Dame, IN.

13. Mehandru, R., B. Luo, J. Kim, F. Ren, B. P. Gila, A. H. Onstine, C. R. Abernathy, et al. 2003. AlGaN/GaN metal-oxide-semiconductor high electron mobility transistors using Sc_2O_3 as the gate oxide and surface passivation. *Applied Physics Letters* 82(15):2530–5232.

14. Nagahara, M., T. Kikkawa, N. Adachi, Y. Tateno, S. Kato, M. Yokoyama, S. Yokogawa, et al. 2002. Improved intermodulation distortion profile of AlGaN/GaN HEMT at high drain bias voltage. 2002 IEEE International Devices Meeting (IEDM), December 8–11, 2002, San Francisco, CA.

15. Nicollian, E. H. and J. R. Brews. 2002. *MOS (Metal Oxide Semiconductor) Physics and Technology.* John Wiley & Sons, Hoboken, NJ.

16. Oh, C. S., C. J. Youn, G. M. Yang, K. Y. Lim, and J. W. Yang. 2004. AlGaN/GaN metal-oxide-semiconductor heterostructure field-effect transistor with oxidized Ni as a gate insulator. *Applied Physics Letters* 85(18):4214–4216.

17. Palacios, T., S. Rajan, L. Shen, A. Chakraborty, S. Heikman, S. Keller, S. P. DenBaars, and U. K. Mishra. 2004. Influence of the access resistance in the rf performance of mm-wave AlGaN/GaN HEMTs. Device Research Conference, June 21–23, 2004, Piscataway, NJ.

18. Pozzovivo, G., J. Kuzmik, S. Golka, W. Schrenk, G. Strasser, D. Pogany, K. Cico, et al. 2007. Gate insulation and drain current saturation mechanism in InAlN/GaN metal-oxide semiconductor high-electron-mobility transistors. *Applied Physics Letters* 91(4):043509-1.

19. Romero, M. F., A. Jimenez, J. Miguel-Sanchez, A. F. Brana, F. Gonzalez-Posada, R. Cuerdo, F. Calle, and E. Munoz. 2008. Effects of N2 plasma pretreatment on the SiN passivation of AlGaN/GaN HEMT. *IEEE Electron Device Letters* 29(3):209–211.

20. Selvaraj, S. L., T. Ito, Y. Terada, and T. Egawa. 2007. AlN/AlGaN/GaN metal-insulator semiconductor high-electron-mobility transistor on 4 in. silicon substrate for high breakdown characteristics. *Applied Physics Letters* 90(17):173506–1.

21. Sze, S. M. 2008. *Physics of Semiconductor Devices*, 3rd ed. Translated by Li Geng and Ruizhi Zhang. Xi'an, China: Xi'an Jiaotong University Press.

22. Winslow, T. A. and R. J. Trew. 1994. Principles of large-signal MESFET operation. *IEEE Transactions on Microwave Theory and Techniques* 42(6):935–942.

23. Wong, M. H., Y. Pei, D. F. Brown, S. Keller, J. S. Speck, and U. K. Mishra. 2009. High performance N-face GaN microwave MIS-HEMTs with 70% power-added efficiency. *IEEE Electron Device Letters* 30(8):802–804.

24. Ye, P. D., B. Yang, K. K. Ng, J. Bude, G. D. Wilk, S. Halder, and J. C. M. Hwang. 2005. GaN metal-oxide-semiconductor high-electron-mobility-transistor with atomic layer deposited Al_2O_3 as gate dielectric. *Applied Physics Letters* 86(6):63501-1.

25. Yue, Y.-Z., Y. Hao, and J.-C. Zhang. 2008a. AlGaN/GaN MOS-HEMT with stack gate HfO_2/Al_2O_3 structure grown by atomic layer deposition. 2008 IEEE Compound Semiconductor Integrated Circuits Symposium, October 12–15, 2008, Piscataway, NJ.

16

Double Gate High Electron Mobility Transistors

Ajith Ravindran

CONTENTS

16.1 Introduction

The need of the hour is power devices that have high power and efficiency. These devices having gained great importance, propelling extensive research and development in the recent years. Over years of research and with silicon-based switching devices nearing its limits of performance, GaN HEMT is the appropriate stand-in for silicon power devices. GaN HEMTs have explored new applications in the semiconductor industry due to their ability to switch high voltages and currents at faster transients. The HEMT, otherwise known as High Electron Mobility Transistor, is a type of Field Effect Transistor (FET) that exhibits low noise figure and very good performance characteristics at microwave frequencies. A HEMT or a heterojunction FET is a key device for high-speed digital circuits and low-noise microwave circuits. These applications include computing, telecommunications, and instrumentation on a broad perspective. The device also shows high performance in RF design. The major advances and utilization of wide bandgap semiconductor devices along with group-III nitrides have led to the development of AlGaN/GaN HEMTs. Comparing the properties of different materials, GaN-based HEMTs have many advantages over other technologies (e.g., GaAs). To be more specific, the high output power density of GaN allows the fabrication of much smaller size devices compared to GaAs with the same output power. Operation at high voltage is possible for HEMTs due to their high breakdown electric field and this characteristic also increases the efficiency of these devices as amplifiers. The fact that wide bandgap materials are employed makes it possible for these devices to operate at high temperatures. Better noise performance is one of the leading parameters that make HEMTs to stand out from all the other kinds of Field Effect Transistors.

The structure of a HEMT device is formed by joining two or more different bandgap materials. Due to the formation of a hetero-structure junction, a two-dimensional electron gas (2DEG) is induced at its junction and is enhanced by the polarization strain. The 2DEG is intrinsic in GaN-based heterostructures. Conduction in HEMT is facilitated by the newly formed pool of electrons. As lattice mismatches exist between the materials Gallium nitride and Silicon, the former cannot be directly grown over a substrate made of the latter material. Due to the scarcity of the bulk GaN, a substrate of the same is hardly a possibility. Hence, the cost of manufacturing GaN HEMTs is reduced by the growth of a buffer layer such as AlN in between the silicon substrate and the GaN layer to avoid excessive lattice structure mismatches. The most important part of HEMT devices is the 2DEG that results from the band gap difference between Al_xGa_xN/GaN. The higher bandgap of Al_xGa_xN that allows free electrons to diffuse from the Al_xGa_xN layer to the lower bandgap of the GaN layer causes the creation of the 2DEG (2 Dimensional Electron Gas) near the interface. A potential barrier that arises as a result, then confines the electrons to a thin layer of charge known as the 2DEG. The formation of the 2DEG layer in AlGaN/GaN HFETs is also influenced by the strong polarization that occurs at the AlGaN/GaN interface. The total polarization results as the sum total of the spontaneous (P_{sp}) as well as the piezoelectric (P_{pi}) polarization components. The polarization effects in the HEMT device further enhance the 2DEG density without using any doping in the AlGaN barrier layer. This is uniquely referred to as "polarization doping" in group III-Nitride heterostructures.

DG-HEMTs, in their extrinsic form, exhibit an evidently superior frequency performance and a higher immunity to short-channel effects due to the reduced parasitic resistances and lower drain conductance. GaN HEMTs are promising power electronic devices with very high-power density and efficiency as compared to their Silicon and Gallium Arsenide counterparts.

In Single Gate HEMT (SGHEMT), as shown in Figure 16.1, when the gate length (L_g) reduces to below 100 nm, it leads to a fall in Transconductance (g_m), resulting in a reduction of voltage gain (g_m/g_d) of the device and thereby reducing the maximum frequency of oscillation (f_{max}). The most important reason for the reduction in g_m with gate length in Single Gate HEMT is the inability of the device to maintain the desired channel aspect ratio (α). Double Gate HEMT (DG-HEMT), which is the emergent solution for the shortcomings of SGHEMT by virtue of its double gate structure and the requirement of the

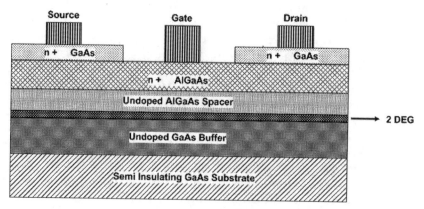

FIGURE 16.1
Single gate HEMT structure.

same supply voltage. The double heterojunctions, virtually increases the value of aspect ratio for very small gate lengths. Short-channel effects are less prominent in DG-HEMTs, leading to a better intrinsic dynamic performance.

16.2 AlGaN/GaN DG-HEMT Structure

GaN is a wide bandgap material and thus it can withstand extreme environmental conditions such as high temperatures and voltages. The material properties of GaN compared to other materials are given in Table 16.1. GaN devices stand at an advantage over other materials with respect to many characteristics. This includes wide bandgap, high breakdown electric field and high electron mobility. Semiconductors with large energy bandgaps are used to build transistors that operate at much higher temperatures, maintain greater voltage levels and handle higher signal power levels than is possible with smaller bandgap conventional materials such as Si, GaAs and InP.

In the Double Gate HEMT Structure as shown in Figure 16.2, an additional top gate electrode covers the normal gate and extends to the source and drain electrodes with overhangs. According to its directions (pointing to the source or the drain), the overhangs'

TABLE 16.1

Material Properties of Some Semiconductors

Material	Bandgap, (eV)	Breakdown Electric Field, (MV/cm)	Thermal Conductivity k, (W/cm-K)	Electron Mobility, $\mu_{n(cm^2/(V \cdot s))}$	Hole Mobility, $\mu_{p(cm^2/(V \cdot s))}$	Relative Dielectric Constant
Si	1.12	0.3	1.3	1400	450	11.7
GaAs	1.42	0.4	0.55	8500	400	12.9
GaN	3.4	3.3	1.3	1000	200	8.9
4H-SiC	3.27	3	3.7	900	120	9.7
6H-SiC	3.02	3.2	4.9	400	90	9.66

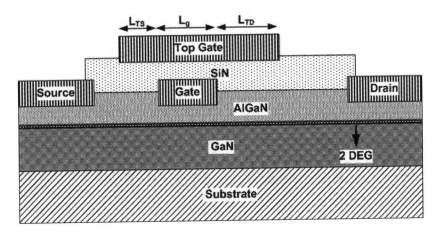

FIGURE 16.2
Double gate HEMT structure.

lengths are expressed as LTS and LTD, respectively. Therefore, the gate adjacent regions, where charging and discharging usually occur, are under the control of the top gate. When the pulse dynamic characterization is carried out, an additional pulse signal is applied on the top gate. If this additional signal is biased constantly at 0 V, the device's operation mode equals the one with the Source Field Plate (SFP) mode. While if this additional signal is the same as that applied on the normal gate, the device's operation mode equals to the one with the Gate Field Plate (GFP) mode. As a result, more information of dynamic performances can be found in the same AlGaN/GaN HEMT, which allows the direct comparison of the effects of the SFP and the GFP on the dynamic characteristics, excluding the material and process variations between different devices, and furthermore distinguish the mechanism differences between the SFP and the GFP. The dynamic characteristics of DG-HEMT working in the SFP-mode, GFP-mode and normal mode are given in Figure 16.3.

When looking at the fabrication, AlGaN/GaN Double Gate HEMT Structure is grown on a 2-inch (0001) sapphire substrate by using metal-organic chemical vapor deposition (MOCVD). The epitaxial structure consists of a 2 µm-thick GaN buffer layer, a 1 nm-thick AlN spacer and a 30 nm-thick $Al_{0.3}Ga_{0.7}N$ barrier layer. Hall measurement results show a 2-D electron gas (2DEG) density of 1.44×10^{13} cm^{-2} and an electron mobility of 1080 cm^2/ Vs, yielding a sheet resistance of 400 Ω.

The device fabrication process begins with mesa isolation by using Cl_2/BCl_3-plasma-based dry etching. Then, source/drain ohmic contacts are formed by depositing Ti/Al/ Ni/Au (20/160/50/100 nm) metal stacks, followed by annealing at 850°C for 30s in a N_2 atmosphere. The contact resistance of ~0.7 Ω·mm can be derived by measuring linear

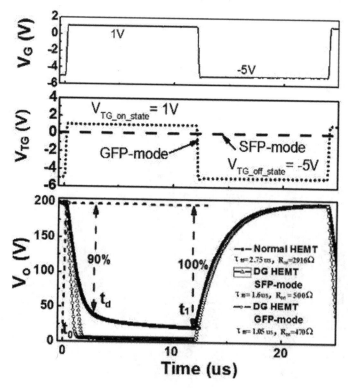

FIGURE 16.3
Dynamic characteristic of DG-HEMT working at SFP-mode, GFP-mode and Normal mode.

transmission line method patterns. Then, the gate is made by depositing Ni/Au (50/200 nm) and lift-off technique. A dielectric layer of 165 nm SiN was deposited at 350°C in a plasma-enhanced CVD system. The top-gate electrode is formed with the same processes as the normal gate. Finally, the electrode pads are opened by reactive-ion dry etching the SiN layer. The gate-source spacing is 5 μm, the gate-drain spacing is 12 μm, and the gate width and length are 100 and 2 μm, respectively.

Both GFP and SFP modes improve the dynamic performance of Double Gate HEMT and among this, GFP mode is the best because of the following two reasons: 2DEG compensation during on-state and the less negative charges trapping during off-state.

Figure 16.4 shows the cross-sectional construction of a double-finger AlGaN/GaN HEMT with a symmetric lateral structure. The silicon-nitride (SiN) passivation layer serves to reduce surface defects and isolation boost electrode isolation. A single-finger gate width of 50 μm was utilized based on the principle that the phase difference will be no more than π/16 at W-band frequencies. The distance between the source and drain is 2 μm. All these physical features contribute toward the lower series resistance in the channel and improved high-frequency characteristics of the device.

The hetero-structure device consists of a 1.5 μm thick GaN buffer layer and a 23 nm thick AlGaN barrier layer with 24% Al content. To enhance the 2DEG channel characteristics, the AlN layer is inserted between the barrier layer and the 1 nm thick buffer layer. The mobility of the 2DEG layer and the sheet carrier concentrations are 2000 cm²/Vs and 1.1×10^{13} cm⁻² respectively. A Silicon Carbide (SiC) substrate is used for getting excellent thermal conductivity.

The 2DEG layer of the AlGaN/GaNHEMTs is determined by polarization, and the polarizationis related by Equation 16.1.

$$P_t = P_{sp} + P_{pi} \tag{16.1}$$

where P_{sp} represents spontaneous polarization for the GaN materials and P_{pi} is the piezo-electric polarization between AlGaN and GaN materials which is given by Equation 16.2.

$$Ppi = 2\frac{as - ao}{ao}\left(E31\frac{C13}{C33}\right) \tag{16.2}$$

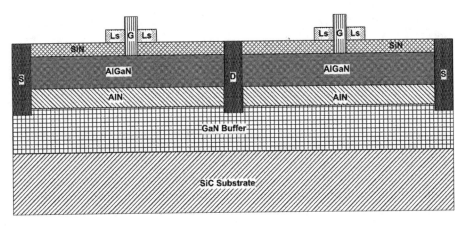

FIGURE 16.4
Structure of AlGaN/GaN HEMT.

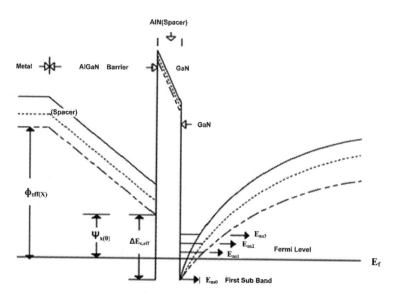

FIGURE 16.5
Conduction energy band diagram of AlGaN/AlN/GaN HEMT.

where α, C, and E represent lattice constants, elastic constants, and piezoelectric constants, respectively. The values can be obtained according to the material statement so that polarization can be calculated.

The conduction energy band diagram of AlGaN/AlN/GaN HEMT is shown in Figure 16.5. The wider bandgap AlGaN barrier and AlN spacer layer thickness gives the effective Conduction band offset $\Delta E_{c,eff}$. Mobility is increased further by using a spacer layer AlN which separates the two-dimensional electron gas (2DEG) from ionized donors near heterointerface. There are two factors that improve the electron mobility of the carrier in the 2DEG quantum well. One is the reduction of alloy scattering because of binary nature of AlN spacer and secondly, the introduction of spacer AlN also reduces the penetration of the electron wave function into the AlGaN barrier layer. The thin AlN spacer interlayer improves transport properties of the 2DEG, and the mobility linearly increases due to suppression of the alloy scattering. The AlN spacer layer is incorporated to spatially separate the heterostructure interface from the doped AlGaN wider bandgap barrier (donor layer) and ensures higher carrier mobility. The high carrier concentration along with a higher mobility can thus be achieved simultaneously. Therefore, the behavior of 2DEG in the new structure can be attributed to the larger $\Delta E_{c,\,eff}$ which is caused by the insertion of the thin AlN layer.

The thin AlN spacer interlayer improves transport properties of the 2DEG, and the mobility linearly increases due to suppression of the alloy scattering. The AlN spacer layer is incorporated to spatially separate the heterostructure interface from the doped AlGaN wider bandgap barrier and ensures higher carrier mobility. Hence the high carrier concentration along with a higher mobility can be achieved.

AlGaN barrier and AlN spacer layer thickness give the effective conduction band offset $\Delta E_{c,eff}$. E_0 and E_1 are the potential of the two allowed energy band in the triangular well. Mobility of the electrons increases because of their separation from a parent ionized donor. Mobility is increased further by using a spacer layer AlN which separates the 2DEG from ionized donors near heterointerface.

The channel, which is a thin layer, is formed in the buffer layer closer to the surface of the barrier layer because of the polarization effect.

16.2.1 Materials Used for Fabrication

GaN HEMTs are not grown on GaN substrates since the growth of high quality GaN substrate is still under development and very expensive. Although native substrate for GaN is not commonly available to date, several other substrate types are practically used in literature such as Sapphire (Al_2O_3), Semi-Insulating Silicon Carbide (SI SiC), and silicon (Si). There is also a newly developed substrate type SopSiC (Silicon on poly-crystalline Silicon Carbide). Traditionally, Sapphire single crystal Aluminium Oxide (Al_2O_3), is the most commonly used substrate for GaN heteroepitaxy. The materials are as follows:

Sapphire Substrate: Sapphire is an interesting choice because of its semi-insulating property that enables the material to be able to with stand the required high growth temperatures. It is also relatively very cheap. However, its very low thermal conductivity (0.47 W/cm·K at 300 K), large lattice mismatch (16%), and large thermal expansion coefficient (TEC) mismatch (34%) with the GaN epilayers make it the worst choice for high power applications. Nevertheless, the power results for GaN HEMTs with sapphire substrates are surprisingly more than 10 times as high as those that can be achieved by GaAs HEMTs.

Silicon (Si) substrate: Silicon preserves its importance and is still the most common and widely used semiconductor material for electronics applications. Despite the very large lattice mismatch (17%) and enormous TEC mismatch (56%), the advantages of low substrate cost and exceptional access to large diameters of substrate, acceptable thermal conductivity (1.5 W/cm·K at 300 K) and integration possibilities with Si electronics make Si substrates interesting candidates for GaN heteroepitaxy. Very promising results have been reported in the literature for GaN HEMTs on a Silicon substrate.

Silicon Carbide (SiC) substrate: The high thermal conductivity (3.7–4.5 W/cm·K at 300 K), low lattice mismatch (3.4%), and relatively low TEC mismatch (25%) are the main reasons for the superior material quality of GaN epilayers grown on S.I. SiC compared to those grown on Sapphire and Silicon. As a consequence, the 2DEG transport properties of GaN epilayers on SI SiC are also far better and it is very clear that at the moment S.I. SiC is the best substrate choice for GaN HEMT-based microwave high power applications. On the other hand, it is very expensive compared to its counterpart materials. The increasing demand for high power GaN HEMTs will obviously push the market to enhance mass production which in turn will bring down SiC substrate prices.

Silicon on poly-crystalline Silicon Carbide (SopSiC) substrate: Silicon on poly-crystalline Silicon Carbide substrate structure combines the advantages of Si and SiC materials. When compared to the conventional SiC single crystal approach, this approach offers a larger diameter substrate and a comparatively lower cost solution. The Sop SiC is expected to show thermal capabilities close to those of polycrystalline SiC. When compared to bulk Silicon, this substrate exhibits major improvement in terms of heat dissipation. SopSiC substrates are fully compatible with the various other compounds used for semiconductor device manufacturing. SopSiC material is expected to be the solution toward bringing in high performances necessary for high power and high frequency devices at very low costs.

Nucleation layer: The nucleation layer stays between the substrate and GaN buffer layers. Nucleation layer can be made of GaN, AlN, AlGaN or graded AlGaN layers. This layer is required due to the high mismatch in lattice constant and dissimilarity between GaN and the substrate materials which results in a low-quality material with high surface roughness. Nucleation layer reduces the density of threading dislocations in the buffer layer. GaN buffer should be highly resistive; nucleation conditions also influence the GaN buffer resistivity. For different substrate types, different materials are chosen for nucleation layer and they have different optimization values such as growth temperature and thickness. For instance, low-temperature GaN or AlN nucleation layers can be used for Sapphire and SiC substrates (Figure 16.6).

GaN buffer layer: In this part, we will examine the GaN buffer layer as the next layer in the structure. In order to attain optimal device performance it is crucial that the GaN buffer layer is of a very high quality, i.e. low defect density and high resistivity (semi-insulating behavior) to avoid both charge trapping of the 2DEG electrons, which causes drain current collapse and hence output power reduction and high buffer leakage, which makes it very difficult or even impossible to pinch-off a device and achieve a high breakdown voltage. The inability to effectively pinch off a device also reduces the available current swing and therefore the achievable microwave output power. In addition, buffer leakage can cause interaction between active devices on the same chip even if they are isolated by mesa structures. Furthermore, the surface of the GaN buffer layer should be smooth to render a good interface between itself and the AlGaN barrier layer. A smooth and sharp interface is required to obtain a high mobility and good confinement of the 2DEG electrons. Good carrier confinement also enables effective device pinch off.

Increasing GaN buffer thickness provides improvements including smoother growth fronts (improved interface roughness of the heterostructure), reducing dislocations, and further removal of the active regions of the device from the defective nucleation layers and substrate. 2DEG mobility values also increase with buffer layer thickness due to improvements in the AlGaN/GaN interface quality.

Spacer layer: In the epitaxial design of GaN HEMT structures, there is a thin improvement layer between GaN buffer layer and AlGaN supply layer, which is called spacer layer. This layer, made of AlN or non-intentionally doped (n.i.d.) AlGaN with the same Al alloy composition as AlGaN supply layer, is inserted. Spacer layer strongly decreases coulomb scattering between the 2DEG electrons and their

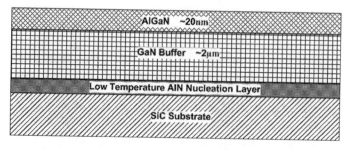

FIGURE 16.6
Structure with low-temperature AlN nucleation layer.

ionized parent atoms in the supply layer. Hence the mobility of the 2DEG electrons is significantly increased to typical values ranging from 1200 to 2000 cm²/Vs at room temperature (RT). The thickness of the spacer layer has a critical importance on the 2DEG properties and sheet resistance so it needs to be very well controlled. It is reported that the growth duration of AlN spacer layer results in 10 s under approximately 0.5 nm/s growth rate condition in order to attain optimum mobility of the 2DEG layer and sheet resistance levels.

AlGaN supply layer: AlGaN layer is used to supply sufficient electrons to the 2DEG and it is usually Si-doped. Al alloy concentration and thickness of AlGaN layer affect the sheet carrier concentration. As we increase the Al alloy concentration, 2DEG density also increases. Moreover, as the thickness of the supply layer increases, the 2DEG density increases much slower than the previous case. Therefore, to achieve a high 2DEG density, maximum Aluminium must be incorporated without causing relaxation in the AlGaN layer. An additional restriction on the maximum thickness of the AlGaN layer is imposed by a very important device structural design concept, which is known as the high-aspect-ratio design concept that puts a limit on the maximum AlGaN thickness.

Cap layer: With the intention to reduce the gate leakage and increase the Schottky barrier height, thin AlGaN cap layer can be introduced to the structure. Conduction band profiles with and without Cap Layer is shown in Figure 16.7. The enhanced barrier height increases the flat band voltage. If we use thick cap layer then we will not need surface passivation. Alternatively, in order to accommodate lattice mismatch between AlGaN and GaN, we can use graded AlGaN cap layer instead of GaN. Furthermore, AlGaN can sustain larger electric fields and has higher Schottky barrier than GaN.

Surface Passivation layer: There are several options for surface passivation layer such as Si_3N_4, SiO_2, Sc_2O_3, MgO, Ni, NiO and Ozone. Among these, Si_3N_4 is the most commonly used material for passivation. Sample structure of AlGaN/GaN HEMT with SiN passivation layer is shown in Figure 16.8. Accurate passivation prevents the surface states from being neutralized by trapped electrons and thereby maintaining the positive surface charge and preventing the formation of a virtual gate. Consequently, better transport properties, higher peak value of transconductance, higher peak drain current, higher gate-to-drain breakdown voltage, reduced DC-RF dispersion, and higher output power density has been achieved using passivation techniques compared to unpassivated devices.

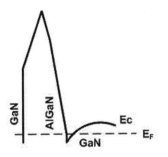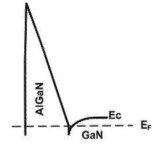

FIGURE 16.7
Conduction band profiles with and without Cap layer.

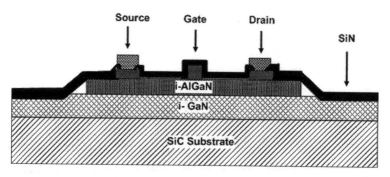

FIGURE 16.8
Epilayer structure with a surface passivation layer of SiN.

On the other hand, passivation technique critically depends on the deposition conditions of the dielectric, as well as the surface cleaning techniques. Care should be taken to reproduce the same surface conditions with varying AlxGa1-xN contamination levels resulting from different processing steps, exposure to air, etc. Special care must be taken to reproduce the same surface condition, with varying Al_xGa_xN contamination levels resulting from different processing steps, exposure to air, etc. If the passivation process is imperfect, then electrons, leaking from the gate metal under the influence of a large electric field present during high power operation, can get trapped.

16.2.2 Device Fabrication

Device fabrication of the AlGaN/GaN HEMT, as shown in Figure 16.4, commences with the definition of the area of the active device. This can be determined either by Cl mesa-etching or by ion implantation. The ohmic contacts are then formed by an initial partial etching of AlGaN in the source and drain regions, depositing the ohmic metals and annealing at 900°C. Though Ti/Al/Ni/Au has been the ideal metallurgy, Ta-based ohmic contacts are also being investigated for their improved morphology. Next, the gate is defined by liftoff of Ni/Au metallurgy. The efficiency of a gate recess, which is frequently employed in compound semiconductor technology, is presently being investigated in the GaN system. Device fabrication is completed with a deposition of a SiN passivation layer. This layer serves the crucial purpose of eliminating dispersion between the large-signal alternating current (AC) and the direct current (DC) characteristics of the HEMT. The maximum AC is less than the DC, the difference being referred to as dispersion. When the device is biased into pinch off, electrons from the gate are injected into the empty surface donors, required to maintain a 2DEG. Compensation of these donors reduces the 2DEG. Under AC drive, the electrons cannot respond due to the long-time constant of the donor traps resulting in a reduced channel current and higher on-resistance (sign of dispersion). SiN passivation eliminates this effect.

16.2.3 Device Operation

The operation of GaN/AlGaN is based on the principle of the two-dimensional electron gas, polarization and charge control. Figure 16.9 shows charge transport mechanism in these devices with help of energy band diagram.

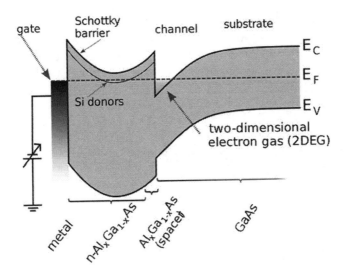

FIGURE 16.9
Transport mechanism in GaN/AlGaN HEMT.

The band diagram consists of a wide gap semiconductor (AlGaN) and a semiconductor with narrower gap (GaN). At the interface, a triangular quantum well is formed in the undoped narrow gap material. Electrons from the wider bandgap material fall into this potential well and are confined within the well. Because of such quantum mechanical confinement in a very narrow dimension, they form a high density of electron gas in two dimensions. Electrons can move freely within the plane of the heterointerface, while the motion in a direction perpendicular to the heterointerface is restricted to a well-defined space region by energy, momentum, and wave function quantization, thus formed is the so-called 2DEG. As the material with a narrow gap is undoped and these electrons are away from the interface, the electron mobility can be simultaneously increased with high concentration of carriers (Figure 16.10).

Figure 16.11 shows the polarization vectors in the AlGaN and underlying GaN. Within the AlGaN layer there are two polarization vectors P_{pi} AlGaN and P_{sp} AlGaN for the piezoelectric and spontaneous polarizations, respectively. The polarization in the AlGaN causes dipole charges to form at the borders of the material with a negative sheet charge at the surface and an equal positive sheet charge at the AlGaN/GaN junction. The polarization in the GaN layer causes a negative sheet charge at the AlGaN/GaN junction and an equal

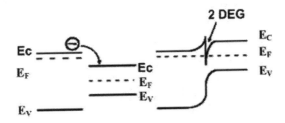

FIGURE 16.10
2DEG formation.

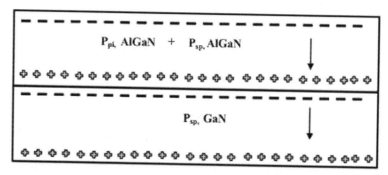

FIGURE 16.11
Polarization vectors in the AlGaN and GaN.

positive sheet charge on the bottom surface. Since the total polarization in the AlGaN is larger, the overall result is a net positive sheet charge at the AlGaN/GaN interface.

The bottom GaN and AlGaN/GaN interfaces are both positive while the charge density at the top AlGaN interface is negative. The critical point to remember is that the interface sheet charges here are not free carriers. They are induced charges as in a typically polarized dielectric. However, it is the presence of these charges and the polarization induced electric field in the AlGaN which allows for 2DEG formation without barrier doping.

There are different effects present in the operation of HEMT devices such as Polarization effects and Trapping Effects. It is now widely recognized that built-in electric fields due to polarization-induced charges play an important role in the electrical and optical properties of Nitride heterostructures grown in [0001] orientation. These fields also provide a source of a 2-dimensional electron gas in AlGaN/GaN heterostructures. In HFETs, the understanding and controlling of the source of electrons is significant for the optimization of their performance. Polarization is mainly divided into Spontaneous Polarization and Piezoelectric Polarization.

Spontaneous Polarization: Al-N and Ga-N bonds are highly ionic and each carries a strong dipole. For example, because the electronegativity of N is much higher than that of Ga, the electron wave function around the Ga-N pair is offset to the nitrogen side. The effect is even more exaggerated in the Al-N pair. This is a special feature of the III-nitrides as the degree of spontaneous polarization is more than five times greater than in most III-V semiconductors.

Figure 16.12 shows GaN grown in the Ga-face and N-face orientation which is the norm for high performance AlGaN/GaN heterostructures. The c-axis polarization vector points from the Nitrogen to the Gallium, as indicated, and creates an internal electric field pointing in the opposite direction; this is referred to as "spontaneous polarization." In an AlGaN/GaN HEMT both the GaN and AlGaN layers have spontaneous polarization vectors which point in the same direction, from the N to the Ga(Al) towards the substrate in Figure 16.11.

Piezoelectric Polarization: The lattice constants a0 and c0 for GaN are slightly larger than those for AlN. As a result, thin AlGaN layers grown on GaN are tensile strained (the GaN is relaxed due to the thick buffer grown on the

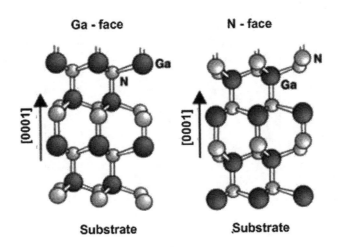

FIGURE 16.12
Ga-face and N-face orientation.

chosen substrate). In the nitride system the piezoelectric constants are more than ten times greater than those typical in most III-V semiconductors and this creates very large polarizations. In Ga-faced material under tensile strain the piezoelectric polarization due to the deformation of the AlGaN layer points parallel to the spontaneous polarization vectors, e.g. towards the substrate in Figure 16.12.

AlGaN/GaN HEMTs show strong current slump which is widely considered to be caused by electron trapping. Traps are very often surface related. So it is not unreasonable to suppose they may contribute to current slump effects in AlGaN/GaN HEMTs. Aside from surface effects, traps may also be formed by dislocations, point defects or impurities.

Charge density is controlled by gate voltage. In AlGaN layer, acceptor states compensate silicon doping thus decreasing charge in the channel. In GaN layer, acceptor states lower Fermi level, hence reducing the 2DEG. Acceptor state addition decreases saturation current and the sheet charge density in the channel.

16.3 AlGaN/GaN DC and RF Characteristics

Figure 16.13a shows the V_{gs} versus transconductance (g_m) and Figure 16.13b shows drain-current characteristic as a function of drain-source voltage. Saturation current exhibits a negative conductance at large V_{ds}; it is due to self-heating and results in a decrease in electron mobility. Beside the self-heating phenomenon, deep traps are also seen at the AlGaN/GaN heterojuction which results in the reduction in the performance of HEMTs. Such trapping effects occur both at the surface and in bulk of GaN epilayer, Therefore, the role of passivation layer is significant to reduce these trapping effects.

Generally, the transistor HEMT is characterized in dynamics by two important parameters: the cut-off frequency (f_T) and the maximum oscillation frequency (f_{max}). RF characteristic of AlGaN/GaN DG-HEMT is given in Figure 16.14.

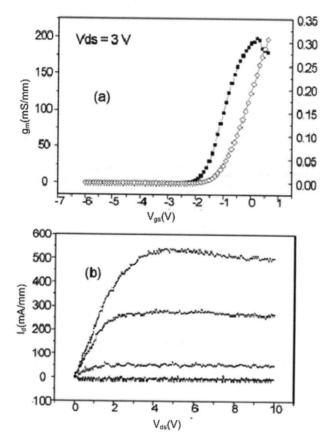

FIGURE 16.13
DC Characteristics of AlGaN/GaN DG-HEMT (a) g_m-V_{gs} curve (b) I_d-V_{ds} curve.

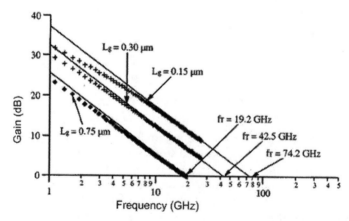

FIGURE 16.14
RF characteristic of AlGaN/GaN DG-HEMT.

16.4　InGaAs/InAs DG-HEMT Structure

The cross-section of an idealized composite-channel DG-HEMT device, is shown in Figure 16.15.

The structure of DG-HEMT is on InP substrate. In this structure, a multicap layer of 5 nm InGaAs (2×10^{19} cm^{-3}), 15 nm InGaAs and 15 nm InAlAs (2×10^{18} cm^{-3}) are introduced to decrease the contact and source/drain parasitic resistance. This is followed by 84 nm Si$_3$N$_4$ passivation layer. Below the cap layer, 2 nm InP etch stop layer is used. Next an InAlAs barrier layer with T_B varying from I to 5 nm is introduced, followed by a thin InAlAs Si-delta doping layer (1×10^{18} cm^3) to provide excess carriers to the channel. To overcome the effect of alloy and impurity scattering, an undoped 2 nm thin spacer layer of InAlAs is placed between the barrier and channel layer. To overcome the effect of alloy and impurities scattering, an undoped 2 nm thin space layer of InAlAs is placed between the barrier and channel layer. The device contains a sub-channel, which includes

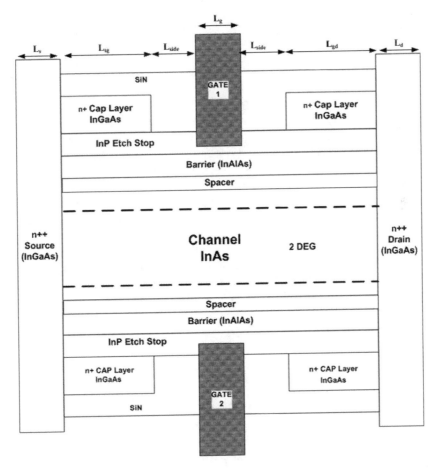

FIGURE 16.15
Structure of InGaAs/InAs DG-HEMT.

thin 8 mm InGaAs sub-channel layer with 2 nm/3 nm of InGaAs upper/lower-channel layers to increase the carrier transport properties thereby improving the device analog/RF performance. The gate is a recessed up to etch stop layer to increase the gate control towards the channel and L_g is swept from 10 to 150 nm. The gate to source ($L_{sg} + L_{side}$) and gate to drain ($L_{side} + L_{gd}$) region spacing's are 0.5 and 1 um, respectively. The contacts of source and drain (S/D) are expanded up to spacer layers, so that the electrons can drive immediately into the channel. The device with highly doped InGaAs source and drain ohmic contacts were formed with an arsenic active concentration of 1×10^{20} cm³. The top and bottom Schottky contact control gate is formed by Gold (Au). All the above active layers are symmetrical with respect to the channel to form a DC structure. These DG-HEMTs works under a common mode condition, which means the voltages applied at both gate terminals are equal. ($V_{gs1} = V_{gs2}$)

The HEMT fabrication is based on both optical and electron beam lithography (EBL). Firstly, the device isolation is achieved through mesa formation by means of a phosphorus acid-based wet chemical etching to expose the InAlAs buffer layer. Secondly, source and drain Ohmic contact are spaced 2.4 μm apart by a lift-off process. Then, Ti/Pt/Au is evaporated by an electron beam evaporator to achieve Ohmic metallization without annealing. Transmission Line Method (TLM) measurements reveal the contact resistance of 0.032 Ω·mm and the specific contact resistivity of 1.03×10^{-7} Ω·cm⁻² on linear TLM patterns. Thirdly, in order to measure on-wafer DC and RF characteristics, the coplanar waveguide bond pads are formed using photoresist AZ5214, and Ti/Au connection wires are evaporated.

The final and most important process in the HEMT fabrication is the gate process, which includes gate lithography, recess, and metallization. Firstly, three layers of electron beam (EB) resist of PMMA/Al/UVIII are coated on the surface. Then EB exposure and the development of each EB resist layer are carried out in turn. Secondly, the gate recess is formed by a phosphorus acid/hydrogen peroxide wet etching till the InP etching-stopper layer, which serves as a surface passivation layer to avoid the kink-effect in DC characteristic. The reduction of side-etched region length (L_{side}) is helpful to suppress the kink-effect, besides, it enhances the electric field in channel and reduces the access part of Rs and R_d, thus increasing the extrinsic transconductance, and finally improving f_T and f_{max}. However, the reduction of Lside is not beneficial to f_{max} due to the enlarged gate-to-drain capacitance (C_{gd}), and also it degrades breakdown voltage due to the increased electric field between gate and drain. In order to reach a balance between the DC and RF characteristics, L side can be optimized through controlling the gate recess etching time. Finally, a Ti/Pt/Au gate metal is evaporated and lifted off.

16.5 InGaAs/InAs DG-HEMT DC; RF Characteristics

Figure 16.16 shows the DC I_{ds} versus V_{ds} characteristics of DG-HEMT device. The saturated drain current (Ids, sat) of 1.1 A/mm is obtained at $V_{gs} = 0.8$ V and $V_{ds} = 1.4$ V for devices with $TB = 2$ nm. This high value of Ids, sat is attributed to the superior electron mobility and conductivity in the InAs channel.

Figure 16.17 shows the I_{ds}–V_{gs} characteristics of an InAs DG-HEMT for V_{gs} from −0.5 to 1 V and $V_{ds} = 0.5$ V with TB as the third parameter. The simulation data show that for TB = 2, 3, 4, 5, and 6 nm the values of $I_{ds,}$ sat are 1.12, 1.14, 1.16, 1.34, and 1.42 A/mm, respectively.

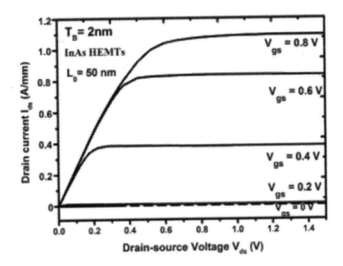

FIGURE 16.16
I_{ds}–V_{ds} characteristics of DG-HEMT.

The devices with thin TB induce higher electric field on the barrier leading to higher depletion of the channel 2DEG, increase sheet resistance and consequently, reduce $I_{ds,}$ sat. From Figure 16.17, the extracted value of VT is 0.26 V for TB = 2 nm measured at a $Ids = 1$ mA/mm and $V_{ds} = 0.5$ V.

While looking at the RF characteristics, the f_T and f_{max} are critical parameters for high-speed applications, which can be respectively extrapolated from the maximum

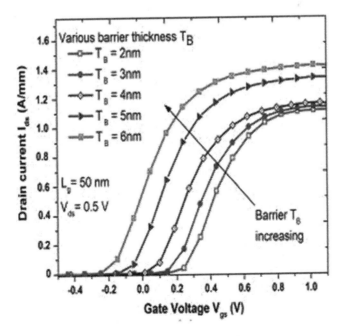

FIGURE 16.17
I_{ds}–V_{gs} characteristics of an InAs DG-HEMT.

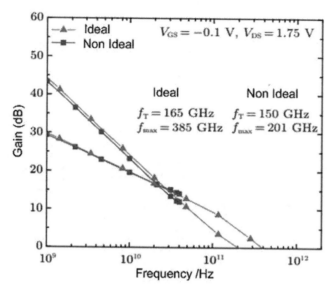

FIGURE 16.18
Frequency characteristic of InP-based HEMT.

available power gain (MAG) and maximum available stable power gain (MSG) using a least-squares fitting with a –20-dB/decade slope. This MSG and MAG are computed from the S-parameters by using Equations 16.4 and 16.5 respectively (Figure 16.18).

$$MSG = \left| \frac{S_{21}}{S_{12}} \right| \tag{16.3}$$

$$MAG = \left| \frac{S_{21}}{S_{12}} \right| \left(K - \sqrt{(K^2 - 1)} \right) \tag{16.4}$$

$$K = \frac{1 - |S_{21}|^2 - |S_{22}|^2 + |\Delta S|^2}{2|S_{12} S_{21}|} \tag{16.5}$$

where K is the Rollett stability factor, $K > 1$ and $\Delta S < 1$ both constitute a primary condition for the stability of the device. Figure 16.17 demonstrates the frequency characteristics of the simulated and measured InP-based HEMTs, which are biased at the peak transconductance points of $V_{GS} = -0.1$ V and $V_{DS} = 1.75$ V. The no ideal f_T and f_{max} are extrapolated to be 150 and 201 GHz, where as the ideal values are 165 and 385 GHz. Notably, the ideal and non ideal gain performances go downward theoretically as frequency increases with a slope of –20 dB/decade. Moreover, S-parameters of the InP-based HEMT are measured over frequencies from 0.1 GHz up to 40 GHz in steps of 0.1 GHz. However, the device is still potentially unstable ($K < 1$) at the instrumentation limit of 40 GHz, therefore, extrapolating the gain at this point with the –20-dB/decade slope indicates a real f_{max} more than 201 GHz. Consequently, the simulation compensates partly for the limitation of test equipment, and gives a more accurate f_{max} of 385 GHz.

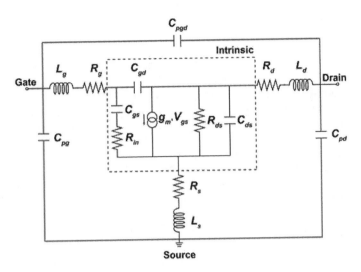

FIGURE 16.19
HEMT small-signal equivalent circuit model.

16.6 Parasitics in DG-HEMT

The small-signal equivalent circuit model is shown in Figure 16.19 and it is based on GaN-based HEMT small-signal models. This is a physically based small-signal equivalent circuit model. The extrinsic elements include the pad capacitances C_{pg}, C_{pgd} and C_{pd}, the pad inductances L_g, L_d, and L_s, and the gate and access resistances R_g, R_d, and R_s.

They are bias independent except for the source resistance R_s. The other parameters are intrinsic elements and are bias dependent. Independent extrinsic components includes pad capacitances C_{pg}, C_{pd} and C_{pgd}, pad inductances L_g, L_s, L_d, and gate and access resistances R_g, R_s and R_d.

Bibliography

Akazaki, T., Enoki, T., Arai, K. & Ishii, Y. Improving the characteristics of an InAlAsInGaAs inverted HEMT by inserting an InAs layer into the InGaAs channel. *Solid State Electron*, **38**, 997–1000 (1995).

Asar, T., Özçelik, S. & Özbay, E. Structural and electrical characterizations of In$_x$Ga$_{1-x}$As/InP structures for infrared photodetector applications. *J. Appl. Phys.*, **115**, 104502 (2014).

Bhat, K. M. et al. Fabrication of double recess structure by single lithography step using siliconnitride-assisted process in pseudomorphic HEMTs. *Microelectron. Eng.*, **127**, 61–67 (2014).

Bhattacharya, M., Jogi, J., Gupta, R. S. & Gupta, M. An accurate charge-control-based approach for noise performance assessment of a symmetric tied-gate InAlAs/InGaAs DG-HEMT. *IEEE Trans. Electron Devices*, **59**, 1644–1652 (2012).

Bhattacharya, M., Jogi, J., Gupta, R. S. & Gupta, M. Impact of temperature and indium composition in the channel on the microwave performance of single-gate and double-gate InAlAs/InGaAs HEMT. *IEEE Trans. Nanotechnol.*, **12**, 965–970 (2013).

Bhattacharya, M., Jogi, J., Gupta, R. S. & Gupta, M. Scattering parameter based modeling and simulation of symmetric tied-gate InAlAs/InGaAs DG-HEMT for millimeter-wave applications. *Solid. State. Electron.*, **63**, 149–153 (2011).

Endoh, A., Watanabe, I., Kasamatsu, A. & Mimura, T. Monte Carlo simulation of InAs HEMTs considering strain and quantum confinement effects. *J. Phys. Conf. Ser.*, **454**, (2013).

Hiyamizu, S. High electron mobility transistors. *Surf. Sci.*, **170**, 727–741 (1986).

Matsuzaki, H. et al. Lateral scale down of InGaAs/InAs composite-channel HEMTs with tungsten-based tiered ohmic structure for 2–S/mm gm and 500-GHz f T. *IEEE Trans. Electron Devices*, **54**, 378–384 (2007).

Mazumder, P., Kulkarni, S., Bhattacharya, M., Sun, J. P. & Haddad, G. I. Digital circuit applications of resonant tunneling devices. *Proc. IEEE*, **86**, 664–686 (1998).

Mohanbabu, A., Saravana Kumar, R. & Mohankumar, N. Noise characterization of enhancement-mode AlGaN graded barrier MIS-HEMT devices. *Superlattices Microstruct.*, **112**, 604–618 (2017).

Nguyen, L. D., Brown, A. S., Thompson, M. A. & Jelloian, L. M. 50-nm Self-aligned-gate pseudomorphicAlInAs/GaInAs high electron mobility transistors. *IEEE Trans. Electron Devices*, **39**, 2007–2014 (1992).

Parveen, Supriya, S., Jogi, J. & Gupta, D. A novel analytical model for small signal parameter for separate gate InAlAs/InGaAs DG-HEMT. *IEEE Reg. 10 Annu. Int. Conf. Proceedings/TENCON*, 1–6 (2012). doi:10.1109/TENCON.2012.6412183

Saravana Kumar, R., Mohanbabu, A., Mohankumar, N. & Godwin Raj, D. In 0.7Ga 0.3As/InAs/In 0.7Ga 0.3As composite-channel double-gate (DG)-HEMT devices for high-frequency applications. *J. Comput. Electron.*, **16**, 732–740 (2017).

Saravana Kumar, R., Mohanbabu, A., Mohankumar, N. & Godwin Raj, D. Simulation of InGaAs sub-channel DG-HEMTs for analogue/RF applications. *Int. J. Electron*, 105, 446–456 (2018).

Sun, S.-X. et al. Physical modeling of direct current and radio frequency characteristics for InP-based InAlAs/InGaAs HEMTs. *Chinese Phys.*, B **25**, 108501 (2016).

Waldron, N., Kim, D. H. & Del Alamo, J. A. A self-aligned InGaAs HEMT architecture for logic applications. *IEEE Trans. Electron Devices*, **57**, 297–304 (2010).

Wong, Y. Y. et al. Growth and fabrication of AlGaN/GaN HEMT on SiC substrate. *2012 10th IEEE Int. Conf. Semicond. Electron. ICSE 2012—Proc.*, 729–732 (2012). doi:10.1109/SMElec.2012.6417246.

Yu, G. et al. Dynamic characterizations of AlGaN/GaN HEMTs with field plates using a double-gate structure. *IEEE Electron Device Lett.*, **34**, 217–219 (2013).

Zhong, Y.-H. et al. 0.15 μm T-gate In 0.52 Al 0.48 As/In 0.53 Ga 0.47 As InP-based HEMT with f_{max} of 390 GHz. *Chinese Phys.*, B**22**, 128503 (2013).

Index

Note: Page numbers in italic and bold refer to figures and tables, respectively.

423